Reliability and Maintainability of Electronic Systems

DIGITAL SYSTEM DESIGN SERIES

ARTHUR D. FRIEDMAN, Editor
George Washington University

S. I. AHMAD AND K. T. FUNG
Introduction to Computer Design and Implementation

MICHAEL ANDREWS
Principles of Firmware Engineering in Microprogram Control

J. E. ARSENAULT AND J. A. ROBERTS
Reliability and Maintainability of Electronic Systems

JEAN-LOUP BAER
Computer Systems Architecture

MELVIN A. BREUER
Digital System Design Automation: Languages, Simulation and Data Base

MELVIN A. BREUER AND ARTHUR D. FRIEDMAN
Diagnosis and Reliable Design of Digital Systems

ARTHUR D. FRIEDMAN
Logical Design of Digital Systems

ARTHUR D. FRIEDMAN AND PREMACHANDRAN R. MENON
Theory and Design of Switching Circuits

Reliability and Maintainability of Electronic Systems

Edited by:
J.E. Arsenault and J.A. Roberts

COMPUTER SCIENCE PRESS

Printed in the United States of America.

Computer Science Press, Inc.
11 Taft Court
Rockville, Maryland 20850

2 3 4 5 6 86 85 84 83 82 81

Library of Congress Cataloging in Publication Data
Main entry under title:
Reliability and maintainability of electronic systems.

Includes bibliographies and index.
1. Electronic apparatus and appliances—Reliability. 2. Electronic apparatus and appliances—Maintainability. 1. Arsenault, J. E., 1941- II. Roberts John Alvin..1943-
TK7870.R43 621.381 79-10543
US ISBN 0-914894-24-2

Distributed by Springer-Verlag
Berlin Heidelberg New York Tokyo
ISBN 3-540-12581-7

PREFACE

The complexity of electronic systems is increasing rapidly and this book attempts to illustrate the practical means that can be used to provide reliability and maintainability in such complex electronic systems. The principles of how to design for sustained performance (reliability) and the problem of how to design for the rapid diagnosis and removal of faults (maintainability) have not changed. However the growth of digital systems, software problems and complexity in general has spawned a large number of new techniques. These attempt to achieve, economically, the reliability and maintainability levels formerly achieved by much simpler systems.

The book is intended for engineers, managers and academics engaged in system engineering and concerned with reliability and maintainability. Accordingly it takes a broad approach to the subject. The editors have tried to select sufficient theoretical and practical information to solve those reliability and maintainability problems frequently encountered. In addition, through a comprehensive set of reference material, the reader will be able to explore, for himself, aspects of the techniques required by those special problems which inevitably appear. To meet the limitations of space, some filtering has taken place and solutions looking for problems, together with obsolete approaches, have been eliminated. The editors have, however, included some techniques which are regarded as new or novel by some specialists.

In order to present a truly practical view of the subject the editors have selected authors who are currently engaged in using the techniques described. This approach is of course not entirely without difficulties and the form and level of treatment varies inevitably from chapter to chapter. However the basic instruction given to the authors was that examples and practical techniques should be emphasized. This is properly reflected in the content and gives the reader an insight into how the techniques are applied.

This book has been organized into three parts: Part One-General, Part Two-Reliability, Part Three-Maintainability; these parts are further subdivided into chapters. As far as possible, hardware and software considerations are discussed concurrently in each chapter.

Chapter 1 offers an overview by emphasising the cost aspect over the entire life of the equipment. This approach has become known as life cycle costing and is an important factor in the evaluation procedure of any system proposed to agencies. The chapter covers procedures and methods, cost factors, management and models. Chapter 2 provides a view of reliability and maintainability management applicable to a variety of organizational arrangements. Each phase of a product cycle and the applicable R-M requirement is described. Training and organization outline requirements are also described. Chapter 3 considers the role of design evaluation and review. Without the use of such techniques a poor or costly design may be committed to production. The chapter describes the various stages of product design at which the technique may be applied. Chapter 4 describes how the problem of design complexity and information proliferation can be brought under control using design automation. The treatment particularly emphasizes those techniques that may simplify the reliability and maintainability aspects of the design process. Chapter 5 is concerned with the problem of ensuring that the configuration is controlled so that what is intended to be producted is actually produced. The concepts of identification, item and baseline are explained and management plan is described. Chapter 6 reviews the current software reliability and maintainability problems and solutions. A duality between software and hardware is proposed and used to clarify a set of techniques. Testing and documentation methods are described which simplify the problems of dealing with the complexity of the task.

In Chapter 7 the various functions applicable in reliability modelling are reviewed. This chapter provides a convenient source of reference for those reliability engineers who are not usually exposed to the theoretical basis of the discipline. Chapter 8 begins with a review of the basic configuration concepts for the achievement of a required reliability performance. The chapter then provides a description of the various estimation techniques so that reliability of each section of a unit can be calculated. Chapter 9 provides practical treatment of the thermal design problems found in electronic equipment. The three modes of heat transfer are examined and the role of each is considered. This chapter can provide a sound basis for thermal design evaluations and reviews. Chapter 10 reviews the various factors in the environment that degrade reliability. Factors considered include nuclear, mechanical considerations, electrical interference and contamination. Chapter 11 describes fault tree analysis techniques and their applicability to design review. Construction methods, the role of the computer and relationship of fault tree methods to those of failure mode and affects analysis are also described. Chapter 12 reviews sneak circuit analysis methods. Unplanned modes of operation are as undesirable as defective intended operational modes. The chapter describes actual cases and covers digital circuit and software aspects of the technique. Chapter 13 provides a unique view of information integrity as applied to system problems. Introductory theoretical aspects and practical examples are provided which clarify the value of applied queing theory and error correction techniques in system design. Chapter 14 is concerned with the problems of ensuring that components are correctly chosen and correctly used in a system. Techniques such as checklists, design reviews, preferred parts lists and derating guidelines are described. A detailed and well documented derating guide is provided. Chapter 15 describes the problems resulting from poor parts control and provides advice for the correct selection and specification of parts. While concentrating on military and related areas the techniques described can be applied to other critical procurements. Chapter 16 introduces the techniques that can be used to screen out defective parts, subassemblies and systems. Practical results from screening programs are given. Chapter 17 covers the general areas of failure reporting and corrective action systems. However the importance and difficulties associated with IC failures is given emphasis. The practical implementation of such systems is described in detail. Chapter 18 describes the role and methods of systems testing to assess the achieved reliability. The recording and statistical significance of failures together with the analysis and corrective action aspects are covered. Comprehensive descriptions with examples is provided for modelling, test plan formulation and reliability growth.

Chapter 19 provides an introduction to mathematical modelling for the maintainability of systems. Examples provide insight into the applicability of the models. Models for various maintenance policies are provided with examples of calculations for practical systems. Chapter 20 describes the methods used to allocate and estimate the maintainability of subsystems. The chapter includes a table of complexity factors for units, assemblies and parts. These tables are used in the solution of complete practical problems. Chapter 21 introduces the concept of diagnostic strategies from which system design can proceed. The application of graph theory to system modelling is presented together with an automated approach. The various modes of Built In Test Equipment monitoring, recovery methods and the use of diagnostic software is described. Chapter 22 discusses the unit design for situations where units are replaced and are subject to diagnosis. The use of signatures for ATE and details for practical design factors for maintainability are described. Software aspects for the test procedures are included in the discussion and test languages are also described. Chapter 23 reviews the role of software simulation in automated testing and provides practical design guidelines for enhancing board testability. The use of simulation is discussed in depth with aspects much as component modelling, fault models and timing models included in the considerations. Chapter 24 provides an overview of the practical aspects of handling components classified as limited life items. The storage and wear out of particular component types is described. Light Emitting Diode, Incandescent Lamps, Contacts, etc., are discussed in detail. Chapter 25 describes the problem and solution to the various aspects of holding an inventory of spares. The economic factor in sparing together with maintenance policy effects is considered. Simple models are given and their limitations are discussed. Chapter 26 provides a practical introduction to the role of manuals in the area of maintainability. Advice is given regarding style, presentation and form. The contractual aspects of manual preparation are outlined with emphasis on gaining a clear understanding of the requirements. New concepts are discussed including the generation of software manuals. Chapter 27 covers the basic concepts of training and presents an overview of training methods. The nature of the learning process is described and related to the problem of teaching maintenance skills. Chapter 28 provides a basis for choosing and planning a maintainability demonstration method to suit particular maintainability requirements. The practical aspects of demonstrations, testing and risk are covered in detail.

The Appendices provide lists of standards, specifications and published texts relevant to the reliability and maintainability field. These sources will provide the reader a basis for a further extension into the practical and theoretical aspects of the art/science of reliability and maintainability.

January, 1979
Ottawa, Canada

J.E. Arsenault
J.A. Roberts

ACKNOWLEDGEMENTS

A book of this kind could not be produced without the co-operation of a large number of people. The editors and authors, however, owe primarily a collective debt of gratitude to Dr. Hans Reiche, who provided every possible encouragement and practical assistance toward bringing the book to completion. Thanks are due to Lyn Drapier Arsenault who critically read and corrected the manuscript and also the staff of the publisher for their valuable assistance. For coping with the seemingly endless number of changes to the manuscript, the editors wish to thank Gail Frankland, Gail Turner and Sandra Dadson.

The editors would like to acknowledge the kind permission of the Reliability Analysis Center to reproduce Figure 4-31 and Table 4-37 from Reliability Design Handbook RDH-376 by R.T. Anderson, and Pergamon Press to adapt the following papers from 'Microelectronics and Reliability':

(a) Roberts, J.A., "The Design Error Contribution to Failure", Volume 14, No. 2 April 1975, p. 159 to 162,
(b) Des Marais, P.J. and Williams, D.G., "System Diagnosis with FLIP", Volume 17, No. 1 January 1978, p. 47 to 52.
(c) Des Maris, P.J., and Arsenault, J.E., "Effects of Design Automation on the Reliability and Maintainability of Electronic Systems", Volume 17, No. 1 January 1978, p. 143 to 153.

Acknowledgement is also due to the following organizations who helped in the production of the manuscript:

(a) Society of Reliability Engineers (Ottawa Chapter),
(b) Computing Devices Company (Control Data Canada),
(c) Thompson-Foss Incorporated.

CONTENTS

PART ONE - GENERAL

PART TWO - RELIABILITY

PART ONE

GENERAL

1. LIFE CYCLE COST
2. MANAGEMENT
3. DESIGN REVIEWS
4. DESIGN AUTOMATION
5. CONFIGURATION
6. SOFTWARE

1 LIFE CYCLE COST

H. Reiche

Under the present economic climate item procurement has taken on a different trend. A system which has been fielded must be supported for the total life. The cost of this support over the life cycle is often more than the initial acquisition cost. Usually the disposal value is small. System procurement therefore looks at the total cost which may be incurred over the life cycle rather than only at the initial, non-recurring cost factor. At the same time management can better foresee the cash flow which is required for the system over its life. Governments and large industrial concerns have introduced guidelines in many countries which assure that the procurement process evaluates and states the total Life Cycle Cost (LCC) for high value purchases. The concept of Life Cycle Management with its life cycle phases allows management to deal with the problem of LCC during the appropriate phases, in an orderly manner and gives them a tool for improved decision making purposes.

The purpose of this chapter is to outline the methodology of LCC and its use in various phases of Life Cycle Management. The phases which may be considered for LCC applications are Concept and Definition, Design and Development, Manufacture and Install, and Operate and Maintain as shown in Figure 1. The subject will be introduced by first stating the procedures and methods which are available to a practical user. The appropriate factors which may form the base for the required data are then listed. A few simple models are mentioned and actual applications are described.

PROCEDURES AND METHODS

Procedures

One type of procedure involves trade-offs between various cost parameters in conjunction with effectiveness which may be operational or logistical in nature. The maximum trade-off level is directly related to optimizing acquisition costs with total life cycle costs for an item with maximum effectiveness. The item must not only meet the operational requirements but also the logistic support requirements and all within the constraint of cost. The basic framework of the logistic support must be identified as part of the procedure because its influence on cost can be a major factor much more so than changes in operational parameters. Another type of procedure may concern itself with establishment of reliability and maintainability levels which can be achieved within a given cost frame. This type of procedure can be effective in the design and development phase.

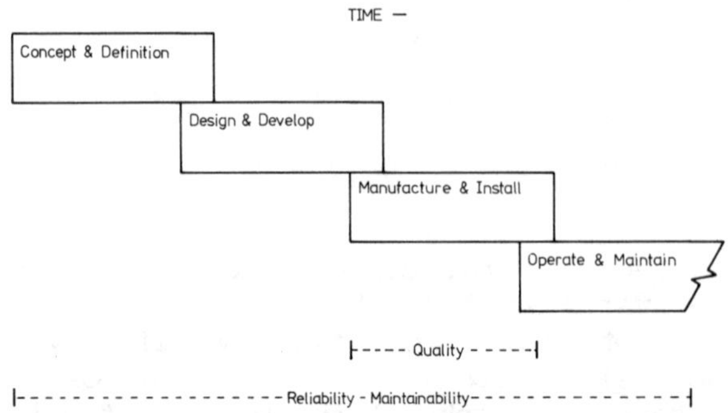

Figure 1. Phases In Equipment Life Cycle

A generalized LCC procedure showing the process steps involved in balancing the user/customer and producer/contractor roles has been developed in a very effective flow chart given in (1) and shown in Figure 2. This process may be repeated during the initial phase with different levels of detail.

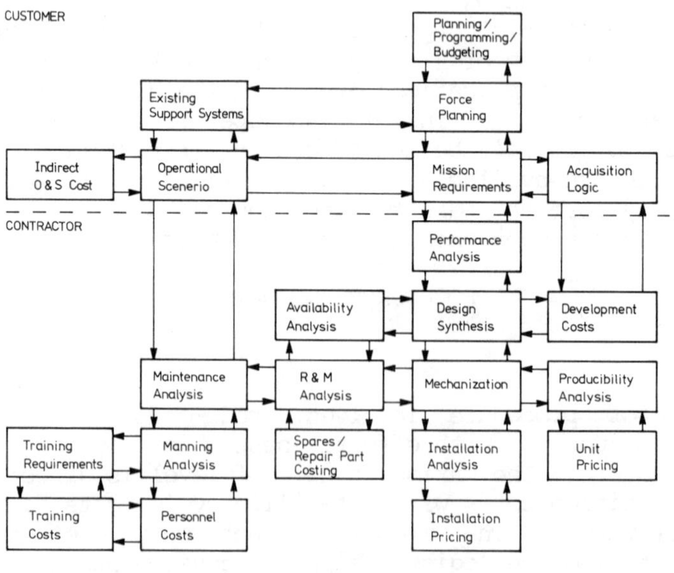

Figure 2. Generalized LCC Procedure

Methods

Before embarking on any LCC study one should ask the following questions:

(a) What is the purpose of the estimate?
(b) What is the impact of accuracy and precision in the estimates?
(c) Who is involved in the cost project?
(d) What are the constraints?
(e) What data are available?
(f) What details are required of the cost structure?
(g) How are the uncertainties to be treated?
(h) What are the general cost fund limitations?
(i) What responsibility has the cost analyst?

Because the data required for any study are important the following should be reviewed:

(a) Are the data oriented towards the problem or not?
(b) Are the data co-ordinated with other information?
(c) Are the data comparable to other data?
(d) Are the data obsolete?
(e) Are the data applicable for this study?
(f) Are the data biased?
(g) Are the data available for the study?

The above type of questions were raised in (2) and represent a fairly comprehensive list.

In order to estimate LCC one must make a comparison between existing costs and new costs. There is little sense in developing LCC for an equipment and then have nothing to compare or relate it to. Historical data are therefore often used as the base for comparison. There are a number of methods which one can use to develop cost estimates.

A simple method is the use of existing catalogue data. Although the price for a single item in a catalogue may not reflect the actual market price, it is a base from which one can operate. In a sense it is historical information which once must have been correct and used. The comparison of one catalogued item and the same from another catalogue may yield a good estimated price. Another method is the use of information which was generated for the purpose of the initial plan. These cost data may be very preliminary but sometimes may be all that are available, especially for a newly developed item. Adjustments to such cost data will have to be made progressively.

The comparative cost method may be used by laying out a simple type of comparative chart which identifies cost relationships of existing items to new ones. For example, one may want to compare the Tactical Air Navigation Equipment (TACAN) produced 15 years ago with that of a new one to be developed. The materials used in the old equipment can be costed and compared with anticipated materials to be used in the development. The type of construction will certainly influence the cost of support and if wired-in units are used for the original equipment, maintenance costs will depend largely on the time to repair. In a functional, modularized equipment this cost will be reduced by a certain percentage. The total chart may then project the two prices and a comparison can be made.

There are other types of methods which may be parametric in nature and deal with specific cost relationships in production, maintenance, performance, spares, etc. Here relationships between costs and certain specific characteristics must be established. For example, the cost of maintaining aircraft engines will depend on the operating hours and the type of engine to be studied. The more complex the engine and the more powerful it is in terms of thrust the more man-hours will have to be spent in maintaining it for a given operating time. That time can be expressed in man-hours per operating hour and a relationship can be established to the cost. The graph for various types of engines against cost will take the form shown in Figure 3. Although often man-hours per flight hour are shown, it has been found that preventive maintenance and training take up a fair portion of the actual operating hours. To leave these out from the cost picture is unrealistic.

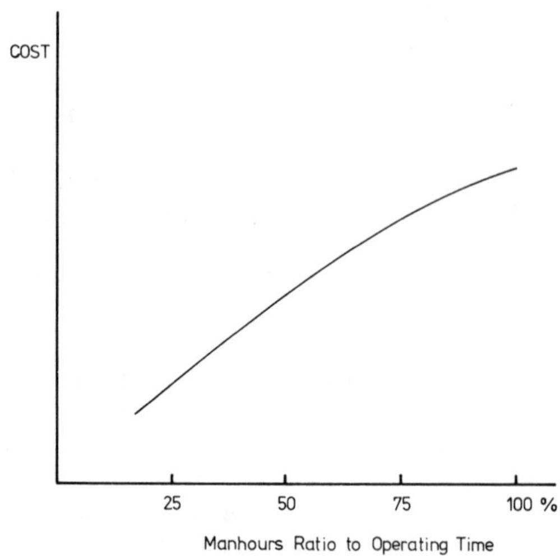

Figure 3. Manhours Ratio to Operating Time

LCC FACTORS

Probably one of the most controversial and difficult subjects of LCC studies is the type of factors which one should include in the evaluation, analysis, estimation, etc. The reason for this controversy lies in the fact that almost everyone has a different view of what constitutes

important factors. It is obvious that for a military aircraft certain factors have a higher priority than others, but these are not necessarily the same for calculating the cost for a staff car.

Basically there are two major types of cost parameters, non-recurring and recurring costs. Regardless of whether we buy the equipment from the shelf or develop it, certain non-recurring cost factors can be identified.

Non-Recurring Cost Factors

The non-recurring LCC factors can usually be related to the following:

Research or Development. This cost will depend on the extent of the research or development specified in the contract. If the extent reaches into the latest state-of-the-art, costs will be high. Costs of minor developments on new equipment may be insignificant in relation to other cost factors. There are few equipments which are bought off-the-shelf without having included some required modifications, especially for the military. These costs are often overlooked but are part of the unit price

Reliability-Maintainability Engineering. The improvement of reliability or maintainability of a design after or before delivery is another factor. Off-the-shelf items may not be fully satisfactory to a customer and the manufacturer may like to offer in his proposal an improvement in these parameters. His cost estimate for such improvement will form part of a non-recurring cost.

Qualification Approval. The approval of an item prior to acceptance or delivery will require tests, facilities and manpower. Often the tests cannot be carried out at the contractor's plant. Shipment of equipments to a site for tests will be necessary, new test equipment, special to the item may be required. In the mechanical field special-to-type test equipment is minimal but in such fields as optics, fluidics, microelectronics, high power, etc. new test methods are often needed. The cost of facilities can be a major one. For example, on a recent test of a reconnaissance drone the facilities required had to be augmented with new sites, protective bunkers, fuel pumps, an improved weather station and a survey of landing areas.

Government Contract Clauses. There are many so called standard clauses in government contracts. Many deal with delivery, packaging, payments, etc. Although these may have some influence on cost they may not be as critical as such clauses as rights regarding information, royalties and patents. Many battles have been fought on these issues but usually the government insists that all information and patent rights resulting from a contract are their property, unless the patent existed prior to the contract. The cost of such factors may be substantial. For example, a

recent purchase of a man portable UHF radio set by the military resulted in a cost for information (drawings, test reports, maintenance data) of almost 1/3 of the unit price.

Life Cycle Management. The management of any project requires manpower at the initial conceptual phase. The development of specifications, requirements and the preparation of the total contract package may be lengthy and costly.

Acquisition. The cost of buying an item includes everything which forms part of the actual item, such as power supplies, cabling or mounts for a radio set, or spare tires, canopy floor mats for a jeep. It has been suggested that taxes and duties should be added to the buying cost, but this may not be easy because each country, province, state and government department has different regulations, one sometimes off-setting the other.

Installation. Once an item has been bought it has to be installed and perhaps tested in its final environment. Additional connectors may have to be purchased, adjustments may have to be made, special mounting brackets may have to be manufactured, interference filters may have to be added externally. This takes time and manpower. Extended field tests prior to taking the item into the final inventory for operational use can be costly.

Test Equipment. Test Equipment for maintenance is a major non-recurring cost item. It must be realized that sometimes an entire so-called "hot set" must be used as the test bed. This practice should be discouraged by use of suitable test equipment, manual, special-to-type or automatic with complex functional capabilities. Unfortunately, many overlook the cost of the software associated with test equipment. There has been the impression that automatic test equipment saves a lot of manpower and money. With few exceptions this has proven to be correct. The cost of software is in many cases the same as that for the hardware or more. The level of maintainers to be used with automatic test equipment, is usually of a higher level than with manual and therefore more costly. Automatic test equipment does provide reduction in test time and improves repeatability. Therefore, when taking this cost factor into account, one must be certain that equipment availability or readiness is a major factor, not necessarily cost reduction.

Training. The introduction of any new item requires training and familiarization. One contributing factor to unreliability remains the human. The problems which can be encountered when introducing a new equipment into the field reflect the extent of training. A brief study was carried out by human behaviour scientists on a new search radar which revealed that a group of maintainers and operators trained for two weeks by the manufacturer introduced 7% of all failures during the first six months of operation of that equipment. Another group trained by the previous

maintainers and operators plus the manufacturer reduced such failures to 1.3% in the next six months. These were failures which can be directly attributed to the human. It is therefore important that proper training be funded and included in the total cost picture.

There are a number of non-recurring costs here. Both in-house and outside training may be required. In-house training, for example by the military, requires manpower. Facilities are normally in existence, but special aids for training may have to be developed. Military training schools have a permanent training staff and the main cost would be in the aids, possible travel and any other support. It is different with training by industry or technical schools. Here special staff may have to be assigned, facilities and overhead become a factor in the cost. Construction of new training support devices may be included and of course the actual cost of lecturer or trainer. With current consulting fees this portion is a considerable one. Total package hardware and training contracts may possibly reduce some of the costs. Nevertheless these training requirements must be detailed, costed and specifically reviewed in light of the benefits which can accrue from proper training.

Transportation. Although this factor is obviously a cost contributing one, the initial conceptual, development and acquisition phases of any program require some travel. Normally the military have their own transportation and saving may be made by using it. But shipment of equipment with commercial carriers, trips to contractors by car or otherwise, mailing of bulk materiel, such as volumes of proposals or contract data packages, all constitute cost factors which occur once.

Documentation. This includes all data and mention has already been made of various data which are required. The estimation of this total cost is difficult because often many of the documents are developed only after the fact. Parts lists, manuals and drawings, today normally supplied in microform, can be costly. Translated information is even more costly and verification can add further costs.

This list does not exhaust all non-recurring costs but gives some insight of their extent. Again it should be pointed out that not all cost items here are of prime importance to all projects and a selection must be made of the high priority factors.

Recurring Cost Factors

The next items to be considered are recurring costs. These are described in the following:

Operating. Under this category one must review the mission requirements of the item. A radar set may require special environmental conditions, such as air conditioning, to operate properly. It may require certain services such as continuous power sources, fuel, and water. A truck

will require fuel, oil, and water. An optical transmission system may require cryogenics for cooling a sensor.

Manpower. There are few equipments which run without manpower for any length of time. The level of manpower required to operate and maintain it will have to be costed and will depend on the type of equipment. A programmer will earn a different salary than a stoker on a ship. The maintainer operating the automatic test equipment will obtain a different income than the maintainer who puts a protective coating on a printed circuit board after repair. The pilot of a 747 will receive more pay than the driver of a tank. The number of persons involved will depend on the type of equipment, the function and mission. In the military, costs for personnel are usually known by some directorate of costing who are part of the finance empire. The Department of Labour normally keeps statistics of manpower costs.

Support. Continuous support in the form of supplies and services is required for any item which is operational. For example, a typewriter or teletype requires paper. The operation of the item may require electricity which must be bought or generated by a local power supply. Operating a commercial airline will require weather forecasting aids and charts on a timely basis. A device which scares birds from airfields may require small amounts of explosives for creating the bangs. There is almost no end to these support requirements which are continuous and recurring costs.

Maintenance. Probably the largest cost in operating any item is the maintenance at various levels. These recurring costs are the major funding problem at the present time for any tight budget. Over the useful life of the item, often ten or more years, these costs may amount to more than 40% of the total life cycle cost. Many believe that with improved or higher reliability of modern parts, the maintenance burden will be reduced. They advocate simplicity but, unfortunately, in order to improve simplicity and equipment reliability one must make use of parts which are extremely reliable; these are very costly. A faster answer to the problem may lie in developing even more complex equipments which have self-maintaining features built in, which take care of at least ten years of service. Although facilities and test equipment for maintenance are usually non-recurring costs, the maintenance itself, such as repair of an equipment, or calibration of a test set, or lubrication of a servo motor, or overhaul and repair after some years of operation, and many more form part of the cost of maintenance. Such costs are often difficult to forecast especially for the total life cycle of the item but because these costs are high an effort must be made to develop a reasonable estimate which can be compared with historical or other data.

Inventory . Not only are spare parts required for support but these are bought on a time scale to replenish the pipeline or stock. There are parts which cannot be stored for a long time without deterioration and

therefore these will create recurring costs. The military have developed shelf life policies for some items and special buying schedules have been instituted to meet the need for operation, without reducing reliability. To the inventory one must add fluids, chemicals, fuel, gaseous materials etc. All of these are recurring costs.

The LCC against some cost factors are shown in Figure 4. The list is incomplete but it would be useless to continue a list of non - or recurring costs without having actual examples to deal with. Nevertheless the above should indicate the broad approach one must take when costing these factors. Another point should be made here. When going out for tender on a contract it is necessary to identify to the bidders all the factors which should be taken into account. If this is not done, the well known examples which have been quoted by many, suggest that the result in many cases is useless information. It is not possible to compare oranges with apples even though all are fruit. The request for bids must require contractors to list their factors dealing with LCC in the same manner, otherwise their tenders should be returned. A recent large European contract for a new military aircraft did exactly this. Companies who are not willing to use the required format should be eliminated. There is no other solution at the present time.

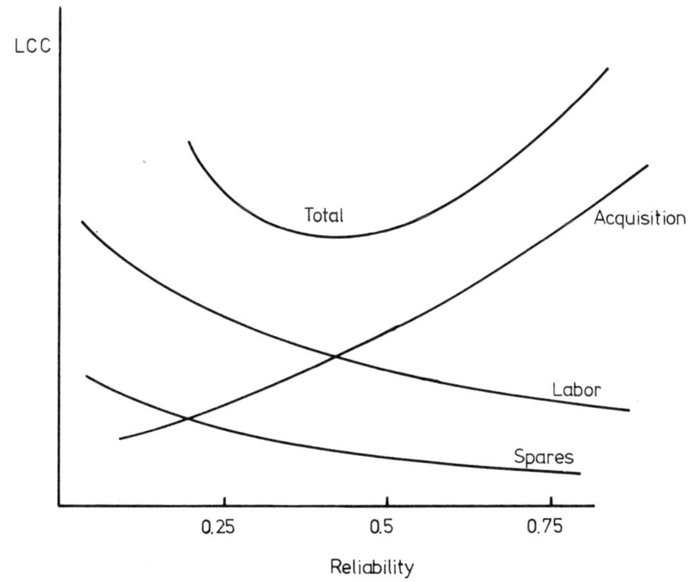

Figure 4. LCC Versus Selected Factors

A practical, simple example of factors required for a LCC calculation follows. It is an example which could be carried out manually. Let us take for this example an aircraft with its equipment and determine what kind of data or factors are required for LCC. The basic data could be the following:

(a) number of sub-systems and equipments per aircraft,
(b) the flying rate, normally stated as monthly flying rates,
(c) the Mean Time Between Failure for each equipment,
(d) the Mean Time to Repair at various repair levels,
(e) the cost per system and equipment,
(f) the cost of one test equipment at each maintenance level,
(g) the cost of qualification/demonstration/approval testing. Various other types of tests may have to be reviewed,
(h) the cost of systems or equipments required for training including training aids and personnel at various maintenance and operational levels,
(i) the cost of any equipment for use on test benches, such as hot sets as standards,
(j) the cost of equipment and parts spares for maintenance support at various levels and to support operating systems, training and tests,
(k) the cost of conferences, meetings, travel,
(l) the cost of publications and drawings,
(m) the cost of field service representatives,
(n) the length of time a system is anticipated to be in service,
(o) the cost of any anticipated overhaul or major modifications, and
(p) the cost of labor.

Some of these relationships to cost are shown in Figure 4.

Estimates

It is seldom possible to evaluate costs accurately at the conceptual phase. Many cost data are not available and others can only be given with a limited accuracy. This does not imply that no LCC can be established at any early phase. Good estimates can be made at any time in the life cycle. Aany government agencies today insist on LCC projections from the start of any large project. Too often they have had to accept an equipment which is not cost effective. Initial estimates certainly can help to clarify certain cost relationships. The data for such estimates must come from various sources. As pointed out before, analog data are a major source. One may lay out a comparative listing of certain features which can then be used as a cost analog estimate. A simple example for an electronic equipment is shown in Table 1.

LCC MANAGEMENT

There is little sense in starting any project unless a proper program and plan have been prepared. An analysis of the LCC requirements set in the concept and definition phase is the basis of a program and plan. Those responsible for life cycle management should prepare an LCC program and plan which may coincide with the development plan. The program and plan identifies the steps which must be taken and the tasks to be completed.

Table 1. Portable UHF Communication Set Cost Comparison

Old Set	New Set	Cost Impact
Frequency range	Extended by 3 MHz	None
Sensitivity	Improved by 1.5 m V	+ 5%
Power output	Same	None
Discrete Parts	LSI	- 10%
PCBs (double sided)	PCBs (Multilayer)	+ 15%
Plug-in	Same	None
Manual module design and Assembly	Automatic	- 20%
Labour Cost		+ 300%
Material Cost		- 10%
Maintenance Cost		+ 10%

A typical example may include the following:

(a) Outline the procedure to be used for an LCC study,
(b) Define the method which is applicable to the particular problem,
(c) List the various factors which should be analysed or quantified and where these factors can be obtained,
(d) Carry out an estimate of the LCC to arrive at a "ball park" value,
(e) Specify the type of LCC model to be used for calculation and the constraints imposed,
(f) Validate the information by carrying out checks against known values and make corrections to the data progressively by updated information,
(g) Prepare conclusions or recommendations in the form of options.

LCC MODELS

There are probably more models in LCC than in any other discipline, The reason is that LCC is coupled in many cases to systems effectiveness, reliability, maintainability, supportability etc. To cover all these models would require a volume like this book alone. It would not serve any purpose. Instead let us write down some of the simpler types of models which leave out the non-sensitive factors. Such models may suffice in many cases and may make it easier for the bidder or contractor. In particular provisions for price adjustment and qualification testing increase the complexity of the model.

Probably the simplest of all models adds the non-recurring and recurring costs. Here LCC are the total LCC costs. NRC are the non-recurring costs and RC the recurring costs, i.e.,

$$LCC = NRC + RC \tag{1}$$

The non-recurring costs can be calculated by adding up the following cost factors:

$$NRC = C_{RD} + C_{RM} + C_{Q} + C_{LCM} + C_{A} + C_{I} + C_{TE} + C_{T} + C_{TR} + C_{S} \tag{2}$$

where,

C_{RD} = Research and development cost,

C_{RM} = Reliability/maintainability improvement cost,

C_{Q} = Qualification approval cost,

C_{LCM} = Life cycle management cost,

C_{A} = Acquisition cost,

C_{I} = Installation cost,

C_{TE} = Test Equipment cost

C_{T} = Training cost,

C_{TR} = Transportation cost,

C_{S} = Support cost.

The RC costs can be added as follows:

$$RC = C_{O} + C_{M} + C_{S} + C_{MT} + C_{IN} \tag{3}$$

where,

C_O = Operating cost,

C_M = Manpower cost,

C_S = Support cost,

C_{MT} = Maintenance cost,

C_{IN} = Inventory.

Another form of this model can be written as the cost of buying and the cost of owning.

$$LCC = CB + CO \tag{4}$$

where,

CB = Buying cost,

CO = Owning cost.

If we make LCC dependent on the system reliability then one may write a simple equation:

$$LCC = C(\lambda_1) + C(\lambda_2) + (Y)(RC) \tag{5}$$

Here $C(\lambda_1)$ are the costs which do not depend on a failure of the system and $C(\lambda_2)$ the costs which are dependent on a failure. Y is the number of years the system is to be in operational use and RC again the recurring costs. The recurring costs usually, but not always, depend on system failure. Figure 5 shows a possible relationship of LCC and failures with Y being 1, 10 or 20 years.

A special model developed by the United States Air Force (USAF) relates Logistic Support to Cost (LSC). It is applicable to aircraft support, but it is not a LCC model as such, as it covers only operation and maintenance. The model is stated as:

$$LSC = \sum_{i=1}^{n} C_i \tag{6}$$

where,

C_i = cost of the ith logistic cost factor

and these cost factors have been identified as:

C_1 = All spares cost,

C_2 = On equipment maintenance cost,

C_3 = Off equipment maintenance cost,

C_4 = Inventory Entry, Supply Management cost,

C_5 = Support Equipment cost,

C_6 = Training, personnel and equipment cost,

C_7 = Management data cost,

C_8 = New facilities cost,

C_9 = Fuel cost,

C_{10} = Spare engines cost.

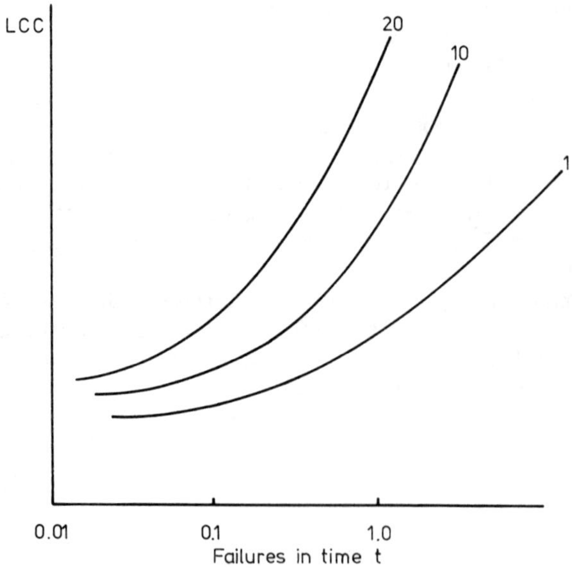

Figure 5. Failures vs. LCC

The management cost (CM) may be calculated by adding up the cost of the specific activities.

$$CM = \sum_{i=1}^{n} CM_i \tag{7}$$

where,

CM_i = The cost of a specific management activity i,

n = Number of activities.

Cost of data (CD) may be calculated in a similar manner.

$$CD = \sum_{i=1}^{n} CD_i \tag{8}$$

Other factors may be evaluated by simple summation such as the cost of training personnel(CT).

$$CT = \sum_{i=1}^{n} CT_i \tag{9}$$

where,

CT_i = Cost of training for student i.

n = Number to be trained.

The maintenance cost (CMT) may be qualified by

$$CMT = C_P + C_S + C_R + C_T + C_{TR} + C_F + C_D \tag{10}$$

where,

C_P = Maintenance personnel cost,

C_S = Support cost,

C_R = Repair materiel cost,

C_T = Test equipment cost,

C_{TR} = Transportation cost,

C_F = Facilities cost,

C_D = Data cost.

A factor which is often overlooked in the LCC is the phase-out cost

(CPO). One may equate the disposal cost to condemnation factor (F_C), the number of corrective maintenance actions (N_{CM}), the actual disposal cost (C_{DIS}) and a disposal value (C_{DV}).

$$CPO = (F_C)(N_{CM})(C_{DIS} - C_{DV}) \tag{11}$$

Whatever the model, one must keep in mind that for the purpose of decision making, options must be developed. The calculated options will then form the basis for trade-offs and in turn, decisions, by management. There must also be a clear indication of the constraints. If the amount of money available for an LCC evaluation of spares is known, for example, there is no sense in calculating spares for a specific life period without this constraint (3). One should also be careful in the apportionment of available funds for spares by not allocating the major portion to low cost spares only.

VALIDATION

Once the details of the design and development phase have been bridged, the technical specifications are formulated for the manufacturing phase. At that stage the cost of validating parameters such as reliability and maintainability enter. The feasibility of the design has been established. At this crossroad many projects have faltered, because the cost of validation against the expected unit price often appears unrealistic. Pressure from management or sales tries to bypass this step as costly and time consuming. The design phase seldom covers the identification of all the support requirements. Most of the major requirements for support are developed during this phase but many items remain obscure. The establishment of cost factors for these items at this stage is often difficult, but if one has to validate the LCC at this stage, a reasonable accuracy of the total cost is expected.

The simulation of these factors by mathematical methods is possible, but the accuracy and confidence one may have in such information may be limited. A close interface between the contractor and the buyer is needed to resolve such cost factors. The prototyping of such support equipment is often not economical. Validation will have to take into account many of these problems and identify the constraints with which the estimates have been made and are being presented. There is no simple solution to this but a full understanding of each other's views, the contractor and the customer, can result in agreeable LCC quantification and validation data, which in turn can be used as contract guarantees. The LCC Contract flow is shown in Figure 6. It indicates in a simplified form the steps which can be taken to arrive at a solution which would allow management to make decisions. The sequence of the steps may be altered as required but the individual steps should not be bypassed.

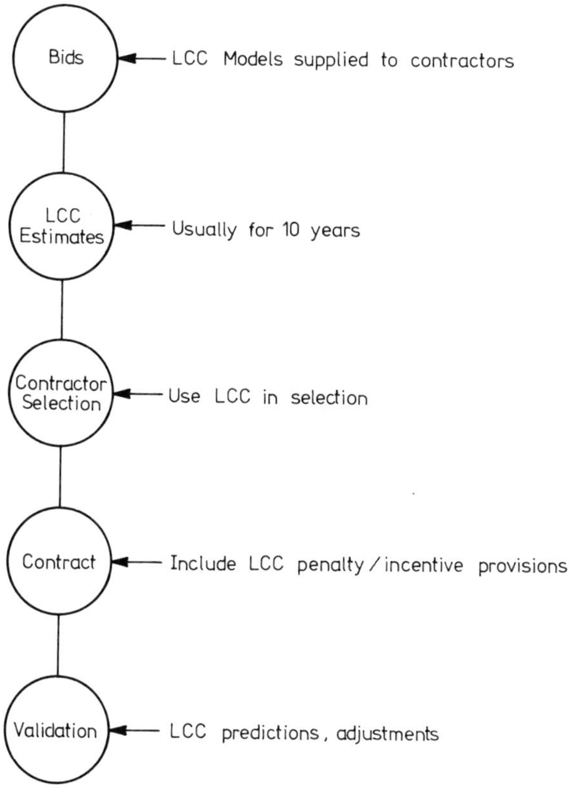

Figure 6. LCC Contract Flow

EXAMPLES

A number of examples are shown here which may help to solve similar or allied problems. Because each examination requires different input factors, quantification of these factors needed for a particular problem should be carried out by consulting historical data if available, and with the help of cost experts who normally have certain basic data at hand which engineers do not have.

Example One

The USAF have been concerned with the cost of aircraft tires. An LCC study indicated that savings could be obtained by making use of LCC data when buying spares. The bid price (C_B) for various companies was added to the shipping cost (C_S) and the maintenance cost (C_{MT}). This value was divided by the index which represented the number of landings achieved for a particular company's tire, i.e.,

$$\frac{C_B + C_S + C_{MT}}{\text{Index}} = \text{C/Landing} \qquad (12)$$

For four different bidders the cost varied from \$6.14 to \$3.47. This information was used to evaluate the number of landings using LCC for purchases and the same without LCC. The increase in landings obtained with the LCC ranged from 46% to 400% for one particular type of tire used on one particular type of aircraft. This kind of information typically indicates the value LCC analysis has to both management and engineering.

Example Two

The Canadian Services have made use of a simple model for costing high value tubes involving tubes cost (C_T), support cost (C_S), and mean time to failure (MTTF), as follows:

$$\frac{C_T + C_S}{\text{MTTF}} = \text{C/Operating Hour} \qquad (13)$$

The support cost includes removal, replacement and transportation. The MTTF was determined in consultation with the manufacturers and a maintenance management information system.

Three bidders were evaluated for the purchase of a magnetron and Table 2 was constructed.

Table 2. Magnetron Evaluation

Manufacturer	Price	Quantity	MTTF
A	4380	98	5000
B	3900	98	7000
C	4750	98	5000

Therefore cost/operating hour for A = 0.89, B = 0.57 and C = 0.85. The preferred manufacturer here is B. It should be stated here that all three manufacturers had military approvals.

Example Three

Before purchasing a portable, multichannel UHF communication set a study of the impact on cost over a ten year period was made. The mean time between failures (MTBF) against cost was calculated, and, as shown in Table 3, there is a direct relationship.

Table 3. MTBF vs. Cost

MTBF	Cost			
	Material	Labour	Manufacturing	Profit
100	6,500	4,000	16,000	2,000
200	8,500	4,200	17,000	2,700
300	12,250	4,800	25,000	3,500
400	20,000	6,600	3,500	5,000

Here the costs were calculated for a single set. Next the maintenance costs per set for one year were calculated based on the following percentages, (a) Personnel 42%, (b) Training 8%, (c) Facilities 21%, (d) Materiel 20%, (e) Test Equipment 4%, and (f) Data 5%. Here, test equipment costs are those for maintenance and depreciation. The percentage values represent the cost contributions to the total maintenance cost.

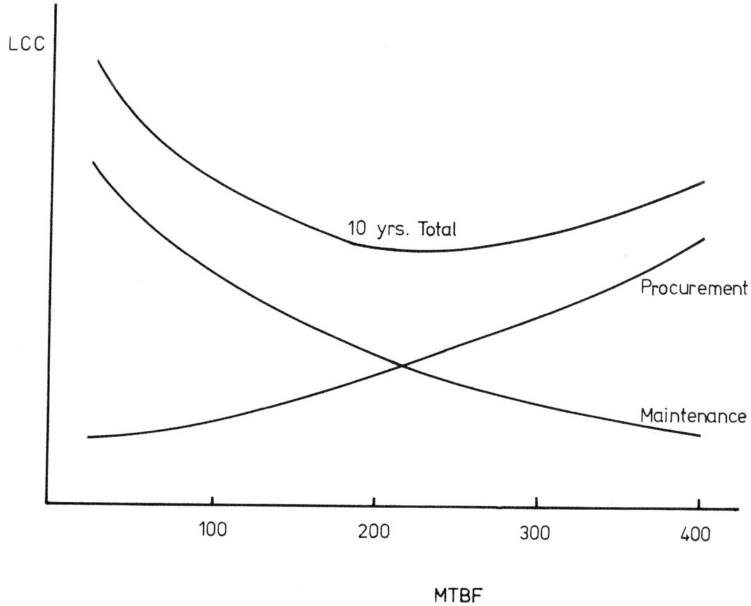

Figure 7. MTBF vs. LCC

If one calculates from all these cost values the cost for the actual number of sets over a 10 year period, using various MTBF figures for comparison, the total LCC can be calculated. The plot of these values is indicated graphically in Figure 7.

The labor cost for corrective maintenance depends on the number of maintainers needed. If the number of systems is (N), the number of operating hours per system per year (H), Mean Time Between Failure (MTBF), Mean Time to Repair (MTTR), then the number of the maintenance man hours per year (MMH) is:

$$MMH = \frac{(N)(H)}{MTBF} \, MTTR \qquad (14)$$

If one now divides equation (14) by the working hours/man year one obtains the number of maintainers needed.

CONCLUSION

LCC is not simply an accounting procedure. Its impact on reliability and maintainability must be understood and a balance of available funds must be maintained to make LCC effective. Because, as in so many management tools, data play a major part in the styling of a task, any available source must be tapped. Expert judgement cannot be replaced with simple mathematical models. The pitfalls encountered should not be discouraging but encouraging engineers to find improved and practical solutions. LCC will not disappear overnight as some may wish, especially in the present economic climate.

REFERENCES

1. Earles, D.R., "LCC - Commercial Application - Ten Years of Life Cycle Costing", Proceedings 1975 Annual Reliability and Maintainability Symposium, IEEE, N.Y., 1975.

2. Jones, M.V., "Estimating Methods and Data Sources Used in Costing Military Systems", Mitre Corp., TM04263.

3. Kiang, T.D., "Life Cycle Management Cost Model", Proceedings 1978 Annual Reliability and Maintainability Symposium, IEEE, N.Y., 1978.

BIBLIOGRAPHY

1. Perdzock, R.C., "Reliability and Maintainability - Keys to Low Total Cost", Fifth Inertial Guidance Test Symposium,

2. Dushman, A., "Influence of Reliability on Life Cycle Cost", ASME, 70-ARC-164, 1970.

3. ____________, "Review of the Application of Life Cycle Costing to the ARC - XXX/ARC-164", Avionic Program Office, Wright Patterson Air Base, Ohio., 1975.

4. ____________, "Life Cycle Costing", Collins Radio Co., Colorado, 10 August, 1966, 533-0758985.

5. Blanchard, B.S., Logistic Engineering and Management, Prentice Hall, Englewood Cliffs, N.J., 1974.

6. Brannon, R.C., Army Life Cycle Cost Model, AD - A021 - 900, Jan., 1976.

7. ____________, Life Cycle Costing Procurement Guide LCC1, LCC2, LCC3, U.S. Department of Defense, Washington D.C., July, 1970.

2 RELIABILITY AND MAINTAINABILITY MANAGEMENT

A. P. Harris

The purpose of this chapter is to provide concepts and guidelines for the successful management of Reliability and Maintainability (R-M) Programs in achieving R-M in products. The general management processes of program definition, organization, control,implementation and training are included. It is recognized that there are many R-M roles and relationships not covered by a contractual arrangement between a customer and a supplier. It attempts therefore to provide useful guidelines for all R-M roles.

This chapter assumes four basic phases during the life cycle of any product and proposes that R-M in the operation of the product is achieved by the implementation of appropriate management and engineering tasks during all four phases. Although demonstration of R-M by test programs is recognized as a task appropriate to some classes of equipment, it is also recognized as inappropriate to others, particularly where large installations are involved.

SCOPE

The following discussion is applicable to the management of all R-M tasks for all products. While the emphasis is on products (systems) the principles contained herein are generally applicable to units, assemblies and parts.

BASIC PREMISES AND PRINCIPLES

This chapter is based on the following premises:

(a) There are four clearly definable phases in the creation of a product, namely, concept and definition, design and development, manufacturing and installation, operation and maintenance. These four phases are illustrated in Figure 1,

(b) There is a special R-M role pertinent to each of these phases. To achieve R-M in the operation and maintenance phase, planned actions and tasks must be carried out during all phases,
(c) The concepts can be applied to any product including those in the electronic, power, nuclear, mechanical and chemical processing fields,
(d) To assemble an effective R-M program during any of the phases requires not only a knowledge of reliability principles but an understanding of the product itself and its technology,
(e) Quality control programs are essential during the manufacturing and installation phase if the desired level of R-M is to be achieved in the operation and maintenance phase,
(f) More effective results can be achieved if R-M activities are managed as part of the product program and not as a separate activity,
(g) The methods and principles can be applied to complete systems, subsystems, equipment assemblies, and components,
(h) The R-M techniques that can be chosen to constitute a R-M program are described herein.

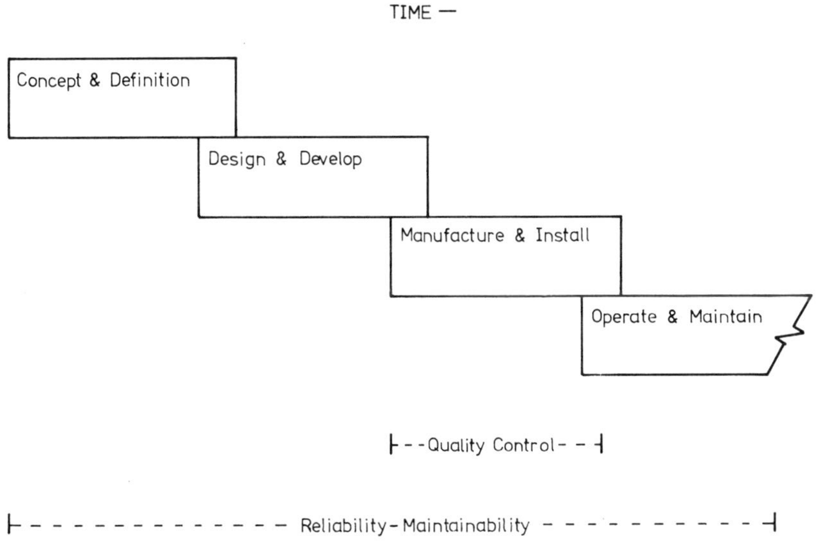

Figure 1. Product Life Cycle Phases

RELIABILITY AND MAINTAINABILITY

R-M methods can be applied to all products and these methods tend to be multi-disciplinary. R-M management is highly dependent on the particular role that organization or R-M group fulfills in the total life cycle of the equipment. It is to a lesser extent also influenced by the technology of the equipment.

Many R-M documents are written on the assumption that in all cases there is a simple customer-supplier relationship, and that the customer will define his requirements specifically and turn his specifications over to a supplier who will then perform the design, development, manufacturing and installation functions. While the customer - supplier relationship is reasonably common, it does not reflect many of the reliability functions and activities pertinent to the consumer product field and other technologies.

The various reliability roles that require management are identified in Table 1. The most common roles are the "owner-user" and the "designer and manufacturer". In addition, some organizations have reliability groups that are solely dedicated to the design and development laboratories or to the manufacturing division. There are also groups with reliability activities who are solely owners and operators of consumer products and these products include domestic appliances and vehicles. The owner/operator or consumer in these cases does not normally have any direct influence over specifications for the concept and definition phase nor have any control over design and development or manufacturing of the product. His control is limited to choice of products or services.

Table 1. Roles in Equipment Life Cycle

Role	Concept and Definition	Design and Develop	Manufacture and Install	Operate and Maintain	Examples
Owner/user	X	-	-	X	Military, Government
Designer/ Manufacturer	-	X	X	-	Industrial Firms
Owner/designer/ user	X	X	-	X	Public Utilities
Manufacturer	-	-	X	-	Industrial
Design and Develop	-	X	-	-	R and D Labs
Owner/Operator	-	-	-	X	Consumer
Conceive/Design/ Manufacture	X	X	X	-	Automobile Co.
Complete service (All phases)	X	X	X	X	Computer Co. Telephone Co.

There are also organizations who fulfill the first three roles but who do not operate and maintain the equipment. These are organizations that provide a complete service to a customer, such as computer companies, vehicle manufactuerers, and most component suppliers.

This chapter therefore examines R-M management on a role basis within each of the four phases of the equipment life cycle. Specifically it proposes management and task guides for those who are, (a) in the concept and definition phase, (b) in the design and development phase, (c) in the manufacturing and installation phase, and (d) in the operation and maintenance phase. Organizations which operate in more than one phase can combine rules and tasks accordingly.

The acquisition of equipment usually involves a customer and a supplier bound together by a contract. It is a routine matter for the customer to specify his R-M requirements and for the supplier to generate plans and take actions to meet these requirements. These requirements become part of the contractual process.

There are, however, many other instances in the electrical field where such an arrangement does not apply but where, R-M programs are still important. An example is consumer products, where requirements are not specified by the customer or consumer, but rather determined by the supplier based on his appraisal of the general market needs. In the four life cycle stages, therefore, the owner or consumer is limited to the operation and maintenance phase.

It is also common to lease computers as a service. In this case, the equipment is designed to meet the need of a certain market segment rather than to the specification of a particular customer. The control of the customer is therefore limited to the operation phase.

THE LIFE CYCLE CONCEPT

All systems and equipment regardless of technology pass through four basic phases from their creation to their removal from service. These four basic phases were illustrated earlier in Figure 1, and are:

(a) Concept and definition phase wherein a need for the product is decided and its basic characteristics defined, usually in the form of a systems or an equipment specification,

(b) Design and development phase wherein the equipment hardware and software is created to perform the functions described in the equipment specification. This phase may conclude with the assembly of a prototype system or equipment and its test to simulate service or an actual service trial,

(c) Manufacturing and installation phase wherein the design is placed into production for service by one or more customers. In the case of very large systems the actual installation on a particular site is regarded as an extension of the manufacturing process. This phase may conclude with a checkout and a process wherein the equipment or installation is turned over to the customer,

(d) Operation and maintenance phase wherein the equipment is operated throughout its useful service life. During this phase the essential repair and maintenance actions are taken and the performance of the equipment is monitored. This phase usually ends when increased cost of repair or other factors render further use of the equipment uneconomic, or when it becomes technologically obsolescent.

It is a basic premise of this chapter that to achieve a desired level of R-M in the operation and maintenance phase, a series of managerial and technical tasks must be coordinated throughout all four phases.

It is a further premise that the managerial and technical tasks necessary in each phase may constitute a manageable program. The choice of tasks that must be performed therefore depends on the phase of activity in which the organization is operating.

It should be noted that in the case of small parts or components, the concept definition and design and development phases may be long past and the supplier is offering a product whose performance is known from much use and service. Occasional recycling back through the design and development phase may be necessary occasionally to meet new requirements or to make improvements in the products.

R-M IN THE CONCEPT AND DEFINITION PHASE

Tasks Involved

In this phase the overall task is to define the functional requirements, R-M, and other related technical constraints (such as environmental requirements and use of standards) and to capture these requirements in a clearly written specification.

The system consideration and economic models that justify particular R-M requirements are not dealt with in this chapter and the reader is referred to Chapter 1.

The specific tasks involved in this phase will include the following:

(a) Definition of terms - agreed terms and definitions should be consistent throughout all phases of the life cycle,

(b) Requirements specification - a recognized and an identifiable specification is required, which defines features and performance requirements, reliability and maintainability requirements, related constraints and management controls placed on contracted designers and manufacturers,

(c) Capability of a product - the purpose of this task is to define the basic performance requirements of the product and its essential features. Departure from the performance stated normally constitutes a failure of the product,

(d) Definition of product R-M target - the purpose of this task is to define the product R-M requirements in quantitative terms and to provide a clear definition of a failure,
(e) Safety - The purpose of this task is to define the safety requirements of the system and control of hazards to personnel and equipment,
(f) Parts control requirements - the purpose of this task is to define component parts control methods to be followed by those in the design and manufacturing phases. It will define the controls to be placed upon component or subsystem suppliers to sustain the reliability and maintainability requirements,
(g) Maintenance philosophy and requirements - in this task both preventive and corrective maintenance actions are considered. The purpose is to indicate the basic approach to the intended maintenance of the product, to indicate the level at which repairs are made, and to describe fault location methods,
(h) Environmental requirements - the purpose of this task is to describe the physical and electrical environments under which the equipment is manufactured, transported, installed and subsequently operated. Guides for defining the environmental requirements are available from documents and publications,
(i) Use of standards - this task involves the definition of design and hardware standards that are to be followed in the design and manufacturing phases with the objectives of supporting the R-M requirements or in controlling the proliferation of parts and materials,
(j) Definition of technical constraints - the purpose of this task is to restrain the use of technologies or design practices that have been found to have a detrimental affect on R-M,
(k) Definition of field study requirements - the purpose of this task is to define the requirements for the collection and analysis of R-M data during a specified period in the operation of the product,
(l) Requirements for operating and maintenance documentation - the purpose of this task is to define the contents of operation and maintenance documentation to be supplied with the product. This documentation may include general engineering descriptions, fault diagnosis techniques, methods of repair, and may include instructions for the original installation of the equipment,
(m) Definition of logistics support - the purpose of this task is to define the intended system of logistic support needed to maintain the equipment in operation. An overall logistic support plan may include elements of tasks defined above,
(n) Training requirements - the purpose of this task is to describe any general or particular training requirements for the operators and maintenance staff. This task must be coordinated with the requirements for technical manuals for operation and maintenance,

(o) Definition of process capability - the purpose of this task is to define the unique combination of machines, tools, test equipment and personnel required to produce the design and its constituent assemblies to specifications and to define the effect of this combination on R-M,

(p) Definition of document control system - the purpose of this task is to describe the system required for the control of the generation, distribution, maintenance, and change of design documentation to assure adherence to R-M.

Management and Control by the Customer

The customer has various controls available to him if he chooses to contract the design and development and the manufacturing of the product to other organizations. These include the following:

(a) Review of requirements with the supplier,
(b) Request for plans and programs from the supplier prior to the design or manufacturing process, (1,2),
(c) Participation in the design review,
(d) Requests for R-M demonstration by tests, (3,4).
(e) Audit of the suppliers programs and results by the customer's representative,
(f) The use of warranties.

R - M IN THE DESIGN DEVELOPMENT AND ENGINEERING PHASE

The implementation of R-M tasks in the design phase is critical. In the total life cycle, the activities which take place in the design phase have the maximum impact on R-M potential. The following tasks are applicable:

(a) R-M Plan - An analysis of the R-M requirement set in the concept and definition phase is the basis of an R-M plan. Those responsible for design of the product are responsible for preparing a R-M plan coincident with the product development plan. This plan is made up of tasks of which the following are typical,
(b) R-M Allocation - Where a system is involved, the R-M goals set in the concept and definition phase may be allocated to subsystems, assemblies, and parts,
(c) R-M Prediction - Prediction of the potential R-M of the system is performed using the appropriate set of failure rates and using appropriate models and procedures,
(d) R-M Analysis - The purpose of this task is to locate R-M weaknesses, by employing such methods as failure mode and effects analysis or fault tree analysis,

(e) Design Reviews - This task is a management task, intended to bring together a group, which reviews a design and points out possible omissions or shortcomings in the equipment that may affect its reliability, its performance or its operation and maintenance. The number of design reviews, their timing and makeup should relate to the complexity and disciplines involved,

(f) Environmental Engineering - Environmental engineering involves deriving an environmental profile for the equipment and its component parts. The environmental profile is a synopsis of the physical and electrical or mechanical stresses that occur during the manufacturing process, during transportation and shipment, and during actual operation of the equipment. Where necessary, evaluation tests simulating these environments are performed on component parts and materials or assemblies to demonstrate that the R-M will not be compromised,

(g) Components Application and Evaluation - This task involves the correct selection of parts and materials to meet the physical, mechanical and electrical stresses imposed by the equipment and therefore must follow the environmental profile analysis above. It is the purpose of this task to ensure that component parts are used within their stress ratings based on known performance data for the part. Where prior knowledge is not available, evaluation testing must be performed. Selection and evaluation of appropriate vendors and their products may be included,

(h) Spares Predictions and Management - This task involves the prediction of spares requirements over given time periods and is based on the maintenance procedures and philosophies expressed in the concept and definition phase,

(i) Safety in Design - This task involves the early identification of hazards and the development of appropriate steps to isolate and control them in the design phase,

(j) Prototype Evaluation - This task involves testing an entire equipment or system or only the unique portion(s) and its interface(s) under conditions which simulate actual operating environments. This simulation may take place in the laboratory, at installations or as a planned experiment in the field,

(k) Preliminary Field Studies - This task involves placing the equipment in actual field service to demonstrate its capability, to see that it performs the intended functions, and to learn of any R-M shortcomings. Accurate record keeping of performance and failures is mandatory,

(l) Parts and Equipment Failure Data - This task involves the collection and analysis of failure information on parts, assemblies, and subsystems which is appropriate for performing the R-M analysis or R-M predictions referred to above.

R-M IN THE MANUFACTURING PHASE

While R-M efforts during this phase do not normally enhance the reliability achieved in the design phase, the design R-M can be degraded by inadequate controls and processes during manufacturing and installation.

During this phase it is necessary to create quality programs with R-M elements whose principal purpose is to control the quality of the manufacturing and installation process and to measure by tests the product at the end of the manufacturing line or after installation. These programs should contain the following tasks:

(a) Quality Control Program - This task involves standard quality control activities. Quality tasks can generally be classified into preventive control measures and corrective control measures,

(b) Vendor Control - This task involves the appraisal of parts, material, and assembly vendors based on a prior examination and analysis of their engineering, manufacturing, and quality control competence. This requirement is normally met by vendors whose product has been accepted under procedures for products of standard quality,

(c) Final Test and Acceptance - This task involves the test of the product to the specification criteria established in the concept and definition phase (as modified by any subsequent agreement) and may be the basis for acceptance of the equipment or system by the customer,

(d) R-M Tests - This task involves the testing of the equipment or system under a set of prescribed conditions for the purpose of determining that the failure rate or mean time between failures meets prescribed levels. This test may be developed from the guides established.

R-M IN THE OPERATION AND MAINTENANCE PHASE

The manager responsible for operating and maintaining the equipment must perform R-M tasks to ensure that the appropriate spares are available for rapid repair and restoration of service, that operator training is established, and that the performance of the equipment is measured and is compared with established criteria. These tasks include the following:

(a) Collection and Analysis of R-M Data - This task involves the record keeping of faults occurring in the system, the diagnosis of primary causes, and the detection of basic patterns of failure that may be correctable by design or manufacturing action,

(b) Spares Prediction and Management - While planning for logistics is set earlier, this task involves ensuring the ready availability of spare parts to replace failed items in the equipment or system. Constraints imposed on this processing include a desire to minimize inventory, a required time for repair and return of the failed item and possible centralizing of the inventory,

(c) Operator Training - This task involves training operators in controlling the equipment, diagnosing its faults, and in updating this knowledge as changes are made to the equipment or system,
(d) Control of Operating and Maintenance Documentation - This task involves ensuring that all operators and their repair crews are provided with current information on the equipment or system so that they can diagnose faults quickly and perform repairs,
(e) Supply of Tools and Test Equipment - This task involves the definition of tools and test equipment required for maintenance, and the development of plans for ensuring that they are available at the appropriate locations when required.

SELECTION OF TASKS FOR AN R-M PROGRAM

In the basic customer-supplier relationship, the R-M program requirements may be completely defined by the customer. While the supplier may wish to negotiate changes during the contractual process or to add tasks for his own benefit, definition of the R-M program has already been accomplished.

For other roles, however, particularly where the product is conceived within the suppliers organization, other guides are necessary. In these instances the selection of tasks for a R-M program for the product will depend on the following general factors:

(a) the level of the R-M requirement,
(b) the phase of the equipment in its life cycle,
(c) the role of the R-M organization,
(d) the nature of the technology and past history of similar equipment,
(e) other project constraints (e.g., schedules and funding available).

R-M groups active in the concept and definition phase, will be primarily interested in establishing a series of constraints or performance goals for design and manufacturing to follow. Their first reliability task, therefore, will be to establish R-M targets or goals for the system or equipment. The technology and its application will influence the choice of the measure of reliability, for example, the measure for a simple communications system may be mean time between failures, the measure for a nuclear power system may be availability, the measure for a component part such as a transformer may be failure rate expressed in terms of failures per million hours or failures per year.

It may also be the role of the R-M group or function in this phase to monitor the activities of those operating in the design and manufacturing phases. For most people a R-M program in the concept and definition phase will include, as a minimum, the following tasks:

(a) Requirements Specification,
(b) Definition of Product R-M Targets.

For most products in the design phase, a minimum reliability program would include:

(a) R-M Prediction,
(b) R-M Analysis,
(c) Design Review,

In the manufacturing phase, a minimum program for most products would include:

(a) Quality Control Program,
(b) Final Test and Acceptance.

In the operation and maintenance phase, a minimum program for most products would involve:

(a) Spares Prediction and Management.

A typical reliability program for systems in the design phase is shown in Figure 2 and a reliability program applicable to new components (electronic or electrical) is illustrated in Table 2.

RELIABILITY ORGANIZATION

Reliability organizations can be classified generally into three types:

(a) Those which establish policy, prgrams, and procedures,
(b) Those which are supportive to the engineering function in any phase,
(c) Those whose function is to monitor, inspect, or audit the activities of others.

Policy and Procedures Organization

These organizations usually operate in, or prior to, the Concept and Definition Phase, and develop broad R-M policies, objectives, and guidelines.

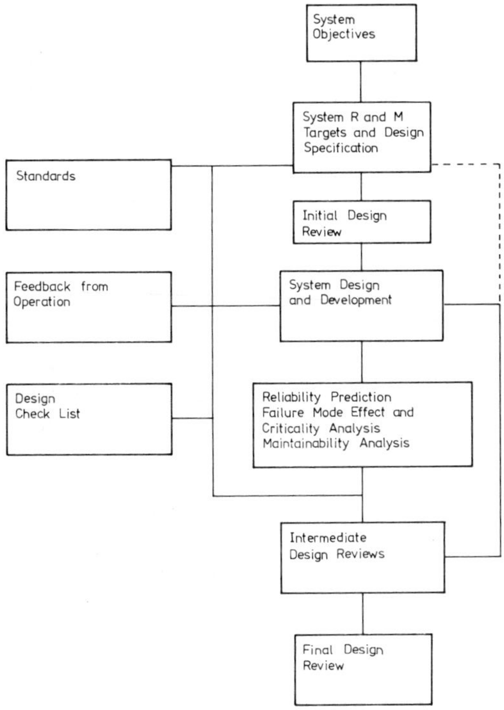

Figure 2. Basic Reliability Program Tasks In System Design

Table 2. Component Reliability Program for New Products

Phase		Task
Concept and Definition Phase	-	Component reliability objectives Component specification
Design Phase	-	Reliability prediction Prototype Evaluation Failure analysis & correction
Manufacturing Phase	-	Qualification tests Final acceptance test Quality control (Process control)
Operation Phase	-	Collection & analysis of rel-iability data. Failure analysis & correction

Support Organization

Support organizations can be directly responsible to the product project manager in some phase of the life cycle. Optionally they can report functionally to the project manager and administratively to a manager of R-M in a so-called "matrix" organization.

In the latter case the manager of R-M is responsible for recruitment, training, and assignment of staff but not for the "line" activity.

Inspection and Audit Organizations

Audit and monitoring organizations are not normally responsible to the project manager in any phase but rather to a senior manager designated as a R-M assurance manager who is preferably on a similar reporting level as the project manager.

When large projects are involved this group may be responsible to the customer (or to another external organization designated by the customer, who is the ultimate owner or user of the equipment) to look after his interests. The customer may also ask the supplier to establish an auditing group.

Preferred Organization

In general, on smaller projects, the supportive organization is preferred as it makes a direct contribution to achieving R-M. On larger projects, combinations may be used; the supportive group established by the supplier and the audit group by the customer. Policy type groups are usually found as part of a customer or user organization. Audit groups not specified by the customer are not recommended.

R-M TRAINING

Because of the scope and interdisciplinary nature of the subject, R-M education and training provide special problems. The kinds of tasks and therefore the training needs of those practicing R-M are highly dependent on the role of the organization in the equipment life cycle and to a lesser degree on the nature of the product. A working knowledge of the technology of the product is essential, but it need not be to the depth required for design. A basic training must include a knowledge of statistics and their practical application. Additional technical training must include environmental engineering, methods of parts control, documentation, and documentation systems. Especially significant are courses in Program Management to promote efficiency in managing large R-M activities. In promoting and gaining acceptance of R-M concepts and program plans, important aids are training in personal communication techniques, including report writing, oral presentations and the use of audio-visual aids.

CONCLUSION

The life cycle concept of R-M allows a systematic approach to any product irrespective of size or technology. At any phase of the life cycle the R-M functions become clear for the organization involved.

If the R-M factors of each phase are tabulated as check lists they provide, a useful tool for both Program Manager and R-M Manager in ensuring that important tasks are not overlooked. Life cycle approaches in R-M are consistent with the total management approach in the creation or acquisition of the product.

For smaller countries required to import technical equipment, the approach will be useful in defining requirements for supplier and in choosing tasks for audit and inspection of the product. Further, it indicates the amount of effort which must be exerted by the purchaser in his main role in obtaining satisfactory service - namely, planned operation and maintenance.

REFERENCES

1. ___________, MIL-STD-785, Reliability Programs for Systems and Equipment Development and Production, U.S. Department of Defense, Washington, D.C.

2. ___________, MIL-STD-470, Maintainability Program Requirements, U.S. Department of Defense, Washington, D.C.

3. ___________, MIL-STD-781, Reliability Tests, Exponential Distribution, Department of Defense, Washington, D.C.

4. ___________, MIL-STD-471, Maintainability Verification/Demonstration/Evaluation, Department of Defense, Washington, D.C.

BIBLIOGRAPHY

1. Harris, A.P., "Reliability Programs", Proceedings of the Northern Electric Reliability Seminar, BNR, Ottawa, Ontario, 7-8 April, 1970.

2. Harris, A.P., "Reliability and Quality Strategy", Canadian Nuclear Association Educational Course, 4 November, 1974.

3. ___________, "Guidelines for Reliability Management", Proposed IEC Standard 56, July, 1977.

3 DESIGN REVIEWS

M. B. Darch

The purpose of engineering design is to develop a basic concept into a piece of equipment that can be reproduced in quantity and placed into use. Each unit is not designed separately but rather a single unit is designed to serve as a template for the rest. An inherent problem with this process is that individual errors made in the design process will be automatically built into each unit during the production run. If these errors are of a nature which affect the ability of the equipment to meet its requirements, corrective action can have a significant impact on the life cycle cost of that equipment.

Design review is a process whereby engineers, independent of the design process, review the design to uncover possible faults. These reviews should occur at the following times:

(a) during the conceptualization of the project,
(b) upon completion of the preliminary systems design,
(c) upon completion of detailed equipment design,
(d) prior to commencement of production, and
(e) periodically during the useful life of the system.

Figure 1 relates these reviews to the design process. This chapter will discuss the design review process from the viewpoint of the Reliability and Maintainability (R-M) engineer. Particular emphasis will be given to the cost implications of design changes at each point in the process.

GENERAL

The Principle of Independent Review

A design review can occur in two different contexts. The first is an internal review by a company of its own products while the second is the review of a design prepared under outside contract. Although the two cases appear to be of a totally different nature, the procedures involved in both should be essentially the same. An underlying problem of any engineering design is that it reflects the education, perceptions and preferences of the design team. The purpose of the design review is to examine the design and ensure that it meets the requirements in a cost effective manner. For the reviews to be conducted properly, the reviewers must be able to stand back from the design to better see any insufficiencies. To do this, the review team must be expert in the field of interest and must also be independent of the design team. This reduces the possibility of the review team missing a design fault due to a familiarity with the design and opens the design process to a greater diversity of opinion and expertise. Regardless of whether the review is being performed internally or externally, it should proceed in a manner which ensures independence of the design and design review teams.

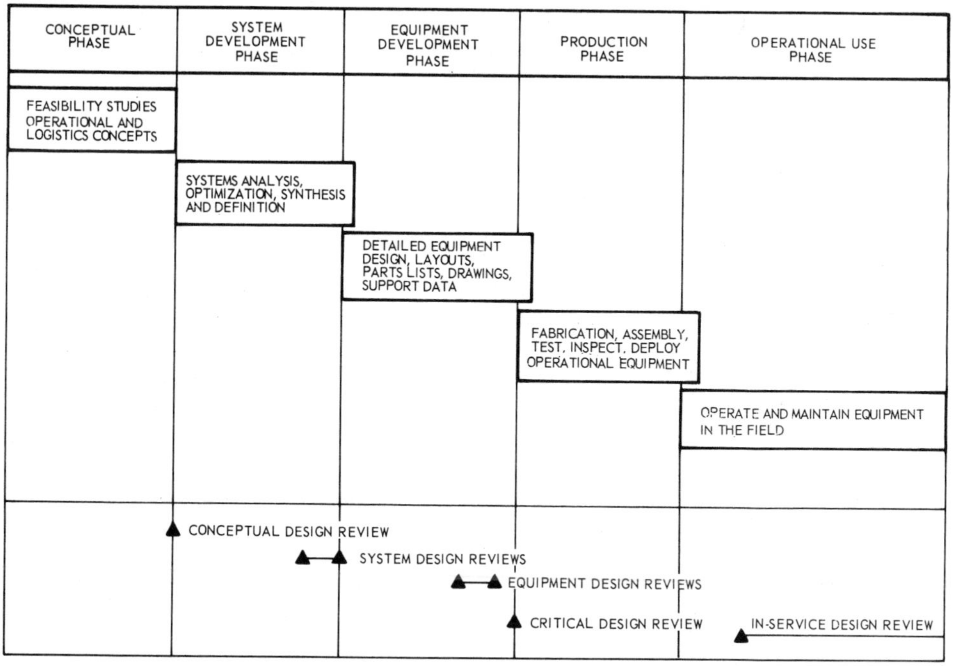

Figure 1. Design Reviews Vs. Development Phase

The Benefits of Design Review

The principle goal of design review is to ensure, at each stage of the design process, that a product is being developed which will meet the stated requirements in the most cost effective manner. The structure of the review process also leads to a variety of secondary benefits.

In the previous section, it was stated that the reviewers should be independent of the design process. This allows an outside perspective to be brought to the design, shedding fresh light on various problems. If the designer knows that he must defend his work at various times, he will take greater care to document the basis for any decision.

The design of large systems requires considerable interface between a variety of engineering disciplines. A design review is a formal opportunity for the design team to meet and examine specifically the various areas of interface. It is inevitable that some problems will arise that are difficult to resolve. The design reviews provide a forum for discussion of these interface problems and their resolution.

Another important benefit is that a formal design review establishes a common base for future work. The design to that point is analyzed, altered as necessary and then documented. All studies, testing, and decisions prior to the review are tabled, discussed and recorded. This provides the design team with a clear indication of what has been accomplished to that point. From there, the design team has a common base from which to proceed with the next stage in the design process.

The Decision Horizon

As a design proceeds towards completion, an increasing amount of the design becomes fixed. The approval of concepts, philosophies and specifications lends a certain character to the final design and therefore to the life cycle costs. If at the end of the conceptualization stage a certain philosophy is accepted, system designers and equipment designers will accept that philosophy and build around it. The nature of the design process is such that after a stage is completed, the results from that stage are accepted as given and the next stage proceeds from there. It is theoretically possible to go back and start again if a fundamental fault is found but the cost in time and dollars makes this a very undesirable course of action.

Figures 2 and 3 relate dollars expenditures and percent of locked-in life cycle costs to the life cycle of a project. These figures are held as being representative for the U.S. Department of Defense. Figure 2 shows that the design and development phase of a project consumed only 15% of the cost of a typical project, as opposed to 35% for the production phase and 50% for the in-service phase. Although only 15% of the expenditures were made prior to production, Figure 3 shows that about 90-95% of the life cycle costs were determined. The design specifications that were approved prior to production determined how it would proceed and, therefore, determined the costs to be incurred in that phase. Similarily the detailed specifications were produced based upon a certain operational, maintenance and supply support policy. These policies and the design dictate such in-service variables as manpower, consumables and spares levels.

The significance of these figures should not be lost on the (R-M) engineer involved in design review. Prior to the conceptual design review, 100% of the design can be altered and 100% of the life cycle cost can be affected. Completion of the conceptual design review gives approval for the basic framework of the design. The concepts approved, although not a written set of specifications, place constraints on the design team, narrow their decision horizon and fix a certain level of the life cycle cost on the project. As the reviews progress, the decision horizon narrows and a greater percentage of the life cycle costs become determined. During the review process, the reviewer must be aware that each stage fixes a greater part of the design and hence of the life cycle costs.

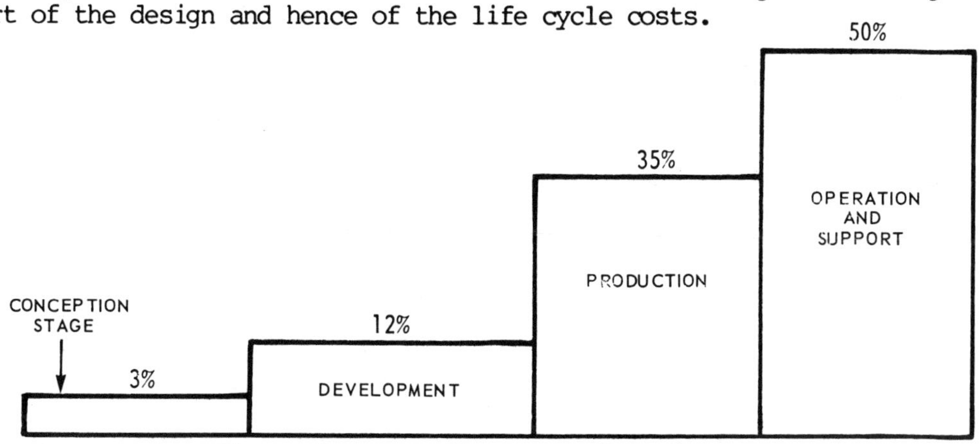

Figure 2. Expenditures During Life Cycle

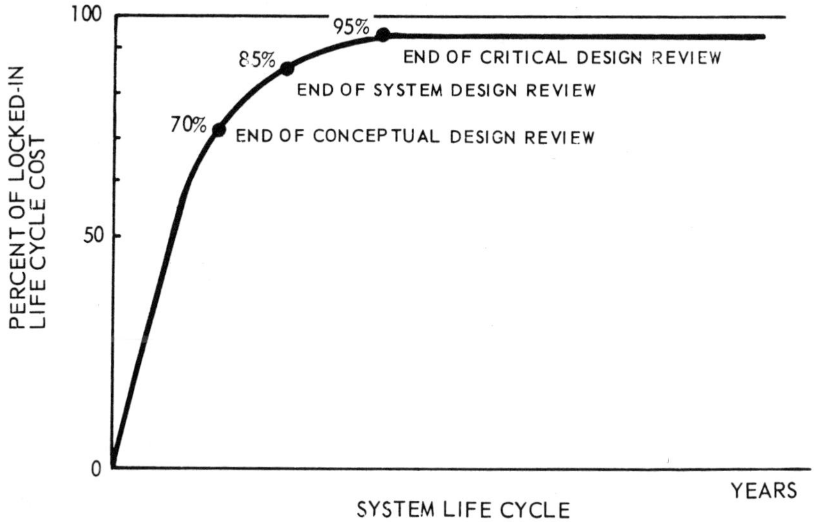

Figure 3. Effect of Early Decision on Life Cycle Cost

The Conceptual Design Review

The first review, and possibly the most important, is the conceptual design review. As the name implies, this review occurs at the end of the conceptual phase. This stage determines the basic approaches, concepts and philosophies that will be used to develop the design. The basic idea or requirement is taken and a strategy for meeting that requirement is developed. Various feasibility studies may have been performed on alternate technologies, different support strategies, etc. The output may include reliability targets, maintenance philosophies, operational parameters and design concepts. The purpose has not been to produce the detailed design but rather to develop the framework within which the detailed design can be built. The importance of this review can be seen in Figure 3. Approval of the design concepts can determine 70% of the total life costs. Its framework determines such things as the maintenance policy, the supply support policy and the basic form of the equipment. These in turn imply a certain technology, certain manpower levels and certain materiel requirements. No hard design has been done but limits and constraints have been imposed that place a definite form on the design. To illustrate this point better, consider the requirement to intercept, identify and possibly destroy aircraft infringing on a country's territorial limits. If the conceptual study determines that this is done best by a manned jet fighter, capable of flying at Mach 2, an immediate estimate could be given of the approximate cost of such a system. The reviewer must therefore be well aware that although the decisions taken at the conceptual design review are not detailed specifications, the approval of basic concepts for the design in fact locks the design process into certain approaches and does have a significant cost impact. At this review, the R-M engineer is concerned with ensuring that sound principles of R-M are being used in the framework. From the reliability viewpoint, the following principles should be adopted:

(a) Reliability goals should be appropriate to the requirement and be realistic,
(b) New technology should be used but the risks should be stated,
(c) Failure of piece parts or sub-systems should have as little effect as possible on total system reliability, and
(d) Proper consideration must be taken of the effect of the operational environment on reliability.

Concerning maintainability the following points apply:

(a) Requirements to disrupt the system for maintenance should be minimized, i.e., minimize preventive maintenance,
(b) Modules should be used to as great an extent as possible,
(c) Components should be grouped functionally,
(d) Skill level requirements for field maintainers should be as low as possible,
(e) Maintenance and supply support policies should be clearly stated and should match the operational environment,
(f) Self-test capabilities should be built into the system, and
(g) Requirements for support equipment should be minimized.

In the conceptual design review, the reviewer should ensure that the above principles have been accepted. If these principles are firmly embedded into the design framework, it becomes much easier to ensure that the principles are adopted into the detail design. Also at this stage, the reviewer must be satisfied that the stated maintenance and supply support concepts are compatible with the requirements and with the basic design philosophy. If R-M targets have been established, the reasoning and supporting data behind the estimates must be evaluated to ensure the validity of the targets.

System Design Review

At this stage, the design is beginning to take on a more concrete form. The overall systems layout has been determined complete with preliminary systems specifications. Initial testing, systems trade-off studies and system optimization studies have been carried out. The framework has been established within which the detailed equipment design can proceed. The perspective of the reviewers has changed from examining general concepts and philosophies to the examination of the application of these concepts to systems design. His concern is beginning to focus on a more detailed design rather than the discussion of ideas.

Concerning reliability, the design must be examined to ensure that system integration is optimized with the interdependencies clearly stated. The design is not the product of an individual but rather of a group. The system configuration must therefore be investigated to determine if all aspects of the design are compatible. Documentation, to support system reliability figures and trade-off studies, should be examined to determine the validity of the conclusions reached.

The concepts of functional grouping and modularity should be prevalent in the system's design. Fault isolation to a module should be possible using either built-in test equipment or commonly available test equipment. The complexity of the system design should be such that the skill level required of technicians to isolate faults is at as low a level as possible. The conceptual design review outlined maintenance and a supply support concept to guide system development. The system design should be reviewed to ensure that it is compatible with the proposed maintenance and supply support concepts.

With the design taking on a firmer shape, the reviewer should be looking for more design work supported by Life Cycle Cost (LCC) studies. R-M is an engineering field in which cost savings are not realized in the design or production stages but rather when the system enters into service. For the production engineer it is rather easy to demonstrate the cost benefits that are derived from design improvements which reduce production costs. These savings will reduce the capital cost of the system and, therefore, are an evident benefit to the system. Improved maintainability and reliability usually necessitate a higher capital cost but offer lower operating and maintenance costs. Life Cycle Cost is a relatively new field and, historically, acquisition decisions have been made primarily on capital cost. For the R-M engineer this means that he must have a strong case to justify the increased capital cost and demonstrate the reduced operating and maintenance costs.

Equipment Design Review

The design has now progressed to the point where the equipment design is complete. Prototypes have been completed as well as testing and trade-off studies. Preliminary production drawings have been prepared together with final specifications. The perspective of the reviewer has narrowed to the fine details of the equipment design. There are no longer abstract ideas and concepts to evaluate but rather detailed drawings, supported by extensive testing and evaluation.

With the completion of the equipment design, reliability targets have been converted to reliability estimates. The test reports and the analysis of those reports must be reviewed to verify the conclusions reached. Clear and concise documentation should be presented which outlines component reliabilities and how these components interact to produce sub-system and system reliabilities. If new technology is being used that has not had extensive testing, a risk analysis should be available outlining the degree of uncertainty associated with the reliability estimate and the effects on system reliability and cost in the event of the estimate not being valid. Particular emphasis should be placed on determining components which have a major effect on system reliability, ensuring that the reliability of such components is as cost effective as possible.

From the maintainability viewpoint, standardization is a key word. Component parts, equipment layouts, packaging and necessary support equipment should be as standardized as possible. This will help to reduce replacement spares, manpower and training costs, as well as assuring better availability of replacement parts. Equipment fault isolation should be as simple as possible and require as low a skill level as possible. Adequate consideration should be taken as to the level to which repair will occur, i.e., at what level do you replace rather than repair. Another important point is access and ease of removal of repairable components. The design should be scrutinized to ensure that low reliability components requiring scheduled maintenance are readily removable and do not require extensive dismantling of other components in the system.

During the detailed design, it is inevitable that some trade-offs between reliability and operational performance will have to be made. These trade-offs must be well documented with a clear statement of the available alternatives, the cost factors involved and the implications of acceptance of the recommended course of action. Life Cycle Cost studies are again important because of the long term nature of the benefits to be derived from improved R-M.

Critical Design Review

The critical design review should be performed after the final production drawings have been completed but prior to their release. The design is essentially frozen with the review being a "last look" to ensure that the optimum design has been developed. The conceptual design review examined the basic concepts and philosophies to be used in the design. The system design review narrowed the scope to examine the system design within the context of the concepts approved at the end of the conceptual phase. The equipment design further narrowed the scope to ensure that the equipment design fit within the system design. The critical design review once again broadens the scope of the review to determine if the final design best meets the operational, maintenance, and supply support requirements.

The importance of the critical design review can be seen in Figures 2 and 3. Figure 3 shows that the design development work has expended an average of 15% of the total life costs for a system while Figure 2 demonstrates that the design produced fixes 90-95% of the life cost, due to materials. Approval of the design is a commitment to a certain cost level and future programs for cost savings have little impact relative to the total life costs.

The critical design review must, therefore, represent a comprehensive analysis of the design versus the intial requirement. All supporting material must be reviewed to ensure that the design is cost effective and that it will meet its stated reliability and performance estimates. It is the last chance before total commitment.

In-Service Design Review

A basic principle of sound system design is feedback to uncover and correct errors. Although this principle is well accepted for errors that have occured up to release of final production drawings, it is rarely applied after that. For the R-M engineer, the most important feedback is from the in-service stage. As mentioned earlier, improvements in R-M do not appear during development or production but rather when the system enters into service.

An in-service review cannot change the design of the system since it has already been manufactured. The review is an analysis of the predicted reliability and support costs versus those that are being experienced. The reviewer must examine the documentation supporting the design to extract the basis for in-service estimates. Actual usage data must then be examined to determine how the theory translated into reality. If predictions were not met, an analysis should be done to determine why this happened. Besides poor design such uncontrollable factors as parts shortage, inflation, or manpower shortages can be the cause of wide variance in predicted versus actual costs.

As stated earlier, a design reflects the knowledge, preferences and prejudices of the designers. The knowledge learned during in-service reviews will not be reflected in the current system but rather in the design of future systems. The designer can determine if the new concepts and ideas built into the design have met their expectations. Any data collected for this review provides a good data source for use in Life Cycle Cost studies in the design phase of related systems.

CONCLUSION

The formal design review provides the opportunity for an independent, objective look at critical points during the design process. These reviews introduce new perceptions and viewpoints to the process and encourage more self review on the part of the designer. The formal meeting of the design team and the reviewers enhances the interchange of ideas between team members and therefore assists in the co-ordination of the diverse design elements. A well-planned review that uses technically competent, independent reviewers greatly decreases the risk of design flaws in the system as it enters production. Such action plays a key role in ensuring that systems are designed, produced and operated with the lowest possible life cycle cost.

BIBLIOGRAPHY

1. Ankenbrandt, F.L., ed., Electronic Maintainability, Electronic Industries Association, Reinhold Publishing, New York, 1960.

2. ________, Configuration Management During Definition and Acquisition Phases, AFSCM 375-1, Wright-Patterson AFB, Ohio 45433.

3. Blanchard, B.S., Logistics Engineering and Management, Prentice-Hall Englewood Cliffs, N.J., 1974, p 376-389.

4. Cunningham, C.E. and Cox, W., Applied Maintainability Engineering, J.Wiley and Sons, New York, 1972, p 147-176.

5. Smith, D.J. and Babb, A.H., Maintainability Engineering, Pitman Publishing, London, 1973, p30-40.

6. ________, MIL-STD-1521 Technical Reviewers and Audits for Systems, Equipment, and Computer Program, U.S. Department of Defense, Washington, D.C.

4 RELIABILITY, MAINTAINABILITY AND DESIGN AUTOMATION

J. E. Arsenault and P. J. DesMarais

The point has nearly been reached in electronic system design where what can be conceived can, in fact, be realized practically. This state of affairs has evolved because of, (a) the progress of the semiconductor industry in increasing the level of hardware integration, (b) the efforts of system workers in optimizing system architectures, (c) the growing repertoire of mathematical techniques for information processing. So it is normal to expect complexity in systems, as the above factors are accommodated to achieve higher and wider performance requirements. However, complexity introduces its own set of related problems, i.e., how to design for sustained performance (reliability) and how to design for detection and isolation of faults (maintainability). The problem of complexity requires an approach whereby the designers can get above the problem of complexity and let the finer design details largely take care of themselves within economic and time constraints. Design Automation (DA) systems are conceived with this purpose in mind. This chapter first discusses DA systems in general and then their influences on system reliability and maintainability design.

DESIGN AUTOMATION SYSTEMS

General

Design automation has been aptly defined (1) as "the art of utilitizing digital computers to help generate, check, and record the data and the documents that constitute the design of a digital system". Its objective is straightforward: to reduce the cost and development time required in engineering complex electronic systems, from conceptual design to finalized production. This goal is basically achieved by relieving the design and manufacturing engineers of tedious, error-prone, and time-consuming manual tasks which, as will be outlined, involve the use of a wide variety of computer-based techniques.

Rapid advances in technology, in particular the introduction of Large Scale Integration (LSI), has led to a level of sophistication such that increasing reliance on computers proves necessary to surmount the complexity barrier. Today, a DA system is structured typically as illustrated in Figure 1, supporting a wide range of engineering activities from design and its evaluation, to hardware implementation, to maintenance and documentation.

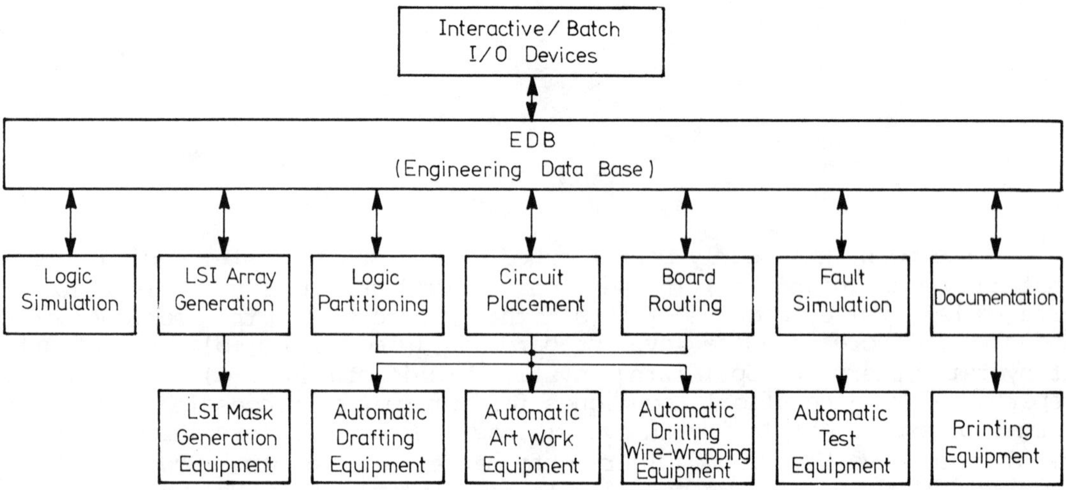

Figure 1. Typical DA System

The Engineering Data Base (EDB) and Its Management

As depicted in Figure 1, the Engineering Data Base (EDB) plays a central role within a DA system. In addition to storage of component model libraries, it holds design and production data such as simulation results, interconnection data, parts and wiring lists, etc., as well as various management information, effectively providing the basis for a fully integrated DA system. As such, the EDB constitutes a key element towards maximizing the cost-effectiveness of the system. On the one hand, it promotes efficient use of the system via the compatibility of the various DA modules whereby information repeatedly used by different programs need be entered only once into the EDB. Extensive consistency and cross-reference checking is usually performed on input data to ensure greater accuracy. On the other hand, the EDB provides a readily accessible yet well-controlled source of documentation throughout the design and production phases, a critical requirement towards assuring fast and orderly development. However, the benefits of the EDB can be fully realized only if its management and its interaction with DA program modules and Input/Output (I/O) devices is supported by an adequate operating system. Ideally, the latter should provide, in addition to straightforward batch procedures, highly flexible interactive capabilities that promote ease of use, hence faster design turnaround.

Design Verification

The design of a system generally goes through two principal phases: (a) the conceptual phase (or system specification), where the architecture of the system is defined in terms of major building blocks and their interfaces, and (b) the logic implementation phase, where the functions of each subsystem are detailed down to the logic component level.

Similarly, logic simulators offer two levels of analysis: (a) high level design language (also called functional or register transfer) simulation, which is aimed at providing a detailed evaluation of the conceptual design while remaining efficient of both computer time and storage, (b) gate level simulation which provides a more extensive verification of the logic design in terms of predicted logic behaviour and detailed timing analysis to flag race and hazard conditions.

Immediate benefits of logic simulation include significantly reduced debugging time and reduced likelihood of uncovering costly design errors late in the development process. This is most valuable in LSI design where repair is impossible once the circuit is fabricated. In addition, the engineer is able to explore new ideas and techniques by using the computer to obtain fast and accurate results, and he can quickly acquire a better understanding of the problems at hand while gaining valuable experience of different design procedures.

However, as outlined above, the participation of the computer in the actual design process is still limited to a large extent, to that of a passive, verification role. Indeed, major problems have yet to be solved in the area of complex system design, namely the need for: (a) more active computer assistance in high-level specification of logic circuits, (b) implementation techniques based on subsystem modules, and (c) the development of a methodology for complex system design and evaluation.

Physical Design

Upon completion, the system logic design must be implemented with hardware building blocks. Again, DA plays a very important role in assuring that the various physical design tasks are performed rapidly and accurately.

Electronic systems generally consist of subsystem units, each holding a number of circuit card assemblies, each in turn holding a number of Integrated Circuit (IC) components. Endeavouring to minimize total assembly costs while providing adequate reliability and maintainability standards, physical design deals with three basic problems: (a) logic partitioning, which involves, first, the selection of ICs from a library of available components so as to minimize both the number of ICs used and the number of interconnections required, and secondly, the assignment of ICs to various circuit cards so as to minimize the number of cards needed while promoting testability of the system via a functionally oriented partitioning scheme, (b) placement, both of components on a card, and cards on a backplane, where optimization aims generally at minimizing interconnection wire length. Note however, that several other (sometimes conflicting) goals must also be satisfied, such as avoiding large wire buildup in the routing channels, eliminating signal cross talk and echoes, providing well-distributed heat dissipation in order to avoid severe heat source concentrations, and, all the while, reducing manufacturing and maintenance costs by promoting ease and neatness of wirability; (c) routing, which consists of defining, again both at card and backplane level, the precise conductor paths required to interconnect system components properly. In this case, the main problem is that conductors cannot be allowed to cross, while satisfying other requirements such as minimum wire size and spacing, number of board layers etc.

Given the complexity of present-day systems, a purely manual approach to solving these problems becomes totally inadequate. On the other hand, DA systems offer a variety of efficient techniques (1) yielding heuristically good solutions which may not be minimal but provide an acceptable compromise between all-too-often conflicting goals.

Maintenance and Documentation

Much in the same way that design simulation verifies that a given logic design behaves as intended, fault simulation predicts how faulty versions of this same design can be expected to respond, thereby providing a very strong basis upon which to diagnose faulty circuits, ideally through the use of Automatic Test Equipment (ATE). In fact, at all system levels, the task of testing logic networks has become so complex that only through the use of computerized techniques (2) can it be achieved efficiently to an adequate level of diagnostic effectiveness. In return, fault simulation exercises a strong influence on logic design testability requirements (3) which, however, proves to be a very small price to pay for its tremendous benefits.

Fault simulation is aimed primarily at supporting production testing and field maintenance, and for that reason, it is generally used only at late stages of system development when logic design has stabilized. However, recognizing the importance of designing testability early in the development process, it becomes debatable whether, in spite of its relatively high cost in computer time, fault simulation should not be used during the design phase as a means of evaluating the diagnosability of the system.

Another important aspect of maintenance is the need for complete and accurate documentation. In that respect, DA systems generally include programs which can quickly produce a wide variety of documents throughout system development, such as parts lists, wiring lists, circuit-location charts, system diagrams, maintenance schedules, etc. More importantly however, DA provides the capability of readily and accurately maintaining up-to-date documentation throughout the life of the system.

RELIABILITY AND DA

General

The most significant effect a full DA system has on electronic system reliability, results from the rapid throughput of this approach to design. This permits, (a) several significantly different design solutions to be evaluated (e.g., different part technologies, different packaging approaches) within the usual development cycle and thus provides for a final selection for production after extensive testing, (b) minimization of extensive re-design associated with development programs, thus allowing the development of a design solution approaching the inherent reliability without a long period of reliability growth.

The effects of DA on system reliability are discussed below under four topics, i.e., (a) Part, (b) Circuit Design, (c) System, and (d) Manufacturing Considerations.

Part Considerations

As parts play a critical role in determining system reliability, various techniques have evolved over the years to (a) ensure that only parts with a proven reliability history are used, (b) that they are second sourced and (c) that they are readily available at competitive prices. Some organizations approach the problem by preparing a Preferred Parts List (PPL) which is continuously updated and issued as a standard. Large projects will sometimes have a PPL imposed by contract. Large projects also may employ a Parts Control Board whereby the prime contractor holds regular meetings at each of his major subcontractors in order that part commonality is ensured to the maximum extent possible. Despite the above efforts, control is still an elusive goal for reasons due mainly to project pressures.

With DA the designer is constrained, at least initially, to implement his design only with those parts available in the EDB. Since from the start of the design, the designer is limited automatically in the choice of parts he may use, then, an ideal PPL situation has been achieved. If new parts are found to be required, for any reason, first they must satisfy criteria (a) to (c) mentioned above, to the maximum extent possible, and thereafter be placed in the EDB for use.

The complexity of parts in the EDB can range from the simplest to the most complex available in the industry. Modelling at the gate level is a feasible solution for parts up to about 500 gates, beyond which functional simulation can be used. Table 1 illustrates the range of past EDP complexity.

Circuit Design Considerations

Some of the tasks now performed manually and/or individually by reliability engineers can be integrated within a DA system to give the design rapid reliability assessments.

Table 1. EBD Complexity Range

SSI (G < 10)	MSI (10 < G < 100)	LSI (100 < G < 1000)	VLSI (G > 1000)
5400 (NAND)	5480 (ADD)	2901 (uP)	2107 (RAM)
5404 (HEX INV)	54150 (MUX)	8080 (uP)	2708 (EPROM)
5474 (FF)	54195 (SR)		9900 (uP)

G = Equivalent Gates

Stress Analysis. For purely digital systems, stress analysis is usually confined to ensuring that integrated circuit loading rules are not exceeded, in some cases ensuring that the number of loads is some percent of the rated loading. This activity is quite manageable when dealing with, for example, a single area of circuitry bounded by a schematic where all lines between parts are immediately apparent. However, such is not the situation for those cases where circuitry is represented by multiple schematics or where a wiring list must be consulted to check interconnection, thence loading. For all of the above cases, a program can be devised to check device loading and report those cases where overloading has occurred. For extensive analog circuitry some programs contain features where component stress is computed and provided to the user.

A case in point is SYSCAP (4) which consists of a number of computer programs with an extensive choice of options. Two useful options are DICAP and TRACAP, which when used together can analyze electrical stresses on the parts comprising a circuit. DICAP can be used to obtain DC stresses for nominal and worst-case conditions and TRACAP can simulate parts stresses from transient and/or duty-cycle circuit operations. The SOPSTO option of DICAP also detects any potential overstress arising from multiple power supplies that increase from zero to rated voltage in a disproportionate manner. A nominal DC analysis is given in Figure 2.

Worst Case Analysis In digital systems many problems are related to timing, such as clock skew, hazard, race and oscillation conditions. With an accurate EDB, i.e., accurate timing relationships for the parts being used in a detailed design, these conditions can be predicted. Figure 3 shows an example of a timing analysis.

Programs are available for analog circuitry which provide worst case analysis where the effect of parameter drift caused by aging, for example, can be taken into account. Design tolerance analysis also may be provided by using Monte Carlo techniques.

INITIAL CONDITIONS

MATRIX SOLUTIONS= 19

NODE VOLTAGE AND AUXILIARY ANSWERS

1	NODE	1	0.	13	NODE	13	-.52861E+00	25	NODE	51	-.20459E+00
2	NODE	26	.10000E+02	14	NODE	14	.26001E+00	26	Q1	PT	.48097E+01
3	NODE	3	-.54723E+00	15	NODE	15	-.29964E+02	27	Q2	PT	.48097E+01
4	NODE	4	-.40664E-02	16	NODE	16	-.29964E+02	28	Q4	PT	.10230E+00
5	NODE	5	-.70498E-04	17	NODE	17	.83234E-02	29	Q6	PT	.12797E+00
6	NODE	6	.89151E+00	18	NODE	18	-.29149E+02	30	Q3	PT	.34551E+01
7	NODE	7	-.11791E+01	19	NODE	19	-.30000E+02	31	Q5	PT	.49031E+00
8	NODE	8	.14264E+02	20	NODE	20	.25140E+00	32	Q7	PT	.13035E+00
9	NODE	28	-.15000E+02	21	NODE	29	-.14264E+02	33	D1	IDT	.17197E-01
10	NODE	10	-.20458E+00	22	NODE	47	.53850E-02	34	D2	IDT	.17316E-01
11	NODE	11	-.57269E-01	23	NODE	49	.15000E+02	35	D3	IDT	.61755E-04
12	NODE	12	.93038E+01	24	NODE	50	-.15041E+01	36	D4	IDT	.42964E-05

Figure 2. DC Nominal Analysis

```
THE FOLLOWING SIGNALS WILL BE DIAGRAMMED FROM START TIME      0 TO END TIME
   A
   B
   N1
   N2

                          -----------------------------------------
                          I                                       I
                ----------                                         ----------
A               0    5    10   15   20   25   30   35   40   45   50   55   60

                          -----------------------------------------------------
                          I
                ----------
B               0    5    10   15   20   25   30   35   40   45   50   55   60

                XXXX-----------   -----   -----   -----   -----   ----------
                XXXX           I   I   I   I   I   I   I   I   I   I
                XXXX           -----   -----   -----   -----   -----
N1              0    '    10   15   20   25   30   35   40   45   50   55   60

                XXXX-----------   -----   -----   -----   -----   -----
                XXXX           I   I   I   I   I   I   I   I   I   I   I
                XXXX           -----   -----   -----   -----   -----   -------
N2              0    5    10   15   20   25   30   35   40   45   50   55   60

   END TIMING DIAGRAM OPTION
END OF TEGAS3 RUN.
```

Figure 3. Timing Analysis

System Considerations

This equipment level represents the highest level of concern at which the designer can evaluate the various system trade-offs. Although there has been some progress in high level system simulation through the use of Register Transfer Language (RTL), these languages are most useful in system conceptual design. RTL type languages are used to declare proposed register structures and logical operations within them. Microprograms are then devised to control the data-processing operations between the registers. The bulk of simulation for design continues to be done using part models as the basic elements.

Mean Time Between Failure (MTBF). Typically, reliability engineers calculate MTBF based on failure rate sources (5) either completely by manual methods or by a stand-alone computer program. This approach also applies to system reliability modelling which is accomplished manually or via a stand-alone computer program.

Because a designer eventually will implement his design at the part level, a set of programs can be interfaced with the EDB to compute the MTBF for any design specified. An additional set of programs can be used to perform system modelling, for example, redundancy given in MTBFs of the individual designs involved. In addition, it will be possible to vary parameters such as the environment in which the system is required to perform, part quality, temperature, and so on, all of which impact MTBF. Figure 4 shows an example of the output from such a program which can be interfaced to an EDB for generating MTBF estimates leading to design trade-offs.

R ESTIMATE COMP RUN NO. 5 18 FEBRUARY 1977 PAGE 19

*** COMPONENT SUMMARY ***

BLOCK 7 I/P CCA 147 1 NO REV

MISSION TIME 24.0 HOURS ENVIRONMENT: GROUND, MOBILE

COMPONENT	TEMP DEG. C	STRESS RATIO	REL. GRADE	APPL. FACTOR	FRS. /MILLION HRS.	QTY.	TFRS. /MILLION HRS.
RN STYLE RESISTOR PACK U5	75.	0.00	SM	4.0	.0853	1.	.0853
RN STYLE RESISTOR PACK U5	75.	.10	SM	1.0	.0242	1.	.0242
MIC OP AMPLIFIER LM101 U6	75.	N/A	B-2	1.0	.8891	1.	.8891
MIC HEX INVERTER 54L04	75.	N/A	B-2	1.0	.3401	1.	.3401
MIC 12 BIT A/D CONVERTER (APPOR)	75.	N/A	B-2	1.0	5.0000	1.	5.0000
PRINTED WIRING BOARD - 4 LAYER	75.	N/A	N/A	1.0	.4740	1.	.4740
SOLDERING CONNECTIONS WAVE	75.	N/A	N/A	1.0	.1043	1.	.1043
CONNECTOR PCB 52 PINS USED	75.	N/A	SM	1.0	2.1231	1.	2.1231

FAILURE RATE 31.6494 MTBF 31596. HOURS NUMBER OF PARTS 60. RELIABILITY .999240703

Figure 4. MTBF Estimate

Sneak Circuit Analysis. This type of analysis (6) is growing in popularity as a useful tool at the system level. It seeks to identify a latent path or condition which inhibits a desired action or initiates an unintended or unwanted action. A sneak circuit is not caused by part failure but is a condition that has been inadvertently designed into a system. Automation has been used in analysis of this type since 1970, using schematics, wiring lists and operational scenarios (operational and maintenance sequences) as a data base. So far, this analysis has been accomplished as a separate activity. Obviously, using the EDB generated for a system and appropriate scenarios, the analysis can be achieved by simulation. Sneak circuit analysis is concerned basically with the topology of a system and in essence so is Fault Tree Analysis. Therefore, the generation of Fault Trees from the system EDB should be relatively straight forward.

Manufacturing Considerations

In manufacturing a system several important benefits can accrue from DA systems.

PCB Repeatability. Because automatic artwork generation equipment can hold tolerance as close as (a) $\pm$ 0.002 inch, (b) registration between layers to $\pm$ 0.0015 inch and (c) drill holes positioned on a grid as fine as 0.002 inch, excellent repeatability and uniform quality can be achieved in PCB manufacturing (7). A description of a program for automatic artwork generation is given by Allum (8).

Effects of Design Change. Because of the many and varied consistency checks associated with DA systems, before changes are implemented, their effects on electronic system performance can be forecast accurately and assessed for suitability. When changes are implemented, they are propagated accurately into the manufacturing process and preclude reliability problems due to poor documentation.

MAINTAINABILITY AND DA

As electronic systems grow in performance they grow also in complexity, leading to correspondingly increased problems in the area of fault detection and fault isolation. This is true for all equipment levels, i.e. part, assembly, unit and system. Ultimately the solution is to design in features at all equipment levels, bearing in mind that a balance between hardware and software approaches to the problem must be struck. DA helps to make this goal a reality, in that it makes it possible to design large systems and small complex ones as well, so that they will exhibit optimum fault detection and isolation characteristics. The main DA tool in accomplishing this is simulation, the starting point of system design. This is discussed under five topics, i.e., (a) Part, (b) Circuit Design, (c) System, (d) Manufacturing, and (e) Field Considerations.

Part Considerations

Designing maintainability into parts is normally not the task of the system designer. However, this is changing, as in the case of the TMS-1000 microprocessor noted by Falk (9) where additional circuitry is added to the chip, along with external test pins for testing purposes. A designer who is aiming to meet specified maintainability targets will exploit any features provided at the part level, in fact, this could influence the choice of parts to some degree.

Circuit Design Considerations

Broadly speaking, the designer will be faced with the problems of, (a) where to locate test points (hardware and/or software) and, (b) what stimulus/response regime is necessary to provide adequate or specified maintainability characteristics.

Circuit Card Assembly. We will assume first that the design has been partitioned logically onto Circuit Card Assemblies (CCAs) and now we wish to design each CCA so that it is testable and also fulfills the required performance characteristics. With a DA system, each CCA is modelled, using the part models in the EDB and connecting them into networks representing the circuits necessary to meet the specified performance. We can assume further that the various simulation possibilities have indeed proven that the CCA meets the performance requirements and is free of logical errors. At this point a set of test inputs can be generated, which are combined with the CCA logic network description and loaded for running with a fault simulator such as CCTEGAS3 (10). A fault list is compiled, which in essence leads back to the faulty part on the CCA (based on the observed output from the CCA with the specified input) and fault detection and isolation percentages are computed. Before the CCA is released for layout it is highly testable and it has developed for it a test program for use on ATE when the CCA is manufactured. Figure 5 shows an example of the output from a fault simulator.

```
INPUT EQUIVALENCE CLASS  =  FIRST MACHINE =    1 ,   NO OF MACHINES =  283
  INPUT OPERAND    19 = 321664
                OUTPUT EQUIVALENCE CLASS  =  FIRST MACHINE =    1 ,   NO OF MACHINES =  282
                  OUTPUT STATES = 3105473247703664
                OUTPUT EQUIVALENCE CLASS  =  FIRST MACHINE =  805 ,   NO OF MACHINES =    1
                  OUTPUT STATES = 3105477247703764
INPUT EQUIVALENCE CLASS  =  FIRST MACHINE =    1 ,   NO OF MACHINES =  282
  INPUT OPERAND    20 = 321666
                OUTPUT EQUIVALENCE CLASS  =  FIRST MACHINE =    1 ,   NO OF MACHINES =  271
                  OUTPUT STATES = 7505477227703764
                OUTPUT EQUIVALENCE CLASS  =  FIRST MACHINE =  477 ,   NO OF MACHINES =    6
                  OUTPUT STATES = 7505476227443764
                OUTPUT EQUIVALENCE CLASS  =  FIRST MACHINE =  584 ,   NO OF MACHINES =    1
                  OUTPUT STATES = 7505476227643764
                OUTPUT EQUIVALENCE CLASS  =  FIRST MACHINE =  791 ,   NO OF MACHINES =    4
                  OUTPUT STATES = 7505473227703664
INPUT EQUIVALENCE CLASS  =  FIRST MACHINE =    1 ,   NO OF MACHINES =  271
  INPUT OPERAND    21 = 321664
                  OUTPUT STATES = 3505477267703764
    NO PARTITIONS
```

Figure 5. Fault Simulation Output

Failure Mode and Effects Analysis (FMEA). FMEA produces a catalogue of potential part failures (opens, shorts and power supply over-voltages) and the related secondary overstress, secondary failure and circuit failure symptoms which would result. This catalogue is useful in fault isolation studies and system level failure mode analysis. FMEA is particularly useful for analog circuits whereas digital circuits can have a practically limitless number of machine states. This makes exhaustive analysis of this type of failure mode, through fault simulator programs, expensive. An economical approach is to be satisfied with fault detection and isolation percentages of the order of 98% each. A computer generated FMEA for an analog case is given in Figure 6.

FAILURE NUMBER 9
Q4 B-E SHORT AND B-C OPEN

MATRIX SOLUTIONS= 25

NODE VOLTAGE AND AUXILIARY ANSWERS

1	NODE	1	-.75000E+00	13	NODE	13	.26447E+01	25	NODE	51	.11108E+02
2	NODE	26	.10000E+02	14	NODE	14	.80555E+01	26	Q1	PT	.30020E-05
3	NODE	3	.10528E+02	15	NODE	15	-.30000E+02	27	Q2	PT	.30020E-05
4	NODE	4	-.42945E-02	16	NODE	16	-.30000E+02	28	Q4	PT	.51566E-03
5	NODE	5	-.70605E-04	17	NODE	17	.11699E-02	29	Q6	PT	.36575E+00
6	NODE	6	.10694E+01	18	NODE	18	-.30000E+02	30	Q3	PT	.12498E-05
7	NODE	7	.98294E+01	19	NODE	19	-.30000E+02	31	Q5	PT	.50187E+00
8	NODE	8	.14125E+02	20	NODE	20	.10150E+01	32	Q7	PT	.54774E-02
9	NODE	28	-.15000E+02	21	NODE	29	-.14268E+02	33	D1	IDT	.93690E-01
10	NODE	10	.11108E+02	22	NODE	47	.75659E-03	34	D2	IDT	.16066E-01
11	NODE	11	.75889E+00	23	NODE	49	.15000E+02	35	D3	IDT	-.39100E-08
12	NODE	12	.10000E+02	24	NODE	50	-.15042E+01	36	D4	IDT	.75134E-03

Figure 6. FMEA Analog Network

System Considerations

The system level represents the highest equipment level with which the designer is concerned. The objective will be to fault detect and isolate in some cases to the faulty replaceable unit, or to the faulty replaceable assembly. In the former case, the system is repaired by unit replacement. The faulty unit will be fault isolated to the faulty replaceable assembly by the use of generalized ATE.

System. In many cases systems are designed so that, during system startup, central computing circuitry can run test sequences, via Built-In-Test (BIT), to check that it is operating normally. When correct operation of this circuitry is confirmed, then the system's peripherals are checked, in turn, under control of the central computing circuitry using BIT associated with each peripheral. To a large degree the development of BIT can be accompanied by a high level simulation, especially when the system is partitioned carefully into function areas. Once the total system is operational, background tests monitor the system continuously in a non-interruptive fashion.

Unit Level. This equipment level is exemplified by assemblies grouped together both physically and electrically to perform a function(s). Typically, a faulty unit can be removed from the system and replaced by a good unit, thus restoring the system to normal operation. In many instances the unit is then connected to ATE for fault detection and isolation to the faulty assembly. In general, test connectors, bringing out the necessary testpoints, are used to accomplish this task and are strategically placed, based largely upon intuition and knowledge of the design. However, programs such as described by DesMarais and Williams (11) are available. These can be adapted to read an EDB and are capable of assigning a minimum number of testpoints to permit isolation to the faulty assembly.

Mean Time To Repair (MTTR). Programs can be developed to compute the MTTR in association with the EDB, with the user supplying only that information which is not readily available in the EDB. An example of a computer generated MTTR Estimate is shown in Figure 7.

Manufacturing Considerations

DA will have considerable impact on manufacturing processes in that when a design is released for manufacturing it will be testable at all equipment levels. This will result in faster throughput from design to finished product as maintainability characteristics will have been designed in.

M ESTIMATE XM-1 BC CCP RUN NO. 3 7 APRIL 1977 PAGE 1

TEST NO	AM SQ	DESCRIPTION OF SRA/PART	REFERENCE DESIGNATION	FAILURE RATE	SET UP TIME	ISOLAT TIME	ACCESS TIME	INTCHG TIME	ADJUST TIME	CHECK TIME	TOTAL TIME	MM/OH
10	0 0	ZERO SW.	1A1S1	.300	5.00	6.00	40.60	6.30	N/A	9.00	66.90	20.07
20	0 0	ONE SW.	1A1S2	.300	5.00	6.00	40.60	6.30	N/A	9.00	66.90	20.07
30	0 0	TWO SW.	1A1S3	.300	5.00	6.00	40.60	6.30	N/A	9.00	66.90	20.07
40	0 0	THREE SW.	1A1S4	.300	5.00	6.00	40.60	6.30	N/A	9.00	66.90	20.07
50	0 0	FOUR SW.	1A1S5	.300	5.00	6.00	40.60	6.30	N/A	9.00	66.90	20.07
60	0 0	FIVE SW.	1A1S6	.300	5.00	6.00	40.60	6.30	N/A	9.00	66.90	20.07
70	0 0	SIX SW.	1A1S7	.300	5.00	6.00	40.60	6.30	N/A	9.00	66.90	20.07
80	0 0	SEVEN SW.	1A1S8	.300	5.00	6.00	40.60	6.30	N/A	9.00	66.90	20.07
90	0 0	EIGHT SW.	1A1S9	.300	5.00	6.00	40.60	6.30	N/A	9.00	66.90	20.07
100	0 0	NINE SW.	1A1S10	.300	5.00	6.00	40.60	6.30	N/A	9.00	66.90	20.07
110	0 0	DECIMAL SW.	1A1S11	.300	5.00	6.00	40.60	6.30	N/A	9.00	66.90	20.07
120	0 0	MINUS SW.	1A1S12	.300	5.00	6.00	40.60	6.30	N/A	9.00	66.90	20.07
130	0 0	ENTER SW.	1A1S13	.300	5.00	6.00	40.60	6.30	N/A	9.00	66.90	20.07
140	0 0	CLEAR SW.	1A1S14	.300	5.00	6.00	40.60	6.30	N/A	9.00	66.90	20.07
150	0 0	ZERO SW.	1A1S15	.300	5.00	6.00	40.60	6.30	N/A	9.00	66.90	20.07
160	0 0	BORE SIGHT SW.	1A1S16	.300	5.00	6.00	40.60	6.30	N/A	9.00	66.90	20.07
170	0 0	CROSS WIND SW.	1A1S17	.300	5.00	6.00	40.60	6.30	N/A	9.00	66.90	20.07
180	0 0	CANT SW.	1A1S18	.300	5.00	6.00	40.60	6.30	N/A	9.00	66.90	20.07

Figure 7. MTTR Estimate

When hardware is built in manufacturing, test programs will be available for use on ATE. Figure 8 depicts typical factory equipment used for both unit and circuit card production testing. In many cases the ATE used for unit testing will be completely different from ATE used for circuit card testing. Another factor is the availability of a set of complete and accurate documentation generated by the DA system at the detailed level.

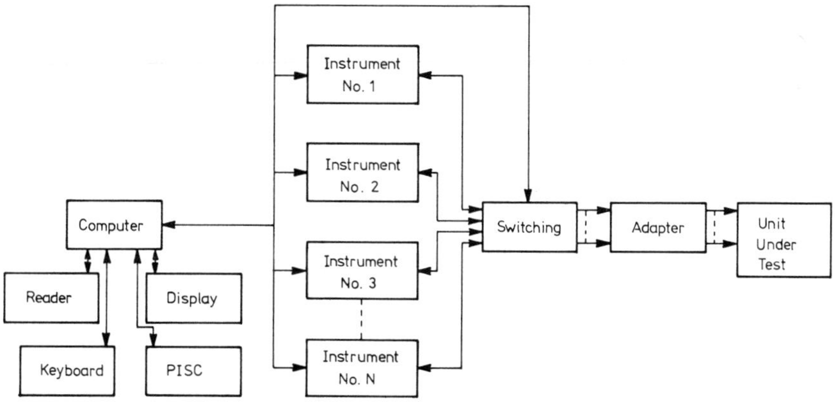

Figure 8. Typical ATE Complex

Field Considerations

DA systems provide significant benefits when electronic systems are placed in the operational environment and are, therefore, subject to maintenance actions. One of the real benefits is a complete and accurate set of documentation for use by Field Service Engineers (FSEs) as an integrated DA system is capable of producing documentation such as the following, (a) Wiring Lists, (b) Simulation Results, (c) Logic Schematics, (d) Printed Wiring Board (PWB) Artwork, (e) Test Programs and (f) Parts Lists. Advances in low-cost portable testers (12) permit FSEs to confirm the failure of suspect CCAs and even repair products on site. Most manufacturers make available translators, capable of interfacing diagnostics generated under a simulation system, to the tester. Otherwise, it is possible to write a translator program to link diagnostics developed under a DA system to the particular tester. Test programs for portable testers can be kept on cassettes.

CONCLUSION

A greater awareness of DA techniques and their introduction and practical application provides the key to designing electronic systems with required reliability and maintainability performance. DA will play an increasingly important role with respect to the global problems of system

Life Cycle Cost (LCC). Larger organizations tend to choose and/or develop an integrated approach to DA to minimize manual input to the design process. Smaller organizations will usually be constrained by economics to select a mix of DA programs and traditional approaches that together provide an economical solution to their needs.

REFERENCES

1. Breuer, M.A., Ed., Design Automation of Digital Systems, Vol. I Theory And Techniques, Prentice-Hall, Inc., Englewood Cliffs, N.J., 1972.

2. Kaplan, G., "Computer-Aided Design", IEEE Spectrum, IEEE, N.Y. October, 1975, p40-47.

3. DesMarais, P., and Krieger, M., "Fault Simulation and Digital Circuit Testability", Microelectronics and Reliability, Vol.15 Supplement, June 1976, p5-13.

4. ___________, SYSCAP II - System Of Circuit Analysis Programs, User Information Manual, Control Data Corporation, 1 September, 1975.

5. ___________, Reliability Prediction Of Electronic Equipment, MIL-HKBK-217B, U.S. Dept. Of Defense, 20 September, 1975.

6. Clardy, R.C., "Sneak Circuit Analysis Development and Application", 1976 Region V IEEE Conference, April 14-16, 1976, IEEE Catalog No. 76CH1068- REG 5.

7. ___________, "Computer-Aided Drafting and Documentation Systems", GTE Lenkurt Demodulator, January 1974.

8. Allum, B., et al, "Computer System Accelerates Design of Printed Circuit Boards", Telesis, Bell-Northern Research, 1976/1, p144-150.

9. Falk, H., "Design For Production", IEEE Spectrum, IEEE, N.Y. October, 1975, p52.

10. ___________, CCTEGAS3-Test Generation and Simulation System, User Information Manual, Control Data Corporation, 4 January, 1975.

11. DesMarais, P. and Williams, S., "System Diagnosis with FLIP", Microelectronics and Reliability, Vol.17, No.1, Pergamon Press, New York, 1978, p47-52.

12. Franson, P., "Portable Tester Checks Complex Logic", Electronics, July 11, 1974, p125.

5 CONFIGURATION MANAGEMENT

H. I. March

The Reliability and Maintainability of an equipment starts with the basic design concept, achieved in most cases on the drawing board during the development phase. This design concept is later described by the appropriate specifications and engineering drawings. As we start referring to the design, we constantly refer to the item's configuration. Once a satisfactory degree of item design identification is achieved, one may choose to freeze the design at that specific configuration level. This type of procedure will require the establishment of a baseline.

There are numerous good reasons for the establishment of baselines and there are various stages in a development program where baselines are necessary. This area will be more thoroughly described in this section when we review the military system of Configuration Management.

It would be naive to believe that established item configuration is set in concrete. There are always changes to be implemented no matter what the reasons for these changes. Therefore, after the establishment of a baseline, there must be a sound and flexible system of implementing changes and maintaining a system of records identifying the numerous changes to a baseline.

The following is a brief description of a Configuration Management System which is employed in almost everything produced. On small projects, the importance of having a Configuration Management Plan is often not recognized, because most of the elements of Configuration Management (CM) are reasonable, practical and imposed by a company as routine procedures. However, on complex designs and large programs a Configuration Management Plan (CMP) is a very necessary data item to ensure that there are no assumptions of required control procedures and that identification and accounting practices are precisely defined and understood.

A large program without a definite Configuration Management Plan would, no doubt, end up as the proverbial Tower of Babel. Since Governments are known to be the buyers of the most sophisticated equipment and usually award contracts of the largest production scope, all our examples will relate to the existing military systems of Configuration Management.

DEFINITIONS

Following are several definitions which should be absorbed and thoroughly understood by the reader before proceeding any further.

Configuration Management

This is a discipline applying technical and administrative direction and surveillance to identify and document the functional and physical characteristics of a Configuration item, control changes to those characteristics, and record and report change processing and implementation status.

Configuration (as defined by the Military)

The functional and/or physical characteristics of hardware/software as set forth in technical documentation and achieved in a product.

Configuration Item (CI)

An aggregation of hardware/documentation, or any of its discrete portions, which satisfies an end use function and is designated by the Buyer (contract) for Configuration Management. The CI's may vary widely in complexity, size and type, from an aircraft, to a computer. During development and initial production, CI's are only those specification items that are referenced directly in a contract. During the operation and maintenance period, any repairable item designated for separate procurement is a configuration item.

Three major areas govern all elements of Configuration Management, these are:

(a) Configuration Identification - this is the current approved or conditionally approved technical documentation for a configuration item as set forth in specifications, drawings and associated lists, and documents referenced therein,

(b) Configuration Control - this is systematic evaluation, coordination, approval or disapproval, and implementation of all approved changes in the configuration of a contract item after formal establishment of its Configuration Identification.

(c) Configuration Status Accounting - the recording and reporting of the information that is needed to manage configuration effectively, including a listing of the approved configuration identification, the status of proposed changes to configuration, and the implementation status of approved changes.

Baseline

The Baseline is a configuration identification document or a set of documents formally designated and fixed at a specific time during a contract item's life cycle. Baselines, plus approval changes from those baselines, constitute the current configuration identification, as follows:

(a) Functional Baseline - the initial approved functional configuration identification,
(b) Allocated Baseline - the initial approved allocated configuration identification,
(c) Product Baseline - the initial approved or conditionally approved product configuration identification.

Although most manufacturers of complex equipment and systems have their own Configuration Manuals, the government in most cases would request the manufacturer to present a Configuration Management Plan (CMP).

CONFIGURATION MANAGEMENT PLAN

The contract bid set either requests the CMP as part of the bid set or approximately 30 days after contract award. The purpose for requesting such data, is to ensure that the manufacturer's CMP is responsive to the customer/user requirements and the system is able to integrate with the customer's controls, both during the item's development cycle and their future product support activities.

The Department of Defense (DOD) ensures that manufacturer's products and systems integrate with their system successfully by issuing several standards which lay down the fundamental requirements of Configuration Management elements; and are implemented via the contract work statement.

Reference (1) is the standard which identifies requirements to the contractor as to what shall be done to implement an effective system. The plan must describe project responsibilities and procedures for implementing CM, and includes schedules for major CM activities and events. A general outline of a CM plan is given in Table 1.

Table 1. Configuration Management Plan Outline

Section	Section Title
I	Introduction
II	Project Organization
III	Configuration Identification
IV	Configuration Control
V	Configuration Status Accounting
VI	Subcontractor/Vendor Control
VII	Program Phasing
VIII	Reviews and Audits
IX	Software/Firmware Documentation and Configuration Control

Assuming that a company must present a CMP, the characteristics identified in Table 1, if properly and adequately described, should suffice as a plan. Let us now further expand and detail the essential characteristics outlined in Table 1.

Introduction (Section I)

State the purpose of CMP and briefly describe its contents - list applicable documents and management approach.

Project Organization (Section II)

Describe the project organization and its relationship to program/ project management. Use organization chart for the illustration of authority/responsibility of key personnel and identify the organizational level of the project Configuration Management Office (CMO). Describe the organization of the Change Control Board (CCB). This Board is separately treated further on in this chapter.

In this section the customer will have to be convinced that management levels and expertise associated with the contract are acceptable. The following points should be considered:

(a) Does the Configuration Manager and his staff have an acceptable background?,
(b) Is the Change Control Board Chairman at the correct management level to perform this function?,
(c) Does the Configuration Management Office have sufficient scope and resources to perform the required task? Are the responsibilities clearly defined?

Figure 1 illustrates an organizational structure which would be quite effective to manage the engineering development phase with limited production quantities. Of course, the structure concerns itself with the top level organization elements with emphasis on engineering. This emphasis may be shifted depending on the kind of development phase the contract has authorized.

Configuration Identification (Section III)

The US Department of Defense (DOD) lists six major categories of identification with which the DOD are concerned, (2). The manufacturer must review and address these identifications accordingly.

Let us now itemize and describe these identification areas in detail:

(a) Functional Configuration Identification (FCI) - this identification, once established, shall serve throughout a CI's life cycle as a description of its required functional characteris-

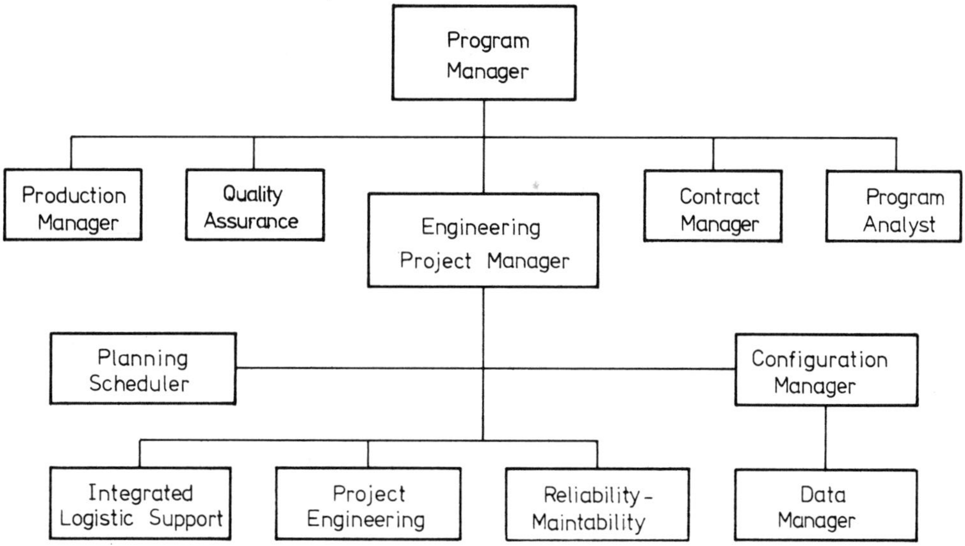

Figure 1. Organization Structure for Development Phase

tics. The FCI must be documented by a performance oriented specification, prepared in accordance with (3). Their types and forms should be selected from (4). The purpose of (3) is to establish uniform practices for specification preparation, to ensure the inclusion of essential requirements, and to aid in the use of analysis of specification content,

(b) Allocated Configuration Identification (ACI) - this identification is optional. Requirements for this type of identification may exist in order to identify contractual division of a total task or the complexities of a CI may warrant identification of selected CI that are part of a higher level CI. These specifications will also be prepared in accordance with (3) and each specification must include the necessary interface requirements with other associated CI's which comprise the higher level CI,

(c) Product Configuration Identification (PCI) - this identification is used to prescribe the necessary "build-to" or form fit and function requirements for a configuration item and the acceptance test for these requirements.

The level of detail to be considered in the PCI should be determined by consideration of requirements for anticipated procurement for configuration audits, and for logistic support. Alternatively potentially repairable items which are part of a CI documentation for the PCI should be prepared in accordance with (3) and (5),

(d) Identification Compatibility - all the above identifications should be mutually consistent and compatable. However, should conflict arise between such identifications, unless otherwise directed, the order of precedence should be:
   (1) 1st Function Configuration,
   (2) 2nd Allocated Configuration,
   (3) 3rd Product Configuration,

(e) Identification Numbers - this configuration identification requirement clearly shows DOD's concern with ensuring that supplier's identifications are designed to interface with the DOD system. The items controlled here, are:
   (1) Documentation Numbering in accordance with (3,5,6),
   (2) Item Numbering, where CI's and portions thereof must be identified with suitable numbering in order to perform the configuration identification, change control and status accounting functions,

(f) Configuration Identification Record - a permanent copy of configuration identification shall be held in the custody of contractor or DOD. If the contract specifies these records must be maintained throughout the life cycle of a CI, and shall include all configuration identification, beginning with the baseline documents and continuing with the addition of proposed and approved changes from those baselines.

Configuration Control (Section IV)

The object of this section is toexplain the method used in the establishment of configuration identification of CI's, controlling all changes from that identification and then ensuring that the hardware/ software matches the configuration identification, except for approved waivers or deviations. To control all change action, the Government has provided two military standards of which either one is mandatory in most contracts (7,8).

The established procedures must ensure that each engineering change is properly coordinated and documented to determine the total impact on all phases of the program involved.

The above mentioned references (7,8) classify the type of changes and provide standard forms for review and approval by DOD of:

(a) Engineering Change Proposals (ECP's),
(b) Request for Deviation/Waiver,
(c) Notice of Revision, (NOR).

Reference (7) defines requirements for engineering changes and classifies all changes into two classes. The details of these classifications are too numerous for this chapter, but may be summarized as follows:

(a) Class I Engineering Changes - this is a noninterchangeable change to an equipment, assembly, or part in which an item is changed to such an extent that the form, fit, or function has been modified,
(b) Class II Engineering Changes - this is an engineering change in which complete equipments, assemblies, subassemblies, or parts are capable of being freely exchanged one for another irrespective of application. Obviously reliability must not be affected.

A convenient checklist for the preparation of the "Configuration Control" section is as follows:

(a) Does the CMP include an adequate change Control System for the various classes of changes established?,
(b) Is a method provided for control during the interfaces between elements of the software product during development and testing?,
(c) Does the CMP identify methods for complying with contractor's requirements and controlling interfaces with external software and hardware items?,
(d) Is the change control process clearly described with flow diagrams and sample forms?,
(e) Is there a uniform system of identifying, analyzing, and implementing change orders as they pertain to depicting the engineering design and development effort?

Configuration Status Accounting (Section V)

The writer must state plans of application of configuration index and status accounting records. He must state his understanding and outline the approach which will be taken in order to discharge his duties and responsibilities to,

(a) Submit data to an integrating agency, who will collate, prepare and distribute reports,
(b) Collate, prepare and distribute some specific reports himself,
(c) Accept inputs from other contractors and collate such data with his own inputs, and prepare and distribute reports.

A very thorough understanding of the customer's requirements relative to the type of development phase in which the contract is awarded will help in arriving at the above (a), (b) and (c) requirements. In most cases, however, the contract specifies the required data via the Contract Data Requirements List (CDRL).

It should be noted that there is latitude with respect to status accounting reporting, particularly if data processing methods are implemented. Therefore, define the method to be used, and define the interfaces, with the software control activity or center. A checklist for the preparation of this section is as follows:

(1) Does the CMP identify the data required for configuration status accounting, and the method for recording it?,
(2) Does it identify the status accounting reports required and their distribution as well as the frequency of distribution?,
(3) Is integration of data as items (a), (b), (c) outlined above included in the status accounting procedures?,
(4) Are the forms for status accounting data adequate for reporting, and are these examples included?

Subcontractor/Vendor Control (Section VI)

The author of CMP must indicate his proposed method for controlling subcontractors and vendors, in so far as it impacts his configuration management commitments to the procuring activity. The methods used to determine their capability and monitor their ability to support the requirements of configuration management must be explained.

Program Phasing (Section VII)

Program phasing includes establishing configuration controls on hardware/software from the definition phase through to production, continuing support by means of Preliminary Design Review (PDR), Critical Design Review (CDR), product configuration baseline, specifications, drawings, Engineering Change Orders (ECO's), approved Engineering Change Proposals (ECP's), Field Change Orders (FCO's) and status accounting reports.

Figure 2 illustrates the hardware/software phasing for a specific program as an example. It must be understood that each program will have its own specific phasing plan. However, Figure 2 may be of considerable assistance to the reader in viewing the interrelationship of various events.

Reviews and Audits (Section VIII)

In this section the writer must describe plans for conducting and supporting the various reviews and audits over the hardware/software development cycle. Some of the major items to be described are the Functional Configuration Audit (FCA) and Physical Configuration Audit (PCA).

If the contractor is requiring privately developed items, (items whose development was not paid for by the contractor) a method for validation and approval of such items should be discussed. Also, products and documents which are required for review and approval, should be identified in the plan and their approval method described.

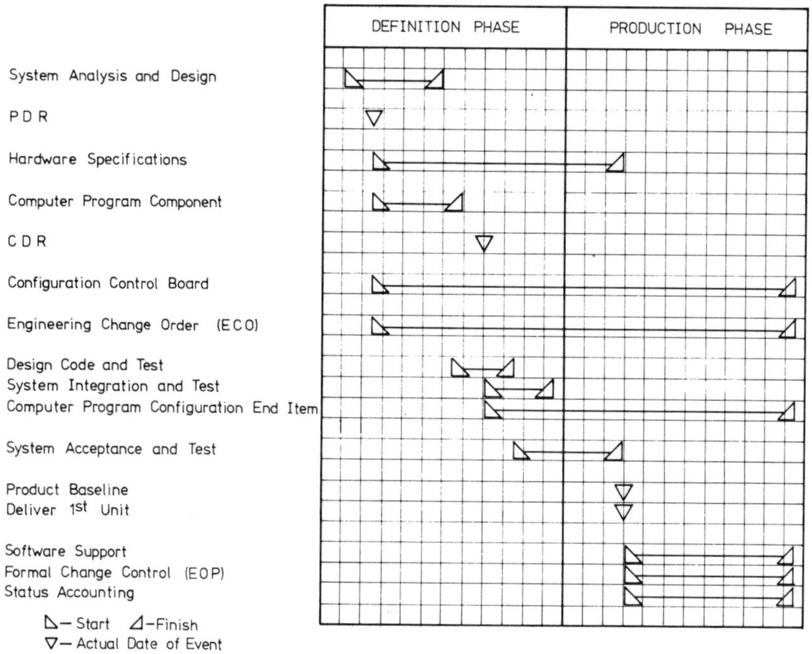

Figure 2. Typical Program Phasing of Hardware/Software

Software/Firmware Documentation and Configuration Control (Section IX)

Each year the manufacturers of sophisticated weapons systems make advances in computer and sensor technology, increasing demands for greater and more complex systems. This is a recognized fact of life by the DOD procuring agencies, but since it is the natural outgrowth of advances in computer technology, their major effort is not in dictating, precisely, the form of development but rather to obtain control of all software development. Because the nature of software is such that modification without physical changes to the system can be made, the proper software control systems must be developed and identified very early. Depending on the software content, the contractor may wish to prepare a separate Software Configuration Management Plan. If this is the case, all elements which are pertinent to software and identified in the foregoing sections should be described.

It is most likely that projects which will have their own Software Configuration Management Plans, will have some form of central software activity center. When describing techniques and facilities which control the software, ensure that safe storage of master tapes and decks is described. Identify responsibilities for withdrawing and changing this programmed media. Describe reproduction and destruction capability, library services, protection against physical loss and/or damage, disaster file and purging policy.

This completes the description of those elements which should be described in a Configuration Management Plan. This area has been treated extensively because the plan covers all requirements of Configuration Management and it is possible to relate the system and management structure of any organization to the required elements of the Plan. Figure 3 on page 71 lists all the components of Configuration Management.

CHANGE CONTROL BOARD (CCB)

One of the major objectives of Configuration Management, is to apply technical and administrative direction and surveillance to properly identify and control design changes to CI, comprising equipment or system or computer programs. The Engineering Change Control Board must be adequately organized, staffed and administered. Figure 4 illustrates typical structure of this board which approves or disapproves proposed equipment changes, with the chairman recording the action taken in the form of minutes.

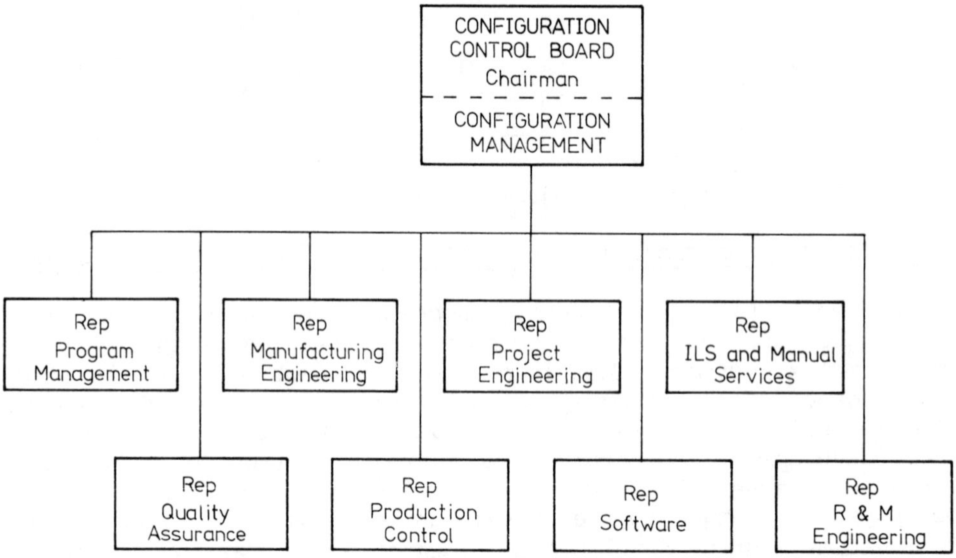

Figure 4. Change Control Board Structure

Change Control Board (CCB) Advantages

The advantages of a CCB can be summarized as follows:

(a) Change processing and change incorporation are a scheduled effort. Dates are monitored by Configuration Management for performance of the agreed plan. Through this method, critical areas are identified prior to imposing shortages or excessive costs,

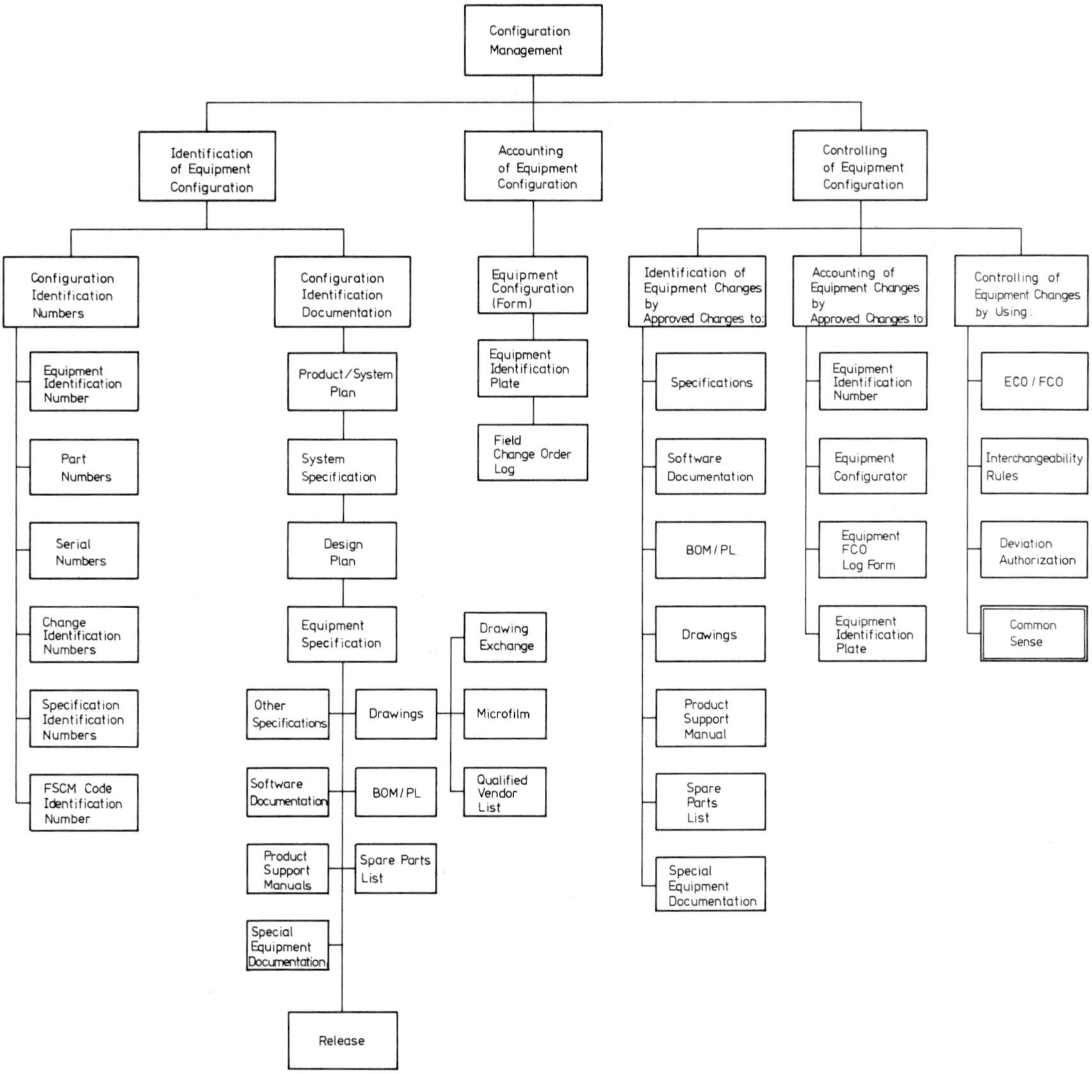

Figure 3. Ingredients of Configuration Management

(b) Configuration Management identify any holding factors and critical areas and coordinate for immediate resolution as a function of change control,
(c) The change board chairman and its members are project/contract oriented and will investigate all changes affecting a specific program or product,
(d) Departments affected are represented in analyzing and scheduling all changes and, therefore, have the opportunity to participate in the final decision,
(e) Board meetings are confined to decisions on presented change packages, based on total facts and impacts presented, thus eliminating lost time for functions not involved in detail analysis,
(f) The change control office is responsible for assigning ECO numbers and maintaining a log book of all actions taken. This maintains controlled visibility of revision level effectiveness and documentation for historical purposes regarding the action taken by the board,
(g) Change orders are discussed with the benefit of having complete cost, schedule, and technical analysis as to the impact on the company's capabilities.

Effective management of the CCB is one of the most important elements of an efficient and accurate change control system. Each member organization/department of the CCB has a specific task and responsibilities. These responsibilities may be slightly differently distributed among the CCB members in each Company. However, the cumulative responsibilities of the CCB should not be altered or diminished.

Responsibilities

Following is a list of responsibilities of the Configuration Control Board and Configuration Control Board Chairman.

(a) Configuration Control Board Cumulative Responsibilities:
  (1) Accurately describe each change,
  (2) Document reason(s) for each change,
  (3) Validate change classifications,
  (4) Schedule each change and establish firm effectivity for that change based on established "Go-ahead" date,
  (5) Approve or disapprove each change order,
  (6) Assign proper disposition on the change order for parts, subassemblies, or assemblies affected by each change,
  (7) Determine if the change is warranted,
  (8) Ensure costs do not exceed benefits,
  (9) Determine all change effects to drawings, specifications, test equipment and test procedures,

(10) Provide the capabilities for handling crash programs in a sophisticated and economical manner,
(11) Must have knowledge of the total Company/division task and individual department status within the total task,
(12) Provide an instantaneous feedback of change status to concerned departments,
(13) Be able to provide expeditious action if schedule dictates,

(b) Configuration Control Board Chairman Responsibilities:
(1) Provides chairmanship of all meetings and publishes the minutes,
(2) Controls the conduct of all board meetings,
(3) Co-ordinates proposed and approved changes when such changes affect two or more different programs and production schedules,
(4) Ensures expeditious and satisfactory processing of all change documents,
(5) Verifies that all decisions rendered by the board members are based on sound and positive analysis.

BASELINE MANAGEMENT

One of the more important aspects of Configuration Management is the concept of baseline management which is formally required by Governments on all military equipment at the beginning of a system configuration item acquisition program. Therefore, it is appropriate that we end this chapter with baseline management.

Establishment

Baselines may be established at any point in a program where it is necessary to define a formal departure point for control of future changes in performance and design. System program management normally employs three baselines for the DOD validation and acquisition of systems:

(a) Functional Baseline,
(b) Allocated Baseline,
(c) Product Baseline.

Program management has the option to employ all three or only the functional and product baselines depending upon complexity or peculiar requirements. These baselines are documented by approved configuration identification item requirements.

Costs

Baselines are the basic requirements from which contract costs are determined. Once defined, changes in these requirements are formally approved and documented to provide an equitable way to adjust contract costs (references (7,8) are the military standards describing the details).

Essentially, configuration management is oriented towards management of, change and use of these separate baselines and it provides the necessary latitudes for defining changes. Initially, most changes may be made within the scope of the functional baseline for the total system/subsystem segment requirements, and ultimately, changes shall be defined as they affect individual configuration items.

Standard Criteria Format

The DOD insists that all descriptions of baselines (functional, allocated and product) of a system, or other configuration items, used to state product performance and design requirements, must be contained in a standard criteria format. In order to achieve an acceptable degree of standardization, the military has published criteria (3), for a uniform specification program for all contractor prepared documents. This is also compatible with criteria for those specifications which may subsequently be prepared and issued under the military series of documents. This specification is intended to be used as a common reference by the industry and government. The reference (3) specification requirements are incident to configuration management and are organized in content, format, and use to recognize the environment imposed by configuration management principles. They recognize the specific periods of design evolution reflected in the baseline principles. Further, they are organized for progressive definition of technical requirements and change control. They are a means of relating performance requirements to the design definition expressed in drawings. Figure 5, from reference (9), illustrates the various baselines and specifications development in accordance with military baseline management plans.

CONCLUSION

The application of configuration management is required in any product development. The degree varies from a common sense approach by individuals to a master submission of a documented program plan. The DOD, which generally funds large system developments of military hardware, makes configuration management disciplines a mandatory requirement. Companies bidding for contracts must remember that once they are awarded a contract, identification, control and accounting of all changes must be done on each contracted item.

One of the major problems regarding this subject is the lack of understanding by development engineers and company management about the specifics of configuration management. Therefore, it is always good practice to hold a configuration control briefing at the start of each project. Such briefings should identify the specific project requirements relative to baselines, hardware/software identification, contents of technical parameters, change control and data submission requirements.

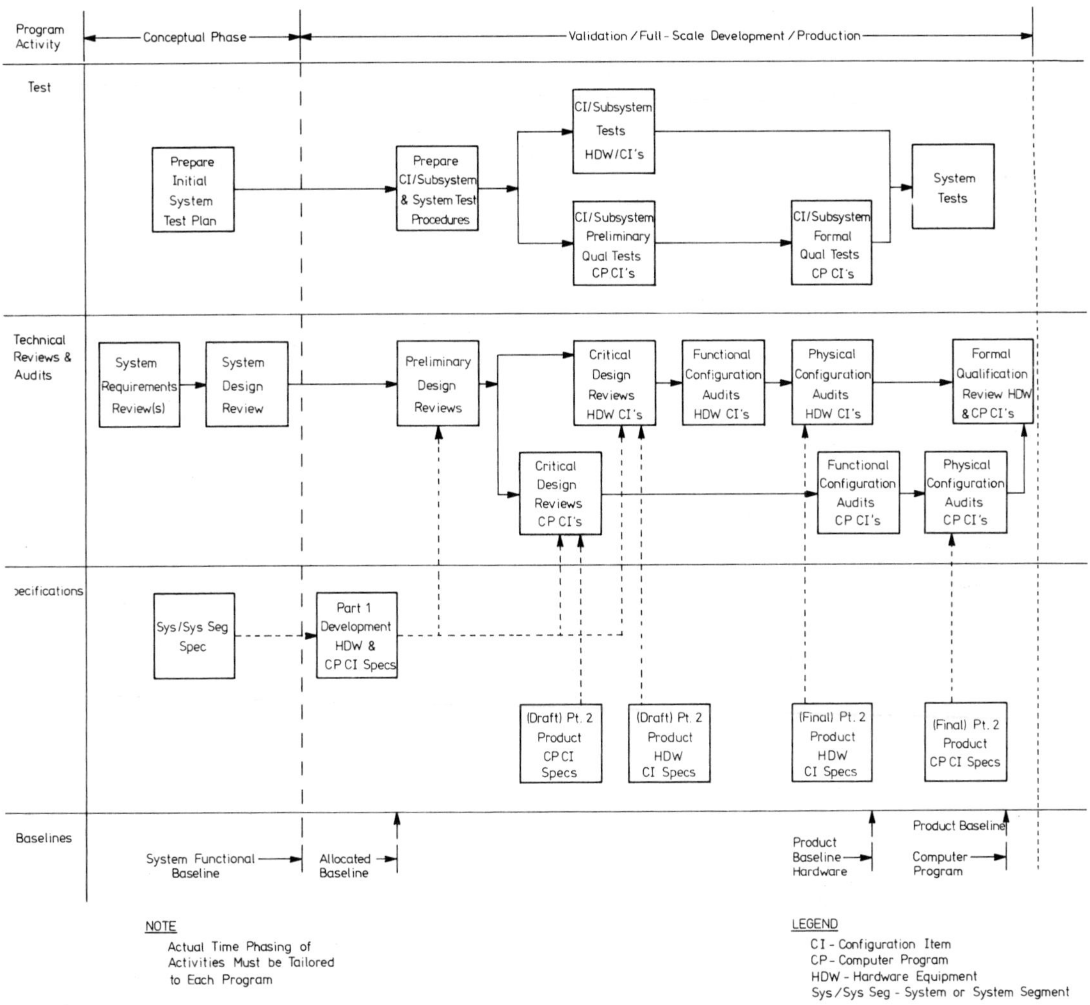

Figure 5. Engineering and Test Flow

Occasionally management does not understand the discipline and consequently does not either budget sufficient funds or provide adequate support and facilities for the performance of configuration management. This situation was recognized by the government procurement agencies some time ago. To rectify this, the military has published several documents and standards (see references). These documents force the manufacturers who wish to supply equipment to the government, to fall in line with the formally described practices.

It is recommended that any company, wishing to supply government agencies with their equipment, make a point to study, in detail, all requirements of configuration management and then generate their own policies on the subjects. These should comply with the military requirements and be flexible enough by adaptation to various in-plant programs.

REFERENCES

1. ______________, MIL-STD-483, Configuration Management Practices For Systems, Equipment, Munitions and Computer Programs, U.S. Department of Defense, Washington, D.C.

2. ______________, "Instruction No. 5010-21 Configuration Management Implementation Guidance", U.S. Department of Defense, Washington, D.C.

3. ______________, MIL-STD-490, Specification Practices, U.S. Department of Defense, Washington, D.C.

4. ______________, MIL-S-83490, Specifications, Types and Forms, U.S. Department of Defense, Washington, D.C.

5. ______________, MIL-STD-100, Engineering Drawing Practices, U.S. Department of Defense, Washington, D.C.

6. ______________, Standardization Manual 4120.3M, U.S. Department of Defense, Washington, D.C.

7. ______________, MIL-STD-480, Configuration Control-Engineering Changes, Deviations and Waivers, U.S. Department of Defense, Washington, D.C.

8. ______________, MIL-STD-481, Configuration Control-Engineering Changes, Deviations and Waivers (Short Form), U.S. Department of Defense, Washington, D.C.

9. ______________, MIL-STD-1521, Technical Reviews and Audits for Systems, Equipment, and Computer Programs, U.S. Department of Defense, Washington, D.C.

6 SOFTWARE RELIABILITY AND MAINTAINABILITY

J. E. Arsenault

The problem of system software Reliability and Maintainability (R-M) is by now generally known and well documented (1). However, it seemingly refuses to be brought to heel. An illustration of this is provided in (2) as follows:

"The world's most carefully planned and generously funded software program was that developed for the Apollo series of lunar flights. The effort attracted some of the nation's best computer programmers and involved two competing teams checking the software as thorough as the experts knew how to make it. In the aggregate, about $600 million was spent on software for the Apollo program. Yet, almost every major fault of the Apollo program, from false alarms to actual mishaps, was the direct result of errors in computer software."

Progress in the R-M field has led to various techniques which, when brought to bear on system hardware, have proved very effective in achieving the desired R-M characteristics. Some of the more effective techniques developed are burn-in, redundancy, standardization, functional partitioning and the use of modularity in electronic packaging to name a few. These techniques were developed to solve a recognized problem and so it would be natural to expect to see a similar development of techniques aimed specifically at the software R-M problem. This has happened with a vengeance, leaving a reviewer of the field with so many opinions as to render him breathless in the search for cost effective solutions to the problem. Fortunately the required techniques are emerging, however slowly, and they are evolving along fairly clear lines. A good appreciation of them can be found in (3, 4). It is also clear that the common element is discipline, i.e., the underlying element in all of the hardware oriented techniques that have proven useful. It may be rather surprising to some that this is the case and more surprising still that there is an extremely strong connection between the proven hardware techniques and the emerging software techniques. It can be argued that there exists a duality between the successful hardware approaches and the emerging software approaches.

Once this is accepted, the whole problem is simplified because the hardware and software problems can be approached together, in a somewhat more unified manner than is apparent today. This chapter will draw on the ideas of discipline and duality in presenting those techniques that have been, or are beginning to be, effective in achieving system software R-M characteristics.

We will finish this section by making our first allusion to the duality between hardware and software mentioned earlier and at the same time provide some insight into the life cycle of software systems. Figure 1 from (5) illustrates the various phases through which large software systems generally proceed. It is difficult to see any general difference between this and a combarable figure if drawn exclusively for hardware. The exception is the block entitled CODING, which is another term for programming and is peculiar to software. We will exploit this similarity in the following sections, after taking a look at the nature of software or for that matter any large system.

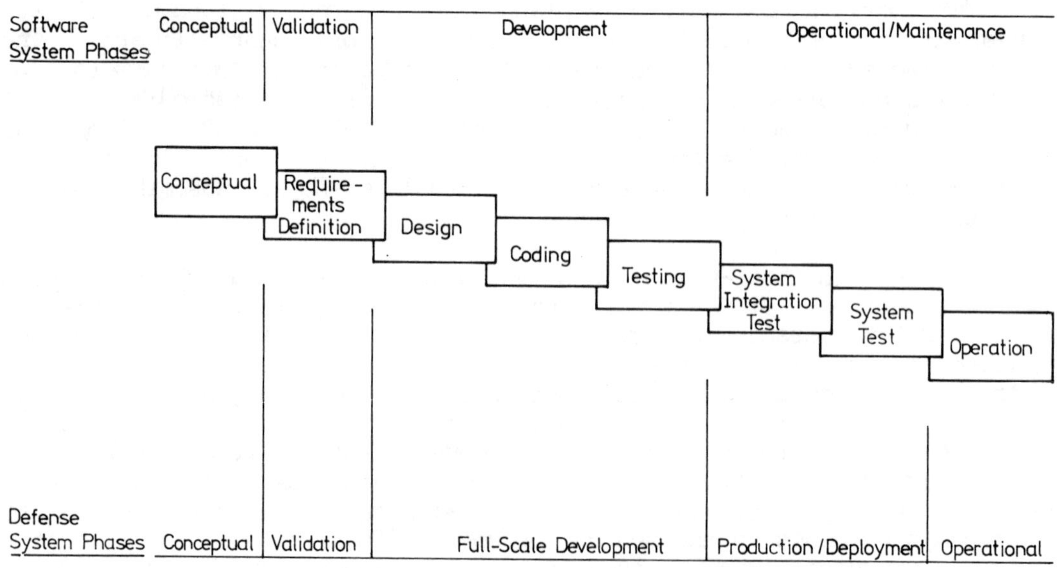

Figure 1. Life Cycle Phases

THE SOFTWARE PROBLEM

Before one can choose and apply the most effective set of tools necessary to solve a problem, the problem must be understood. The basic problem in software is nothing more than the management of complexity. This is aptly described in reference (6) as follows:

"We have learned from our experience with building and managing complex organizations that when the complexity of any level grows beyond a certain range, function becomes impaired, operation becomes inefficient, and reliability declines. We know that ad hoc corrections and local improvements in efficiency can only go so far in correcting the problems, and that sooner or later we must face a total reorganization of the system that must essentially alter the hierarchical control and levels structure."

The idea of hierarchy is further developed in (7) and applied directly to software:

"Now the effect of this relationship, i.e. hierarchy, is profound when you consider system design and reliability. It says that system design time is proportional to the number of levels in the hierarchy used to structure the design. For example, if a designer had to build a system which required 256 elements and he had a choice of building his subassemblies from 16 components and using 2 levels in the hierarchy or using 4 components per subassembly and using 4 levels, he should find that the second structure should take only one half the time to design compared to the first structure. But note the number of specifications that are required. In Figure 2 the first structure using four components per subassembly requires only 17 descriptions of the relationship of the 16 components to make up a subassembly, whereas the second structure requires 85 descriptions describing a simpler relationship of 4 components required to make up a subassembly. This perhaps explains why our intuition fails us and we choose normally to write 17 specifications then we rewrite them over and over again instead of 85 specifications which are each 4 times smaller. If we assume that the relative design times are valid, then the level of effort for each of the 17 specifications would be 10 times greater than for any of the 85 specifications used in the first structure ... Similarly the testing and checkout of each subassembly is 10 times more complex instead of our intuitive guess of 4. This is the reason why testing is grossly underestimated for unstructured systems... The problem is that we do not structure our systems into small enough modules."

Having isolated this core software problem we can now look at the tools which are going to help in the generation of software with the desired R-M characteristics. Each of these tools is listed below and discussed in some detail in the following sections:

(a) Specification,
(b) Design,
(c) Programming,
(d) Testing,
(e) Documentation,
(f) Modelling,
(g) Maintainability,
(h) Management.

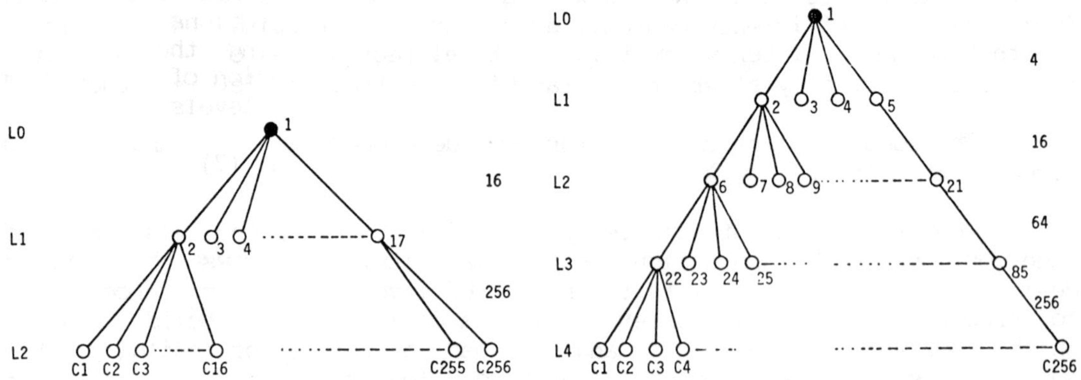

Figure 2. Comparision of Two Structured Systems

SPECIFICATIONS

The thorough specification of a large software system cannot be by a single specification alone. In fact, a series of timely specifications is required and they must be developed in a logical sequence, suggested by Figure 3 of reference (8). The milestones in the process of specification development are a series of design reviews known individually as SRR, SDR, PDR and CDR, some of which are discussed in Chapter 3 and are geared to (9).

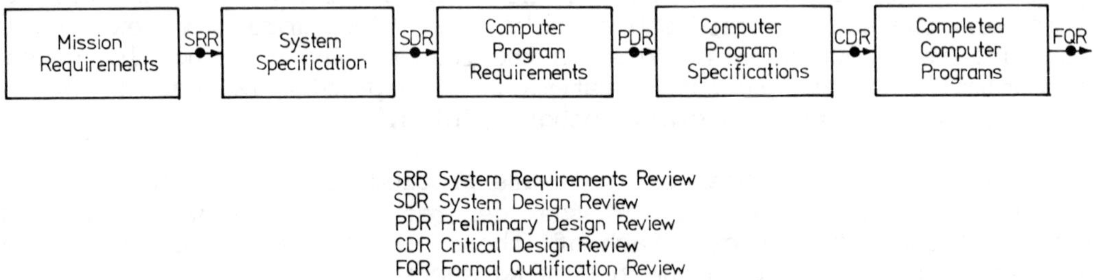

Figure 3. Simplified Specification Model

The system specifications are determined after the System Requirements Review (SRR) and are reviewed at the System Design Review (SDR). Computer program requirements are then developed, refined and reviewed during the Preliminary Design Review (PDR). Computer program specifications are written and discussed at length during the Critical Design Review (CDR). Permission to proceed with the implementation is given, the computer programs are completed and a Formal Qualification Review (FQR) is held.

The important point to note is that at each stage the specifications are reviewed in progressively increasing detail by both the customer and developer so that the direction of development is clear to both parties and always there exists an approved specification.

The contents of specifications will vary with the nature of the software system to be developed. This is discussed in a later section under Documentation.

DESIGN

The most effective design technique to deal with the hierarchial problem is known as "top down design." Proceeding from an approved software specification the top down approach provides a systematic design methodology that leads to the development of a system structure which can cope with inherent system complexity in an effective and readily understandable way. A brief description of the technique is given in (10) as follows:

"(a) Starting with the problem statement (or functional or external specification), design the structure of the entire program or system using one or more forms of analysis. Note that when designing a system (a collection of related programs), an intermediate step is usually necessary. Before the system can be decomposed into modules, it must be decomposed into independent programs, or components...

(b) Review the completed structural design, trying to maximize module strength and to minimize coupling.

(c) Review the design again, using the guidelines of ..... decision structure, input/output isolation, restrictive modules, data access, size, recursion, predictable behaviour..."

Guidelines outlining the methods of decomposition that can be applied to software are given in (10, 11).

Upon identification of the system's various levels of abstraction and of the connections between them, top down design achieves a decomposition of the system into a number of highly independent modules, resulting in a significantly simpler structure.

When the decomposition has been completed a structure similar to Figure 4 should have been achieved.

Arriving at the final structure is a highly iterative technique greatly aided by the Hierarchy, plus Input, Process, Output (HIPO) method fully described in (12). The end product is a HIPO diagram for each module starting, as in Figure 4, with module 1.0 and proceeding downward until all modules have a HIPO diagram. An example of a completed HIPO diagram is shown in Figure 5.

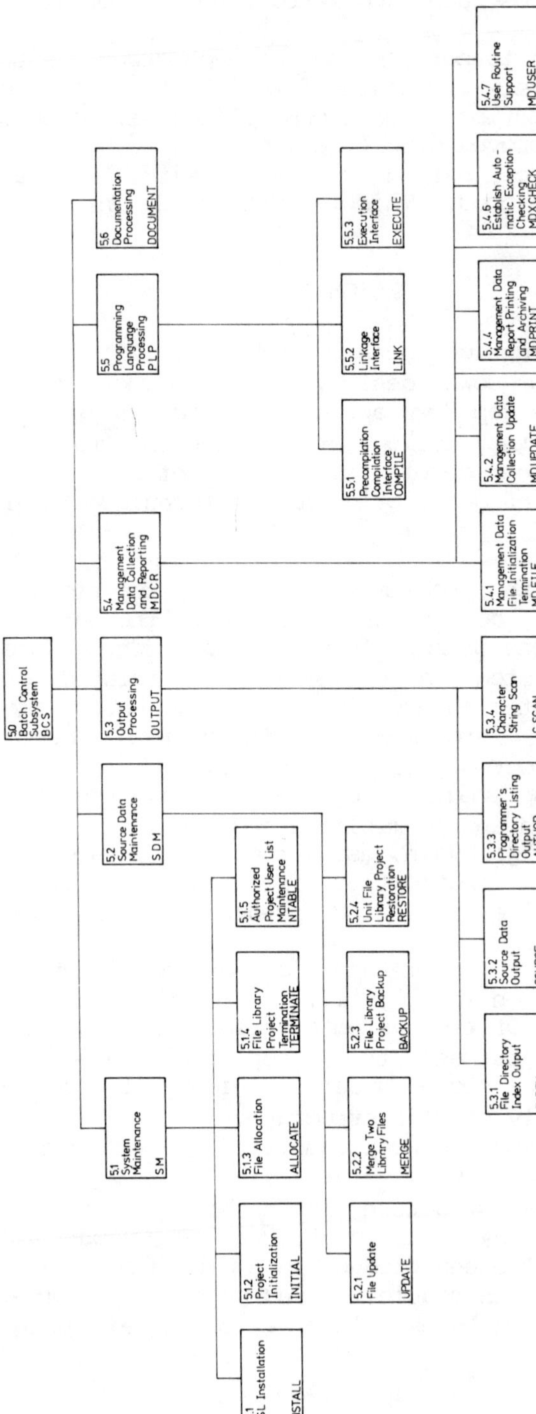

Figure 4. Decomposed Software System

It may appear as though a detailed HIPO chart prepared for a module would be sufficient to present to a programmer for coding. This is not necessarily the case and represents a narrow application of the HIPO chart, which is intended as a design aid in defining a software system via a top-down approach to design. Module specifications based on HIPO charts are still recommended for formally documented systems.

At this point a comment on module size would be appropriate. The recommended finished module size is 10 to 100 high level statements (13) which should fit, approximately, on one page about 11 inches long.

The top down approach may also be used to check out the consistency of module specifications before coding them. This is done by "stubbing" the system with a simulation program which, externally, looks like the real program. Using top down decomposition, one begins by specifying overall functions to be performed; then one determines how these general functions are accomplished by more specific functions. The process is repeated until specific functions, which can be accomplished by a single module, are derived. Module specifications are checked directly from the top-down decomposition.

Of the major advantages attributed to the top-down approach, the following are particularly significant to enhancing software R-M:

(a) Emphasis is no longer placed on program implementation and coding, but rather on system design, by far the most critical software development phase,

(b) Effective modularity of the system can be achieved from the fact that common functions to be performed are identified early enough in the design,

(c) Modules may be tested at each successive level of development, which greatly simplifies debugging and guarantees program correctness as it proceeds, not as an afterthought following completion of the coding effort.

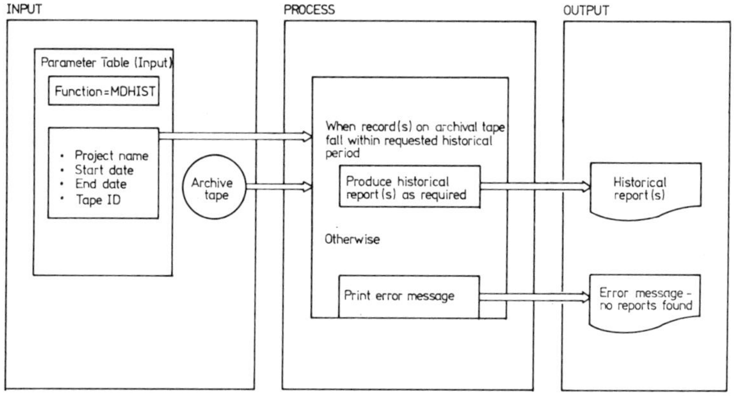

Figure 5. High Level HIPO Chart

PROGRAMMING TECHNIQUES

Assumming that a properly specified module is to be programmed, are there any fundamental guidelines which can be offered as to how the programming should proceed? A technique known as Structured Programming (SP) has evolved and appears to provide what is needed. Basically SP offers a set of fundamental programming constructs which, in general, may be applied to any programming task.

The early concepts of structured programming were based on three basic software constructs, i.e., Sequence, IF THEN, DO WHILE. These basic structures have evolved into the six now in common use, illustrated in Figure 6 and defined as follows:

(a) Concatenation - statements are executed in the order in which they appear, with control passing unconditionally from one statement to the next. This hardly requires explaining, but is necessary for the construction of a block from statements that are to be executed sequentially.

(b) IF c DO s - the condition c is tested. If c is true, the statement is executed; otherwise, s is not executed, control passes to the next statement. Note that the statement s may itself be a block, or it may be a simple statement. This is true in each of the program constructs, wherever a statement can appear.

(c) IF c THEN a ELSE b - the condition c is tested. If c is true, the statement a is executed and the statement b is skipped; otherwise, the statement a is skipped and the statement b is executed.

(d) WHILE c DO s - the condition c is tested. If c is true, the statement s is executed and control returns to the beginning for another test of c. If c is false, then s is skipped and control passes to the next statement.

(e) REPEAT s UNTIL c - the statement s is executed and then the condition c is tested. If c is false, control returns to s for another iteration. If c is true, control passes to the next statement.

(f) CASE i OF $(s_1, s_2 \ldots, s_n)$ - the ith statement of the set $(s_1, s_2 \ldots, s_n)$ is executed, and all other statements of this set are skipped. Control passes to the next statement (following s).

A structured program consists of a sequence of the structures described above and therefore will be highly readable, testable and maintainable. The principles of SP have been developed along similar lines by a number of authors (14, 15, 16, 17) but the basic result is always a disciplined and controlled method of programming. Specific applications of SP to different programming languages are given in (18, 19, 20).

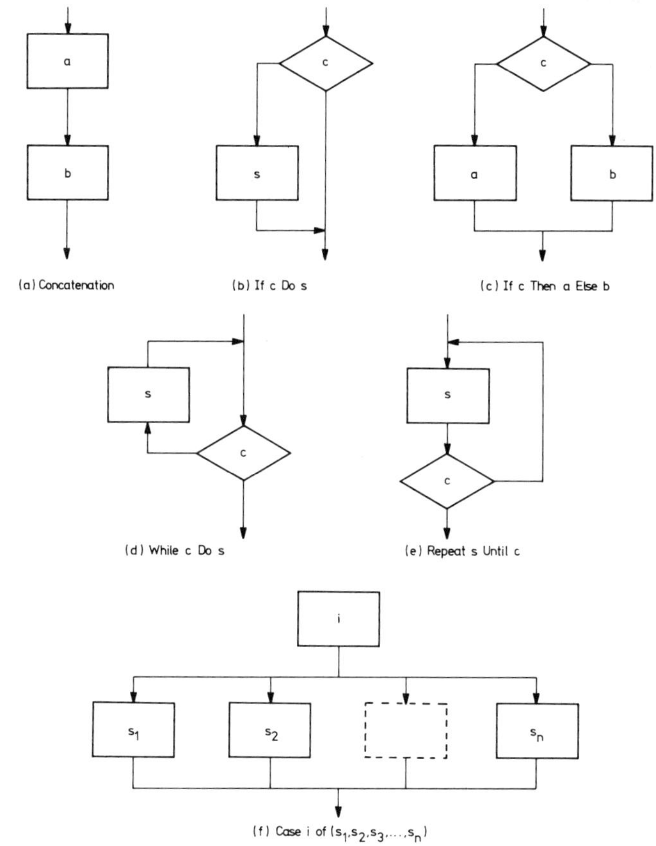

Figure 6. Structured Programming Constructs

TESTING

The problems of testing software, theoretically and practically, are formidable. There are very good reasons for this as explained in reference (21).

"The main reason why it is difficult to establish software reliability is our inability to easily visualize the dynamic behaviour of a program from the text describing it. Consider the program flow chart in Figure 7. It has 4 blocks of sequential code and 8 decision blocks. Also there are 2 nested loops where the inner can be executed up to 12 times and the outer loop, 2 times. Now this module has 1.6×10^{19} possible ways of traversing through the flow chart. If you tested one path every nano-second, it would take you over 500 years. Clearly, testing alone will not prove the the correctness of this module. The only way in which you can gain confidence in the behaviour of this module is to prove the correctness of each nested substructure. There are 13 blocks nested within each other as shown in the figure. For each block, you must satisfy yourself that for all possible inputs it will generate the correct outputs.

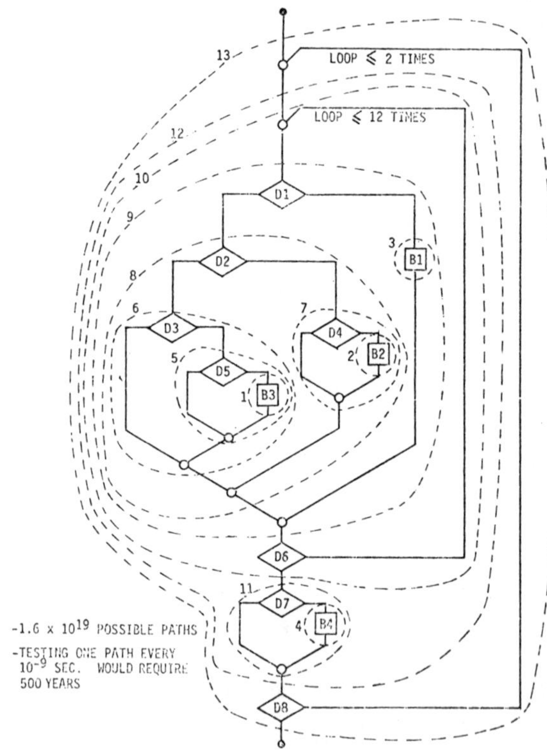

Figure 7. Typical Simple Program

Since each block has a single input and single output, the output from the inner blocks serves as a subset of the outputs or inputs for the outer blocks. Thus only 13 sets of tests have to be derived. Using structured programming techniques the number and difficulty of the proofs is drastically minimized. Only in this way can the reliability of software be achieved in a manageable form."

The testing of programs is divided into two phases, i.e., debugging and integration. In the debugging phase, a module is tested against its specification while in the integration phase modules are tested as a group against a group specification and so on, until finally the software system is tested against its system specification.

When structured programming is used, a significant degree of program correctness is built in before debugging begins. Module testing is therefore oriented primarily toward testing that all branches of each program can be properly entered and exited. Results are precalculated and tested for both sensible and non-sensible input data. Test cases for module level testing are defined using a simple networking technique, as shown in Figure 8. This ensures that all paths between decision points are tested at least once, so that when delivered to integration the routine will not fail under any circumstance. The total number of individual tests required is a function of the program complexity.

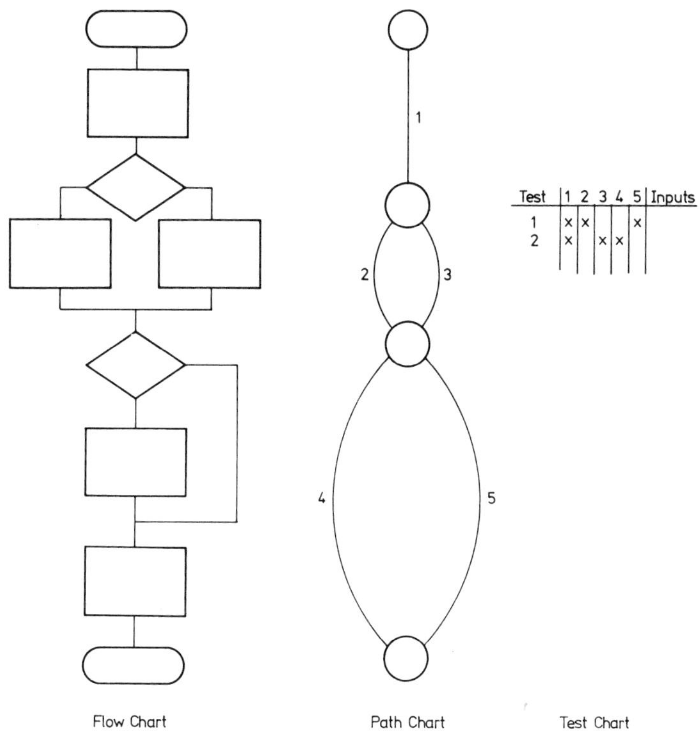

Figure 8. Test Path Tracing

Top-down design assures that proper interfaces and program module compatibility are built into the software design. The purpose of top-down integration is to validate that part of the design. The top-level modules are integrated first, using simple program stubs to represent lower-level modules; lower-level modules are then successively integrated, again using the program stubbing technique. Test cases are designed to test each module in its operating environment. At the end of integration testing, all program units operate compatibly in their operational environment. Each stub replaced by a deliverable module represents a measurable milestone.

Based upon analysis of system and software requirements, a set of test scenarios, which exercise the integrated software in all operating modes and ranges of input data, is defined. The software is then qualified by testing with those scenarios to ensure that all performance requirements are satisfied.

Further information on program test methods will be found in (22, 23).

DOCUMENTATION

As has been pointed out, software systems evolve in phases from concept through to operation. Reference (24) provides a global approach to the problem of documentation, throughout the various phases and the stages within these phases. The three phases applicable to the software life cycle are: Initiation, Development and Operation. The Development phase is further sub-divided into four stages, i.e., Definition, Design, Programming and Test. In addition each stage will have associated with it one or more documents. Table 1 summarizes the above and definitions follow:

Table 1. Documentation Within the Software Life Cycle

Initiation Phase	Development Phase				Operation Phase
	Definition Stage	Design Stage	Programming Stage	Test Stage	
	Functional Requirements Document	System/Subsystem Specification	Users Manual		
		Program Specification	Operations Manual		
	Data Requirements Document	Data Base Specification	Program Maintenance Manual		
		Test Plan	Test Plan	Test Analysis Report	

Phases

Initiation - the objectives and general definition of the requirements for the software are established. Feasibility studies, cost-benefit analyses, and the documentation prepared within this phase are determined by agency procedures and practices.

Development - the requirements for the software are determined and the software is then defined, specified, programmed, and tested. Documentation is prepared within this phase to provide an adequate record of the technical information developed.

Operation - the software is maintained, evaluated, and changed as additional requirements are identified.

Stages

Definition - the requirements for the software and documentation are prepared. The Functional Requirements and the Data Requirements Document may be prepared.

Design - the design alternatives, specific requirements, and functions to be performed are analyzed and a design is specified. Documents which may be prepared include the System/Subsystem Specification, Program Specification, Data Base Specification, and Test Plan.

Programming - the software is coded and debugged. Documents which may be prepared during this stage include the Users Manual, Operations Manual, Program Maintenance Manual, and Test Plan.

Test - the software is tested and related documentation reviewed. The software and documentation are evaluated in terms of readiness for implementation. The Test Analysis Report may be prepared.

Some of the documents listed in Table 1 specify the use of flowcharts. There is an argument that with top-down design, structured programming and commented code, flowcharts may no longer be required.

Document Types

Functional Requirements Document - the purpose is to provide a basis for the mutual understanding between users and designers of the initial definition of the software, including the requirements, operating environment, and development plan.

Data Requirements Documents - the purpose is to provide, during the definition stage of software development, a data description and technical information about data collection requirements.

System/Subsystem Specification - the purpose is to specify, for analysts and programmers, the requirements, operating environment, design characteristics, and program specifications (if desired) for a system or subsystem.

Program Specification - the purpose is to specify, for programmers, the requirements, operating environment, and design characteristics of a computer program.

Data Base Specification - the purpose is to specify the identification, logical characteristics, and physical characteristics of a particular data base.

Users Manual - the purpose is to describe the functions performed by the software in non-ADP terminology, such that the user organization can determine its applicability and when and how to use it. It should serve as a reference document for preparation of input data and parameters and for interpretation of results.

Operations Manual - the purpose is to provide computer operation personnel with a description of the software and of the operational environment so that the software can be run.

Program Maintenance Manual - the purpose is to provide the maintenance programmer with the information necessary to understand the programs, their operating environment, and their maintenance procedures.

Test Plan - the purpose is to provide a plan for the testing of software; detailed specifications, descriptions, and procedures for all tests; and test data reduction and evaluation criteria.

Test Analysis Report - the purpose is to document the test analysis results and findings, present the demonstrated capabilities and deficiencies for review, and provide a basis for preparing a statement of software readiness for implementation.

It is rather obvious that all the documents listed in Table 1, for economic reasons, only apply to the largest of projects. Consequently reference (25) also provides scoring criteria to determine those that should apply to any software project. In addition, detailed guidelines are offered for each of the ten document types listed above. Other documents (25, 26) offer guidance in the area of software systems documentation.

With respect to the top-down approach, the resulting system modularity leads to well structured documentation whereby design structures and program modules can be documented accurately and fully as development progresses. This ensures that, at the design level, none of the original intent of the system purpose is lost during the development process; in addition, good communication is established from the design to the program implementation process, such that better software reliability can be achieved. Furthermore, complete and accurate documentation that can communicate a good understanding of each module, within the framework of the overall system, together with a detailed description of its internal logic, will effectively guarantee the attainment of adequate software maintainability.

MODELLING

Software reliability modelling deals with the subject of predicting how many more errors there are likely to be in a software system and what the time between error occurence is likely to be at any time in the life cycle. Many models have been postulated: (27 - 32), some are speculative in nature and others are based on the reduction of data compiled from software projects. At this time the set is concerned largely with obtaining sufficient data collected under controlled conditions in order that the field of competing models may be narrowed. Collected data takes on a somewhat general shape as shown in Figure 9, from reference (33), for different phases in the software life cycle. It should be noted that comprehensive data collection usually begins when a module meets its specification and is brought under strict configuration control. From this point a data base is built up during integration and may continue throughout other software life cycle phases.

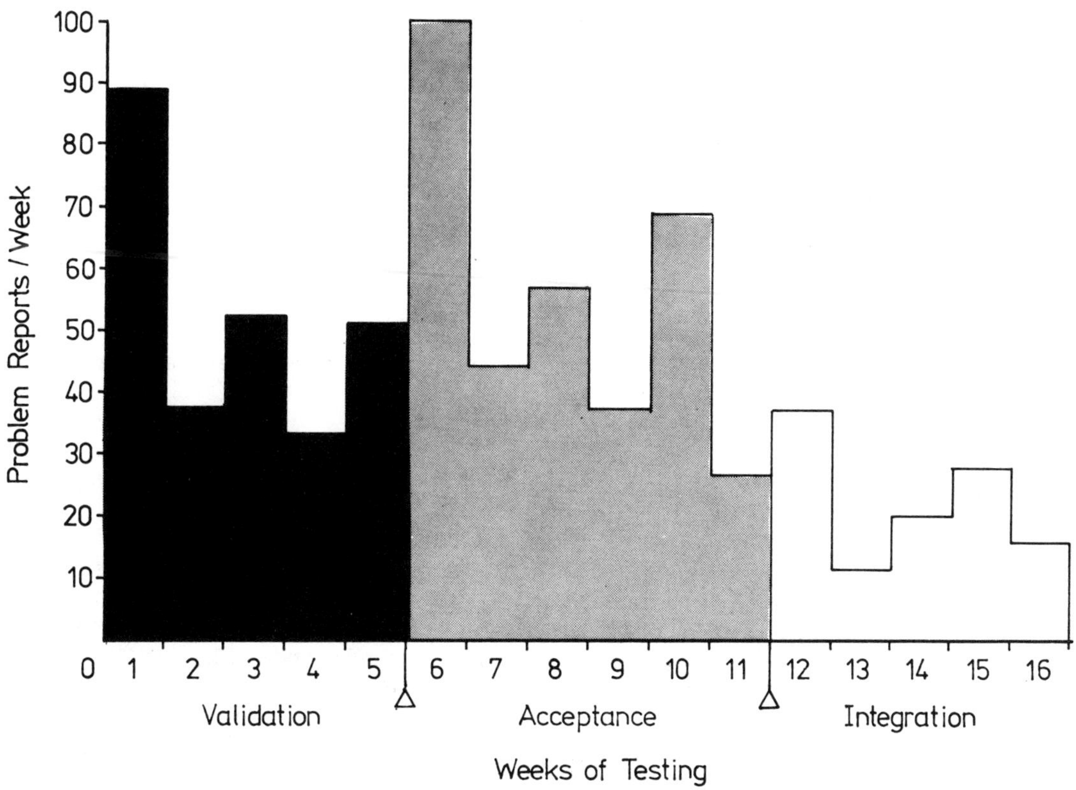

Figure 9. Software Problems Encountered During Testing

It is interesting that a form of the exponential model widely applied to hardware has been adapted for use in software reliability modelling (34). The exponential model is simple to manipulate, allows for adjustment as data accumulates, and has been proven in practice on development projects. As with all models, the fundamental assumption of this model is that a software system has associated with it, at any time, an error rate which will remain constant unless effort is applied to reduce it further. At some point in time, provided that sufficient effort with respect to error removal is applied, system failures due to software errors will be a very small percentage of system failures due to hardware, i.e., negligible. At about the same time the economics of error removal begins to place a practical constraint on the error removal effort.

The exponential software model can be described by:

$$N_D = N_0 (1 - \exp(-\lambda t)) \tag{1}$$

where,

N_0 = initial number of faults,

N_D = detected number of faults,

λ = fault rate,

t = time.

Differentiation of N_D with respect to t gives the expected number of faults per unit time, i.e.,

$$\frac{dN_D}{dt} = \lambda N_0 \exp(-\lambda t)) \tag{2}$$

and the reciprocal of (2) therefore gives the mean time between software faults, i.e.,

$$MTBF = 1/\frac{dN_D}{dt} \tag{3}$$

The resulting curves for (2) and (3) are shown in Figure 10. Figure 10 (a) will be recognized as the classic 'debugging curve' compiled for many projects and widely published in the literature. Thus any project can, using the unique curve peculiar to the project and by performing a curve fitting analysis, derive values for,λ, N_D, and since dN_D/dt will be known, $MTBF_S$ can be estimated at any point in time.

As a result of this approach the reliability parameters of the software will be measured constantly via a failure reporting and analysis system. This will permit the debugging effort to be adjusted as required to ensure that in the final analysis projections will yield the desired $MTBF_S$ at the required point in time.

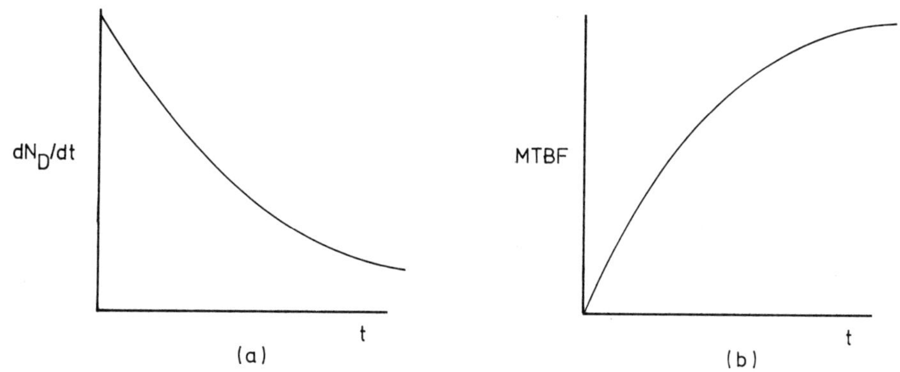

Figure 10. Plots of (a) dN /dt and (b) MTBF vs. Time

A rather unique approach to the general problem of software modelling has been provided in (35) and has been summarized in (36). The overall approach can be thought of as providing a "physics of software" based on the general properties of algorithms. Only a few definitions form the underlying basis for the complete characterization of computer programs and they are:

n_1 = number of unique or distinct operators appearing in an implementation,

n_2 = number of unique or distinct operands appearing in an implementation,

N_1 = total useage of all of the operators appearing in an implementation,

N_2 = total useage of all of the operands appearing in an implementation,

From this the vocabulary is defined as,

$$n = n_1 + n_1 \tag{4}$$

and the length N is defined as,

$$N = N_2 + N_2 \tag{5}$$

With respect to the error content (E) of modules delivered for integration, this is given as:

$$E = \frac{N\log_2(n_1 + n_2)}{2n_2/n_1N_2} \tag{6}$$

and the estimated number of bugs (B) is given by:

$$B = E^{2/3}/3000 \quad (7)$$

Other related topics included in the physics of software are program length modularity, volume, language, and programming effort.
With respect to data collection is general, Chapter 17 should be conculted.

MAINTAINABILITY

Maintainability implies designing for controlled change, in which some parts or aspects remain the same while others are altered, all in such a way that a desired new characteristic is achieved. Changes are required to remove errors from software, to add new capabilities, and to improve a system's performance. Software maintainability is, therefore, a broad sugject and not easy to achieve because changes are required for many resons.

Maintainable software can be obtained most readily by the introduction of software design disciplines outlined previously and appropriate management concern. Special emphasis should be placed on documentation and configuration management. If the rules of structured programming are used to the maximum extent possible, the result, will be virtually, a self-documenting software system. Structured programming demands the use of simple basic programming structures which are implemented by small blokcs of code. Each block or structure should be commented separately and commenting should be used to describe departures from the basic structures, as necessary. Configuration management ensures that the software currently running on a machine is traceable back to listing, decks, cassettes, tapes, etc.

MANAGEMENT

Computer programming, not surprisingly, offers an opportunity for a psychological study of the process (37). In any case the objective is to obtain reliable and maintainable software.

Organization

A team approach to a project has been found to be an effective way to detain reliable and maintainable software. This idea has been developed (38) and is known generally as the Chief Programmer Team (CPT) approach. This approach offers a disciplined alternative to the seemingly unorganized generation of software. It moves the associated problems from the private to the public domain where they can be reconized early and solved appropriately. Figure 11 shows one such organization sturcture and is repeated as often as required to match the demands of the system under development.

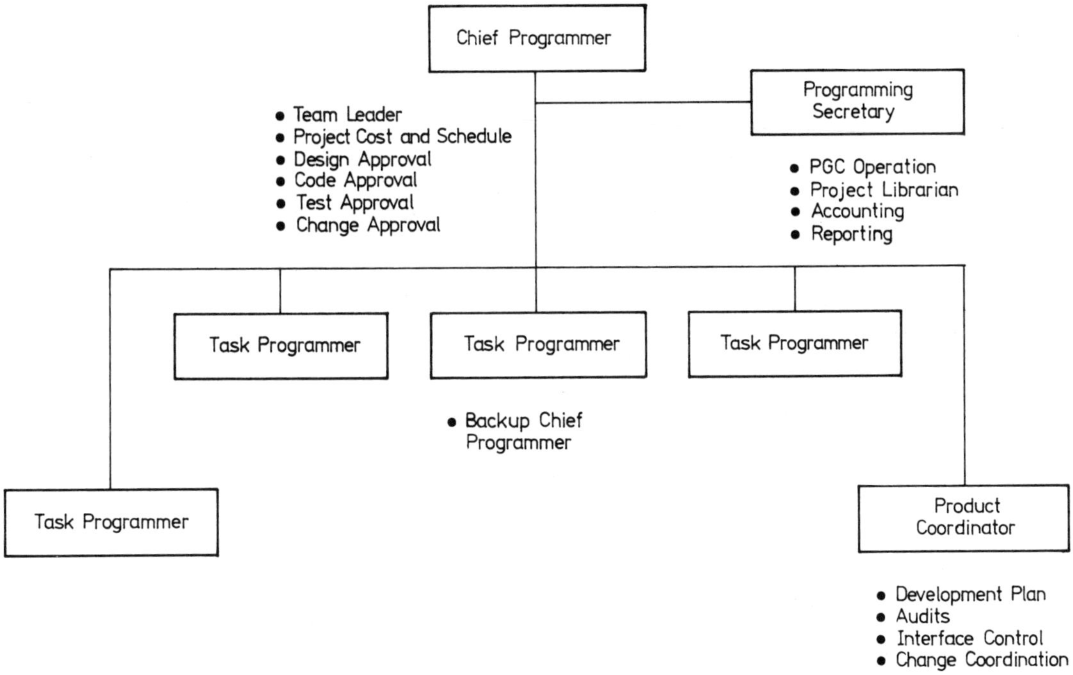

Figure 11. Chief Programmer Team Organizational Structure

The Chief Programmer is the key person on the team. He is the team leader, both administratively and technically. He is responsible for completing the software within cost and within schedule, and has approval authority for all aspects of software development. His administrative role is lessened by the Programming Secretary and the Product Coordinator. He has technical responsibility for the overall software design and for design, programming and testing of the software system. His job input is the overall Software System Requirement Specification. His first output is a group of subprogram requirement specifications.

As the software requirements are allocated to specific application areas, a Task Programmer is assigned technical responsibility for a specific application. The Task Programmer is assigned a schedule and budget for his application area; he is expected to exercise cost and schedule control and to provide clear visibility to the Chief Programmer on cost and schedule performance. Each Task Programmer is responsible for design, programming and testing of his application; the only constraints being the requirement specification, schedule and cost.

The Product Coordinator is responsible for maintaining a Software Development Plan and for ensuring adherence to the plan. As part of that responsibility, he participates in software reviews and audits to verify that standards and conventions are followed both in software and in documentation. He also verifies that defined internal and external interfaces are maintained and coordinates software changes. He is the software project interface with the program-level configuration management function.

The Programming Secretary acts as the operator for the Program Generation Centre (PGC) for making controlled program changes, and is the project librarian. He is also responsible for keeping accurate accounts of all problems reported, changes made or in progress, and he provides regular reports on review status and change status to provide management visibility of development progress.

One of the Task Programmers acts as a Backup Chief Programmer. This is an understudy role; the Backup Chief Programmer must assume the role of Chief Programmer, either temporarily or permanently, when called upon. Because of this, he maintains familiarity with the overall software design and the current administrative and technical status. In this role, he performs administrative and technical review functions as assigned by the Chief Programmer.

Design Reviews

Apart from the usual Design Review process described in Chapter 3, which may involve the use of a formal review as given in (9), there are techniques emerging which are oriented toward software. One such technique is "structured workthroughs" (39). A structured workthrough procedure involves a peer group review, at the informal level, for each module and the system as a whole. To achieve this a review kit is prepared containing all the documentation associated with the module, such as the module specification, listing, test results, etc. The kit is then reviewed by the programming team. The process is summarized in Figure 12.

Configuration Control

The fundamentals of configuration control discussed in Chapter 5 also apply no less to software. A well functioning configuration control system requires that the software currently running on a machine is traceable back to a set of documents. This ensures that controlled software changes are carried out with maximum benefit.

Figure 13 indicates a configuration control system associated with the Chief Programmer Team kind of organization.

CONCLUSION

The importance of discipline in the design and production of software systems has been emphasized in the foregoing and the whole process

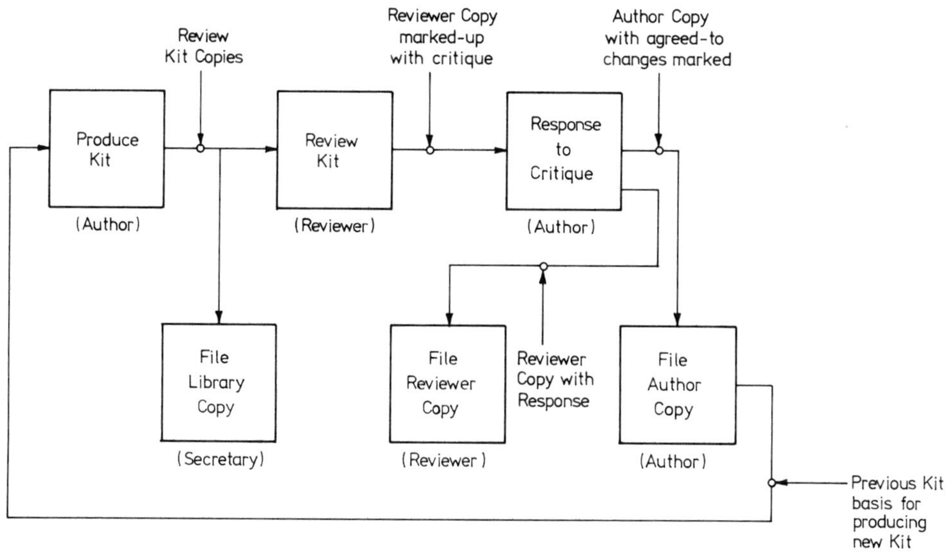

Figure 12. Review Kit Flow

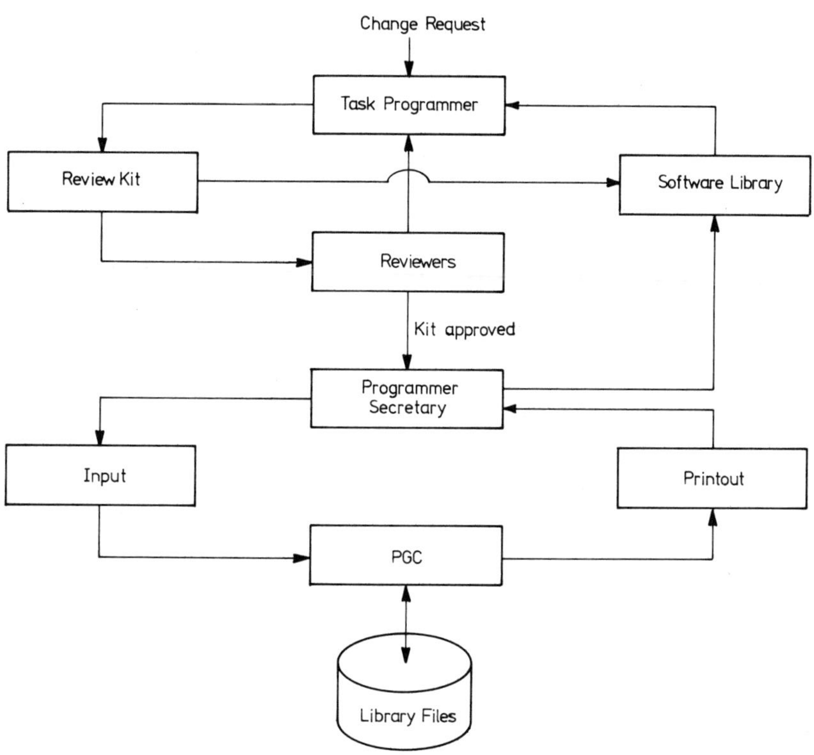

Figure 13. Configuration Control Flow

has become known as Software Engineering. As such the field is taking on a task not a great deal different from Hardware Engineering and it would seem that the process will continue to develop.

REFERENCES

1. _________, Software Engineering, NATO Science Committee Conference, Garmish, Germany, October 1969.

2. _________, ___________, Air Force Magazine, July 1973, p46 - 57.

3. Myers, G.J., Software Reliability, Wiley-Interscience, John Wiley and Sons, New York, 1976.

4. Myers, W., "The Need for Software Engineering", Computer, February 1978, p12 - 26.

5. _________, Software Life Cycle Management Standards Volume 1, Department of National Defence, Ottawa, Canada, 1978.

6. Pattee, H., "Postscript: Unsolved Problems and Potential Applications of Hierarchy Theories", pp. 129 - 156 in Hierarchy Theory, the Challenge of Complex Systems, ed. H. Pattee, Publ., George Braziller Inc. New York 1973.

7. Ronback, J.A., "Software Reliability - How It Affects System Reliability", Proceedings, 1975 SRE Canadian Reliability Symposium, Pergamon Press, New York, p 125 - 127.

8. Reifer, D.J., A New Assurance Technology for Computer Software, The Aerospace Corporation, NTIS Report No. AD/A-020 483, September 1975

9. _____________, MIL-STD-1521, Technical Review and Audits for Systems, Equipment, and Computer Programer, US. Department of Defense, Washington, D.C.

10. Myers, G.J., Reliable Software Through Composite Design, Petrocelli/Charter, New York, 1975.

11. McGowan, C.C. and Kelly, J.R., Top-Down Structured Programming Techniques, Petrocelli/Charter, New York, 1975.

12. Katzan Jr., H., System Design and Documentation - An Introduction to the HIPO Method, Van Nostrand Reinhold, New York, 1976.

13. Ibid (3), p 96.

14. Dijkstra, W.D., A Discipline of Programming, Prentice-Hall, Englewood Cliffs, N.J., 1976.

15. Wirth, N., Systematic Programming - An Introduction, Prentice-Hall, Englewood Cliffs, N.J., 1973.

16. Dahl, O.J., et al., Structured Programming, Academic Press, New York, 1972.

17. Knuth, D.E., Structured Programming with GO TO Statements, Stanford University, NTIS Report No. PB-233 507, May 1974.

18. Kessler, M.M., et al., Programming Language Standards, IBM Corp., NTIS Report No. AD/A-016 771, March 1975.

19. Tenny, T., "Structured Programming in Fortran", Datamation, July 1974, p 110 - 115.

20. Rieks, G.E., "Structured Programming in Assembler Language", Datamation, July 1978, p 79 - 84.

21. Ibid (7), p 123.

22. Hetzel, W.C., ed., Program Test Methods, Prentice-Hall, Englewood Cliffs, N.J., 1972.

23. Miller Jr., E.F., ed., "Program Testing", Computer, April 1978, p 10 - 12.

24. ____________, FIPS Publication 38 - Guidelines for Documentation of Computer Programs and Automated Data Systems, U.S. Department of Commerce, February 1976.

25. Landon, K.R., Documentation Standards, Petrocelli Books, New York, 1974.

26. Krehne, R.S., et al., Handbook of Computer Documentation Standards, Prentice-Hall, Englewood Cliffs, N.J., 1973.

27. Sukert, A.N., "An Investigation of Software Reliability Models", Proceedings, 1977 Annual Reliability and Maintainability Symposium, IEEE Pub. No. 77CH1161-9RQC, January 1977, p 478 - 488.

28. Hecht, H., et al., "Reliability Measurement During Software Development", The Aerospace Corporation, NTIS Report No. N78-10487, September 1977.

29. ___________, A Methodology for Producing Reliable Software, Volume 1, McDonnell Douglas Astronautics Co., NTIS Report N76-29945, March 1976, Appendix B.

30. Wolverton, R.W. and Shick, G.J., "An Analysis of Competing Software Reliability Models", IEEE Transactions on Software Engineering, Vol., SE-4, No. 2, March 1978, p 104 - 120.

31. Moranda, P.B., "Prediction of Software During Debugging", Proceedings, 1975 Annual Reliability and Maintainability Symposium, IEEE Pub. No. 75 CHO 918-3RQC, January 1975, p 327 - 332.

32. ___________, Quantitative Methods for Software Reliability Measurements, McDonnell Douglas Astronautics, NTIS Report AD/A-035 585, December 1976.

33. Craig, G.R., et al., Software Reliability Study, TRW Systems Group NTIS Report AD787784, RADC Report RADC-TR-74-250, October 1974.

34. Krten, O.J., "Improving Existing Software Reliability Models to Provide Early Prediction of Trends", Society of Reliability Engineers, Software Reliability Lecture Series, Ottawa, Canada, 1978.

35. Halstead, N.H., Elements of Software Science, Elsevier North-Holland Inc., New York, 1977.

36. Fitzsimmons, A. and Lone, T., "A Review and Evaluation of Software Science", Computing Surveys, Vol. 10, No. 1, Association of Computing Machinery, March 1978, p 3 - 18.

37. Weinberg, G.M., The Psychology of Computer Programming, Van Nostrand Reinhold, New York, 1971.

38. Baker, F.T. and Mills, H.D., "Chief Programmer Teams", Datamation, December 1973, p 58 - 61.

39. Yourdon, E., Structured Walkthroughs, Yourdon Inc., 1978.

PART TWO

RELIABILITY

7 MATHEMATICAL MODELLING

A. H. K. Ling

As electronics technology becomes more and more complex, it is almost impossible to perform reliability analysis unless global approaches are adopted. Reliability engineers have concentrated increasingly on the systems approach of modelling, as shown in Figure 1. Large complex equipment, or interconnections of equipment, are analyzed by subdividing the whole into functional units, assemblies, components, or until the subdivision is convenient for analysis. Each item can exhibit either a good or bad state. This subdivision generates a system's functional block diagram. Having generated the required data, mathematical models are formulated to fit this logical structure and the calculus of probability is used to compute and predict the system reliability in terms of subdivided units. Using the analytical results, reliability engineers and managers can make decisions as to the significance of these results and make the necessary prognosis.

In this Chapter, some of the most commonly occurring failure distribution functions, associated with electronics, are discussed. Using these distribution functions, reliability expressions are derived for simple structures where repair and failure dependence are ignored. These are analyzed by means of set theory. The more complicated models of systems, where repair and dependency is allowed, are developed in Chapter 19. The models and techniques of analysis are applied by example to show their efficacy.

The treatment of reliability mathematical modelling is by no means an innovation, and the reader is advised to refer to the frequently quoted references, (1-9).

The elementary relationships of failure rate, failure distribution functions, failure density functions, reliability, mean time between failure (MTBF), mean time to repair (MTTR), and availability should be clearly understood and are summarized in Table 1. For their derivation see (1-9).

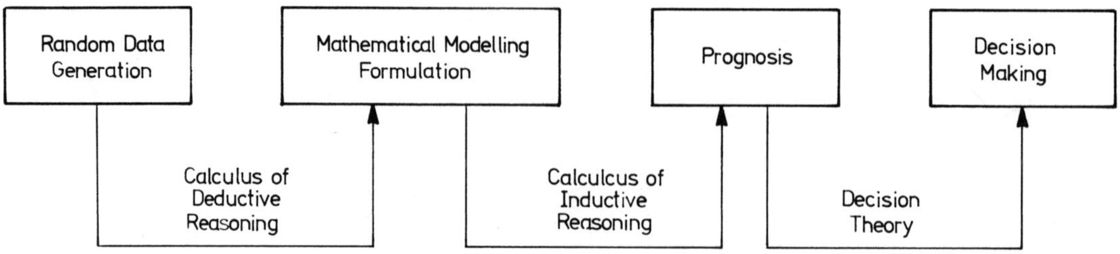

Figure 1. The Modelling Process

Table 1. Relationships Used in Reliability Modelling

$$f(t) = \frac{(n(t_i) - n(t_i+\Delta t_i))/N}{\Delta t_i} \qquad t_i < t \leq t_i + \Delta t_i \tag{1}$$

$$z(t) = \frac{(n(t_i) - n(t_i+\Delta t_i))/n(t_i)}{\Delta t_i} \qquad t_i < t \leq t_i + \Delta t_i \tag{2}$$

$$R(t) = 1 - F(t) \tag{3}$$

$$F(t) = \int_0^t f(x)\, dx \tag{4}$$

$$R(t) = \exp\left(-\int_0^t Z(x)\, dx\right) \tag{5}$$

$$f(t) = Z(t) \exp\left(-\int_0^t Z(x)\, dx\right) \tag{6}$$

$$Z(t) = \frac{f(t)}{R(t)} \tag{7}$$

$$MTBF = \int_0^\infty t\, f(t)dt = \int_0^\infty R(t)dt \tag{8}$$

$$MTBF = \lim_{s \to 0} R^*(s) \tag{9}$$

$$A(t) = 1 - P(\bar{x}) \tag{10}$$

$$A(0,T) = \frac{1}{T}\int_0^T A(t)\, dt = \frac{1}{T}\sum_{t=0}^{t=T} A(t) \tag{11}$$

$$A = \lim_{t \to \infty} A(t) = \frac{MTBF}{MTBF + MTTR} \tag{12}$$

The notations used in Table 1, are defined as follows:

f(t) : Failure density function, defined as the ratio of the number of failures occurring in the intervals $t_i < t \leq t_i + \Delta t_i$, to the size of the original population, divided by the length of the time interval.

Z(t) : Failure rate is defined as the ratio of the number of failures occurring in the time interval to the number of survivors at the beginning of the time interval divided by the length of the time interval. It is a measure of the instantaneous speed of failure.

R(t) : System reliability, is defined as the probability that a given system (component, etc.) operates without failure until a given time when used under specified environmental conditions.

F(t) : Failure distribution function, is defined as the unreliability of the system at time t.

MTBF : Mean-time-between-failure, is defined as the mean of all time periods over an infinite length of time when a system or equipment enters an inoperative state.

MTTR : Mean-time-to-repair, is defined as the mean of all time periods over an infinite length of time, taken to restore the failed system to an operative state.

$R^*(s)$: Laplace Transform of R(t).

P(x) : Probability of success (same as R(t)).

$P(\bar{x})$: Probability of failure 1-P(x).

A(t) : Pointwise (or instantaneous) availability, is defined as the probability that a system is operational at the time t, regardless of how often it has failed during the period of time.

A(0,T) : Interval availability, is the fraction of time, the system is operational during $t \varepsilon (0,T)$.

A : Steady state availability, is the probability that the system will be operational at any random point of time.

PROBABILITY DISTRIBUTION FUNCTIONS

The time between successive failures is a continuous random quantity. This random quantity from the probabilistic standpoint can be fully determined if the distribution function is known. These failure models are related to life test results and failure rate data via probability theory.

In constructing a failure model, test data based on specific conditions is necessary. On the basis of tests, physical failure information (10-15), engineering judgment and time statistical tests, a failure rate model is chosen. The parameters of the models are estimated directly from graphs or computed using theory of estimation (16, 17).

Figure 2 shows a typical time versus failure rate curve. This is the well- known "bath-tub" curve. Zone ab is the infant mortality period. The failure rate in this region is high. This is attributed to gross built-in flaws which soon cause the parts to fail. After this zone, the failure rate under certain circumstances remains constant, zone bc. This is the useful operating life and in this zone failure data is published in reliability parts manuals (18) and it is these part failure rates which are usually summed up to calculate the inherent system reliability. Finally, whatever wear or aging mechanisms are involved, occur in the wear out time, zone cd, i.e., the failure rate increases rapidly.

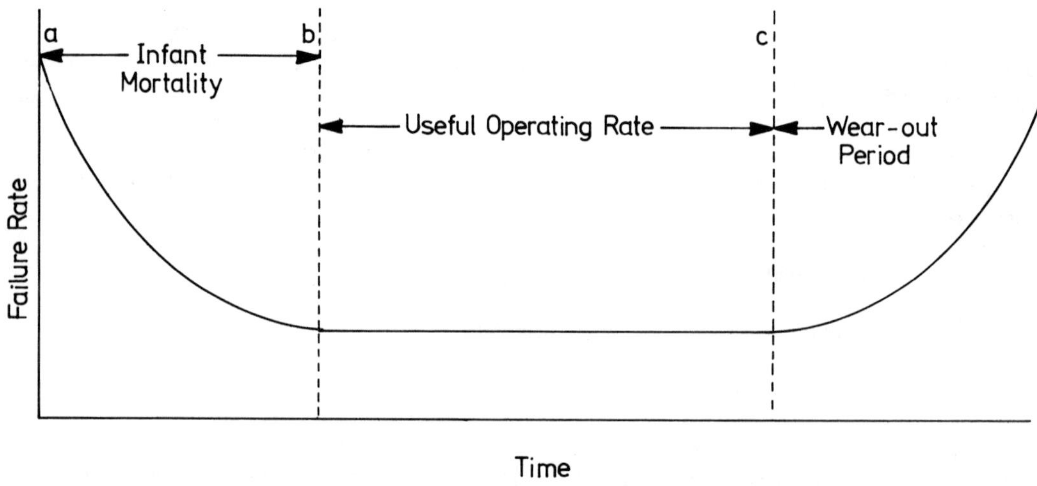

Figure 2. The "Bath-Tub" Failure Curve

The "bath-tub" failure curve gives a good insight into the life cycle reliability performance of an electronic system. Depending on the physical meaning, the random quantities obtained can have different probability distributions. A great many of these distributions have been discussed in the references (1, 7, 13, 19 - 22). They are repeated here since they are useful in reliability mathematical modelling, and provide better understanding of reliability. In practice, the time between failure of a complex system or a single component tends to obey only a few well-defined probability distributions, discussed in the next sections:

(a) Exponential,
(b) Rayleigh,

(c) Normal,
(d) Gamma, and
(e) Weibull distribution laws.

Exponential Law

In the case of an exponential law of distribution for the times of occurrence of failure, the failure rate of the electronic devices is a constant. This is given by:

$$Z(t) = \lambda \tag{13}$$

Then from the basic probability theory given in Table 1, the qualitative characteristic of reliability can be expressed by

$$f(t) = \lambda \exp(-\lambda t) \tag{14}$$

$$F(t) = 1 - \exp(-\lambda t) \tag{15}$$

$$R(t) = \exp(-\lambda t) \tag{16}$$

$$MTBF = 1/\lambda \tag{17}$$

The typical characteristics of the exponential law are plotted in Figure 3. The constant failure rate of the exponential law implies that the probability of a component's failure at any given time interval remains constant until the component fails. This is the component's failure rate, operating under normal conditions in the region bc of the "bath-tub" curve. The interval between individual failures is random. In particular, for a NAND gate, which has a failure rate of $0.2456/10^6$ hour, we can expect at most 3 failures, in a batch of 10×10^6 components under test, during the first hour. The same number of failures occur during the next hour and so on.

Rayleigh Law

When aging and excessive stresses are present, causing wear and deterioration of electronic devices, the failure rate will increase with time. The most appropriate failure rate to describe this is,

$$Z(t) = \frac{t}{\sigma^2} \qquad t \geq 0 \tag{18}$$

where,

σ is the constant of the Rayleigh distribution.

The associated density, reliability and MTBF functions, as in Table 1, are given by,

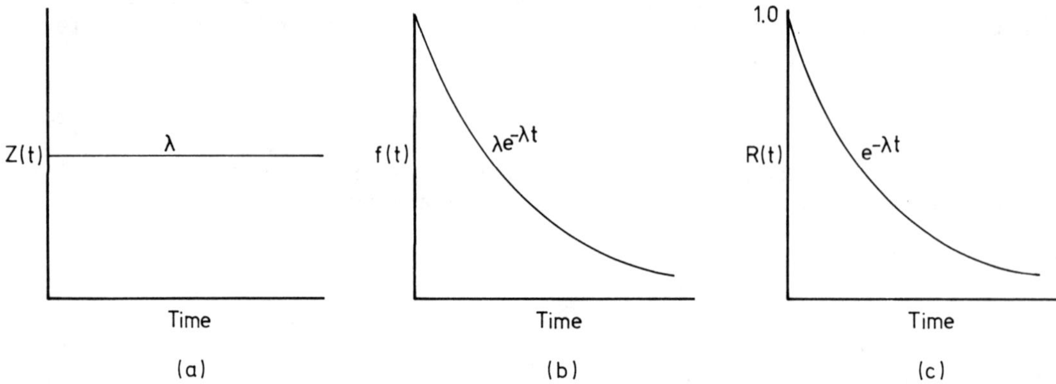

Figure 3. Exponential Law Distribution for (a) Z(t), (b) f(t), (c) R(t).

$$f(t) = \frac{t}{\sigma^2} \exp(- \frac{t^2}{2\sigma^2}) \qquad (19)$$

$$R(t) = \exp(- \frac{t}{2\sigma^2}) \qquad (20)$$

$$MTBF = \sigma \sqrt{\pi/2} \qquad (21)$$

Equation (19) is recognized as the Rayleigh density function. The typical characteristics of equations (18) to (20) are plotted in Figure 4, for different values of $1/\sigma^2$.

It is worthwhile to observe from equations (18) to (20) that the reliability of the Rayleigh law decreases at a much higher rate, than in the case of the exponential law, as time increases. However, in the region of early time, where failure rate is insignificant, the reliability of the components decreases more slowly with time than in the case of the exponential law. This implies that a complex electronic system designed to operate for a short time only should be constructed using components whose time between failures is Rayleigh distributed.

The Normal Distribution Law

One of the most familiar terms of statistical distribution functions is the normal distribution law. Its failure density function is given by:

$$f(t) = \frac{1}{\sigma\sqrt{2\pi}} \exp (- \frac{1}{2} (\frac{t-t_0}{\sigma^2})^2) \qquad (22)$$

where σ is the standard deviation, α is a shaping factor, and t_0 is a scale location factor. The duration of a component's successful operation cannot be negative. Therefore it is meaningful only to discuss the quantitative characteristics of reliability when the normal law of distribution is restricted to $t \geq 0$.

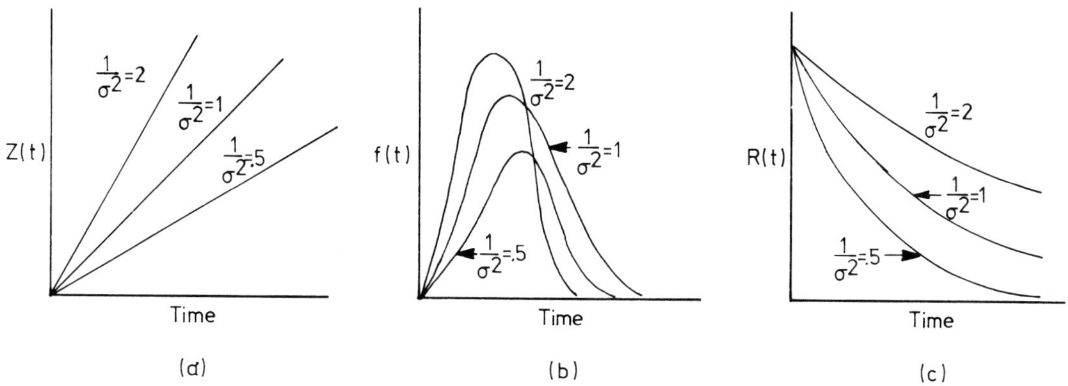

Figure 4. Typical Rayleigh Law Distribution for (a) Z(t), (b) f(t) and (c) R(t).

For the normal law of distribution, the probability that a random quantity x occurs in the interval from α to β is given by the expression:

$$p(\alpha < x < \beta) = \int_{\alpha}^{\beta} f(x)dx = \frac{1}{2}\left(\operatorname{erf}\left(\frac{\beta-\mu}{\sigma\sqrt{2}}\right) - \operatorname{erf}\left(\frac{\alpha-\mu}{\sigma\sqrt{2}}\right)\right) \tag{23}$$

where μ is the mean (expected value, E (x)) and $\operatorname{erf}\left(\frac{\beta-\mu}{\sigma\sqrt{2}}\right)$, $\operatorname{erf}\left(\frac{\alpha-\mu}{\sigma\sqrt{2}}\right)$ are probability integrals of the form:

$$\operatorname{erf}(x) = \frac{2}{\sqrt{\pi}} \int_0^x \exp(-t^2)dt \tag{24}$$

which is the error function. Usually equation (24) is tabulated.

Then,

$$F(t) = \int_0^t f(x)dx = \frac{1}{2}\left(\operatorname{erf}\left(\frac{t-\mu}{\sigma\sqrt{2}}\right) + \operatorname{erf}\left(\frac{\mu}{\sigma\sqrt{2}}\right)\right) \tag{25}$$

From (25) and Table 1, the reliability function for the normal law is given by:

$$R(t) = 1 - F(t) \tag{26}$$

or equivalently,

$$R(t) = \frac{1}{2}\left(\operatorname{erfc}\left(-\frac{t-\mu}{\sigma\sqrt{2}}\right) + \operatorname{erfc}\left(-\frac{\mu}{\sigma\sqrt{2}}\right)\right) \tag{27}$$

where,

$$\operatorname{erfc}(x) = 1-\operatorname{erf}(x) \tag{28}$$

and,

$$Z(t) = \frac{f(t)}{R(t)} = \frac{\frac{1}{\sigma\sqrt{2\pi}} \exp\left(-\frac{1}{2}\left(\frac{t-\mu}{\sigma^2}\right)^2\right)}{\frac{1}{2}\left(\operatorname{erfc}\left(-\frac{t-\mu}{\sigma\sqrt{2}}\right)+ \operatorname{erfc}\left(-\frac{\mu}{\sigma\sqrt{2}}\right)\right)} \tag{29}$$

The MTBF for the normal law is a complex expression, and will not be derived here.

<u>The Gamma Distribution Law</u>

In the case of the Gamma distribution law, the failure density function is given by:

$$f(t) = \lambda_0 \frac{(\lambda_0 t)^{k-1}}{(k-1)!} \exp(-\lambda_0 t) \tag{30}$$

where λ_0 is a scaling parameter and k is a shaping parameter.

The corresponding relationship of Table 1 is given by

$$R(t) = 1-\frac{\lambda_0}{(k-1)!} \int_0^t (t^{k-1} \exp(-\lambda_0 t))dt \tag{31}$$

$$= \exp(-\lambda_0 t) \sum_{i=0}^{k-1} \frac{(\lambda_0 t)^i}{i!}$$

Using equations (30) and (31) we obtain,

$$Z(t) = \frac{\lambda_0 (\lambda_0 t)^{k-1}}{(k-1)! \sum_{i=0}^{k-1} \frac{(\lambda_0 t)^i}{i!}} \tag{32}$$

and,

$$\text{MTBF} = \frac{k}{\lambda_0} \tag{33}$$

The typical characteristics are plotted in Figure 5 for different values of k and $\lambda_0 = 0.5$.

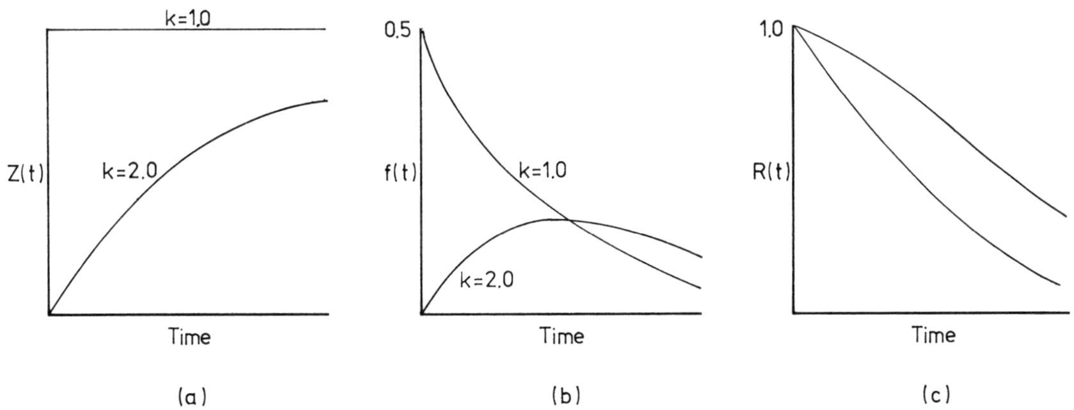

Figure 5. Typical Characteristic Plot for (a) Z (t), (b) f(t), (c) R (t) for $\lambda_0 = 0.5$

It should be obvious from equations (30) and (31) that for k=1, the Gamma distribution changes to an Exponential distribution. For $k > 1$, failure rate increases. Meanwhile for $k < 1$, it decreases.

The Gamma distribution law of failure density represents the distribution of failures of components subjected to wear (19) and of redundant systems where there is a renewal process based on component replacement. The parameter k, then is equal to the number of all devices. For $k < 1$, the Gamma distribution gives the failure density distribution during the burn-in period.

The Weibull Distribution Law

In many cases exponential failure data curves obtained cannot be approximated by any of the distribution laws previously discussed. The most appropriate failure rate model to fit the curve is generally the Weibull Model (15). The failure rate is given as,

$$z(t) = \alpha\beta\, t^{\alpha-1} \quad \alpha \geq 0 \tag{34}$$

The parameter β is the scaling factor and α is the shaping factor. Now, from Table 1, the associated functions are:

$$f(t) = \alpha\beta\, t^{\alpha-1} \exp\,(-\beta t^{\alpha}) \tag{35}$$

$$R(t) = \exp(-\beta t^{\alpha}) \tag{36}$$

and,

$$\text{MTBF} = \frac{\Gamma(1/\alpha + 1)}{\beta 1/\alpha} \;. \tag{37}$$

Note: Γ is the Gamma function defined by

$$\Gamma(x) = \int_0^{\infty} \exp(-t)\, t^{x-1}\, dt \tag{38}$$

equivalently,

$$\Gamma(x+1) = x! \simeq \sqrt{2\pi x}\, (x/e)^{x} \tag{39}$$

where,

e = the base of the natural logarithm.

$\Gamma(x)$ is generally tabulated in advanced mathematical tables.

Equation (35) is the Weibull distribution. The typical characteristics of this model are plotted in Figure 6 for various values of α and $\beta = 0.5$.

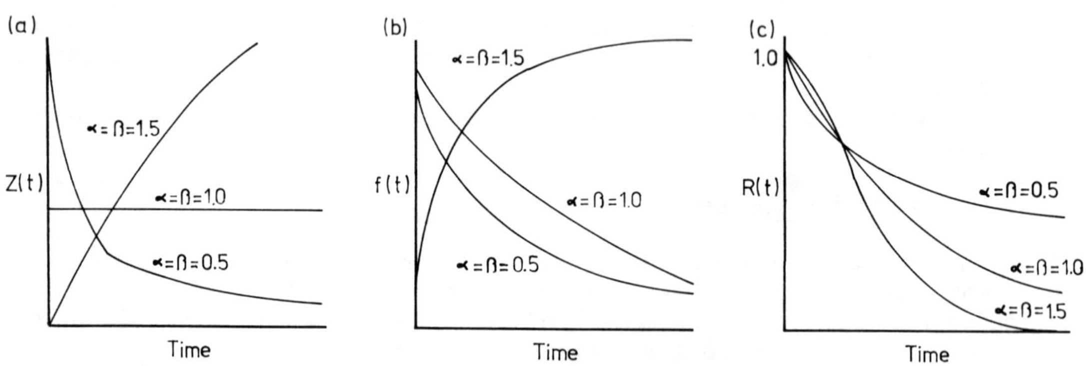

(a) (b) (c)

Figure 6. Weibull Model Distribution For (a) Z (t), (b) f(t), and (c) R (t)

From equation (34) and Figure 6, it can be observed that by appropriate choice of the two parameters α and β a wide range of failure rate models can be generated. For a fixed value of α, a change of β alters the vertical amplitude of the Z(t) curves.

The Weibull model includes all the previously discussed models. In particular for $\alpha=1$, the Weibull distribution changes to an exponential distribution. In general, for $\alpha>1$, the failure density function starts from 0 and increases with time. For $\alpha<1$, failure density starts from $+\infty$ and tends to 0 as time increases. It can be inferred that, like the Gamma distribution, the Weibull distribution represents the failure distribution of components which are subjected to wearing due to aging and stresses (20).

Superposition of Distribution

The distributions discussed only fix the failure density function over a limited period of operation in general. The "bath-tub" curve of Figure 2, over the burn-in period, can be represented by Gamma and/or Weibull; over the normal period of operation by the Exponential distribution; over the wear-out region by Gamma and Normal distributions. Thus, most component failure patterns involve a superposition of different distribution laws. Consequently, with the aid of the above laws, a failure density function, a reliability and MTBF expression can be obtained. In practice, this is a very difficult task. Hence, approximation and much judgement is involved. Each observer may consequently give a different solution to any distribution.

RELIABILITY MODELLING OF SIMPLE STRUCTURES

In this section, the reliability functions of some simple, well-known structures will be derived. These functions are based on a distribution law. However, for simplicity of explanation, only the exponential law is shown. Of course, other distributions are valid. First the model is analyzed through set theory (3, 9). In general, set theory is easy to apply when the structure is elementary. It fails, or becomes complicated, for complex structures. In such cases, we can resort to either the joint density approach, Markovian modelling or Linear Flow Graph technique (24 - 26).

Series Configuration

The simplest, and perhaps the most commonly occurring, configuration in reliability mathematical modelling is the series configuration. The functional operation of this system depends on the proper functioning of all the system's components. A component failure represents total system failure. A series reliability configuration is represented by the block diagram as shown in Figure 7, with n components. All the n components must be functioning for the system's success.

The block diagram of Figure 7 can represent a group of n components, specially integrated to perform a specific operational function or functions. For example, a shift register will require the proper functioning of all the gates and flip-flops. The subsystems of an aircraft (submarine, space shuttle, etc.) include the electronics, powerplant, armament, etc., which must all function for system success. An information processing system, which consists of an input device, a processor, a display unit and a memory unit, can operate successfully only if each of the subsystems perform as designed and do not fail.

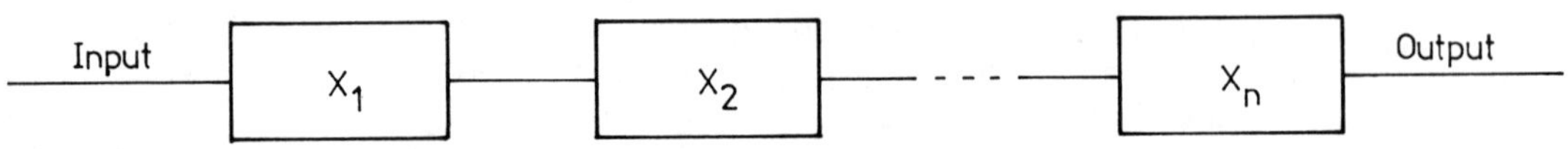

Figure 7. A Series Configuration Of n-Components

Let us now consider a system consisting of n subsystems as in Figure 7. From elementary probability theory (2, 5), the probability of X_1 and X_2 and X_3 and...X_n operating successfully can be represented using Boolean Algebra as:

$$R = P(X_1 \cap X_2 \cap X_3 \cap \ldots \cap X_n) \tag{40}$$

where,

$\cap$ = intersection or logical AND.

An equivalent expression is:

$$R = P(X_1)\ P(X_2/X_1)\ P(X_3/X_1 X_2)\ \ldots\ P(X_n/X_1 X_2 \ldots X_{n-1}) \tag{41}$$

where,

$P(X_2/X_1)$ = conditional probability that X_2 is good, given that X_1 is good. Thereby, the failure rate of one is dependent on the failure rate of the other. However, for most practical purposes it is safe to assume that component failure is statistically independent. Equation (41) then reduces to,

$$R = P(X_1)P(X_2)P(X_3)\ \ldots\ P(X_n) \tag{42}$$

or,

$$R = \prod_{i=1}^{n} P(X_i) \tag{43}$$

If the components exhibit a constant failure rate given by $Z(t)=\lambda_i$, i=1, 2, 3, ...n, then from equations (5) and (16),

$$R(t)= \prod_{i=1}^{n} \exp\ (-\lambda_i t) \tag{44}$$

which is equivalent to:

$$R(t) = \exp(- \sum_{i=1}^{n} \lambda_i t). \tag{45}$$

When all the n components are identical,

$$R(t) = \exp(-n\lambda t). \tag{46}$$

The MTBF can be obtained very easily from equations (8) and (17) as,

$$\mathrm{MTBF} = \int_0^{\infty} \exp(- \sum_{i=1}^{n} \lambda_i t)dt = 1/ \sum_{i=1}^{n} \lambda_i \tag{47}$$

or for n identical components,

$$\mathrm{MTBF} = 1/n\lambda \tag{48}$$

The reliability of the series sytem is observed to be the product of the reliability of each subsystem. The MTBF is the reciprocal of the sum of all the failure rates of the components.

The reliability and MTBF graph are plotted in Figure 8. It should be observed that the reliability and MTBF of a series system are always worse than the poorest components of the subsystems.

Parallel Configuration

The next commonly occurring configuration encountered in reliability mathematical modelling of electronics systems is the parallel configuration. It can be represented by the functional block diagram shown in Figure 9.

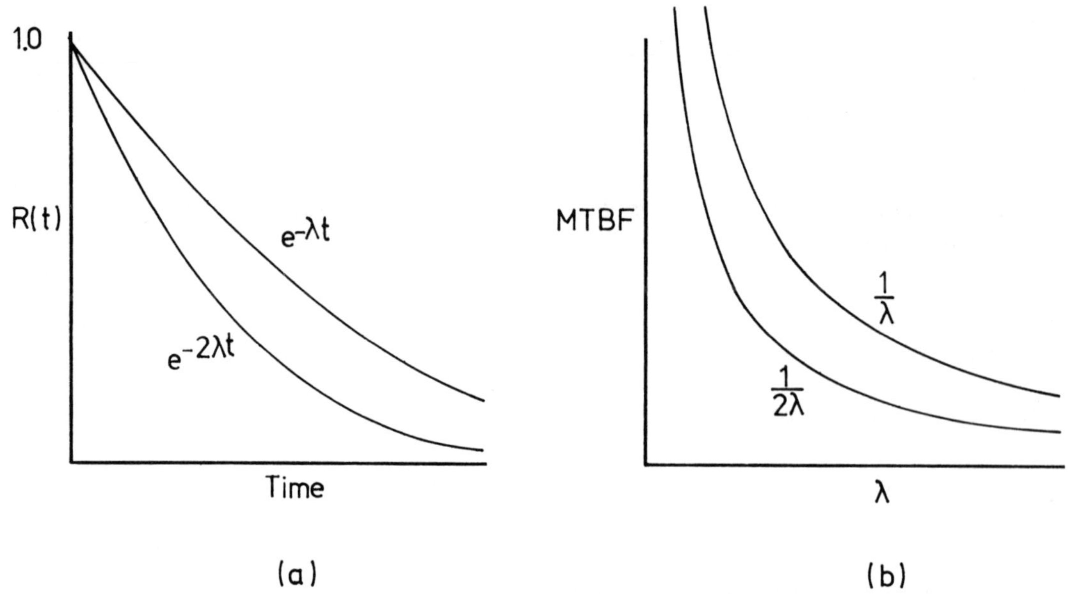

Figure 8. (a) Reliability and (b) MTBF Plot for a Single Component and Two Components in Series $\lambda = 0.25$/hour.

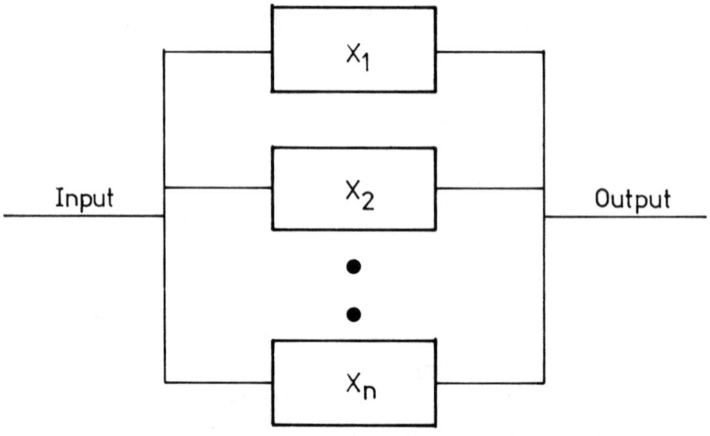

Figure 9. Reliability Block Diagram of Parallel Configuration of n Components

Figure 9 illustrates n units, performing the same operation, such that if one or more units are still operating , irrespective of how many have failed, the system is still operational.

Examples of parallel units, or what is called redundancy, are legion: for example, the display consoles of computer communication terminals, power generators in aircraft, buffers in a processor, computer memories. By introducing the n-1 redundant units, the system reliability will improve tremendously. (This will be illustrated by example in a later section.) The system will be successful if any one of the redundant units survives. The probability of success is the probability of the union of the n successful components. This is given by:

$$R = P(X_1 \cup X_2 \cup X_3 \cup \ldots \cup X_n). \tag{49}$$

Expansion of equation (49) by Boolean Algebra gives

$$R = P(X_1)+P(X_2)+P(X_3)+\ldots+P(X_n)-(P(X_1X_2)+P(X_1X_3)+\ldots+P(\underset{i\neq j}{X_iX_j}))$$
$$+\ldots+(-1)^{n-i}P(X_1X_2X_3\ldots X_n). \tag{50}$$

Irrespective of whether the failure is dependent or not, equation (50) is difficult to solve. Therefore, an alternative must be used. This is approached with the help of the probability of failure. The parallel system fails if all the n units fail. Thus, the probability of the system failure is the intersection of the probability of unit failures. This is given by,

$$P_f = P(\overline{X}_1 \cap \overline{X}_2 \cap \overline{X}_3 \cap \overline{X}_4 \cap \ldots \cap \overline{X}_n) \tag{51}$$

which from equation (40) to (43), with X's replaced by $\overline{X}$'s, gives

$$P_f = \prod_{i=1}^{n} P(\overline{X}_i) \tag{52}$$

since,

$$R = 1-P_f = 1 - \prod_{i=1}^{n} P(\overline{X}_i) \tag{53}$$

or similarly,

$$R = 1 - \prod_{i=1}^{n} (1-P(X_i)) \tag{54}$$

In the case of constant failure rate,

$$R(t) = 1 - \prod_{i=1}^{n} (1-\exp(-\lambda_i t)) \tag{55}$$

The computation of the parallel system MTBF is in general a difficult task. However, for a constant failure rate it can be obtained from equation (8) as,

$$\text{MTBF} = \int_0^\infty R(t)\,dt = \int_0^\infty \left(1 - \prod_{i=1}^{n} (1 - R_i(t))\right) dt \tag{56}$$

$$= \int_0^\infty \left(1 - \prod_{i=1}^{n} (1 - \exp(-\lambda_i t))\right) dt \tag{57}$$

$$= \int_0^\infty \sum_{i=1}^{n} (-1)^i \exp(-(\lambda_{j_1} + \ldots + \lambda_{j_i}))t\,dt \tag{58}$$

$$\begin{Bmatrix} j_1, j_2, \ldots, j_i \\ j_1 > j_2 > \ldots\ j_i \end{Bmatrix}$$

Finally,

$$\text{MTBF} = \sum_{i=1}^{n} \Sigma (-1)^{i+1} (\lambda_{j_1} + \lambda_{j_2} + \ldots + \lambda_{j_i})^{-1} \tag{59}$$

$$\begin{Bmatrix} j_1, j_2, \ldots, j_i \\ j_1 > j_2 > \ldots\ j_i \end{Bmatrix}$$

where the second sum is taken over all possible i-tuples of the form $j_1, j_2, \ldots j_i$ for which $j_1 > j_2 > \ldots\ j_i$. For n=2, we obtain,

$$\text{MTBF} = \frac{1}{\lambda_1} + \frac{1}{\lambda_2} - \frac{1}{\lambda_1 + \lambda_2} \tag{60}$$

and for n=3,

$$\text{MTBF} = \frac{1}{\lambda_1} + \frac{1}{\lambda_2} + \frac{1}{\lambda_3} - \left[\frac{1}{\lambda_1+\lambda_2} + \frac{1}{\lambda_1+\lambda_3} + \frac{1}{\lambda_2+\lambda_3}\right] + \frac{1}{\lambda_1+\lambda_2+\lambda_3} \tag{61}$$

It should be clear from the two examples, how the MTBF can be generalized to any larger value of n for constant failure rate. As n increases, so does the mathematical complexity. Consequently equation (59) can be expressed also as,

$$\text{MTBF} = \frac{1}{\lambda_1} + \frac{1}{\lambda_2} + \ldots + \frac{1}{\lambda_n} - \left[\frac{1}{\lambda_1+\lambda_2} + \frac{1}{\lambda_1+\lambda_3} + \ldots + \underset{i \neq j}{\frac{1}{\lambda_i + \lambda_j}}\right] + \ldots + (-1)^{i+1} \frac{1}{\lambda_1+\lambda_2+\ldots+\lambda_n} \tag{62}$$

In comparing a parallel system, with n independent units, with a series system containing the same n independent units, we can make several observations:

(a) The parallel system has a higher reliability than any of its respective components,
(b) The MTBF of a parallel system is much higher than the series system.

This infers to the designer that, by providing redundancy, the system designer can obtain a reliability and MTBF much higher than that of the individual units. This is shown in Figure 10. However, the extent of redundancy is constrained by such factors as cost, weight, volume, etc., (27, 28).

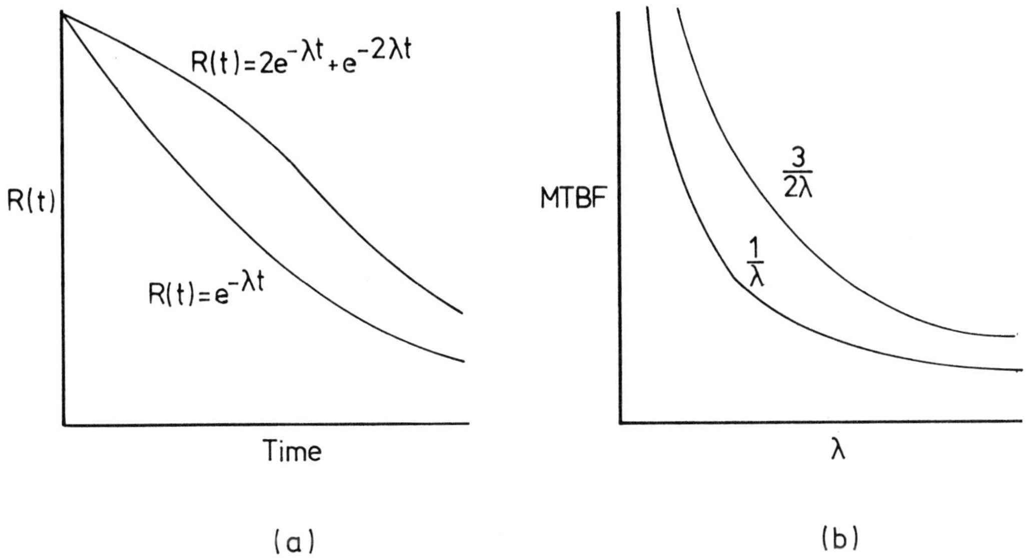

Figure 10. (a) Reliability and (b) MTBF Plot for a Single Unit and Two Units in Parallel

A k-out-of-n Configuration

A system consisting of n subsystems, of which only k need to be functioning for system success, is called a k-out-of-n configuration. Here k is less than n. An example of such a system is the computer system which consists of n terminals of which k need to be operational for success, or an army consisting of n tankers, surveillance aircraft, submarines, etc., of which k of each must be operational for success; more examples are found in (27). There are two categories of the k-out-of-n configuration, (a) an active (on-line) standby and, (b) a cold (off-line) standby system.

Active (On-Line) Standby System

In an Active (On-Line) Standby System all n subsystems are operational. If all subsystems are identical and failures are statistically independent, the probability of exactly k subsystems out of a total of n being successful is represented by,

$$B(n,k;p) = \binom{n}{k} p \quad (1-p)^{n-k} \quad \text{for } k=0,1,2,\ldots,n \tag{63}$$

This is the well-known Binomial distribution function. Now, if at least k out of the n subsystems are required to be simultaneously operative for system success then the probability of system success is given by,

$$R\,(n,k;p) = \sum_{i=k}^{n} \binom{n}{i} p^{i}(1-p)^{n-i} \tag{64}$$

For constant failure rate equation (64) gives

$$R\,(n,k;p) = \sum_{i=k}^{n} \binom{n}{i} \exp\,(-\lambda_i t)^{i}\,(1-\exp(-\lambda_i t))^{n-i} \tag{65}$$

and,

$$MTBF = \frac{1}{\lambda} \sum_{i=0}^{n-k} \frac{1}{(n-i)} \tag{66}$$

It is of interest to note that if k=1, the system becomes a parallel configuration, whereas, if k=n, the system is a series configuration just as given earlier.

If all the n subsystems are not identical, equation (65) is no longer valid. To compute the system reliability, we have to enumerate the system's successful paths. This is aided by a reliability signal flow-graph and application of set theory. The graph will have $\binom{n}{k}$ parallel paths. Each path, containing k elements, corresponds to one of the n events taken k at a time. The success of each path depends on the success of all the k elements. Therefore, by set theory, the reliability is given as:

$$R(t) = P(X_a \cup X_b \cup X_c \cup \ldots \cup X_n) \tag{67}$$

where,

$X_a, X_b, \ldots, X_n$ are sets of path elements. This can be solved as in equation (50).

To understand this better, let us look at an example. For a 3-out-of-4 configuration, the reliability graph is shown as in Figure 11. The probability of success is given by the probability of success of the path $X_1 X_2 X_3$ or $X_1 X_2 X_4$ or $X_1 X_3 X_4$ or $X_2 X_3 X_4$. This is given as:

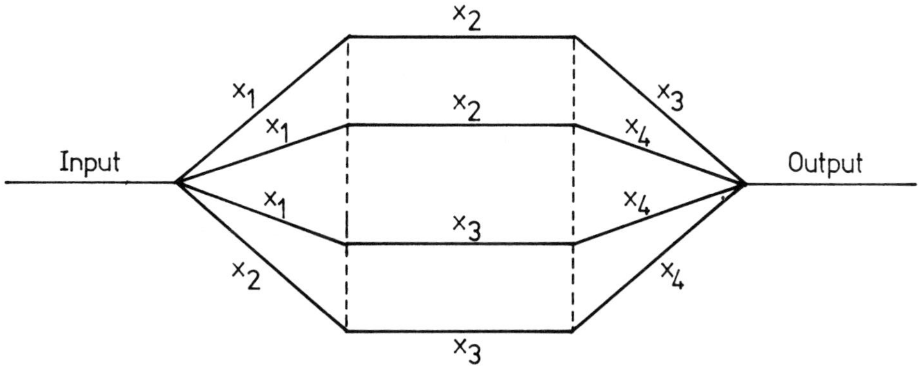

Figure 11. Reliability Graph for a 3-out-of-4 Configuration

$$R = P(A_1 \cup A_2 \cup A_3 \cup A_4) \tag{68}$$

where,

$A_1 = X_1X_2X_3$, $A_2 = X_1X_2X_4$, $A_3 = X_1X_3X_4$ and $A_4 = X_2X_3X_4$

Similarly as for equation (50),

$$R = P(A_1) + P(A_2) + P(A_3) + P(A_4) - \sum\sum P(A_iA_j) + \sum_i \sum_j \sum_{\substack{k \\ i \neq j \neq k}} P(A_iA_jA_k) - \sum_i \sum_j \sum_k \sum_{\substack{l \\ i \neq j \neq k \neq l}} P(A_iA_jA_kA_l) \tag{69}$$

All redundant terms are ignored, for example, $P(A_1A_2) = P(X_1X_2X_3)\, P(X_1X_2X_3)$ becomes $P(X_1X_2X_3X_4)$. The redundant X's are dropped. Equation (69) becomes:

$$R = P(A_1) + P(A_2) + P(A_3) + P(A_4) - \binom{4}{2} P(B) + \binom{4}{3} P(B) - \binom{4}{4} P(B) \tag{70}$$

where,

$B = X_1X_2X_3X_4$.

or,

$$R = P(X_1X_2X_3) + P(X_1X_2X_4) + P(X_1X_3X_4) + P(X_2X_3X_4) - 3P(X_1X_2X_3X_4) \tag{71}$$

If all components are identical, the same reliability expression is obtained by either using equation (64) or (71). In general, the MTBF expression is very difficult to derive for a large system. For constant failure rate, it is the reciprocal of the sum of the failure rate of its

$$R(3,4,P)= \exp(-(\lambda_1+\lambda_2+\lambda_3))t + \exp(-(\lambda_1+\lambda_2+\lambda_4))+ \exp(-(\lambda_1+\lambda_3+\lambda_4))t \quad (72)$$

$$+ \exp(-(\lambda_2+\lambda_3+\lambda_4))t- 3\exp(-(\lambda_1+\lambda_2+\lambda_3+\lambda_4))t$$

Therefore,

$$MTBF = \frac{1}{\lambda_1+\lambda_2+\lambda_3} + \frac{1}{\lambda_1+\lambda_2+\lambda_4} + \frac{1}{\lambda_1+\lambda_3+\lambda_4} + \frac{1}{\lambda_2+\lambda_3+\lambda_4} \quad (73)$$

$$- \frac{3}{\lambda_1+\lambda_2+\lambda_3+\lambda_4}$$

The same solution could have been obtained for the MTBF from equation (66) if all subsystems are identical.

Cold (Off-Line) Standby System

A system consisting of n identical, statistically independent subsystems is said to be a k-out-of-n Cold (off-line) Standby System, provided the system works in the following manner. Initially all the n subsystems are good, of which only k are on line. The other n-k are on standby. The failed subsystem is instantaneously replaced by the cold standby subsystems, requiring negligible time. The probability that at least k subsystems out of the n subsystems will be successful is given in references (27-29):

$$R(k,n,p) = \sum_{i=0}^{n-k} \frac{p^k}{i!} (-k \ln p)^i \quad . \quad (74)$$

The MTBF is given as,

$$MTBF = \frac{1}{\lambda} \sum_{i=0}^{n-k} \frac{(i)^k}{k} \quad . \quad (75)$$

It is worthwhile to note, at this point, that for a fixed k and p, the reliability of both the hot and cold standby k-out-of-n configuration increases with n.

For a comparison of the three configurations discussed so far, let us look at the reliability and MTBF of each. This is plotted in Figure 12, for three series, three parallel and 2-out-of-3 configuration.

Figure 12 illustrates that to improve the reliability and MTBF of a system, we can utilize the parallel and/or k-out-of-n configuration. However, the number of redundancies for both cases are constrained by such factors as cost, weight, volume, etc. Thus the choice of configuration is up to the system designer, subject to the above constraints.

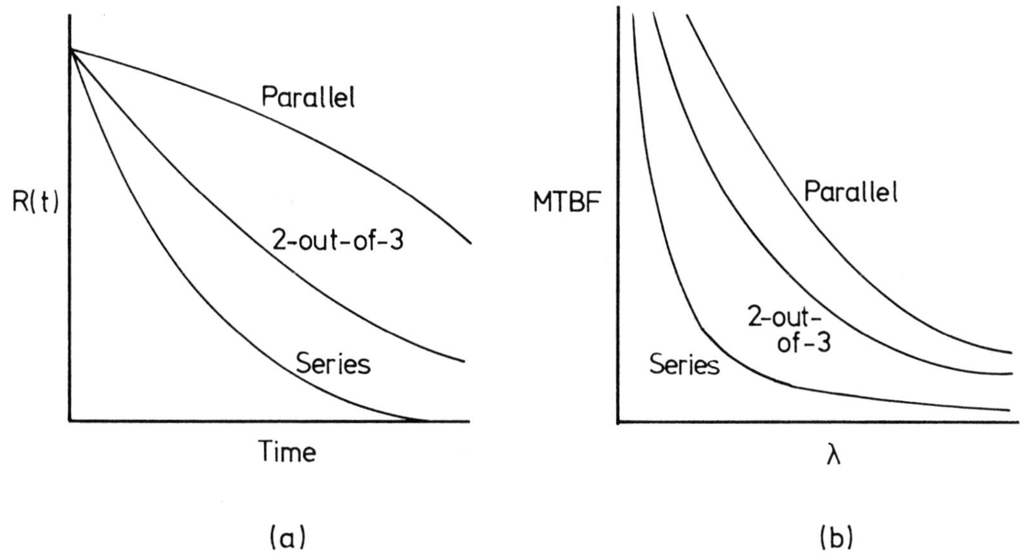

Figure 12. Plot of (a) Reliability and (b) MTBF for a Series, Parallel and a k-out-of-n Configuration

Triple Modular Redundancy (TMR) And n-Triple Modular Redundancy

The use of redundancy to enhance reliability has found wide acceptance as a fundamental procedure. Majority voting systems are an extension of the parallel and k-out-of-n configuration. In this case, the voter makes the decision as to whether the system satisfies the desired operational function. TMR (29) was the earliest example of this system. The simplex unit is triplicated and each of the three independent units feeds into a 2-out-of-3 selector. The system fails if more than one unit fails. The reliability function for this system is identical to equation (64).

The TMR can be expanded to the NMR. The simplex unit is replicated N times (N=2n+1). At least n+1-out-of-N have to be operational for the structure to survive. Thus the reliability is given by,

$$R(NMR) = \sum_{i=n+1}^{N} \binom{N}{i} p^{i} (1-p)^{N-i} \qquad (76)$$

which is another form of the k-out-of-n configuration. The same observations are made for NMR, as for k-out-of-n configuration, in all aspects.

APPLICATIONS

To understand the application of the theories that have been used in the previous modelling, let us look at a few practical problems.

Example 1

Let us consider a computer interface Circuit Card Assembly (CCA). Table 2 gives the interconnected components, quantities and the failure rate per million hours.

Table 2. Interface CCA Components and Failure Rates

Components	Failure rate/ 10^6 hour	Quantities	Total failure rate/10^6 hour
Capacitor tantalum	0.0027	1	0.0027
Capacitor ceramic	0.0025	19	0.0475
Resistor	0.0002	5	0.0010
J-K,M-S Flip Flop	0.4667	9	4.2003
Triple Nand Gate	0.2456	5	1.2286
Diff line receiver	0.2738	3	0.8214
Diff line driver	0.3196	1	0.3196
Dual Nand gate	0.2107	2	0.4214
Quad Nand gate	0.2738	7	1.9166
Hex invertor	0.3196	5	1.5980
8-bit shift register	0.8847	4	3.5388
Quad Nand buffer	0.2738	1	0.2738
4-bit shift register	0.8035	1	0.8035
And-or-inverter	0.3196	1	0.3196
PCB connector	4.3490	1	4.3490
Printed wiring board	1.5870	1	1.5870
Soldering connections	0.2328	1	0.2328
	Total	67	21.6720

The components are in an airborne, inhabited environment at a temperature of 50°C. It is given that for the computer interface to be successful, all of the interconnected components must be operational. Therefore, the components will be modelled as a series configuration, as shown in Figure 13.

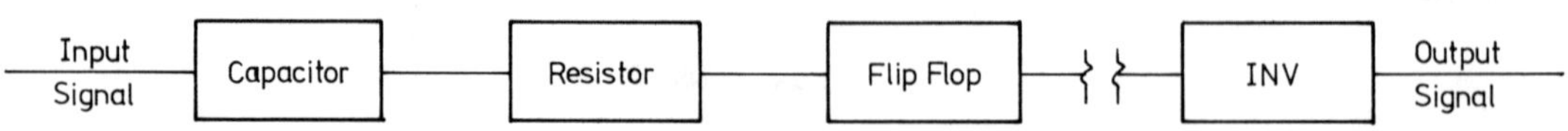

Figure 13. Reliability Block-Diagram of Computer Interface

The reliability of this CCA is thus given as the product of the reliability of each individual component, namely,

$$R(t) = \prod_{i=1}^{n} P_i(t)$$

with constant failure rate (for most practical purposes, this is true):

$$R(t) = \exp(-\sum_{i=1}^{n} \lambda_i t)$$

For a mission time of 12 hours, the CCA has a reliability of

$$R(t) = \exp(-21.6720 \times 10^{-6} \times 12)$$

$$= 0.9997$$

which implies 99.97% of the time in the 12 hours the CCA is successful. The associated MTBF is simply,

$$MTBF = 10^6/21.6720$$

$$= 46142.48 \text{ hours}$$

As observed earlier, the reliability and MTBF of the system is less than the reliability of the worst components in the CCA.

Example 2

A unit consists of the CCAs as given in Table 3, with their corresponding failure rate/10^6 hours, MTBF and reliability. The CCAs are to be designed for an airborne inhabited environment with a temperature of 50°C. Determine the unit reliability and its MTBF.

Table 3. Unit CCAs and Failure Rates

CCA No.	Failure rate 10^6 hour	MTBF (hours)	Reliability
1	20.0504	49874	0.99975942
2	16.5871	60288	0.99980097
3	17.6010	56815	0.99978881
4	30.0516	33276	0.99963945
5	17.8623	55984	0.99978568

Table 3 (Cont'd)

6	21.7848	45904	0.99973862
7	15.7005	63692	0.99981161
8	19.9872	50032	0.99976018
9	141.8336	7051	0.99829944
Total	301.7583		

The CCAs 1-9 must all be functioning for the unit to be operational. Therefore, like Example 1, this is a series configuration as given by Figure 14.

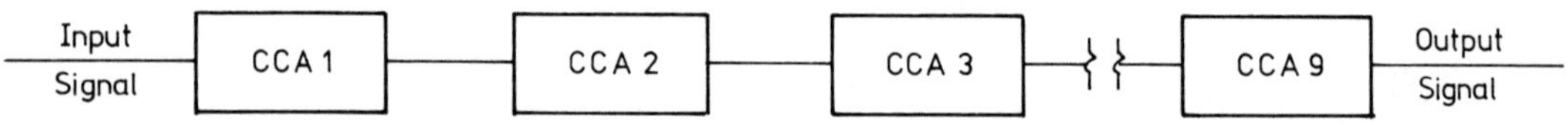

Figure 14. Reliability Block-Diagram for Example 2

Thus assuming a constant failure rate, for a mission time of 12 hours, the unit reliability is given as,

$$R(t) = \exp(-301.7583 \times 10^{-6} \times 12)$$

$$= 0.9963$$

and the corresponding MTBF is given as

$$MTBF = 10^6/301.76 = 3313 \text{ hours.}$$

It can be observed again that the system reliability and MTBF are less than the reliability and MTBF of the worst of the unit's CCAs.

Example 3

In an aircraft electrical system, it is generally wise to include a redundant power supply. This is arranged so that if one fails, the other will take over the load, thus preventing a disaster. Consider the case when both generators are identical, with a constant failure rate of 300 $\times 10^{-6}$/hour. We want to determine the system reliability and MTBF for t=12 hours. Since only one generator is necessary to sustain a successful flight, we can treat this as a parallel model. The other generator is a redundant unit. This is shown in Figure 15.

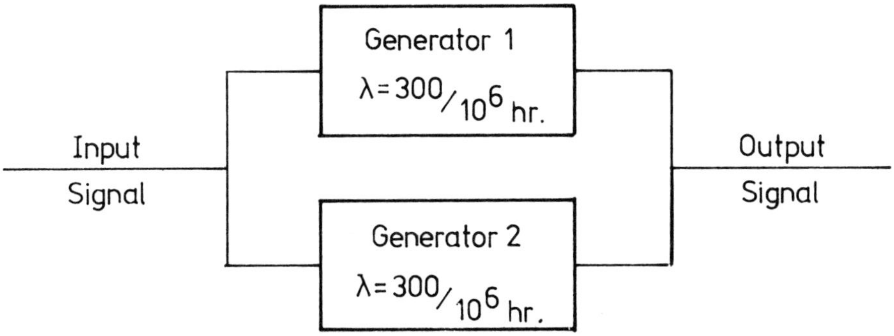

Figure 15. The Reliability Block-Diagram for Power Generator

Therefore, using equation (55), we have,

$$R(t) = 1-(1-\exp(-300\times10^{-6}\times12))^2$$

$$= 0.9999$$

and,

$$MTBF = \frac{2}{300\times10^{-6}} - \frac{1}{2\times300\times10^{-6}} = 5000 \text{ hours}$$

The reliability of each generator for t=12 hours is R(t)=0.9964. Thus, we can infer that for a parallel structure, the system reliability and MTBF are better than the best subsystem reliability and MTBF. Therefore, there are improvements by introducing redundancy.

Example 4

Given three identical satellites, if any two of these satellites are in operation with an earth antenna (assumed perfect), the system is considered successful. Assuming a constant failure rate of $400/10^6$ hours determine the system reliability and MTBF for a period of 24 hours. The system reliability block-diagram is given in Figure 16. Since at least two satellites must be in operation for success, this is a 2-out-of-3 configuration.

For a mission time of 24 hours, the system reliability is given as,

$$R(2,3,p) = \sum_{i=2}^{3} \binom{3}{i} \exp(-\lambda it)\,(1-\exp(-\lambda(3-i)t))$$

$$= 0.9997$$

$$MTBF = \frac{3}{2\lambda} - \frac{1}{3\lambda} = 2917 \text{ hours.}$$

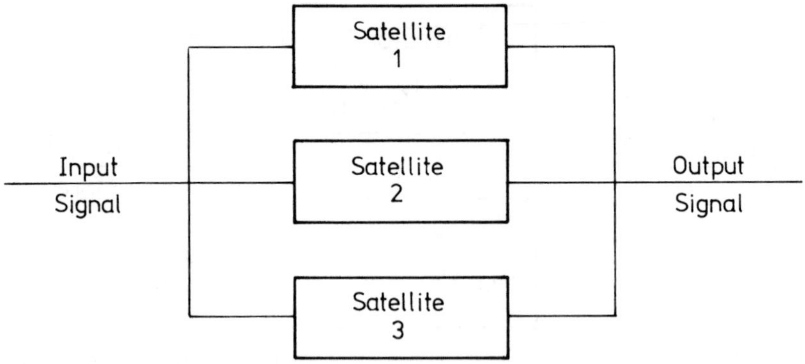

Figure 16. A 2-out-of-3 Satellite Reliability Block Diagram

CONCLUSION

The task of reliability modelling can be difficult and the best a reliability engineer can do is to analyze a system through a simple modelling configuration. Using these simple structures, mathematical formulation can be obtained. The models derived here are by no means complete; it is suggested that the reader refer to the quoted references for information not discussed here.

REFERENCES

1. Barlow, R.E. and Prochan, F., Mathematical Theory of Reliability, John Wiley and Sons, Inc., New York, 1965.

2. Beckmann, P., Elements of Applied Probability Theory, Harcourt, Brace and World, Inc., New York, N.Y., 1968.

3. Hinkle, M.L., "An Engineer's Approach to Reliability Mathematics", 1967 Annual Reliability and Maintainability Symposium Proceedings, pl-16.

4. Feller, W., An Introduction to Reliability Theory and Its Application, John Wiley and Sons Inc., New York, 1957.

5. Lipschutz, S., Theory and Problems of Probability, Schaum's Outline Series, McGraw Hill Book Co., 1965.

6. Parzen, E., Stochastic Process, Holden Day Inc., San Francisco, 1962.

7. Polovko, A.M., Fundamentals of Reliability Theory, Academic Press, New York, 1968.

8. Rau, J.G., Optimization and Probability in Systems Engineering, Van Nostrand Reinhold, New York, N.Y., 1970.

9. Shooman, M.L., Probabilistic Reliability. An Engineering Approach, McGraw-Hill Book Co., New York, 1968.

10. Disney, R.L., Lipson, C. and Sheth, N.J., "The Determination of the Probability of Failure by Stress/Strength Interference Theory", 1968 Annual RAM Symposium Proceedings, p417-422.

11. Glinski, G.S., de Mercado, J. and Thompson, P.M.,"A Diffusion Method for Reliability Prediction", IEEE Transactions on Reliability, Vol. R-18, No. 4, November 1969, p149-156.

12. Kato, Y. and Karasama, H., "Some Approaches to Reliability Physics", 1968 Annual RAM Symposium Proceedings, p607-614.

13. Kirkman, R.A., "Failure Concepts in Reliability Theory", IEEE Transactions on Reliability Vol. R-12, No. 4, December 1963, p1-10.

14. Moyer, E.P., "Device Failure Distribution from Failure Physics", 1967 Annual RAM Symposium Proceedings, p598-611.

15. Weibull, W.A., "Statistical Distribution Function of Wide Application", Journal of Applied Mechanics, Vol. 18, 1951, p293-297.

16. Martz, Jr., H.R., "Pooling Life Test Data by Means of the Empirical Bayes Method", IEEE Transactions on Reliability, Vol. R-24, No. 1, April 1975, p27-30.

17. __________, Censored Sample Size Testing for Life Test", 1973 Annual RAM Symposium Proceedings, p280-288.

18. __________, Reliability Prediction of Electronics Equipment, MIL-HDBK-217B, Department of Defense, Washington, D.C., Sept. 1974.

19. Hameed, M.A., "A Gamma Wear Process", IEEE Transactions on Reliability, Vol. R-24, No. 2, June 1975, p152-153.

20. Malik, M.A., "A Note on the Physical Meaning of the Weibull Distribution", IEEE Transactions on Reliability, Vol. R-24, No. 1, April 1975, p95.

21. Nylander, J.E., "Statistical Distributions in Reliability", IRE Transactions on Reliability and Quality Control, Vol. RQC-11, No. 1, May 1962, p33-43.

22. Robins, R.S., "On Models for Reliability Prediction", IRE Transactions on Reliability and Quality Control, Vol. RQC-11, No. 1, May 1962., p33-43.

23. Welker, E.L. and Lipow, M., "Estimating the Exponential Failure Rate from Data with No Failure Events", 1974 Annual RAM Symposium Proceedings, p420-427.

24. Chow, Y. and Cassignol, E., Linear Signal Flow Graphs and Applications, John Wiley and Sons Inc., New York, 1962.

25. Doyon, L.R., "Solving Complex Reliability Models by Computer", 1968 Annual RAM Symposium Proceedings, p431-447.

26. Htun, L.T., "Reliability Prediction Techniques for Complex Systems", IEEE Transactions on Reliability, Vol. R-15, No. 2, August 1966, p58-69.

27. Ling, A.H.K., An Optimal Reliability Design Technique, University of Ottawa, Department of Electrical Engineering Undergraduate Thesis, Ottawa, 1975.

28. de Souza, A.A.P., "An Optimising Technique for a k-out-of-n Configuration", Microelectronics and Reliability, special issue, Canadian SRE Symposium Proceedings, May 1975, p217-222.

29. Benning, C.J., "Reliability Prediction Formulas for Standby Redundant Structures", IEEE Transactions on Reliability, Vol. R-16, December 1967, p136-137.

8 ALLOCATION AND ESTIMATION

J. E. Arsenault and J. A. Roberts

Most often the designer of an electronic system is faced with the problem of meeting a reliability goal or requirement, or having to specify a reliability goal or requirement. Reliability will normally be stated in one of two ways (1) i.e., (a) probability of mission success R(t) or, (b) Mean Time Between Failure (MTBF). If one is setting goals then a common practice is to review the reliability performance of existing systems and specify goals at least equal to and most likely better than current performance.

It is extremely important to realize that a system design may be substantially different depending on which specified parameter, i.e. R(t) or MTBF, the designer is trying to meet. The mathematics of reliability indicate that only for a short time is it possible to obtain a substantial gain in reliability by parallelling elements into a redundant configuration, while for the longer term the gain in MTBF is only moderate (2). Figure 1 shows the reliability properties just mentioned.

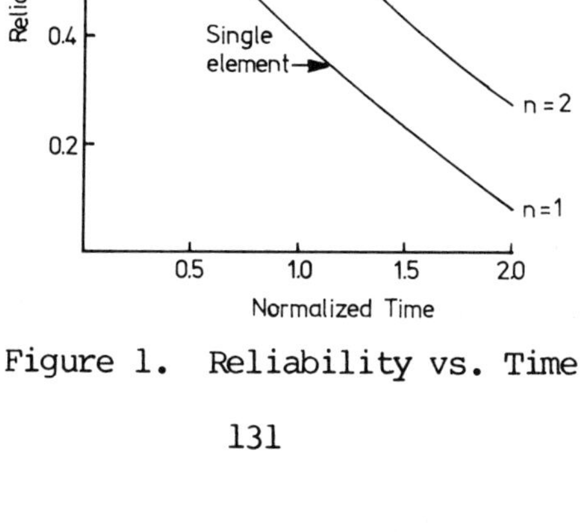

Figure 1. Reliability vs. Time

The underlying basis for all reliability design work is the generation of meaningful failure rates. Without accurate failure rate estimates for a preliminary design, the designer may plan an expensive or heavy configuration where a less expensive and lighter design would suffice. Accurate failure rate estimates are particularly valuable where a design must pass a Reliability Demonstration Test (RDT).

The most common sources for failure rate data are (3) and (4) and actual experience, where (3) is most widely used, as it includes factors to account for the use of a wide range of parts in a design and is readily available.

DESIGNING TO MEET THE REQUIREMENTS

Assuming that the designer is given a system performance specification he will initially proceed to design his system with performance uppermost in mind. Having designed a system that is probably satisfactory from the performance viewpoint, the design will be subjected to a reliability analysis to determine if the specified reliability requirement can be met. Hopefully the requirement can be met without redundancy although this is not always the case. Consideration must also be given to system degraded modes when they occur naturally and the necessity for incorporating them in the design.

Single Unit Case

Here a designer is given a reliability requirement to meet where the function to be implemented will be packaged in a single unit. The same applies to a single Circuit Card Assembly (CCA) as well. In this case the reliability is given by:

$$R(t) = \exp(-t \sum_{i=1}^{n} \lambda_i) \qquad (1)$$

where,

$R(t)$ = probability of mission success or reliability,

exp = natural logarithm,

t = mission duration,

λ_i = failure rate of the i th part,

n = total number of parts.

It is to be noted that this equation is true only if the following conditions are met in essence, i.e., (a) the configuration must be a series one, (b) the parts must be independent, (c) the parts must be governed by a constant failure rate model. Condition (c) is usually achieved during the random failure portion of the life of an item as shown in Figure 2. Curve (a) of Fig. 3(b) is a graph of R(t) for this case.

In addition, the MTBF is given by:

$$MTBF = 1/\sum_{i=1}^{n} \lambda_i \qquad (2)$$

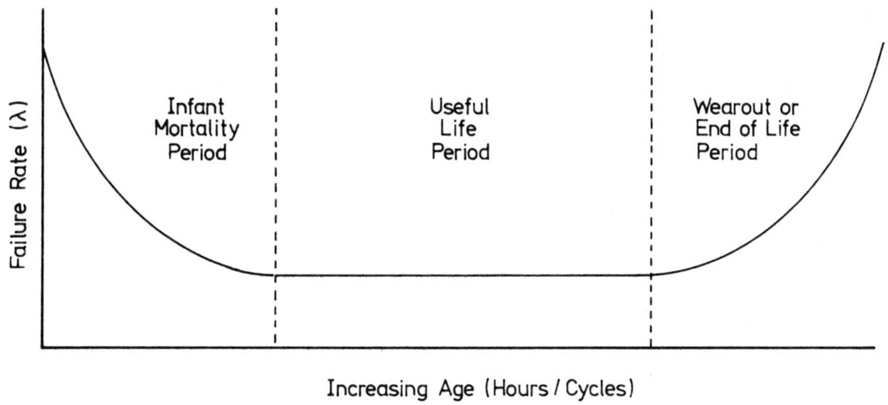

Figure 2. General Form of the Failure Curve

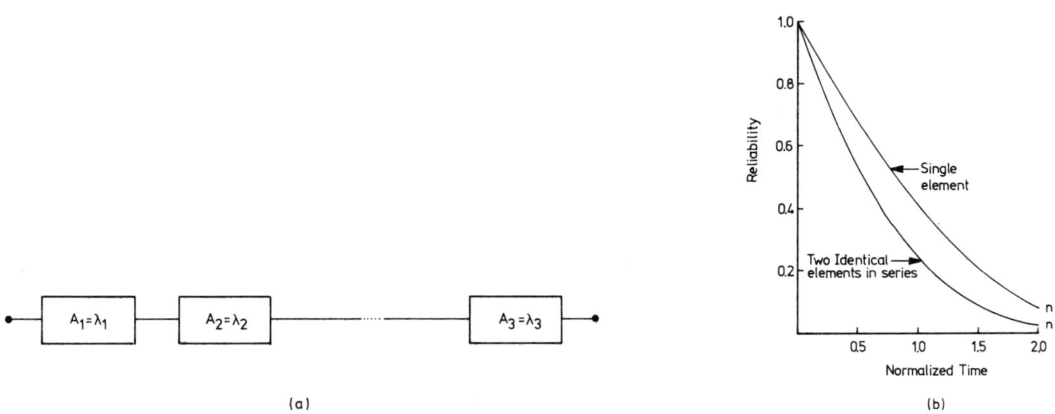

Figure 3. Reliability vs Time for Series Configuration(s)

Assuming that the reliability requirement has been met the designer may try to economize his design by reducing the quality and hence cost of certain types of parts. If he has not been successful in meeting the requirements at this point he can, (a) increase part quality, (b) design in redundancy, (c) try to simplify his design, (d) reduce part electrical stress and, (e) reduce part temperature. Normally (c), (d) and (e) are included as conservative design practice while (a) and (b) are usually addressed until a satisfactory result is achieved. (a) is covered in the next section while (b) is discussed later in this section.

System of Units in Series

The next order of complexity a designer can face is the case where because of packaging limitations it is necessary to design a system consisting of more than one unit. This type of design is known as the series configuration and as shown in Chapter 7 the reliability of the configuration is the product of the individual unit reliabilities. This is shown in Figure 3(a). Curve (b) of Figure 3b illustrates the effect of time on the reliability of a series configuration.

$$R(t) = R_1(t)R_2(t)R_3(t)\ldots R_n(t) \tag{3}$$

where,

$R_j(t)$ = reliability of the j th unit. (j=1,2,3.. n)

The MTBF for this configuration is:

$$MTBF = 1 / \sum_{j=1}^{n} \lambda_j \tag{4}$$

where,

λ_j = unit failure rate,

n = total number of units.

When the designer has completed an evaluation of his design based on the above discussion he should apply the same logic for the single unit case in deciding what the next step should be with respect to trade offs involving part quality, temperature, etc.

Redundant Case

Possibly the designer has exhausted all methods in achieving his system reliability requirement, except for redundancy or the specification calls for redundancy. In either case, redundancy has been assessed as the method to be used in meeting the reliability requirement. Redundancy can take many forms. However, there are two major classes i.e., active and standby, which can be further sub-classified as shown in Figure 4 (1).

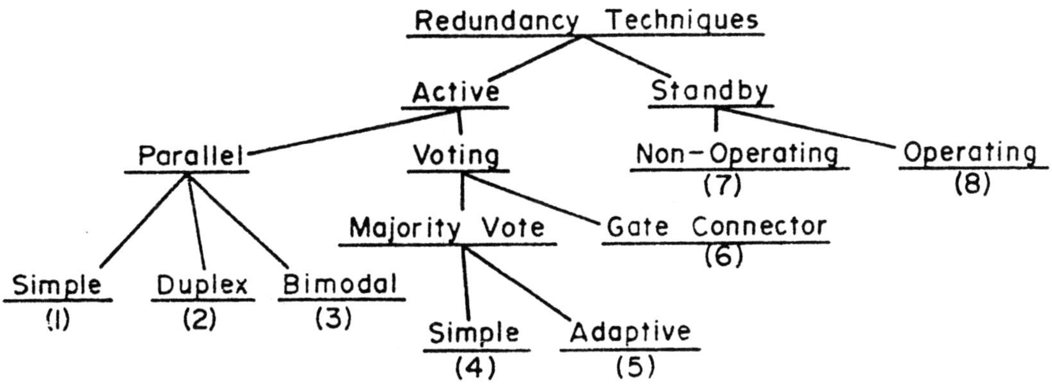

Figure 4. Redundancy Techniques by Type (1-8)

Each of these techniques is summarized in Figure 5(a) and 5(b) as per reference(1). The more common techniques are discussed in the following paragraphs.

Case 1 - Simple. This technique provides protection against irreversible hardware failures for continuously operating systems and is illustrated in Figure 6(a).

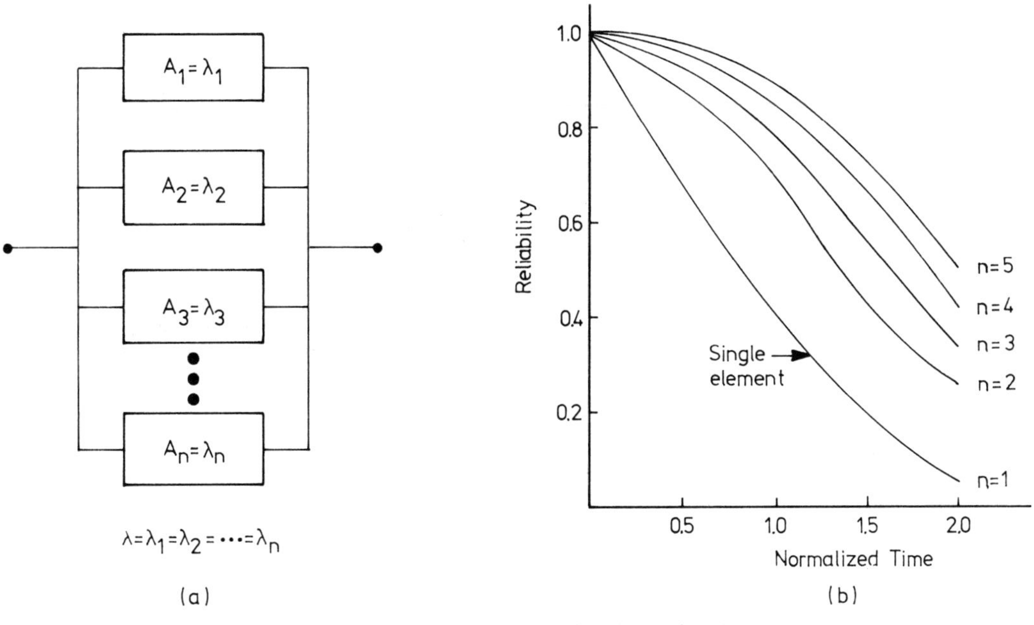

Figure 6. Simple Parallel Redundancy

Simple Parallel Redundancy

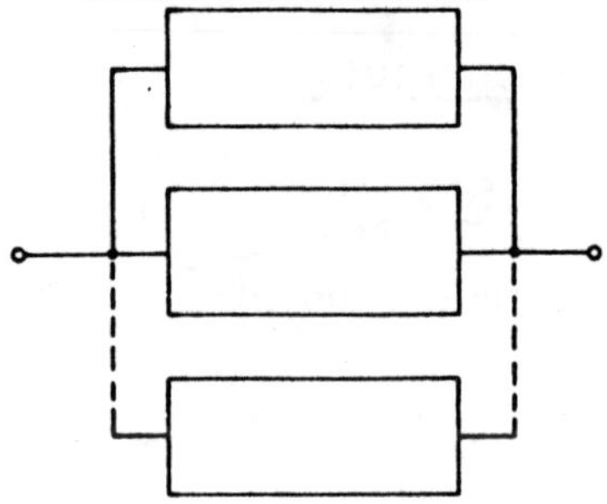

In its simplest form, redundancy consists of a simple parallel combination of elements. If any element fails open, identical paths exist through parallel redundant elements.

Duplex Redundancy

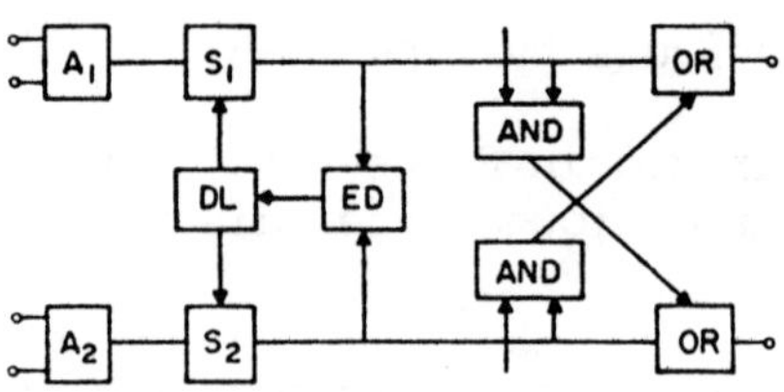

This technique is applied to redundant logic sections, such as A1 and A2 operating in parallel. It is primarily used in computer applications where A1 and A2 can be used in duplex or active redundant modes or as a separate element. An error detector at the output of each logic section detects noncoincident outputs and starts a diagnostic routine to determine and disable the faulty element.

(a) Bimodal Parallel/ Series Redundancy

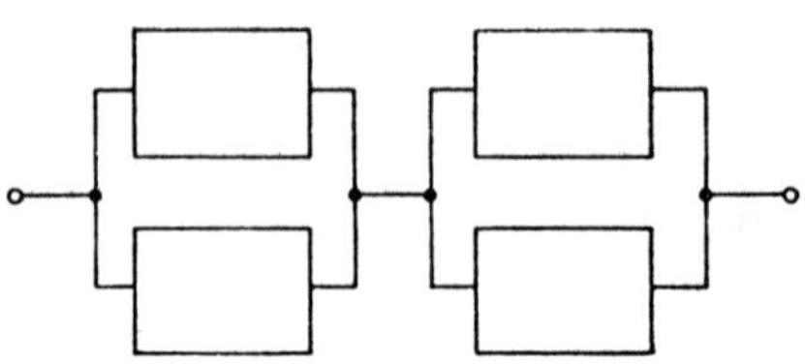

(b) Bimodal Series/ Parallel Redundancy

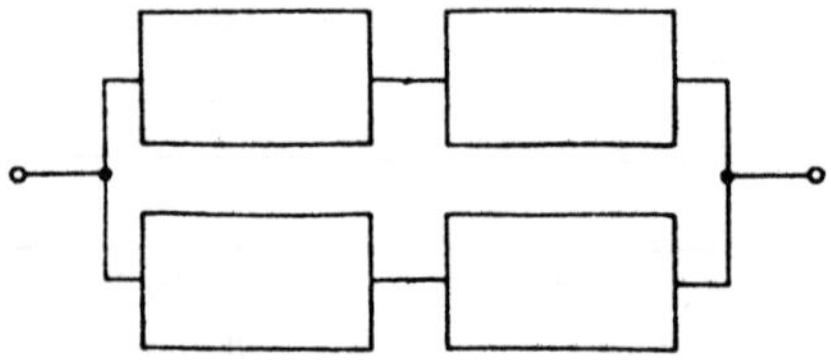

A series connection of parallel redundant elements provides protection against shorts and opens. Direct short across the network due to a single element shorting is prevented by a redundant element in series. An open across the network is prevented by the parallel element. Network (a) is useful when the primary element failure mode is open. Network (b) is useful when the primary element failure mode is short.

Figure 5(a). Redundancy Technique Summary

Majority Voting Redundancy

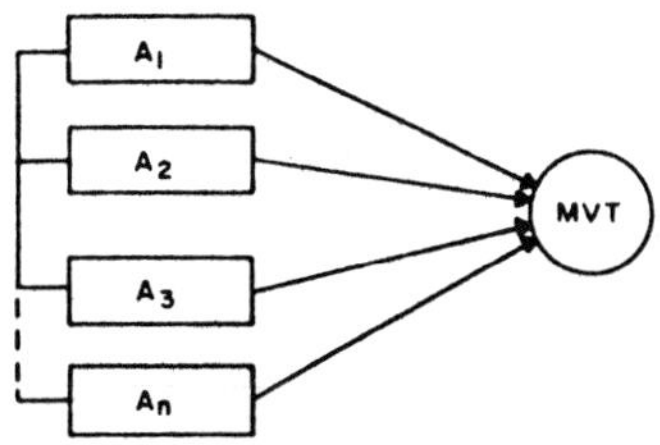

Decision can be built into the basic parallel redundant model by inputting signals from parallel elements into a voter to compare each signal with remaining signals. Valid decisions are made only if the number of useful elements exceed the failed elements.

Adaptive Majority Logic

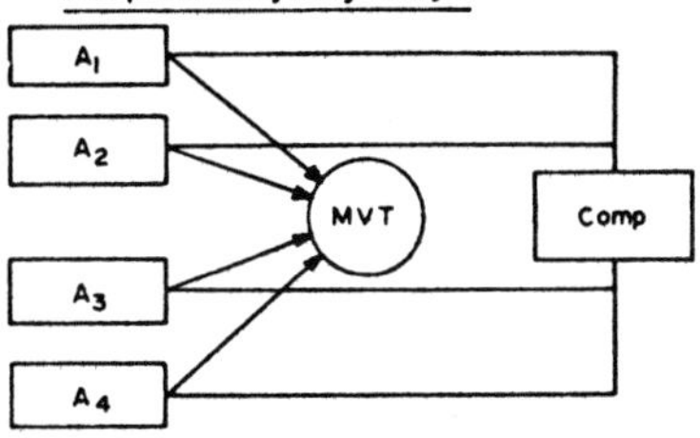

This technique exemplifies the majority logic configuration discussed previously with a comparator and switching network to switch out or inhibit failed redundant elements.

Gate Connector Redundancy

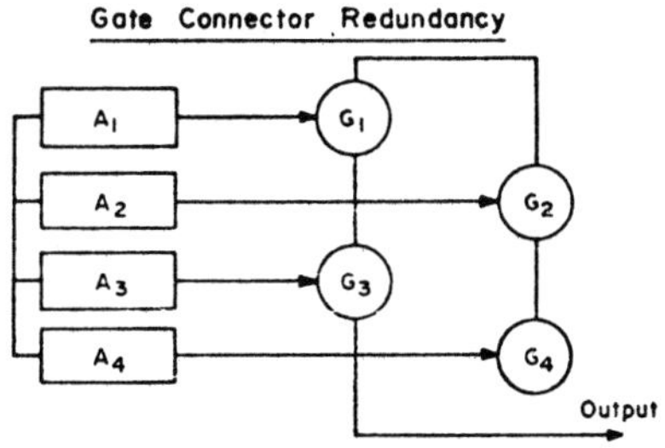

Similar to majority voting. Redundant elements are generally binary circuits. Outputs of the binary elements are fed to switch-like gates which perform the voting function. The gates contain no components whose failure would cause the redundant circuit to fail. Any failures in the gate connector act as though the binary element were at fault.

Standby Redundancy

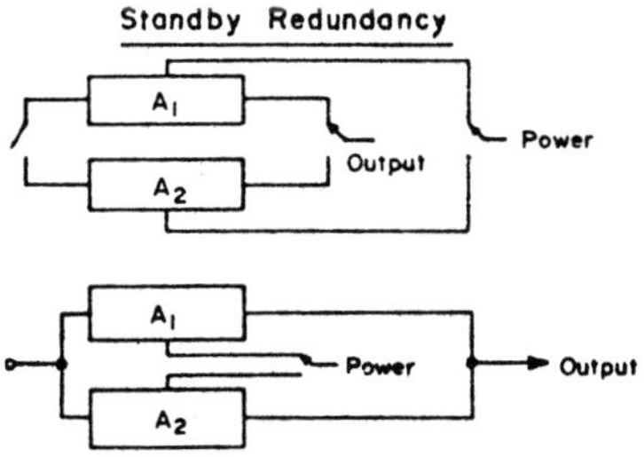

A particular redundant element of a parallel configuration can be switched into an active circuit by connecting outputs of each element to switch poles. Two switching configurations are possible.

1) The element may be isolated by the switch until switching is completed and power applied to the element in the switching operation.
2) All redundant elements are continuously connected to the circuit and a single redundant element activated by switching power to it.

Operating Redundancy

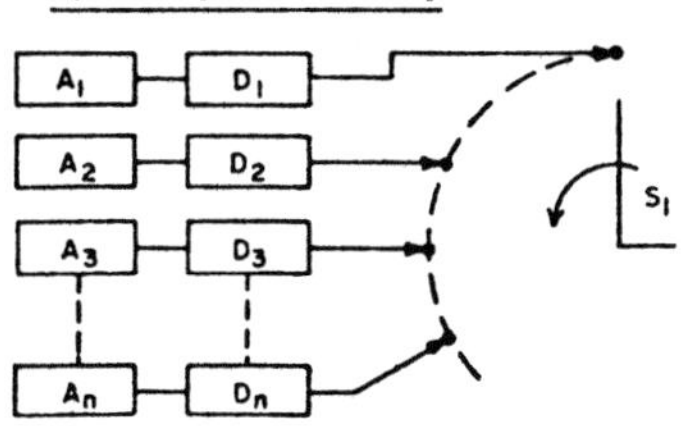

In this application, all redundant units—operate simultaneously. A sensor on each unit detects failures. When a unit fails, a switch at the output transfers to the next unit and remains there until failure.

Figure 5(b). Redundancy Technique Summary (Cont'd)

The reliability for the above configuration is given by:

$$R(t) = 1-(1-\exp(-\lambda_j t))^n \tag{5}$$

which may be further simplified, where $\lambda_j t$ is small, to:

$$R(t) = 1-(\lambda_j t)^n \tag{6}$$

R (t) is shown in Figure 6(b) for various values of n. It is important to note that the gain in reliability as units are added falls off rapidly after the addition of only a few units.

The MTBF for this configuration is given by:

$$\text{MTBF} = \frac{1}{\lambda_j} \sum_{i=1}^{n} \frac{1}{i} \tag{7}$$

A treatment for the case where the units are not-identical is given in (5). In considering this configuration the effect of repair on reliability should be examined as it has a significant impact on system reliability. This is shown in Figure 7.

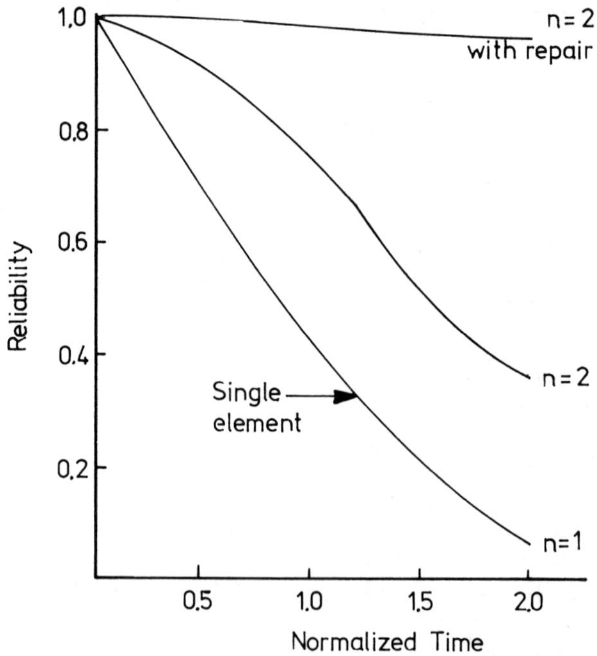

Figure 7. Redundancy with Repair

The reliability for an active simple parallel configuration for identical elements is given by:

$$R_s(t) = \frac{s_1 \exp(s_2 t) - s_2 \exp(s_1 t)}{s_1 - s_2} \tag{8}$$

where,

$$s_1 = \frac{-(3\lambda+\mu)+(\lambda^2 + 6\lambda\mu + \mu^2)^{1/2}}{2} \tag{9}$$

where,

$$s_2 = \frac{-(3\lambda+\mu)-(\lambda^2 + 6\lambda\mu + \mu^2)^{1/2}}{2} \tag{10}$$

μ = unit repair rate

The MTBF for this case is:

$$MTBF = \frac{3\lambda+\mu}{2\lambda^2} \tag{11}$$

A thorough treatment of redundant systems with repair is given in (6).

Case 3 - Bimodal. This technique is applied primarily at the part and assembly level where protection against the short or open failure mode is desired. The two configurations associated with this type of redundancy are shown in Figure 8. The configuration of Figure 8(a) protects primarily against the short failure mode and the expression for reliability is:

$$R(t) = 2 \exp(-2\lambda t) - \exp(-4\lambda t) \tag{12}$$

and the MTBF is:

$$MTBF = \frac{3}{4\lambda} \tag{13}$$

The configuration of Figure 8(b) protects primarily aginst the open failure mode and the expression for reliability is:

$$R(t) = 4 \exp(-2\lambda t) - 4 \exp(-3\lambda t) + \exp(-4\lambda t) \tag{14}$$

and the MTBF is given by:

$$MTBF = \frac{11}{12\lambda} \tag{15}$$

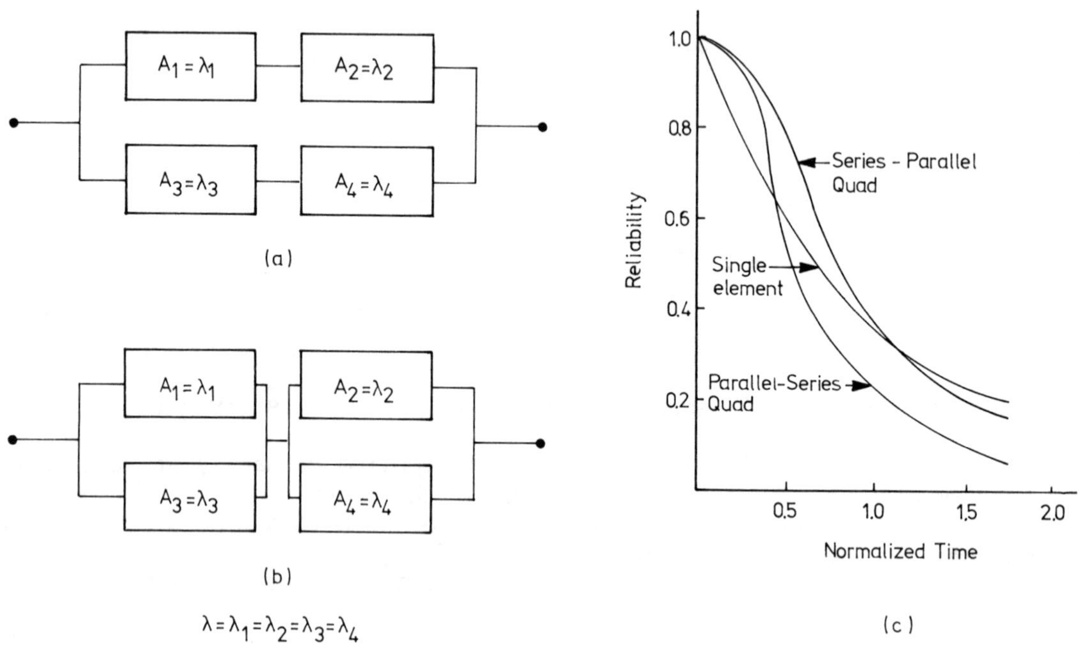

Figure 8. Bimodal Redundant Configurations

The reliability R(t) for each quad configuration is shown in Figure 8(c).

Case 4 - Simple Majority Voting. A common problem in using redundancy is the determination of which redundant unit is in error when the output from each is compared. This is a significant problem in digital computing systems when high reliability is desired. One solution is to use an odd number of parallel circuits and assume the majority are correct in a case of disagreement. For this a majority voter (MV) is required as shown in Figure 9(a). The reliability for this configuration is given by:

$$R(t) = \left(\sum_{i=0}^{n} \binom{2n+1}{i} (1-\exp(-\lambda t)) \exp(-\lambda t(2n+1-i))\right) \exp(-\lambda_m t) \qquad (16)$$

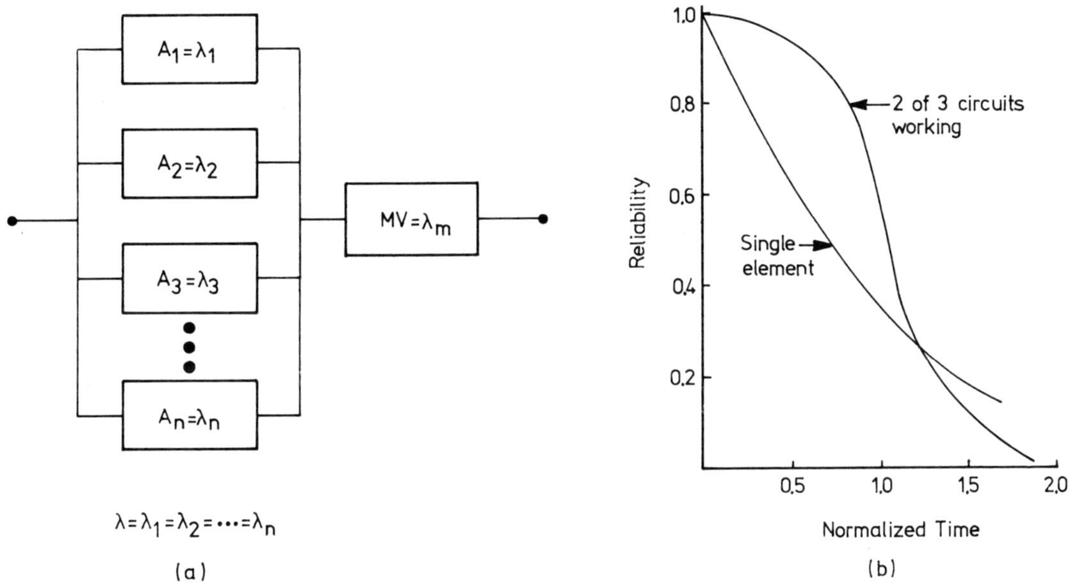

Figure 9. Majority Voting Redundancy

where,

λ_m= failure rate of majority voter.

Equation (16) can be simplified, for small values of λt, to:

$$R(t) = \exp(\lambda_m t) - \binom{2n+1}{n+1}(\lambda t)^{n+1} \quad (17)$$

The MTBF may be calculated for individual cases from:

$$MTBF = \int_0^\infty R(t) \quad (18)$$

The reliability R(t) for the simple case of 2 out of 3 circuits working is shown in Figure 9(b).

Case 7/8 - Standby Redundancy. This technique is such that each redundant unit has associated with it a sensor capable of switching when a failure is detected. This technique may also be used to give a standby redundancy capability by altering the switching arrangement to activate the units as they are switched into the system. The configuration for operating redundancy is shown in Figure 10(a).

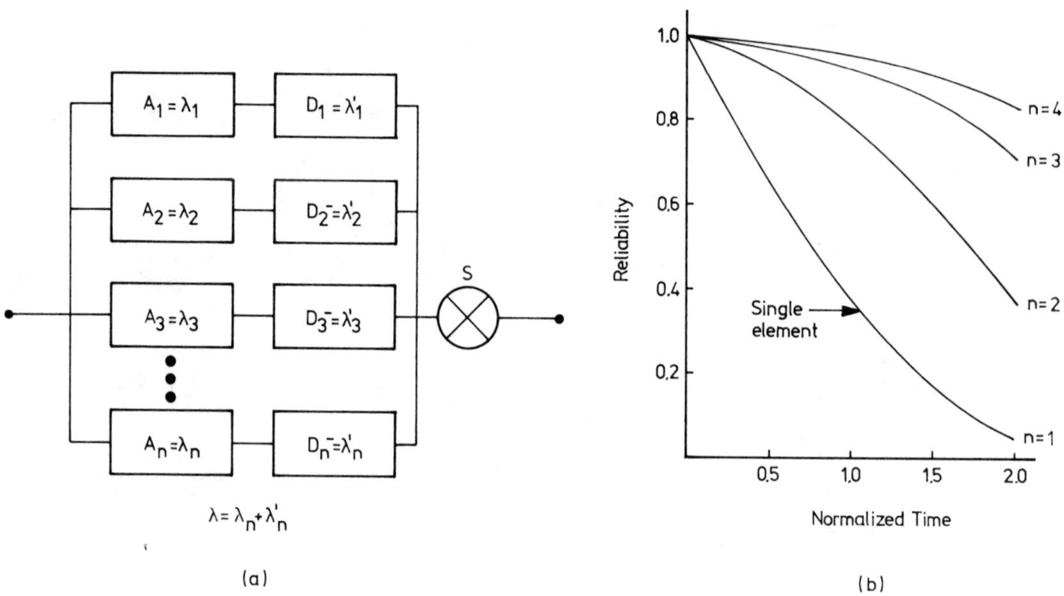

Figure 10. Operating Redundancy

The reliability expression for the configuration of Figure 10(a) assuming error detector and switching reliability to be 1.0, is:

$$R(t) = \exp(\lambda t) \left(\sum_{r=0}^{n-1} \frac{(\lambda t)^r}{r!} \right) \tag{19}$$

where;

n = number of parallel elements.

The reliability expression for the non-operating standby case is given by:

$$R(t) = \exp(-\lambda t) \left(1 + \frac{\lambda}{\lambda_s} (1 - \exp(-\lambda_s t))\right) \tag{20}$$

where,

λ_s = failure rate of switching function.

Hardware to implement the preceding redundancy techniques is described in (1).

r-out-of-n Configuration

In many cases the system operates if at least r-out-of-n units function, e.g., an air traffic control system where 8 of 10 displays are always required. Figure 11(a) shows the configuration.

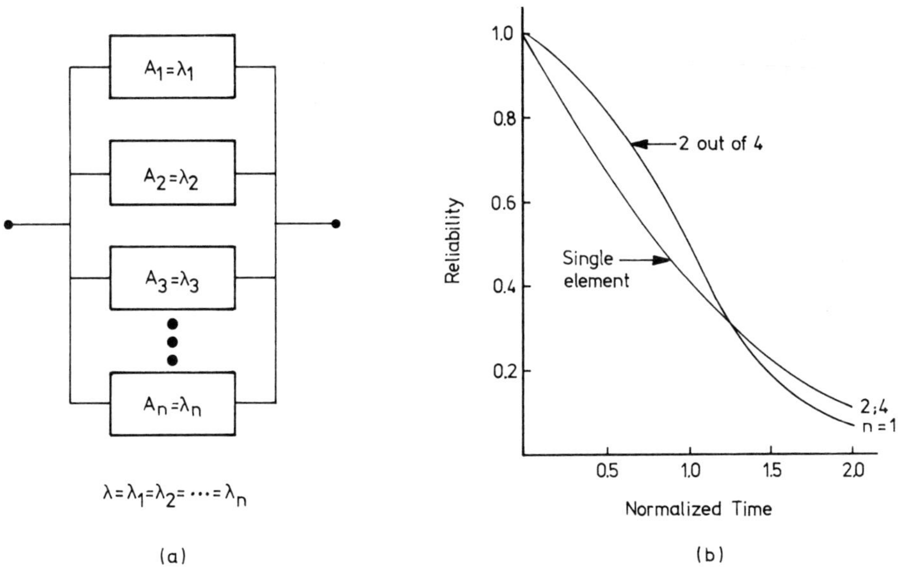

Figure 11. r-out-of-n Configuration

The reliability characteristic for this configuration is shown in Figure 11(b) and is given by:

$$R(t) = \sum_{k=r}^{n} \binom{n}{k} \exp(-k\lambda t)(1-\exp(-\lambda t))^{n-k} \tag{21}$$

where,

r = number of units required for system operation.

RELIABILITY ESTIMATES

Generally, two methods are used to make reliability estimates, i.e., (a) Parts Count Method and, (b) Parts Stress Analysis Method. The Parts Count method requires less information, generally that dealing with the quantity of different part types, quality level of the parts, and the application environment. This method is applicable in the early design phase and during bid/proposal formulation. The Parts Stress Analysis requires the greatest amount of detail and is applicable during the later design phases where actual hardware and circuits are being designed. It should be noted however, that the parts stress analysis method may also be used during any phase provided that enough detailed information is available. Whichever method is used, the objective is to obtain a reliability estimate that is expressed as a failure rate. From this basic figure R(t) and MTBF may be developed.

Calculation of failure rate for an electronic assembly, unit or system requires knowledge of the failure rate of each part of which the item of interest is comprised. If we assume that the item will fail when any of its parts fail, the failure rate of the item will equal the sum of the failure rate of its parts (7). This may, in general, be expressed as:

$$\lambda = \sum_{i=1}^{n} \lambda_i \tag{22}$$

where,

λ_i= failure rate of the i th parts,

n = number of parts.

Parts Count Reliability Prediction (Adapted from (3))

This prediction method is applicable during bid proposal and early design phases. The information needed to apply the method is (a) generic part types (including complexity for microelectronics) and quantities, (b) part quality levels, and (c) equipment environment. The general expression for equipment failure rate with this method is:

$$\lambda = \sum_{i=1}^{n} N_i \, (\lambda_G \, \pi_Q)_i \tag{23}$$

for a given equipment environment where:

λ = total equipment failure rate (failures/10 hr.),

λ_G = generic failure rate for the i th generic part (failures/10^6 hr.),

π_Q = quality factor for the i th generic part,

N_i = quantity of i th generic part,

n = number of different generic part categories.

Information to compute equipment failure rates using equation (23) is given in Tables 1 through 12 and applies if the entire equipment is being used in one environment. If the equipment comprises several units operating in different environments (such as avionics with units in airborne inhabited (A_I) and uninhabited (A_U) environments), then equation (23) should be applied to the portions of the equipment in each environment. These "environment-equipment" failure rates should be added to determine total equipment failure rate. Environmental symbols are as defined in Table 11.

The quality factors to be used with each part type are shown with the applicable λ_G tables and are not necessarily the same values that are used in MIL-HDBK-217C Part Stress Analysis. Multi-quality levels are presented for microelectronics, discrete semiconductors, and for established reliability (ER) resistors and capacitors. The λ_G values for the remaining parts apply, providing that the parts are procured in accordance with the applicable parts specifications and, for these parts, π_Q =1. Microelectronic devices have an additional multiplying factor, π_L (learning factor) as defined in Table 4.

It should be noted that no generic failure rates are shown for hybrid microcircuits. Each hybrid is a fairly unique device. Since none of these devices has been standardized, their complexity cannot be determined from their name or function. Identically or similarly named hybrids can have a wide range of complexity that thwarts categorization for purposes of this prediction method. If hybrids are anticipated for a design, their use and construction should be thoroughly investigated on an individual basis with application of the prediction model of MIL-HDBK-217C.

Parts Stress Analysis Method

This method is applicable when most of the design is completed and a detailed parts list and parts stresses are available. It can also be used during later design phases for reliability tradeoffs vs part quality and stresses. The most commonly used source for the application of this method is (3), para 2.0, which is based on the use of large scale data collection efforts to obtain the relationships (i.e., models) between engineering (temperature, stress, etc.) and reliability (part quality, failure rate, etc.) variables.

Part failure models vary with different part types, but, their general form is:

$$\lambda_i = (\lambda_B)(\pi_E)(\pi_A)(\pi_Q)\ldots(\pi_N) \qquad (24)$$

where,

λ_B = base failure rate,

π_E = environmental adjustment factor,

π_A = application adjustment factor,

π_N = additional adjustment factors.

Base Failure Rate (λ_B). The value is obtained from reduced part test data for each generic part type. The data is generally presented in the form of failure rate against normalized stress and temperature factors. This is shown in Figure 12 where the value of λ_B is generally determined by stress level (current, voltage, etc.) at the expected operating temperature. These values of applied stress relative to the rated stress represent the variables over which design control can be exercised and which influence part reliability.

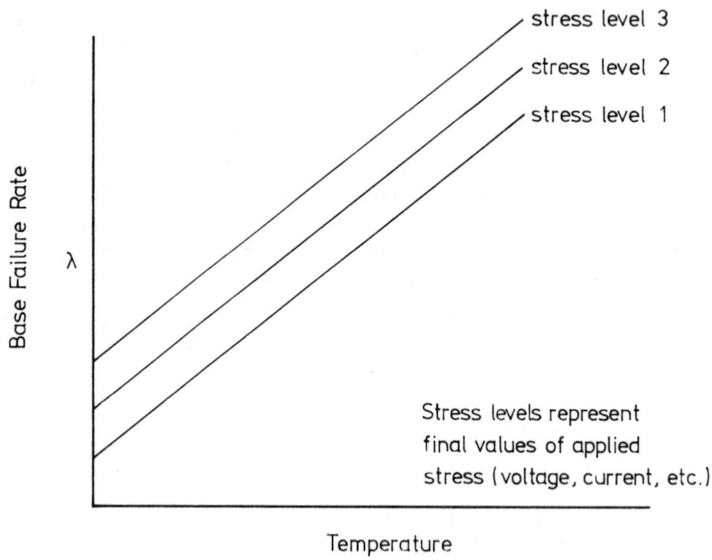

Figure 12. Part Base Failure Rate vs Stress and Temperature

Environmental Factor (π_E). The environmental factor accounts for the influence of environments other than temperature and is related to operating condition (vibration, humidity, etc.). The environments have been defined in Table 12.

Application Adjustment Factor (π_A). This factor depends on the application of the part and takes into account secondary stress and application factors.

Quality Adjustment Factor (π_Q). This factor is used to account for the degree of manufacturing control with which the part was fabricated and tested before shipment to the user. Table 14 identifies those parts with multi-level quality specifications, while Tables 3, 6 and 9 identify quality factors for microcircuits, discrete semiconductors and established capacitors and resistors respectively.

TABLE 1
GENERIC FAILURE RATE, λ_G, FOR BIPOLAR DIGITAL DEVICES (TTL & DTL)
VS. ENVIRONMENT (f./10^6 hr.)*

CIRCUIT COMPLEXITY	G_B&S_F	G_F	A_{IT}	A_{IF}	N_S	G_M	A_{UT}	A_{UF}	N_U	M_L
1-20 gates	.0070	.029	.070	.13	.093	.091	.11	.20	.12	.21
21-50 "	.020	.062	.12	.21	.17	.16	.20	.33	.23	.34
51-100 "	.032	.094	.18	.29	.24	.23	.28	.45	.34	.47
101-500 "	.079	.22	.37	.56	.49	.45	.61	.89	.71	.85
501-1000 "	.13	.34	.56	.82	.73	.67	.92	1.3	1.1	1.2
1001-2000 "	.29	.78	1.3	1.8	1.7	1.5	2.1	2.9	2.5	2.7
2001-3000 "	.81	2.1	3.5	5.1	4.5	4.1	5.8	8.1	6.7	7.3
3001-4000 "	2.2	5.7	9.6	14.	12.	11.	16.	22.	18.	20.
4001-5000 "	5.9	16.	26.	38.	33.	30.	43.	60.	49.	54.
ROM**, ≤ 320 bits	.0083	.022	.036	.053	.048	.043	.060	.085	.070	.078
" 321-576 "	.012	.033	.055	.081	.072	.066	.091	.13	.11	.12
" 577-1120 "	.020	.052	.087	.13	.11	.10	.14	.20	.17	.19
"1121-2240 "	.029	.078	.13	.20	.17	.16	.22	.31	.25	.29
"2241-5000 "	.045	.12	.20	.30	.27	.24	.33	.48	.39	.45
"5001-11000 "	.068	.18	.31	.47	.41	.38	.51	.75	.60	.70
11001-17000 "	.10	.28	.48	.73	.63	.58	.79	1.1	.92	1.1

*See Tables 3 and 4 for Π_Q and Π_L values.

**RAM failure rate = 3.5 X ROM failure rates.

TABLE 2
GENERIC FAILURE RATE, λ_G, VS. ENVIRONMENT FOR BIPOLAR BEAM LEAD, ECL, ALL LINEAR, AND ALL MOS DEVICES (f./10^6 hr.)*

CIRCUIT COMPLEXITY	G_B&S_F	G_F	A_{IT}	A_{IF}	N_S	G_M	A_{UT}	A_{UF}	N_U	M_L
1-20 gates	.010	.048	.099	.16	.14	.12	.21	.30	.25	.24
21-50 "	.048	.19	.31	.40	.43	.34	.73	.86	.92	.52
51-100 "	.076	.31	.48	.59	.68	.54	1.2	1.3	1.5	.78
101-500 "	.19	.82	1.2	1.4	1.7	1.3	3.1	3.4	3.9	1.7
501-1000 "	.32	1.4	2.0	2.3	2.8	2.1	5.1	5.5	6.4	2.6
1001-2000 "	.74	3.1	4.6	5.2	6.4	4.8	12.	13.	15.	6.0
2001-3000 "	2.0	8.4	13.	14.	17.	13.	33.	35.	41.	16.
3001-4000 "	5.4	23.	35.	39.	47.	36.	90.	96.	111.	44.
4001-5000 "	15.	62.	94.	105.	128.	97.	241.	258.	299.	121.
ROM**, ≤ 320 bits	.021	.087	.13	.15	.18	.14	.33	.36	.42	.17
" 321- 576 "	.031	.13	.19	.22	.27	.20	.49	.53	.62	.26
" 577-1120 "	.048	.20	.31	.35	.42	.32	.78	.84	.98	.41
" 1121-2240 "	.072	.30	.45	.52	.63	.48	1.2	1.3	1.5	.61
" 2241-5000 "	.11	.46	.70	.80	.96	.74	1.8	1.9	2.2	.94
" 5001-11000 "	.17	.70	1.1	1.2	1.5	1.1	2.7	2.9	3.4	1.5
"11001-17000 "	.25	1.1	1.6	1.9	2.2	1.7	4.1	4.5	5.2	2.2
Linear, ≤ 32 transistors	.011	.052	.12	.20	.16	.15	.22	.35	.27	.33
Linear, 33-100 transistors	.023	.11	.24	.41	.35	.31	.48	.73	.60	.66

*SEE TABLES 3 and 4 for Π_Q and Π_L values.
**RAM failure rate = 3.5 x ROM failure rate.

TABLE 3

π_Q, QUALITY FACTORS FOR USE WITH TABLES 1 & 2*

Quality Level	π_Q
S	0.5
B	1
B-1	2.5
B-2	5
C	8
C-1	45
D	75
D-1	150

*See Table 13 for descriptions of quality levels. π_Q values shown here are different from those in Table 13.

TABLE 4

π_L, LEARNING FACTORS FOR USE WITH TABLES 1 & 2

The learning factor π_L is 10 under any of the following conditions:

(1) New device in initial production.
(2) Where major changes in design or process have occurred.
(3) Where there has been an extended interruption in production or a change in line personnel (radical expansion).

The factor of 10 can be expected to apply until conditions and controls have stabilized. This period can extend for as much as six months of continuous production.

π_L is equal to 1.0 under all production conditions not stated in (1), (2) and (3) above.

TABLE 5
GENERIC FAILURE RATE, λ_G, (f./10^6 hr.) FOR DISCRETE SEMICONDUCTORS VS. ENVIRONMENT (SEE TABLE 6 FOR QUALITY FACTOR)

PART TYPE	G_B&S_F	G_F	A_{IT}	A_{IF}	N_S	G_M	A_{UT}	A_{UF}	N_U	M_L
Transistors										
Si NPN	.017	.11	.28	.59	.26	.59	.60	1.2	.84	.96
Si PNP	.025	.17	.46	.96	.41	.96	.96	1.9	1.4	1.5
Ge PNP	.025	.25	.75	1.6	.84	1.6	.78*	1.6*	2.1*	2.5
Ge NPN	.072	.66	2.0	4.3	2.2	4.3	3.3*	6.6*	5.4*	6.6
FET	.046	.31	.78	1.6	.70	1.6	1.7	3.4	2.3	2.6
Unijunction	.15	1.0	2.7	5.6	2.4	5.6	6.3	13.	9.0	9.0
Diodes										
Si, Gen. Purpose	.0051	.036	.098	.20	.090	.20	.24	.48	.33	.33
Ge. " "	.0066	.078	.25	.51	.30	.51	.44	.87*	.75*	.81
Zener & Avalanche	.016	.096	.24	.51	.22	.51	.54	1.1	.72	.84
Thyristor	.023	.16	.43	.90	.40	.90	1.0	2.0	1.4	1.4
Si Microwave Det.	.19	2.2	6.0	12.	3.9	12.	7.5	25.	17.	46.
Ge " "	.41	5.6*	18.*	35.*	**	35.*	**	**	**	**
Si " Mix.	.25	3.0	8.0	16.	5.1	16.	17.	34.	23.	64.
Ge " "	.72	10.*	31.*	61.*	**	61.*	**	**	**	**
Varactor, Step Recovery, Tunnel	.24	1.5	3.9	8.1	3.5	8.1	8.6	17.	12.	13.
LED	.034	.14	.25	.49	.45	.35	.91	1.8	1.4	.88
Single Isolator	.051	.21	.38	.74	.68	.53	1.4	2.7	2.1	1.30

*This value is valid only for electrical stress, $S \leq 0.3$.

**Do not use in these environments since temperature normally encountered combined with normal power dissipation are above the device ratings.

TABLE 6
Π_Q, QUALITY FACTORS FOR TABLE 5

PART TYPE	JANTXV	JANTX	JAN	NON-MIL HERMETIC	PLASTIC
Microwave Diodes	0.3	0.6	1.0	2.0	--
All Other Types	0.1	0.2	1.0	5.0	10.

TABLE 7 GENERIC FAILURE RATE, λ_G, (f./10^6 hr.) FOR RESISTORS (see Table 9 for quality factor)

RESISTORS, FIXED			USE ENVIRONMENT									
CONSTRUCTION	STYLE	MIL-R-SPEC.	G_B& S_F	G_F	A_{IT}	A_{IF}	N_S	G_M	A_{UT}	A_{UF}	N_U	M_L
COMPOSITION	RCR	39008	.00051	.0032	.0037	.0075	.0046	.0066	.014	.027	.021	.02
"	RC	11	.0025	.016	.018	.038	.023	.033	.069	.13	.10	.099
FILM	RLR	39017	.0012	.0031	.0043	.0088	.0033	.0062	.016	.032	.021	.028
"	RL	22684	.0061	.015	.022	.044	.017	.031	.079	.016	.10	.14
"	RN	55182	.0014	.0033	.0049	.01	.0037	.007	.018	.036	.024	.032
"	RN	10509	.0073	.017	.025	.05	.019	.035	.09	.18	.12	.16
FILM, POWER	RD	11804	.012	.026	.055	.11	.026	.078	.15	.29	.19	.46
" , NETWORK	RZ	83401	.026	.072	.17	.34	.14	.24	.80	1.6	1.2	1.1
WIREWOUND,	RBR	39005	.0085	.019	.058	.12	.020	.078	.22	.44	.17	.39
ACCURATE	RB	93	.043	.094	.29	.58	.10	.39	1.1	2.2	.85	1.9
WIREWOUND,	RWR	39007	.014	.044	.073	.15	.037	.091	.19	.37	.25	.54
POWER	RW	26	.072	.22	.36	.73	.19	.45	.94	1.9	1.3	2.7
WIREWOUND,	RER	39009	.0079	.021	.045	.090	.024	.056	.11	.22	.16	.34
CH. MOUNT	RE	18546	.040	.010	.22	.45	.12	.28	.56	1.1	.79	1.7
RESISTORS, VARIABLE												
WIREWOUND,	RTR	39015	.014	.034	.078	.16	.077	.15	.19	.37	.24	1.1
TRIMMER	RT	27208	.072	.17	.39	.79	.39	.74	.93	1.9	1.2	5.5
W.W., PREC.	RR	12934	.84	2.1	5.5	11.	4.6	11.	14.	29.	18.	132.
W.W., SEMI-	RA	19	.31	.84	2.4	4.8	2.1	9.5	*	*	*	*
PREC.	RK	39002	"	"	"	"	"	"	*	*	*	*
W.W., POWER	RP	22	.31	.78	2.0	3.9	1.7	7.8	*	*	*	*
NON-W.W.	RJR	39035	.02	.067	.12	.23	.95	.23	.27	.54	.34	1.8
TRIMMER	RJ	22097	.10	.33	.58	1.2	4.8	1.2	1.4	2.7	1.7	9.2
COMPOSITION	RV	94	.12	.49	1.1	2.1	.88	4.1	6.6	13.	6.8	18.
NON-WW. PREC.	RQ	39023	.086	.35	.65	1.3	.58	1.3	2.8	5.7	2.6	10.
FILM	RVC	23285	.096	.34	.6	1.2	.51	1.2	2.2	4.4	2.0	9.6

* - not normally used in these environments

TABLE 8 GENERIC FAILURE RATE, λ_G, (f./10^6 hr.) FOR CAPACITORS
(See Table 9 for quality factors)

CAPACITORS, FIXED			USE ENVIRONMENT									
DIELECTRIC	STYLE	MIL-C-SPEC.	G_B&S_F	G_F	A_{IT}	A_{IF}	N_S	G_M	A_{UT}	A_{UF}	N_U	M_L
Paper	CP	25	.011	.022	.057	.11	.047	.057	.16	.31	.15	.29
"	CA	12889	.012	.031	.087	.17	.083	.087	.47	.90	.53	.44
Paper/Plastic	CZR	11693	.0047	.0098	.025	.05	.021	.025	.072	.14	.064	.13
"	CPV	14157	.0021	.0042	.0088	.018	.0088	.0088	.025	.05	.023	.044
"	CQR	19978	"	"	"	"	"	"	"	"	"	"
"	CHR	39022	.0028	.006	.012	.024	.012	.012	.035	.069	.032	.06
"	CH	18312	.02	.042	.084	.17	.087	.084	.24	.49	.22	.42
Plastic	CFR	55514	.0041	.0086	.022	.043	.018	.03	.067	.13	.075	.13
"	CRH	83421	.0023	.0048	.0096	.019	.010	.0096	.028	.055	.025	.048
MICA	CMR	39001	.0005	.0022	.0059	.012	.0043	.0084	.044	.088	.042	.042
"	CM	5	.003	.013	.035	.071	.026	.050	.27	.53	.25	.25
"	CB	10950	.09	.19	.42	.85	.3	.60	1.9	3.8	1.4	3.0
Glass	CYR	23269	.0003	.0014	.0037	.0075	.0066	.0053	.027	.054	.028	.026
"	CY	11272	.001	.0043	.011	.022	.020	.016	.082	.16	.084	.079
Ceramic	CKR	39014	.0036	.0076	.033	.066	.0098	.016	.068	.14	.032	.12
"	CK	11015	.011	.023	.099	.20	.029	.047	.20	.41	.096	.35
"	CCR	20	.0008	.0032	.008	.016	.0058	.011	.058	.12	.070	.057
TA,SOL.	CSR	39003	.012	.026	.078	.16	.035	.052	.15	.29	.14	.26
TA,Non-Sol	CLR	39006	.0061	.014	.082	.16	.049	.069	.14	.28	.15	.23
" "	CL	3965	.018	.043	.24	.49	.15	.21	.42	.83	.46	.69
AL OXIDE	CU	39018	.074	.23	1.2	2.3	.96	1.6	4.8	9.7	5.3	5.5
" DRY	CE	62	.090	.36	1.9	3.7	1.7	2.6	10.	21.	12.	8.7
CAPACITORS, VARIABLE												
Ceramic	CV	81	.32	1.6	2.5	4.8	4.2	3.5	24.	48.	19.	31.
PISTON	PC	14409	.099	.54	1.2	2.3	1.7	1.5	9.2	18.	22.	7.5
Air,Trimmer	CT	92	.4	3.0	4.8	9.4	8.	6.8	49.	98.	37.	60.
Vacuum	CG	23183	1.2	6.2	15.	29.	15.	21.	140.	270.	94.	*

*-Not normally used in this environment

TABLE 9

π_Q FACTOR FOR RESISTORS & CAPACITORS

FAILURE RATE LEVEL	$*\pi_Q$
L	1.5
M	1.0
P	.3
R	.1
S	.03

*For Non-ER parts (Styles with only 2 letters in Tables 3-7 & 3-8), π_Q = 1 providing parts are procured in accordance with the part specification; if procured as commercial (NON-MIL) quality, use π_Q=3. For ER parts (Styles with 3 letters), use the π_Q value for the "letter" failure rate level procured.

TABLE 10 GENERIC FAILURE RATE, λ_G, (f./10^6 hr.) FOR INDUCTIVE, ELECTROMECHANICAL & MISCELLANEOUS PARTS (SEE TABLE 11 FOR π_Q)

PART TYPE	USE ENVIRONMENT									
INDUCTIVE	G_B&S_F	G_F	A_{IT}	A_{IF}	N_S	G_M	A_{UT}	A_{UF}	N_U	M_L
Low power pulse transformer	.003	.0048	.041	.082	.017	.047	.069	.14	.065	.12
Audio transformer	.006	.0096	.082	.16	.034	.094	.14	.28	.13	.24
High power pulse & power transformer, filter	.019	.053	.31	.60	.13	.35	.46	.92	.98	.86
R.F. transformer	.024	.038	.33	.64	.14	.38	.56	1.1	.52	.96
R.F. coils, fixed	.0016	.004	.021	.042	.0096	.048	.039	.078	.038	.12
" ", variable	.0032	.008	.042	.084	.019	.096	.078	.16	.077	.24
MOTORS	*	15.	19.	19.	24.	19.	41.	41.	49.	*
RELAYS										
General purpose	.13	.30	.65	1.3	.89	.81	2.8	5.6	2.9	16.
Contractor, high current	.44	1.0	2.2	4.5	3.0	2.8	9.6	19.	10.	56.
Latching	.10	.24	.52	1.0	.71	.65	2.2	4.5	2.3	13.
Reed	.11	.26	.55	1.1	.75	.69	2.4	4.8	2.5	14.
Thermal bi-metal	.29	.69	1.5	3.0	2.0	1.9	6.4	13.	6.7	37.
Meter movement	.90	2.1	4.6	9.2	6.3	5.8	20.	40.	21.	*
SWITCHES										
Toggle & push button	.035	.011	.18	.35	.15	.61	1.8	3.5	.84	24.
Sensitive	.15	.44	.74	1.5	.59	2.5	7.4	15.	3.4	100.
Rotary	.22	.67	1.1	2.2	.89	3.8	11	22.	5.1	150.
CONNECTORS (PER PAIR)										
Circular, Rack & Panel	.0062	.029	.12	.24	.053	.12	.17	.34	.23	.18
Printed wiring board	.0031	.028	.060	.12	.036	.060	.090	.18	.11	.090
Coaxial	.0084	.032	.13	.26	.060	.10	.18	.36	.24	.20

TABLE 10 (Cont'd)

PART TYPE	USE ENVIRONMENT									
	$G_B \& S_F$	G_F	A_{IT}	A_{IF}	N_S	G_M	A_{UT}	A_{UF}	N_U	M_L
P.C. WIRING BOARDS										
Two-sided	.0012	.0024	.005	.01	.0048	.0048	.012	.024	.012	.024
Multi-layer	.15	.30	.63	1.3	.60	.60	1.5	3.0	1.5	3.0
CONNECTIONS **	See MIL-HDBK-217C									
TUBES	See MIL-HDBK-217C									
LASERS	See MIL-HDBK-217C									

* Not normally used in these environments.

** Usually negligible compared to system failure rate.

TABLE 11 Π_Q FACTOR FOR USE WITH TABLE 10

PART TYPE	QUALITY LEVEL	
	MIL-SPEC	NON-MIL
Inductive	1	3
Motors	1	1
Relays	1	3
Switches, toggle & sensitive	1	20
Switches, rotary	1	50
Connectors	1	3
P.W. Boards	1	--
Others	--	1

TABLE 12
ENVIRONMENTAL SYMBOL IDENTIFICATION AND DESCRIPTION

ENVIRONMENT	π_E SYMBOL	NOMINAL ENVIRONMENTAL CONDITIONS
Ground, Benign	G_B	Nearly zero environmental stress with optimum engineering operation and maintenance.
Space, Flight	S_F	Earth orbital. Approaches Ground, Benign conditions without access for maintenance. Vehicle neither under powered flight nor in atmospheric re-entry.
Ground, Fixed	G_F	Conditions less than ideal to include installation in permanent racks with adequate cooling air, maintenance by military personnel and possible installation in unheated buildings.
Ground, Mobile	G_M	Conditions more severe than those for G_F, mostly for vibration and shock. Cooling air supply may also be more limited, and maintenance less uniform.
Naval, Sheltered	N_S	Surface ship conditions similar to G_F but subject to occasional high shock and vibration.
Naval, Unsheltered	N_U	Nominal surface shipborne conditions but with repetitive high levels of shock and vibration.
Airborne, Inhabited, Transport	A_{IT}	Typical conditions in transport or bomber compartments occupied by aircrew without environmental extremes of pressure, temperature, shock and vibration, and installed on long mission aircraft such as transports and bombers.
Airborne, Inhabited Fighter	A_{IF}	Same as A_{IT} but installed on high performance aircraft such as fighters and intercepters.
Airborne, Uninhabited, Transport	A_{UT}	Bomb bay, equipment bay, tail, or wing installations where extreme pressure, vibration, and temperature cycling may be aggravated by contamination from oil, hydraulic fluid and engine exhaust. Installed on long mission aircraft such as transports and bombers.
Airborne, Uninhabited Fighter	A_{UF}	Same as A_{UT} but installed on high performance aircraft such as fighters and intercepters.
Missile Launch	M_L	Severe conditions of noise, vibration, and other environments related to missile launch, and space vehicle boost into orbit, vehicle re-entry and landing by parachute. Conditions may also apply to installation near main rocket engines during launch operations.

TABLE 13. π_Q, QUALITY FACTORS

Quality Level	Description	π_Q
S	Procured in full accordance with MIL-M-38510, Class S requirements.	1
B	Procured in full accordance with MIL-M-38510, Class B requirements.	2
B-1	Procured to screening requirements of MIL-STD-883, Method 5004, Class B, and in accordance with the electrical requirements of MIL-M-38510 slash sheet or vendor or contractor electrical parameters. The device must be qualified to requirements of MIL-STD-883, Method 5005, Class B. No waivers are allowed.	5
B-2	Procured to vendor's equivalent of screening requirements of MIL-STD-883, Method 5004, Class B, and in accordance with vendor's electrical parameters. Vendor waives certain requirements of MIL-STD-883, Method 5004, Class B.	10
C	Procured in full accordance with MIL-M-38510, Class C requirements.	16
C-1	Procured to screening requirements of MIL-STD-883, Method 5004, Class C and in accordance with the electrical requirements of MIL-M-38510 slash sheet or vendor or contractor electrical specification. The device must be qualified to requirements of MIL-STD-883, Method 5005, Class C. No waivers are allowed.	90
D	Commercial (or non-mil standard) part, hermetically sealed, with no screening beyond the manufacturer's regular quality assurance practices.	150
D-1	Commercial (or non-mil standard) part, packaged or sealed with organic materials (e.g., epoxy, silicone, or phenolic).	300

Table 14. Parts with Multi-Level Quality Specifications

Part	Quality Designators
Microelectronics	A, B, B-1, B-2, C, C-1, D, D-1
Discrete Semiconductors	JANTXV, JANTX, JAN
Capacitors, Established Reliability (ER)	L, M, P, R, S
Resistors, Established Reliability (ER)	M, P, R, S

Parts Stress Analysis and the Computer

Although, in general, the Parts Stress Analysis Method will yield reliability estimates of the highest confidence, generating them can be a very time consuming and error prone effort. This is because of the amount of detail involved in computing individual part failure rates from their individual models. An obvious approach is to use a computer and program the failure rate model(s) for each generic part type into a sub-routine. It is then only necessary to input the variables associated with a particular design situation and a consistent set of results will be achieved. Assuming such a program is available it is possible to change common variables such as temperature, environment, etc., until the best possible results are achieved for a particular case. Figure 13 is an example of a completed data input form for such a program while Figure 14 is a sample output from the same program.

CONCLUSION

Given certain reliability requirements e.g., R(t), MTBF, a mathematical analysis permits the requirements to be allocated to the different items comprising a system. All reliability calculations are based on part failure rate for which accurate data sources are available. It should be noted that there are certain fundamental limitations associated with reliability estimates: principal of these is that the basic information used in part failure rate models is averaged over a wide data base involving many persons and a variety of data collection methods and conditions which prevent exact coordination and correlation. The user is cautioned to use the latest part failure data available, as part failure rates are continuously improving.

RELIABILITY DATA INPUT FORM

LETTER CODE (LC)	DESCRIPTION	AREA - 1	QUALITY	AREA - 2	AREA - 3	ENVIRONMENT	SPARE	SPACE - 1	SPACE - 2	SPACE - 3	SPACE - 4	SPACE - 5	QUANTITY	TEMPERATURE °C	SPACE - 6	SPACE - 7	APPLICATION FACTOR	LINE NUMBER
NPN	TRANSISTOR Q8 2N2222A		L			GM		50	500		200	1.5	1	40	8	40	1	27
DIODE	CR1 1N914		L			GM		23	75		200	.6	1	40		6.67	1	28
DIODE	/RECTIFIER CR2 1N4002		L			GM		.3	1		200	1	1	40		28	1	29
ZENER	DIODE VR1 1N962B		L			GM		2.8	28		200	1	1	40			1	30
MIC	DUAL D TYPE FF U2 CD4013AE	L	M	D	I	GM		1	16		24		1	40	1	3	1	31
MIC	QUAD NAND GATE U3 CD4011AE	L	M	D	I	GM		1	14		4		1	40	1	3	1	32
MIC	DUAL MONO M/V U6 CD4098BE	L	M	D	I	GM		1	16		62		1	40	1	3	1	33
MIC	OPTICAL ISOLATOR U9		D	D	I	GM		1	8		2		1	40	1	3	1	34
MIC	OP AMP AR1,2 CA3078AT	L	M	L	I	GM		1	8		26		2	40	1	3	1	35
RELAY	DRY REED K1 MRB-1812	R	Q	B	A2	GM		.1					1	40	1	.1	1	36
RELAY	ARMATURE K2 2BN-1B-126	R	Q	B	L5	GM		1					1	40	3	.1	1	37

FORM 1432 PROJECT____ PREPARED BY____ DATE____ PAGE____ OF____

Figure 13. Reliability Estimate Data Input

R ESTIMATE FOR THE HSD4-1 RUN NO. 2 17 JUNE 1977 PAGE 4

*** COMPONENT SUMMARY ***

BLOCK 2 ARITHMETIC PROCESSOR CCA REV/1

MISSION TIME 4.0 HOURS ENVIRONMENT: AIRBORNE, INHABITED

COMPONENT	TEMP DEG. C	STRESS RATIO	REL. GRADE	APPL. FACTOR	FRS. /MILLION HRS.	QTY.	TFRS. /MILLION HRS.
MIC HEX D TYPE FF 54174	90.	N/A	B	1.0	.2143	1.	.2143
MIC DATA SEL/MULTIPLEXER 54LS253	90.	N/A	B	1.0	.1184	1.	.1184
CAPACITOR CERAMIC	90.	.20	M	1.0	.0115	20.	.2298
CAPACITOR TANTALUM	90.	.30	M	1.0	.0530	2.	.1060
RCR STYLE RESISTOR	90.	.30	M	1.0	.0115	1.	.0115
PRINTED WIRING BOARD	90.	N/A	N/A	1.0	2.3280	1.	2.3280
SOLDERING CONNECTIONS WAVE	90.	N/A	N/A	1.0	.0004	776.	.3414
CONNECTOR PCB 116 PINS	90.	N/A	SM	1.0	4.8361	1.	4.8361

FAILURE RATE 39.2585 MTBF 25472. HOURS NUMBER OF PARTS 833. RELIABILITY .999842978

Figure 14. Reliability Estimate Program Output

REFERENCES

1. Anderson, R.T., Reliability Design Handbook, Reliability Analysis centre, Rome Air Development Center, Griffiss Air Force Base, N.Y., 1976, p25.

2. Poppelbaum, W.J., Computer Hardware Theory, The MacMillan Company, New York, N.Y., 1972, p663.

3. ____________, MIL-HDBK-217B Reliability Prediction of Electronic Equipment, Department of Defense, Washington, D.C., Sept, 1974.

4. ____________, Reliability-Maintenance Data Summaries, Government-Industry Data Exchange Program, (GIDEP), Corona, California, 91720.

5. Shooman, M.L., Probabilistic Reliability - An Engineering Approach, McGraw-Hill, New York, N.Y., 1968, p119 and p206.

6. Rau, J.G., Optimization And Probability In Systems Engineering, Van Nostrand Reinhold Company, New York, N.Y., 1970, Chapter 6.

7. Blakeslee, T.R., Digital Design With Standard MSI & LSI, John Wiley & Sons, New York, N.Y., 1975, p325.

9 THERMAL DESIGN

D. Watson

There is a predictable relationship between the operating temperature of electronic parts and reliability. Materials employed in the manufacture of electronic devices have finite temperature limits; when these limits are exceeded, the physical properties are altered and the device ceases to perform its intended function. This physical breakdown of material usually results in irreversible and immediate failure of the device.

Failure also can occur gradually as a result of sustained operation at intermediate temperatures, producing slow but relentless deterioration of materials and finally ending in failure of the device. Statistical failure rate data presented in (1) confirms that the breakdown of electronic parts is closely related to operating temperature. Figure 1 illustrates the very important role which temperature plays in achieving reliable performance of electronic equipment.

What then is the proper operating temperature of an electronic device? What guidelines are available to assist the designer in establishing an acceptable temperature level? These questions are basic to the design of electronic equipment.

WORKING TEMPERATURE OF ELECTRONIC PARTS

Generally speaking, the most reliable working temperature of electronic parts is room temperature. Unfortunately, this is seldom possible to achieve in practice because wherever heat is generated, a temperature rise is created.

When setting a limit on the working temperature it is necessary to consider the reliability objectives of the system. A high Mean Time Between Failure (MTBF) demands low working temperature. The precise relationship between operating temperature and failure rate for a particular system is revealed by parts stress analysis, which calculates failure rate from the known system variables, including temperature, part quality, severity of the operating environment, circuit complexity, and the experience of the manufacturer in making the device.

In an effort to attain consistent thermal performance, equipment users have established guidelines which limit the permissible temperature of electronic parts. Examples of such guidelines are presented below.

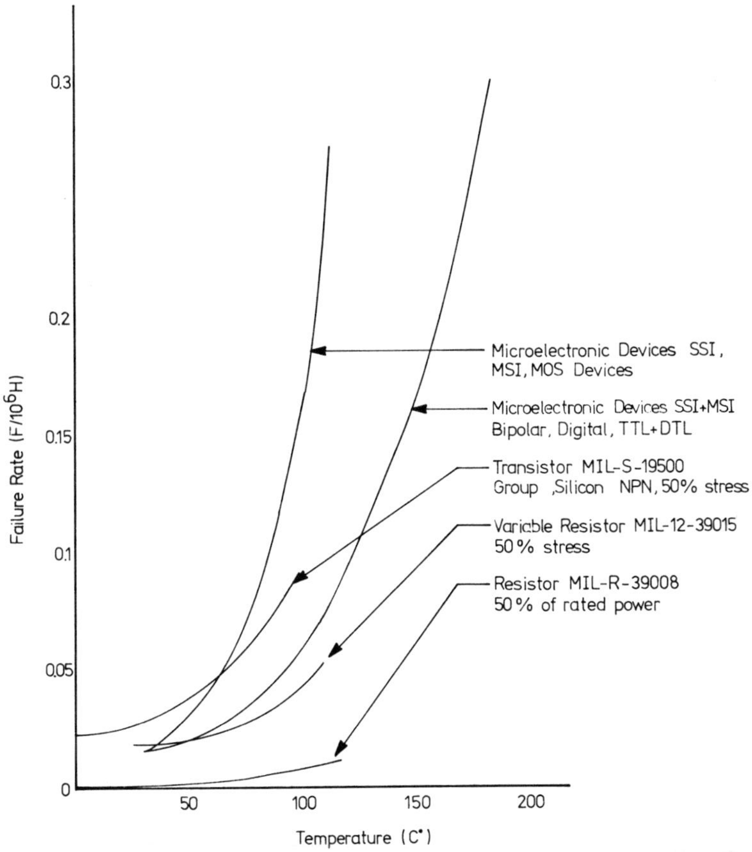

Figure 1. Failure Rate vs. Temperature for Selected Devices

United States Navy

For naval equipment, a limit is imposed on the Critical Component Temperature (CCT) of electronic parts in a Standard Electronic Module (SEM) in reference (2). The CCT for semiconductor devices is 105°C at the junction. For all other devices the CCT is 20°C less than the manufacturer's maximum rated temperature measured on the external surface of the device. The operating conditions that apply to SEMs are presented below in Table 1.

Table 1. Operating Environment for Standard Electronic Modules

	Module Class 1	Module Class 2
Fin or Guide Rib Temperature	0°C to + 60°C	-55°C to +100°C

Commercial Air Lines

The thermal design of electronic equipment for use onboard commercial aircraft is controlled by ARINC Specification 404 which is presented, in part, in Table 2.

Table 2. Component Temperatures for Equipment in Commercial Air Lines

Maximum component temperatures should not exceed the following values for the listed components

Part Type	Maximum Case Temperature
Capacitors	
Tantalum - Solid, Foil and Sintered Anode	$0.67 \times T_M = T_C$
Aluminum Electrolytic, Paper, Ceramic, Glass, Mica.	$0.72 \times T_M = T_C$
Resistors	
All	$0.68 \times T_M = T_C$ or 120°C whichever is lower.
Semiconductors, Integrated Circuits, Diodes.	$0.6 \times T_{MJ} - (\theta_{JC} \times P) = T_C$
Relays and Switches	$0.75 \times T_M = T_C$
Transformer, Coils and Chokes	Min. of 35°C below manufacturer's hot spot temperature.

T_M = Component manufacturer's max. permissible body temperature at zero power in °C,

T_{MJ} = Component manufacturer's max. permissible semi-conductor junction temperature at zero power in °C,

θ_{JC} = Component manufacturer's thermal resistance from junction to case in °C/watt,

P = Normal power dissipation for the specific circuit application in watts,

T_C = External surface or case temperature of component in °C.

Other Design Criteria For Working Temperature

Further discussion on the working temperature of electronic parts is presented in Chapter 14.

HEAT FLOW PROCESSES

The internal temperatures of electronic equipment are completely controlled by the heat transfer processes that carry the heat from the source (the electronic components) to the sink (usually the external atmosphere). There are three ways in which heat flow is accomplished.

(a) Conduction,
(b) Convection,
(c) Radiation.

Conduction

In a solid material, heat flow is considered to take place by molecular interaction, whereby energy is carried from molecule to molecule in a manner resembling the flow of electricity. Basic equations for heat conduction in a solid material are presented in Table 3.

Table 3. Heat Conduction Equations

Description of Heat Flow Path		Heat Flow Equation
Constant cross section	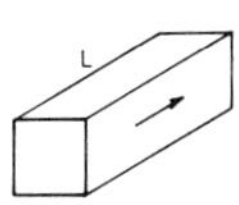	$\Delta T = Lq/Ak$ (1) $\Delta T = Rq$ (2)
Pipe or disk (radial flow)	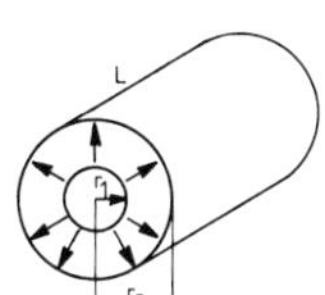	$\Delta T = \ln(r_2/r_1)q/2\pi Lk$ (3)
Hollow sphere (radial flow)	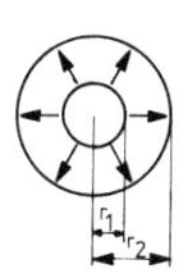	$\Delta T = (1/r_1 - 1/r_2)q/4\pi k$ (4)
Plate or rod with uniformly distributed heat input q. Heat flow to opposite ends.		ΔT_{max} at center $= qL/8Ak)$ (5) ΔT at x from one end $= (qL/8Ak) - (q/2LAk)(L/2-x)^2$ (6)

The similarity between the flow of heat in a solid material and the flow of electricity through a resistor is apparent from equation (2) where the quantity L/Ak has been replaced by R, the thermal resistance. The temperature difference T is analagous to the voltage drop, E, through a resistor and the heat flow, q, is analagous to the current I. This simple analogy aids in the cross-fertilization process between persons practicing in the thermal and electrical fields.

Convection

Convection is defined as heat flow between a solid surface and a fluid. In electronic equipment the fluid is usually air. Convection processes can be divided into two catagories:

(a) Natural Convection - where the air adjacent to the surface is not moving except for natural air currents,

(b) Forced Convection - where the air is forced (usually by a fan) to flow with appreciable velocity across the heat transfer surface.

Natural Convection. Natural Convection is the most commonly employed cooling method in the electronic equipment industry. It is dependent, to a large extent, on the creation of air currents adjacent to the surface (air currents are produced by differences in air density between heated and unheated regions). Since these air currents are affected by the size, shape and orientation of the surface, the heat transfer equations contain parameters which account for the specific surface configuration.

The general equation of heat flow from a heated surface to the surrounding air by natural convection is given by equation (7) as stated in reference (3):

$$q_c = 0.00394CA(\Delta T)^{1.25}/ (L)^{0.25} \tag{7}$$

where,

q_c = heat flow (watts),

C = a constant which is dependent on the orientation of the surface, (see Table 4),

L = characteristic length which is determined by the shape and orientation of the surface, inches, (see Table 4),

A = surface area (inches^2)

ΔT = temperature difference between surface and air (degrees C).

Table 4. Values of C and L for Equation (7)

Surface	C	L, inches
Horizontal plane, facing upward	0.71	$\frac{\text{length x width}}{\text{length + width}}$
Horizontal plane, facing downward	0.35	$\frac{\text{length x width}}{\text{length + width}}$
Vertical plane, rectangular	0.55	vertical height but limited to 24 in. (even if actual dimension is more).
Vertical plane, irregular shape	0.55	$\frac{\text{area}}{\text{horizontal width}}$
Vertical plane, circular	0.55	0.785 x diameter
Horizontal cylinder (pipes, wires)	0.45	diameter
Vertical cylinder	0.45-0.55	vertical height but limited to 24 in. (even if actual dimension is more)
Sphere	0.63	radius
Small electronic parts	1.45	L selected from one of the above surfaces.

Forced Convection. Forced convection takes place when air is moving with appreciable velocity over a heat transfer surface. The rate of heat transfer is greatly enhanced by the moving air stream. Thus forced convection is a much more efficient method of cooling electronic equipment than natural convection. Turbulence in the air stream contributes an additional advantageous factor which further promotes heat transfer between the surface and the moving air stream.

In most electronic equipment applications, the extremely complex nature of the air flow pattern at each point in the system is seldom accurately known. However, considerable research has established the heat transfer properties of commonly encountered shapes. The basic expression for forced convection heat transfer is given by:

$$q = h_c A \Delta T \qquad (8)$$

where,

q = rate of heat flow, between the surface and the moving air stream, watts,

h_c = convective heat transfer coefficient, watts/in^2 °C,

A = surface area available for heat transfer, in^2 ,

ΔT = temperature difference between the surface and air, °C.

Table 5 presents values of h_c which are applicable to various commonly encountered surface shapes and conditions.

Table 5. Forced Convection Heat Transfer Coefficient h_c for use in Equation (8)

Conditions	h_c (watts/in^2 - °C)	
	General Equation	For air at Standard Atmospheric Conditions (see Note 1)
Turbulent Air Flow Parallel to Flat Plate or Large or Cylinder.	$(0.0280k/L)(\rho VL/\mu)^{0.8}$	$0.00037((V)^{0.8}/(L)^{0.2})$ (9)
	or	or
	$(0.055k/L)(\rho VL/\mu)^{0.75}$	$0.0006((V)^{0.75}/(L)^{0.25})$ (10)
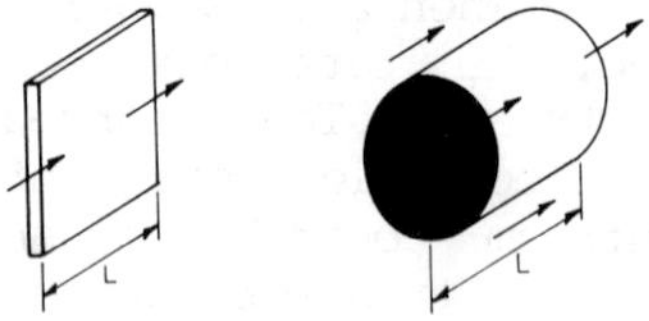	(For the above equation L is limited to 2 ft. even if actual length is greater, reference (2). See Note 5.	

Table 5 (Cont'd)

Laminar Air Flow Parallel to Flat Plate of Large Cylinder 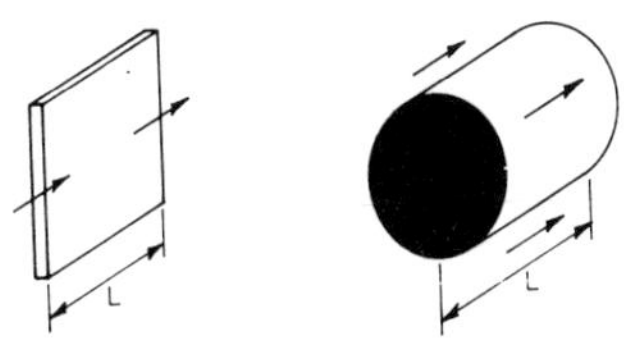	$(0.592k/L)(\rho VL/\mu)^{0.5}$ Also valid for open finned surface, reference (4), see below. See Note 5.	$0.0025((V)^{0.5}/(L)^{0.5})$	(11)
Turbulent Air Flow Parallel to Finned surface (open one side) 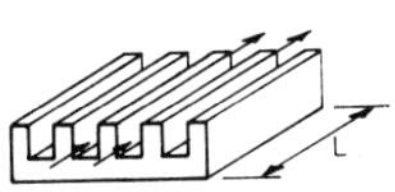	$(0.592k/L)(\rho VL/\mu)^{0.5}$ Re up to 100,000 and Re up to 400,000 if entering air is streamlined, reference (2). See Note 5.	$0.0025((V)^{0.5}/(L)^{0.5})$	(12)
Forced Air Through Cold Plate or Shrouded Finned Surface 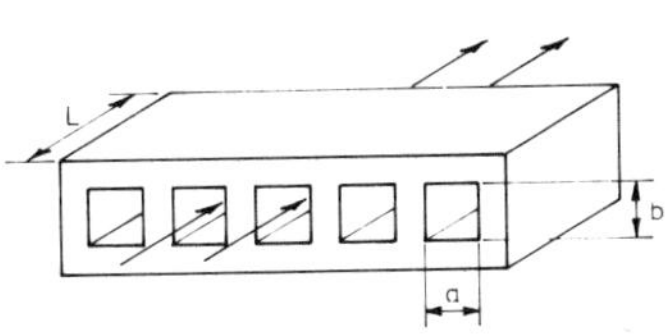	$1.615(\emptyset)^{1/3}k/d$ Reference (2). See Notes 3, 5.	$(0.103/d)(m/L)^{1/3}$	(13)

Table 5 (Cont'd)

Turbulent Air Flow Inside a Round Pipe or Long Passage	$(0.0198k/D)(\rho VD/\mu)^{0.8}$ $0.00026((V)^{0.8}/(D)^{0.2})$ See Note 5.	(14)

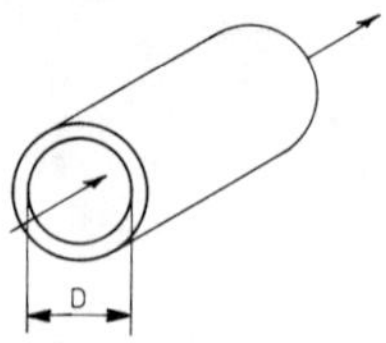

Air Flow Over a Sphere	$(0.37\ k/D)(\rho VD/\mu)^{0.6}$ $0.0023((V)^{0.6}/(D)^{0.4})$ Reference (2). See Notes 4 and 6.	(15)

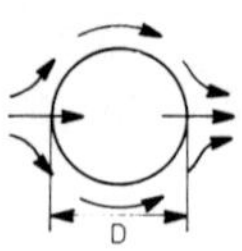

Air Flow Over a Single Cylinder		
Re = 1-4	$(0.891k/D)(\rho VD/\mu)^{0.330}$	(16)
Re = 4-40	$(0.821k/D)(\rho VD/\mu)^{0.385}$	(17)
Re = 40-4000	$(0.615k/D)(\rho VD/\mu)^{0.466}$	(18)
Re = 4000-40,000	$(0.174k/D)(\rho VD/\mu)^{0.618}$	(19)
Re = 40,000-400,000	$(0.024k/D)(\rho VD/\mu)^{0.805}$	(20)
	Reference (3).	

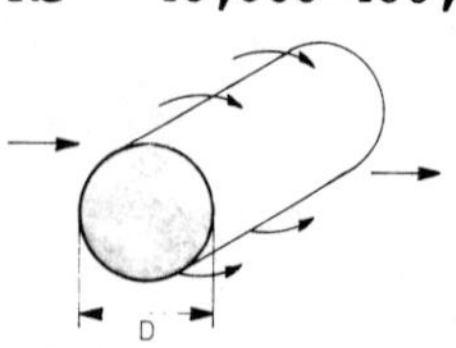

Table 5 (Cont'd)

Notes:

1. k, air at 25°C = 6.7×10^{-4} watt-in./in.^{2}-°C

   Cp, specific heat air at 25°C = 455 watt-sec./pound-°C

   ρ, air at 25°C = 4.3×10^{-5} pounds mass/in.^{3}

   μ, air at 25°C = 1.04×10^{-6} pounds mass/in.-sec.

2. $$d = \frac{4\text{x duct cross section}}{\text{duct perimeter}} = \frac{4ab}{2(a + b)}$$

   $$Re = \text{Reynolds Number} = \frac{\rho VL}{\mu} \quad \text{(dimensionless)}$$

   where,

   V = velocity in./sec.

3. $$\emptyset = \frac{4}{\pi} \times \frac{\text{specific heat x mass flow}}{kL} \quad \text{(dimensionless)}$$

   $$= \frac{4\ Cp\ m}{\pi\ kL} \quad (\emptyset \text{ greater than } 10)$$

   where,

   m = mass flow rate pounds mass/sec.

4. Equation (15) is valid for Reynolds numbers between 17 and 70,000.

5. Fluid properties evaluated at bulk air temperature.

6. Fluid properties evaluated at the film temperature, which is the mean of bulk air temperature and surface temperature.

Radiation

Hot objects emit heat energy in the form of electromagnetic waves which travel through air or a vacuum with very little loss of intensity until intercepted by another solid object. Some materials, such as glass, permit radiant energy to pass through with very little loss of intensity. The majority of substances, however, absorb a portion of the radiation and reflect the remainder.

The ability of a particular surface to emit and absorb radiation is defined by the term emissivity which is the ratio of the amount of radiant energy emitted by a surface to that emitted by a black body.

Radiant heat transfer is governed by the equation:

$$q_r = 0.00368AFeFa((T_1/100)^4 - (T_2/100)^4) \tag{21}$$

where,

q_r = heat flow by radiation between two surfaces which are at absolute temperatures T_1 and T_2, watts,

A = surface area, in.^{2},

Fe = emissivity factor which accounts for the radiating characteristics of the heat transfer surfaces, i.e., the radiating surface with emissivity E_1 and the receiving surface with emissivity E_2, (dimensionless).

Fa = configuration factor which accounts for the portion of receiver surface that is in the direct line of sight of the emitter surface. In most electronic equipment applications, Fa = 1,

T_1 = absolute temperature of heat radiating surface, °K,

T_2 = absolute temperature of heat receiving surface, °K,

Table 6 defines the value of Fe for typical engineering situations.

Table 6. Values of Fe in Equation (21)

Radiating surface small compared to enclosing surface.	$Fe = E_1$	(22)

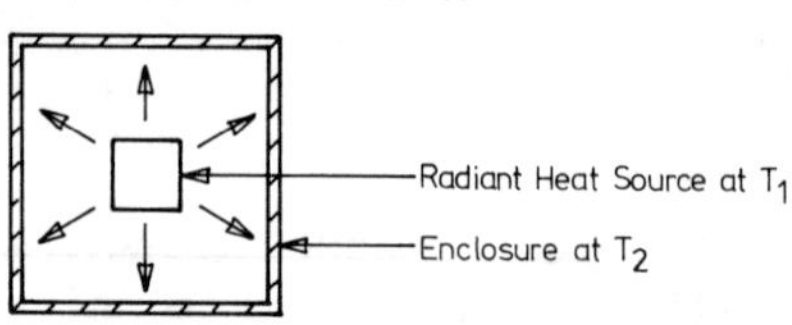

Table 6 (Cont'd)

Radiating surface large compared to receiving surface (separating space small).	$Fe = 1/(1/E_1 + 1/E_2 - 1)$	(23)

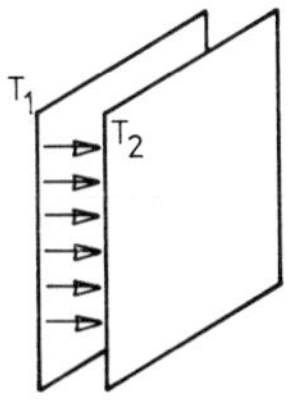

Plates Separated by Small Gap

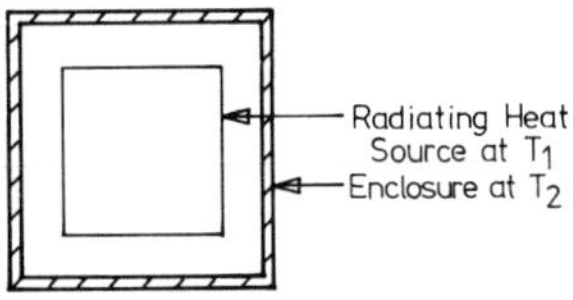

Enclosed Object with Small Separating Space

COOLING METHODS

This section presents the cooling methods most commonly employed: natural, forced air, cold plate and ARINC.

Natural Cooling

Natural cooling is the preferred method of controlling internal temperatures of electronic equipment. It is the simplest, most economical, and usually the most reliable method available to the equipment designer. Because of its complete dependance on natural convection and radiation, this method is limited to relatively low power levels, typically 0.13 watts per square inch of external surface.

The heat flow process for naturally cooled electronic equipment is characterized by two major thermal impedances, (a) the thermal impedance between the external surface and the surrounding air and, (b) the thermal impedance between the heat source and the external surface of the unit. These impedances give rise to two major temperature differentials, i.e.,

(a) from surrounding air to external surface,
(b) from external surface to internal parts.

(a) Temperature Rise, Surrounding Air to External Surface

The temperature rise of the external surface of an enclosure can be computed from equations (7) and (21) which express the heat flow due to natural convection and radiation from the external surface. The total heat flow is expressed by equation (24).

$$q_t = q_c + q_r \qquad (24)$$

$$q_t = 0.00394CA(\Delta T)^{1.25}/(L)^{0.25} + 0.00368AFeFa((T_2/100)^4 - (T_1/100)^4) \quad (25)$$

Example of External Surface Temperature Calculation

The following example illustrates the method of calculating external surface temperature for an enclosure having dimensions as shown in Figure 2. Ambient air temperature is 25°C. Power dissipation is 50 watts. External surface is a paint finish having emissivity of 0.9. The enclosure is located in a large room and is cooled by natural convection and radiation.

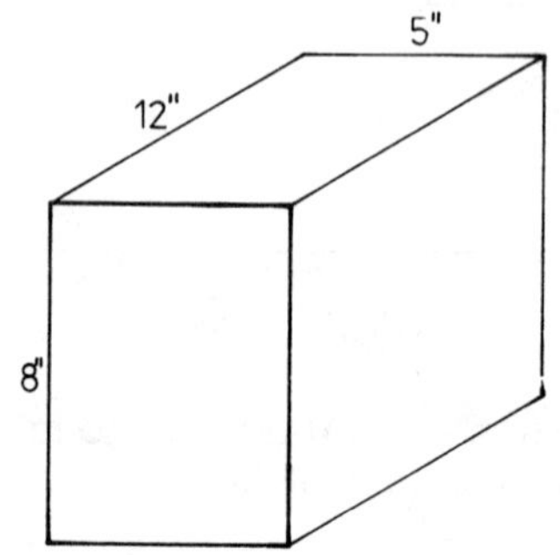

Figure 2. Enclosure Cooled by Radiation and Natural Convection

Heat Loss by Convection. Table 7 presents a tabulation of the known parameters of each surface of the box. The resulting equation (26) expresses the total convection heat loss in terms of temperature rise ΔT.

Table 7. Tabulated Values for Convection Heat Dissipation for Box Illustrated in Figure 2

Surface	A, in.^{2}	C	L, in.	$(L)^{.25}$	$\frac{0.00394\ CA}{(L)^{.25}}$
Top	60	0.71	$\frac{12 \times 5}{12 + 5} = 3.53$	1.37	0.122
Bottom	60	0.35	$\frac{12 \times 5}{12 + 5} = 3.53$	1.37	0.060
Sides (2)	192	0.55	Height = 8.0	1.68	0.247
Ends (2)	80	0.55	Height = 8.0	1.68	0.103
	392				0.532

$$q_c = (0.00394CA/(L)^{0.25})\ (\Delta T)^{1.25} = 0.532\ (\Delta T)^{1.25} \quad (26)$$

$$q_r = 0.00368FeFaA((T_2/100)^4 - (T_1/100)^4)$$

Table 7 (Cont'd)

$$q_r = 0.00368 \times 0.9 \times 1.0 \times 392\ (((273 + 25 + \Delta T)/100)^4 - ((273 + 25)/100)^4)$$

$$q_r = 1.3\ (((298 + \Delta T)/100)^4 - 78.86) \tag{27}$$

$$q_{total} = q_c + q_r$$

$$q_{total} = 0.532\ (\Delta T)^{1.25} + 1.3\ (((298 + \Delta T)/100)^4 - 78.86) \tag{28}$$

Heat Loss By Radiation. Radiant heat loss is derived from equation (21). Equation (27) expresses the radiant heat loss for the chosen example in terms of temperature rise ΔT.

Combined Heat Loss. The total heat loss, sum of convective and radiant heat losses, is expressed by equation (28). Figure 3 is a graphical representation of equation (28) which shows the temperature rise to be 18°C with a power dissipation of 50 watts. Approximately 20 watts are dissipated by convection and 30 watts by radiation.

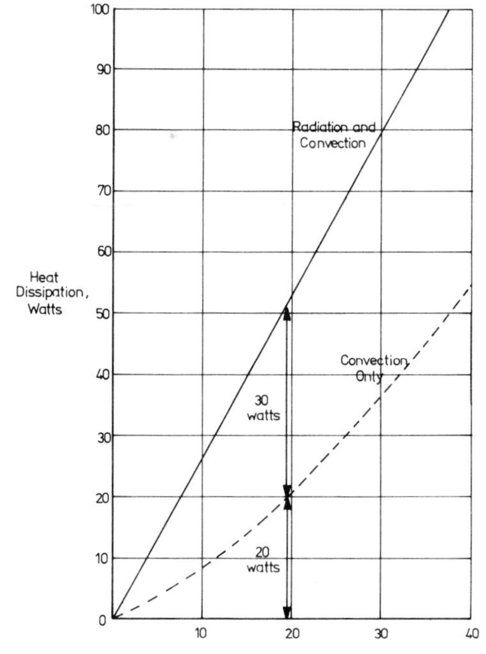

Figure 3. Graph of Heat Dissipation vs. Temperature Rise for Box of Figure 2

(b) Heat Dissipation from an Equal Sided Enclosure

The heat transfer properties of a naturally cooled enclosure having all sides of equal length and a painted exterior surface (emissivity of 0.9) is illustrated in Figure 4. The graph, which is derived from equation (25), shows the relationship between surface temperature and total heat dissipation for enclosure sizes between 2 in. and 25 in. The percentage heat loss by natural convection is noted also.

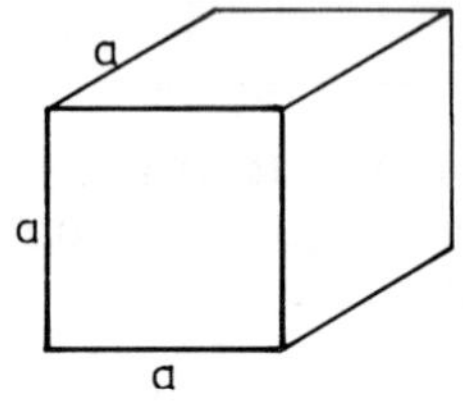

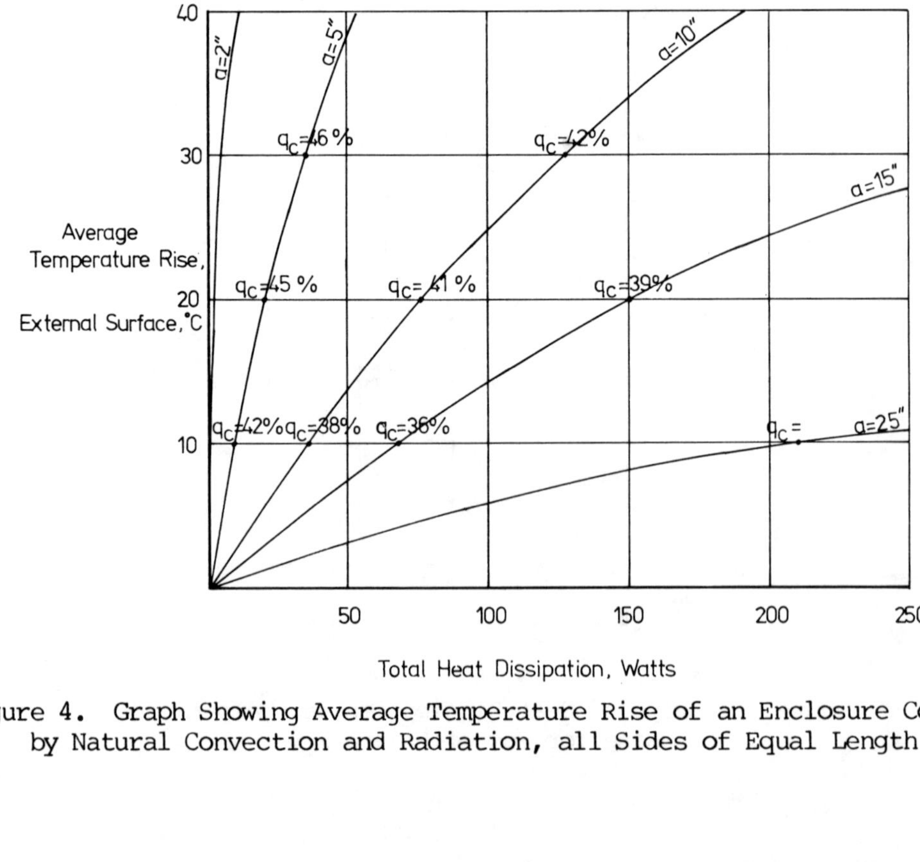

Figure 4. Graph Showing Average Temperature Rise of an Enclosure Cooled by Natural Convection and Radiation, all Sides of Equal Length

Example of Internal Temperature Rise Calculation

The internal temperature rise, i.e., the temperature differential between the wall of an enclosure and the electronic parts, is determined by the thermal impedance of the internal heat paths. Good thermal design practice employs a primary mode of heat transfer, either conduction or convection. However, all three heat transfer processes including radiation are almost always present to some degree. The calculation of internal temperature rise is illustrated by the following example.

A circuit card whose construction is illustrated in Figure 5, contains 42 dual-in-line microcircuits. It is desired to calculate the temperature at the center of the card when the wall temperature is 30°C and the total power consumption of the card is 5 watts. The card is situated in an enclosure which contains numerous other circuit cards, thus all of the heat flow is considered to take place in the plane of the card.

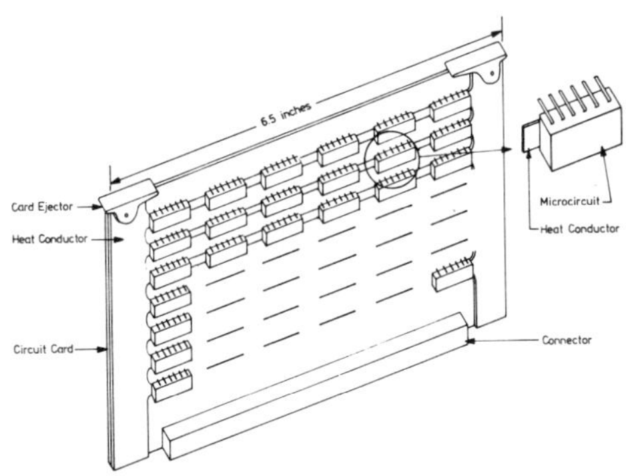

Figure 5. Circuit Card Containing Heat Conductor

The following analysis employs conduction as the primary heat transfer mechanism. Natural convection and radiation will also carry a portion of the heat load and produce an estimated 10% reduction in temperature rise.

Temperature drop, edge of card. The thermal impedance at the edge of the card is created by the interface between the heat conductor and the side wall. Light pressure at the sliding surface is maintained by a spring element. Experimental test data indicates that the thermal impedance at the interface, which contains a contact area 0.2 in. by 3.5 in., is 2°C/watt. Heat flow across interface is 5/2 = 2.5 watts and ΔT_1, temperature drop to the edge of card, 2.5X2 = 5°C.

Temperature drop, center to edge of heat conductor. The heat conductor may be considered as a flat plate which is uniformly heated by the microcircuit devices. The temperature rise is given by equation (5) of Table 4.

$$\Delta T_2 = qL/8AK$$

where,

$q =$ total heat input = 5 watts,

$L =$ length of card = 6.5 in.,

$A =$ cross section area of heat conductor,

$= 7 \times 0.15 \times 0.05 = 0.0525 \text{ in.}^2$,

$K =$ thermal conductivity of aluminum, type 6063,

$$= 4.88 \frac{\text{watt-in.}}{\text{in.}^2 - °\text{C}}$$

ΔT_2, temperature drop center to edge of heat conductor,

$$= \frac{qL}{8AK} = \frac{5 \times 6.5}{8 \times .0525 \times 4.88} = 15.8°\text{C}.$$

If heat conduction in the multilayer printed wiring board is also considered, then the effective area is the sum of the cross section areas of the heat conductor, printed wiring board and copper conductors. The effective thermal conductivity of the composite material is given by

$$AK = A_1K_1 + A_2K_2 + A_3K_3$$

where subscripts 1, 2 and 3 represent aluminum, epoxy glass laminate and copper conductors respectively . It is estimated that the copper conductors are equivalent to a 0.0015 inch thick sheet of copper over the entire heat transfer area.

$$AK = 0.0525 \times 4.88 + 3.5 \times 0.06 \times 0.008 + 3.5 \times 0.0015 \times 9.9$$

$$AK = 0.0017 + 0.256 + 0.052 = 0.31$$

$$\Delta T_2 = \frac{qL}{8AK} = \frac{5 \times 6.5}{8 \times 0.31} = 13°\text{C}$$

The heat conduction of the copper conductors has contributed significantly (17%) to the overall heat transfer process.

<u>Temperature drop, microcircuit case to heat conductor</u>. This temperature drop is calculated on the basis of conductive heat flow through a 0.002 inch gap separating the microcircuit case and heat conductor.

$$\Delta T_3 = \frac{qL}{AK}$$

where,

q = heat dissipation of microcircuit = $\frac{5}{42}$ = 0.12 watts

L = 0.002 in.

A = area of heat flow path = 0.77 x 0.15 = 0.115 in.^{2}

K = thermal conductivity of air = 6.7×10^{-4} = $\frac{\text{watt - in.}}{\text{in.}^2 - °\text{C}}$

$$\Delta T_3 = \frac{qL}{AK} = \frac{0.12 \times 0.002}{0.115 \times 6.7 \times 10^{-4}} = 3°\text{C}$$

Thermal Profile. The thermal profile over the 6.5 in. dimension of the card is illustrated by Figure 6. Maximum temperature occurs at the center and the temperature at intermediate points is defined by equation (6) which gives a parabolic profile.

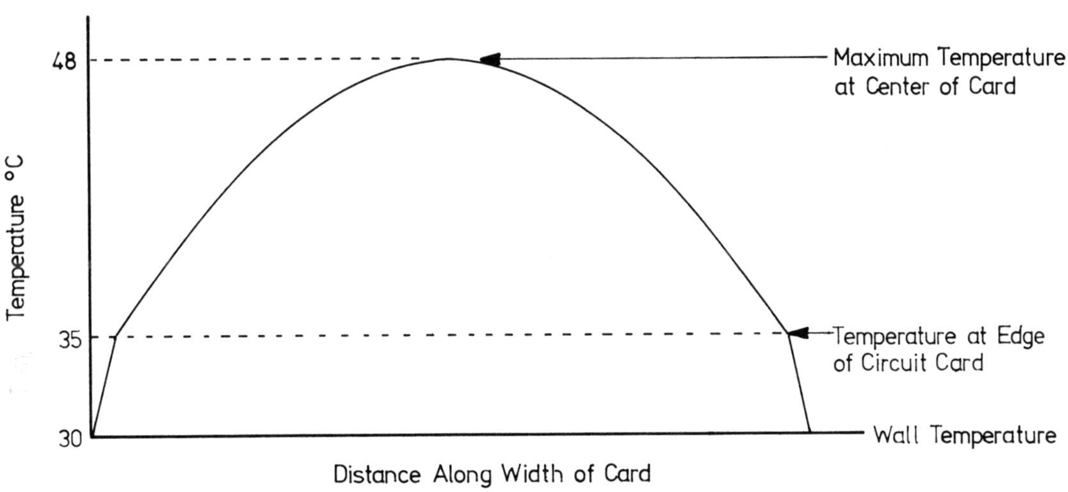

Figure 6. Thermal Profile of Circuit Card

Forced Air Cooling

Forced air cooling is the logical alternative when the power concentration of the equipment exceeds the capability of natural cooling, approximately 0.13 watts per square in. The advantage of forced air cooling is illustrated dramatically by Figure 7 in which the heat dissipating capability of a 6 in. x 6 in. vertical plate is shown with natural cooling (air velocity equals zero) and with a moving air stream.

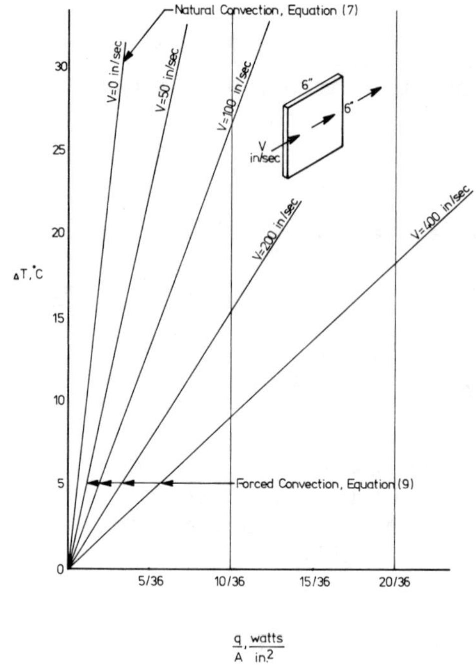

Figure 7. Graph Showing Heat Dissipation from 6 in. x 6 in. Vertical Plate with Natural and Forced Air Cooling

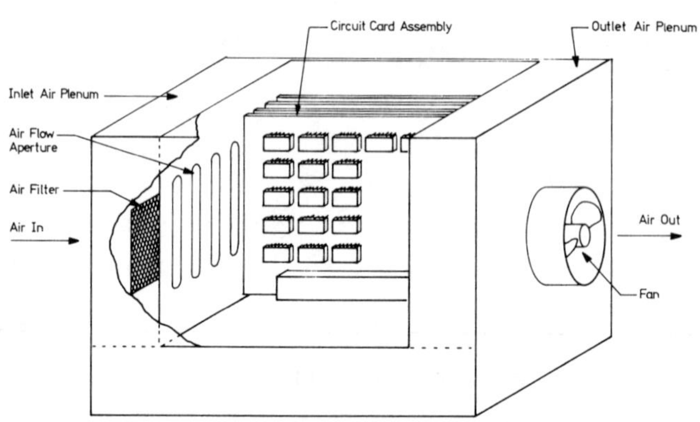

Figure 8. Forced Air Cooling System

Figure 8 illustrates the basic components of a forced air cooled system including an enclosure with inlet and exhaust ports, a fan and a heat source. Heat transfer takes place directly between the heat source and the air stream causing the temperature of the air to rise in accordance with the equation:

$$\Delta T = \frac{q}{mC_p} = T_2 - T_1 \qquad (29)$$

where,

ΔT = temperature rise of the air, °C,

Cp = specific heat of air, $\frac{\text{watt - sec.}}{\text{lb. - °C}}$

q = heat absorbed by air stream, watts,

m = mass flow of air, $\frac{\text{lb.}}{\text{sec.}}$

At normal atmospheric conditions Cp = 455 watt-sec./lb.-°C and the density of air is 0.075 pounds per cubic foot; one pound mass flow is equal to 13.3 cubic feet per second volumetric air flow. Equation (29) can be rewritten in the form of

$$\Delta T = 1.76\ q/CFM = T_2 - T_1 \tag{30}$$

where,

CFM = cubic feet per minute.

The temperature differential between the heat source and the moving air stream, from equation (8) is:

$$\Delta T = T_3 - T_{air} = \frac{q}{hA} \tag{31}$$

where,

ΔT = temperature differential, heat source to air stream,

q = heat produced by the source,

h = heat transfer coefficient $\frac{\text{watt}}{\text{in}^2 - °\text{C}}$,

A = surface area, in^2

Example of Heat Transfer Principles of Forced Air Cooling

An enclosure is cooled by a fan which delivers 20 cubic feet per minute when installed. The total heat load from the electronic parts plus fan is 115 watts and the average air velocity over the surface of the components is 100 inches per second. It is desired to find the average case temperature of a 14 pin dual-in-line (DIL) which is situated in the middle of the enclosure and whose heat dissipation is 250 milliwatts. The ambient air temperature is 25 °C.

The temperature rise of the air stream between inlet and exhaust is given by: $T_2 - T_1 = 1.76q/CFM = (1.76 \times 115)/20 = 10$ °C. At the middle of the enclosure the temperature rise is 10/2 = 5 °C. Air temperature surrounding the DIL is 25 + 5 = 30°C. The temperature differential between the moving air stream and the surface of the DIL could be calculated from equation (9). However the surface area, A, and the heat transfer coefficient, h, are not well defined quantities because of the intricate shape of the device. It is therefore considered more appropriate to employ heat transfer data supplied by the manufacturer of the device. In still air conditions, the thermal impedance, case to air, for a 14 pin ceramic dual-in-line microcircuit is 64°C per watt. The effect of the moving air stream may be estimated by reference to Figure 7, which indicates that the heat transfer capability of a flat surface is approximately tripled when the velocity of the surrounding air is increased from zero to 100 inches per second. Hence the estimated thermal impedance, case-to-air is 64/3 = 21°C/watt. Case temperature = 30 + 21 x 250/1000 = 35 °C.

Advantages and Limitations of Forced Air Cooling. The great advantage of forced air cooling is its superior heat dissipation capability. Compared with natural cooling, forced air cooling has several times the heat removal capacity and is also less influenced by the temperature of adjacent equipment. Its major disadvantage, however, is the contamination of internal parts by dust, oil fumes and cigarette smoke carried into the equipment by the air stream. Air filters are only a partial solution to the problem since only a percentage of the contamination can be removed. Lack of regular cleaning also can destroy the effectiveness of the cooling system. These considerations have led many equipment users, particularly in the military field, to turn to cold plate cooling.

Cold Plate Cooling

Cold plate cooling, illustrated in Figure 9, is characterized by finned air passages which carry forced air within the walls of the unit. This method has two important advantages over natural and forced air cooling methods. Firstly, it provides an efficient and controllable method of heat removal at the walls of the box. Secondly, it prevents airborne contaminants from entering the interior parts of the equipment.

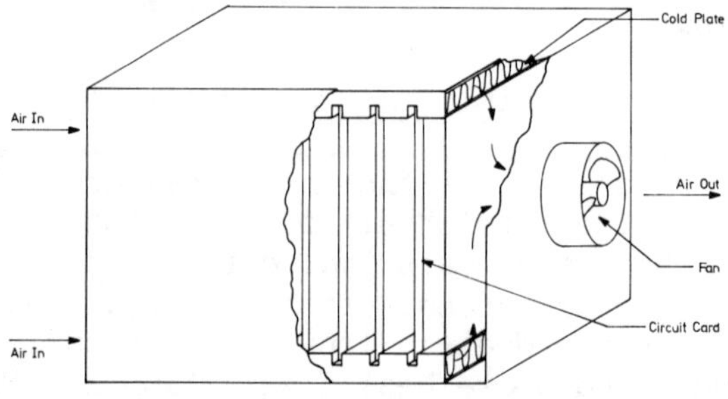

Figure 9. Cold Plate Cooling System

These advantages make cold plate cooling an attractive choice for operation in hostile environments such as those encountered in military applications.

Since cooling air is not permitted to enter the interior parts of the unit the heat transfer properties of the inner parts are controlled by conduction, natural convection and radiation. Conduction is usually the primary method of transferring heat from the electronic parts to the wall of the cold plate. Our previous example of heat transfer analysis inside a natural convection cooled box could apply equally well to a cold plate design. The similarity ends, however, at the wall of the box. Here, the cold plate design offers greatly improved heat transfer properties by virtue of the increased surface area of the fins and the moving air stream. Equation (13) is employed to compute the temperature drop in cold plates.

ARINC Cooling Method

The ARINC cooling system was devised by Aeronautical Radio Inc. (Airlines Electronic Engineering Committee) to fulfill a need for a standardized method of controlling the temperatures of electronic equipment on board commercial aircraft. ARINC specifications have a two fold purpose, i.e., (a) to indicate to prospective equipment manufactures the coordinated opinion of the airline industry and, (b) to channel new equipment designs in a direction which can result in the maximum possible standardization.

The system, which is illustrated in Figure 10, is controlled by ARINC Specification 404 and features a rack structure containing air plenums operating at negative air pressure. The standardized modular units contain an air exhaust aperture on the bottom surface which makes a sealed connection with a corresponding aperture in the rack plenum. Cooling is achieved by drawing air in a downward direction through the equipment and out of the exhaust aperture.

The ARINC cooling system allows considerable latitude in the internal design of the units. The use of cold plates is encouraged to prevent contamination of internal parts, however, the system also permits air to be drawn directly through the unit. Component temperatures will be governed by the effectiveness of the internal heat flow paths and may be computed using the heat transfer concepts discussed previously.

The rate of air flow for each unit of the ARINC System is set at 21.8 Kg. per hour per 100 watts of dissipated energy when the unit inlet air temperature is 38°C at sea level. This flow rate gives an air temperature rise of 16.5°C between inlet and outlet. Air pressure losses through the system must not exceed 1.0 inches of water.

UNITS AND CONVERSION OF UNITS

The large number of different units required to express thermal quantities is a great source of frustration to the heat transfer engineer. Although this chapter employs inch, pound, watt, second and degrees Celcius units, the metric system promises to provide much needed standardization. Table 8 provides a list of units used in this chapter and their corresponding metric form.

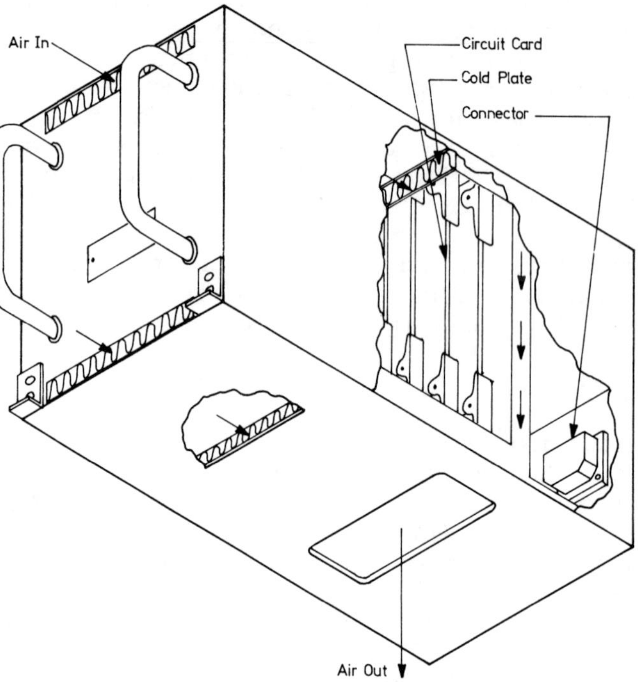

Figure 10. ARINC Cooling Method

Table 8. Definitions and Units

Symbol	Description	Units Used in this Chapter	Metric Unit
A	Area	$in.^2$	$cm.^2$
C	Constant	dimensionless	dimensionless
Cp	Specific Heat	$\frac{\text{watt-sec.}}{\text{pound-}^\circ\text{C}}$	$\frac{\text{watt-sec.}}{\text{kg.-}^\circ\text{C}}$
D	Diameter	in.	cm.
E	Emissivity	dimensionless	dimensionless
Fa	Configuration Factor	dimensionless	dimensionless
Fe	Emissivity Factor	dimensionless	dimensionless

Table 8 (Cont'd)

h	Heat transfer coefficient, convection	$\frac{\text{watt}}{\text{in.}^2 - {}^\circ\text{C}}$	$\frac{\text{watt}}{\text{cm.}^2 - {}^\circ\text{C}}$
k	Thermal conductivity	$\frac{\text{watt-in.}}{\text{in.}^2 - {}^\circ\text{C}}$	$\frac{\text{watt-cm.}}{\text{cm.}^2 - {}^\circ\text{C}}$
L	Distance	in.	cm.
m	Mass flow rate	$\frac{\text{pounds mass}}{\text{sec.}}$	$\frac{\text{kg.}}{\text{sec.}}$
q	Heat flow	watts	watts
R	Thermal resistance	$\frac{{}^\circ\text{C}}{\text{watt}}$	$\frac{{}^\circ\text{C}}{\text{watt}}$
Re	Reynolds number	dimensionless	dimensionless
T	Temperature	°C or °K	°C or °K
ΔT	Temperature difference	°C	°C
V	Velocity	$\frac{\text{in.}}{\text{sec.}}$	$\frac{\text{cm.}}{\text{sec.}}$
ρ	Density	$\frac{\text{pounds mass}}{\text{in.}^3}$	$\frac{\text{kg.}}{\text{cm.}^3}$
μ	Viscosity	$\frac{\text{pounds mass}}{\text{in.} - \text{sec.}}$	$\frac{\text{kg.}}{\text{cm.} - \text{sec.}}$

Table 9 provides a list of conversion factors which can be used to obtain equivalent metric units.

Table 9. Conversion of Units

Symbol	To convert	To	Multiply by
A	in.^2	cm.^2	6.4516

Table 9 (Cont'd)

C_p	$\frac{\text{watt-sec.}}{\text{pound-°C}}$	$\frac{\text{j}}{\text{kg. - °C}}$	2.2046
h	$\frac{\text{watt}}{\text{in.}^2 - \text{°C}}$	$\frac{\text{watt}}{\text{cm}^2 - \text{°C}}$	0.1550
k	$\frac{\text{watt-in.}}{\text{in.}^2 - \text{°C}}$	$\frac{\text{watt-cm.}}{\text{cm.}^2 - \text{°C}}$	0.3937
L	in.	cm.	2.54
m	$\frac{\text{pounds mass}}{\text{sec.}}$	$\frac{\text{kg.}}{\text{sec.}}$	0.4536
ρ	$\frac{\text{pounds mass}}{\text{in.}^3}$	$\frac{\text{kg.}}{\text{cm.}^3}$	0.02768
μ	$\frac{\text{pounds mass}}{\text{in.-sec.}}$	$\frac{\text{kg.}}{\text{cm.-sec.}}$	0.1786

CONCLUSION

The foregoing discussion has presented the thermal engineering aspects of electronic equipment reliability. The three modes of heat transfer, conduction, convection and radiation have been examined and formulae have been presented, which theoretically enable the thermal engineer to compute temperature levels at every point in the system. Unfortunately, mathematical models are never one hundred percent accurate, especially in complex assemblies where the heat transfer modes are almost impossible to analyze with a high degree of precision. This is where thermal testing plays an important role. Thermal mock-ups are a valuable tool in the thermal design of electronic equipment. They are used to verify the thermal performance and to make final adjustments which optimize the design. Reliable thermal design of electronic equipment is therefore a two stage process, theoretical analysis backed up by thermal testing.

REFERENCES

1. ____________, MIL-HDBK-217 Reliability Stress and Failure Rate Data for Electronic Equipment, U.S. Department of Defense, Washington, D.C.

2. ____________, MIL-M-28787 Module, Electronic, Standard Hardware Program, General Specification For, U.S. Department of Defense, Washington, D.C.

3. Reliability/Design Handbook, Thermal Applications, 4 Volumes, Thermal Technology Laboratory, Distributed by National Technical Information Service, U.S. Department of Commerce, 5285 Port Royal Road, Springfield, Virginia, 22161, NTI5 Code AD/A-009013, AD/A-009014, AD/A-009015, AD/A-009016, July, 1973.

4. Kraus A.D., Cooling Electronic Equipment, First edition, Prentice-Hall of Canada Ltd., Toronto, 1965.

BIBLIOGRAPHY

1. McAdams W.H., Heat Transmission, Third edition, McGraw-Hill Book Company Inc., Toronto, 1954.

10 ENVIRONMENTAL FACTORS

J. A. Roberts

It is accepted practice to include temperature and vibration as environmental factors when specifying the reliability of a system. In actual use, a system will, however, experience a wide range of additional hazards and stresses. It is therefore not surprising that although temperature and vibration are major causes of reduced Mean Time Between Failures (MTBF), the actual or "field" reliability may be considerably lower than that experienced in testing or established by calculation (1).

The situation as experienced by the U.S. Department of Defense (DOD) is shown in Figure 1. The wide gaps between the field, specified and demonstrated reliability, illustrate the extent of the problem. In another study which covered 95 subsystems, the mean field MTBF was found to be 1070 hours while the mean demonstrated MTBF was 16,377 hours. There are many reasons for these discrepancies. MIL-STD-721 defines reliablity as "the probability that an item will perform its intended function for a specified interval under stated conditions". Since reliability is defined in terms of probability, the lower bounds of the stated reliability can sometimes encompass the field reliability figure. Here, however, we shall be mainly concerned with the design of systems which achieve the required reliability operating under the usual 'stated conditions' together with other hazards and stresses. It is possible to envision and design for the wider range of environments found in field use. A key element which should encourage such considerations in the design phase, is the current contractual emphasis on stating reliability in terms of field reliability (2). Realism in testing is also a factor being addressed in new issues of MIL-STD-781 (2, 3). Clearly a new emphasis, which will lead to more stringent design considerations, is required and has been set in motion.

THE ENVIRONMENT

The causes of equipment failure in the field are difficult to gather and quantify. The chart shown in Figure 2 displays recently obtained data for naval and avionic equipments (4). This figure not only emphasizes the uncertainty of the available data but also provides a measure of the problem. Clearly, the effect of vibration and temperature are dominant in the area of identifiable failure causes. The large element assigned to 'design factors' and 'other causes' in the charts will be explored as well as those more common and identifiable causes shown.

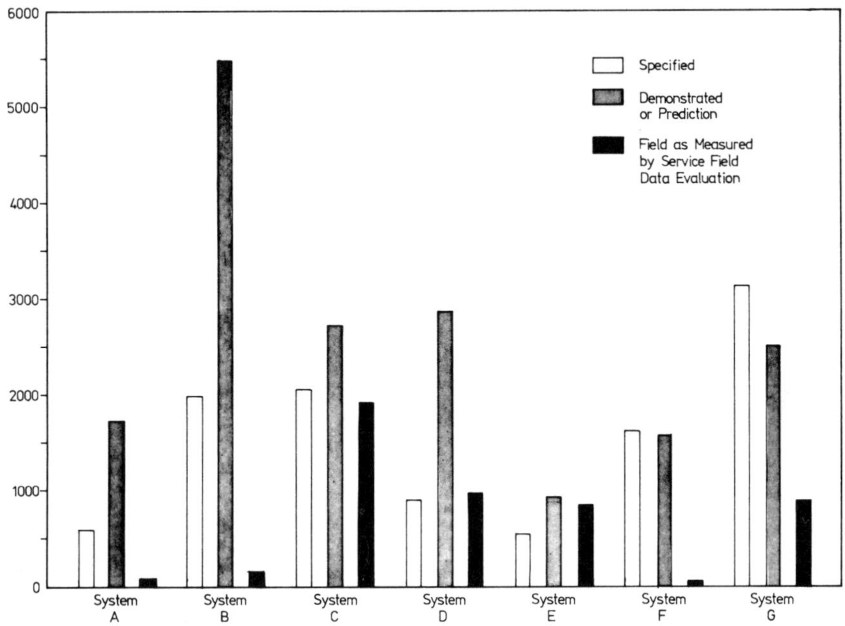

Figure 1. Selected Equipment Reliability Trend for Contemporary Repair Systems (1)

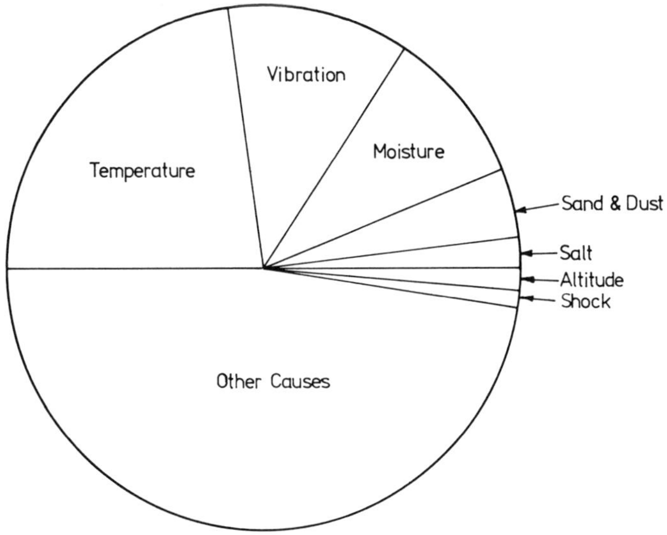

Figure 2. Avionic Equipment Failure Causes (4)

Herein, the position will be taken that all the environmental stresses and hazards can be identified and that through design sophistication, we can eliminate these as dominant factors causing failure. It was, for example, common practice to use leather, wood, fragile tubes and electromechanical relays extensively in equipment. Now, these have been eliminated, partially because of their poor environmental performance, and as a result equipment reliability has been greatly improved. The hazards of the environment are relatively easy to list. Table 1 identifies those common environmental stresses and hazards that are known to cause degradation or failure.

Table 1. Environmental Stresses and Hazards

Thermal	Mechanical	Contamination	Electrical	Nuclear	Other
Extremes	Vibration	Sand and Dust	EMI (1)	EMP (2)	Human
Cycling	Shock	Salt	Surge	Blast	Test Systems
Shock	Stress	Moisture	Static	Neutrons	Maintenance
Radiation	Penetration	Biological	Transients	Ionizing Radiation	Dormancy
	Acoustic Noise	Lubricants	Lightning		Packaging
	Vacuum	Cleaning Fluids	Corona		
	Pressure	Fuel			
		Humidity			
		Atmosphere			
		Fungus			

Note:

(1) EMI - Electromagnetic Interference
(2) EMP - Electromagnetic Pulse

THERMAL EFFECTS

The thermal element is particularly well documented, controllable and important. Chapter 9 of this book covers thermal design. The thermal aspect of failure has been quantified at the equipment level by Duhig and Weaver (4). This suggests that forced cooled equipment shows a 3.8% reduction in MTBF for each degree centrigrade rise in case temperature while free convection cooled equipment showed a 1.4% reduction. Clearly, simple temperature changes or careful thermal design can have valuable benefits (5). Less well understood and quantified are the effects of equipment switch-on at low temperature or of short term, high temperature operation and similar transient thermal problems. For example, it has become clear that the poor reliability of tantalum capacitors in high surge, low temperature, applications may lead to premature failures. In order to observe such failures at the development or test stage, more realistic and comprehensive testing is required. The Combined Environmental Reliability Test (CERT) program offers the possibility of achieving more comprehensive tests (6). These tests attempt to duplicate realistic mission profiles and are part of a general change toward testing that attempts to duplicate real stress.

MECHANICAL EFFECTS

Vibration and shock are regarded as major causes of equipment failure. A large body of test oriented literature has been published and an identifiable industry has been developed to produce the required test equipment. The military agencies have moved toward random vibration testing in order to achieve more realistic testing (3, 7, 8). Changes to MIL-STD-781 have not met with universal approval (3, 9), since there are many unknowns involved in the process of testing and in the evaluation of real environments. The difference between the actual peak limits, and those achieved in testing for one particular system is illustrated in Figure 3 (4). Such results typify the differences between expected and achieved reliability. Recently experiments have shown that gun mounted equipment is actually subject to 1000g shocks while equipment is subjected to a maximum of 100g shock during testing (7). The military specifications and standards are the sources of what tests are required and are being brought into line with actual experienced working limits, (although subject to error). The manufacturers of test equipment are the prime sources of material describing test methods used in practice to demonstrate compliance. While some progress is clearly being made, the prime problem of how to design equipment which can withstand the shock and vibration experienced during testing and in operation has been left to one side. Central to this problem of shock and vibration design is the fact that the equipment has to be designed while models (built for test purposes) are still undergoing tests. Consequently, a race develops between the model evaluation and the main design effort (10). In practice only a design which is conservative and correct the first time will likely meet the specifications. Therefore, design experience is the only basis for success. This is not to say that theoretical design aids do not help; in addition much useful, practical data has been published (11, 12, 13). There is, however, no way to accurately simulate the entire equipment so that a test/change style of development may be adopted.

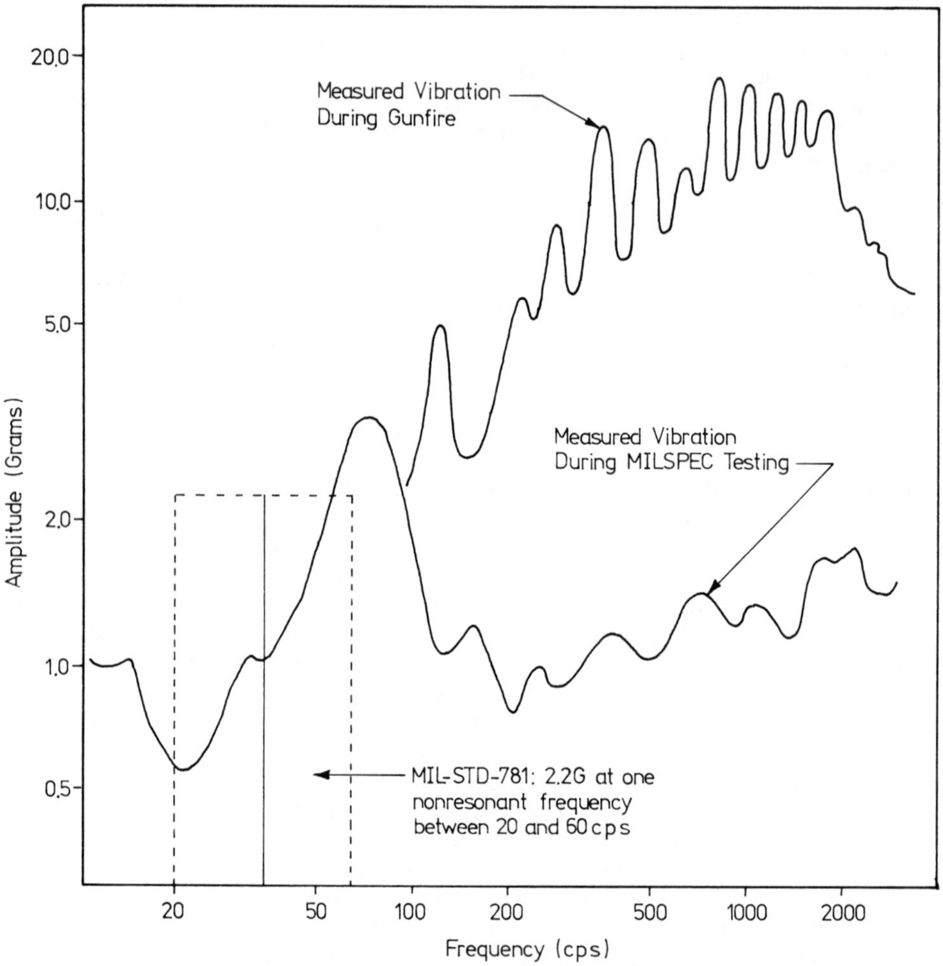

Figure 3. A-7D/E Vibration Levels: Forward-Looking Radar (4)

Avionic equipment is subject to special stresses typified by vacuum, noise and electrical corona requirements. Special care in packaging design and component evaluation is required to avoid problems in this area (14, 15). Recently a renewed emphasis on acoustic testing has been made, as a result of missile, hydrofoil and gun mounted equipment showing sensitivity to the acoustic stress inherent in such environments.

Penetration of equipment is assumed to be absent in test routines. However, in use, covers are often removed, cases are damaged and gun fire will damage equipment. Equipment that has been designed to suffer random short circuits, without damaging circuitry beyond the area directly affected, will be far easier to repair after such events.

CONTAMINATION

Sand and dust requirements have long been part of the environment specified for testing military equipment. Owing to the reduced use of components such as variable capacitors, unsealed switches and high impedance analog circuits the effect of such tests is fortunately reduced. However, in real situations, when humidity is combined with such contaminants, severe problems have reappeared (16). Correct packaging can form the basis of the solution to these problems. In military packaging it has become normal practice to use conduction cooling for individual components and printed circuit board and convection cooling for finned housings. Therefore air and the attendant dust does not impinge directly upon the electrical components (17). Commercial practice still uses filters and fans which direct air over the actual circuits being cooled (16). All is well until the filters become dirt laden. A moderately dirty filter can easily raise the temperature differential to over 30°F.

Where salt spray tests are specified, very severe corrosion can be expected in a short time. The actual circumstances will be very different in practice. Total immersion in seawater is sometimes the more realistic test. Salt-humidity protection and design techniques consist of paying close attention to covers, ventilation ducts, sealing, critical circuit isolation and the avoidance of dissimilar metal combinations. Protective coatings, such as carefully chosen (see MIL-I-46058) conformal coatings on circuit boards, have played a major role in reducing failures induced by dirt, moisture and the salt spray tests. Dessicants can also help but are rapidly overcome if exposure is extended.

Modern materials have eased the problems of protecting against biological and fungal contamination. The specifications called up in agency procurements still require tests to be performed. It is not likely that such tests will be dropped since recent experience has included the discovery of organisms that are able to live in the very hostile environment of aircraft fuel tanks. More probable sources of electronic system contamination include lubricants, fuels and cleaning materials. Recent procurement specifications have included lists of fuels and lubricants that are likely to become, at sometime or other, contaminants of nearby electronic equipment. Oils can have a dramatic effect on electrical contacts. While the contacts of sockets and plugs maintain a gas tight direct seal all is well. However, any maintenance activity may introduce oil between the contacts and reliability falls immediately. Of course relays and switches suffer from rapid wear when contaminated. Oils, fuels and cleaning materials can destroy seals used in capacitors and some cleaning fluids can totally destroy incompatible solid plastics within a few seconds of exposure (17, 18).

Atmospheric gases unhappily include ozone, sulphur dioxide and other severe contaminants (19, 20, 21, 22). Airborne particulates can produce complete open circuits in uncovered, low level relay contacts and considerable study of these effects has been made by the telephone industry. The gases form tarnish films which degrade contacts and degrade the strength of rubber elements. The solutions to these problems can be found in the use of sealed enclosures, active carbon filters and the use of solid state switching elements.

ELECTRICAL EFFECTS

The electrical environment is not benign. Equipment has to contend with lightning, static electricity, ground loops, primary power surges and radiated emissions. Considerable efforts are made, via the applicable Military Standards such as MIL-STD-461 and more specialized documents (e.g., QSTAG 307), to offer standardized tests. Even these standards cannot account for the complete range of environmental risks. Commercial equipment is often subject to special testing involving high energy spark generators, which are operated in close proximity to equipment under test (22). Such generators have been found, by empirical means, to offer risks equivalent to those of real situations. These tests relate to both transient upset problems in working systems and to static electricity induced failures. Some passive circuit elements such as thick film resistors and certain discrete semiconductor devices (notably small signal field effect transistors and VHF bipolar devices) are easily damaged by static electricity. Many integrated circuits (including all MOS and some Schottky bipolar devices) are also easily damaged. Linear circuits such as operational amplifiers with high input impedances and a MOS capacitor compensation element also show high sensitivity (24, 25). While it is economic to provide protection for inputs at risk at the circuit card level, clearly test points and input connector pins should have protective circuits similar to those shown for Electromagnetic Pulse (EMP) protection.

Lightning induces vast current and voltage surges into equipment directly affected. It is impossible to protect equipment which is connected directly to long exposed lines or antenna systems. If such equipment is poorly grounded, the damage can extend further into a system and destroy related equipment. Ground connectors are rarely sufficient to accommodate the kind of massive grounding straps that are required.

Equipment connected to commercial power lines also suffer surges and grounding problems (16, 22, 25). Power drops of 20%, outages of tens of milliseconds, voltage transients 2000% over nominal, lasting 50 µ sec., which happen 10,000 times per year, are experienced (26). Ground and primary power neutral differences of more than 50VAC are also commonplace. Local regulators and uninterruptible power supplies are the solution to transient faults, and the sometimes more difficult ground loop problems (16, 25). Care should be taken, however, to test such devices fully since some types produce truly extraordinary transients even though they smooth long term changes.

Many systems generate large amounts of internal noise and transients. Careful grounding and shielding practices (22, 26), such as single point grounds, ground planes and shielded lines, are vital parts of a reliable system. The selection of the correct logic type for each digital application (22) plays a vital role. Nothing is more convincing to design engineers than the noise difference between Emittor Coupled Logic (ECL) and Transistor-Transistor Logic (TTL) when operated at very high speeds. Where relay and similar electromagnetic devices are used great care and extensive transient suppression are required. Standard low level logic circuits are easily triggered by ground noise and radiated EM interference.

NUCLEAR EFFECTS

The reliability of military and civilian electronic equipment during and after a local nuclear explosion has become a major requirement. Although much of the data on the precise energy levels emitted is classified, and released only to consultants in this field, it is possible (based on the open literature), to provide guidelines sufficient to assess approximately the complexity and cost of modest hardening techniques (27).

We shall be concerned here mostly with conventional components, built into normal structures and intended for direct human operation. For equipment of this type, which might include, for example, communications systems, battery control or detection systems, it is possible to keep non-recurring engineering costs to less than 20% of the total engineering cost and to less than 3% of the recurring production cost.

A nuclear explosion presents several threats to electronic equipment, these include:

(a) neutron effects,
(b) ionizing radiation effect,
(c) electromagnetic pulse (EMP),
(d) soft X-ray effects,
(e) thermal radiation,
(f) blast,
(g) internal EMP (IEMP).

The hardening of electronic equipment must be balanced so that the appropriate level of attention is given to each of the above areas. While the location and function of the equipment determine the balance required it is generally found that EMP together with neutron and ionizing radiation are of prime significance. These will be considered first.

EMP (28, 29)

The nuclear burst produces an electromagnetic pulse which is dispersed according to the inverse square law for the energy density of the propagating field. The radiated fields cause charges to flow in distant conductors, such as cables, antennas, cases and interconnect wiring, in the same manner as those produced by a very high power communication radio transmission. The levels of maximum E fields and H fields can attain values of 10^5V/m and 10^2A/m respectively (27). Clearly large amounts of energy can be induced into systems components, quite beyond their normal design stress. Two general forms of EMP hardening have been considered, circumvention or filtration. Circumvention methods attempt to isolate and close down a system for the duration of the EMP (approximately 2 μ sec). Since these switching operations must occur within about 10 nsec from the first arrival of EMP energy, circumvention is exceptionally difficult. Accordingly most systems use filtration; this includes screening, clipping circuits, grounding balancing lines, cable routing, aperture control, control of loops. Since integrated circuits are damaged at energy levels as low as 1 to 10μJ, protection must be used at both the circuit and system level. Energy over the entire frequency range of 10 kHz to 400 MHz must be considered. Protection at the system level to a large extent should emphasize a "keep it out of the box" philosophy. This means, in practice using:

(a) double screened cables,
(b) closely twisted, balanced lines,
(c) exceptionally heavy and short grounding straps which have full regard for skin effect,
(d) input power lines as short as possible,
(e) EMI screen at meters and displays,
(f) EMI backshielded connectors,
(g) hardened filter optic lines.

In the case of aircraft or similar semicontinuous structures it is often economic to make a massive effort at this level to exclude entry of EMP. In this way the circuit level protection can be eliminated. In most systems, however, circuit level protection is required. The techniques include:

(a) screened modular construction, each screen or case offers some 40 DB of protection,
(b) combined tantalum and ceramic capacitor bypassing of each power line to each circuit board,
(c) short intermodule connections (having no loops) which are held firmly to a ground plane or the case,
(d) exceptional attention to ground loops, bonding technique, grounding point choice and skin effect,
(e) use of spark gaps at antenna, telephone line and unscreened power line entry points with associated screens to prevent re-radiation. Spark-gaps are slow but can absorb large energy pulses,
(f) use of semiconductor high-speed p-n junction transient suppression diodes after each spark gap,

(g) use of arrays of p-n junction diodes on all input logic and small signal input analog lines between modules. Usually the output structures are capable of absorbing the pulse,

(h) use of L-C filter pin or L-C section filters at incoming lines.

Figure 4 shows circuit forms which illustrate practical application of these methods.

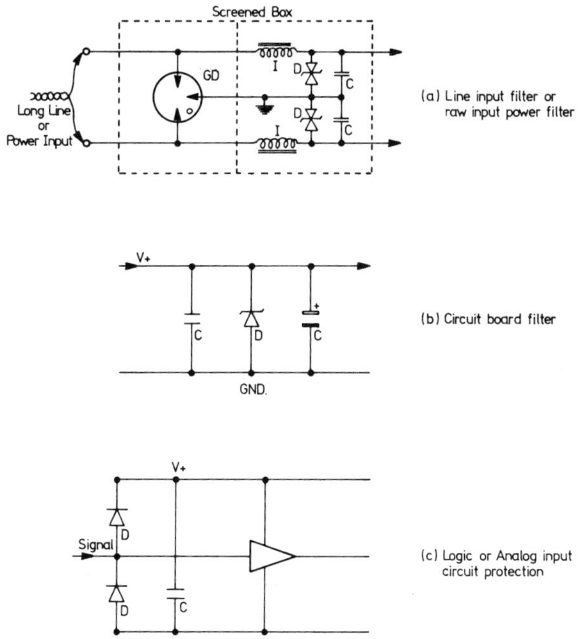

Figure 4. Basic EMP Protection

Neutron Effect (30, 31)

Fast neutrons (energies 10KeV) are of major importance in the degradation of semiconductor performance. These particles collide with the atoms of the lattice structure and produce crystal defects. In terms of macroscopic electrical effects on bipolar devices the degradation takes the form of current gain reductions, increased leakage current and reduced breakdown voltage. Neutron exposure is expressed in terms of neutron energy (MeV) and flux (neutrons/cm^2). It is usual to write in terms of neutrons/cm^2 with all values corrected to 1 Mev equivalent.

In order to design a reliable system which can operate after such exposure, account must be taken of the degradation of each semiconductor device. Bipolar devices are most sensitive to neutron radiation. However, at the levels we are concerned with here, it is possible to rely on a few simple relationships and rules.

The gain reduction can be obtained from:

$$\frac{1}{\beta_\phi} = \frac{1}{\beta_o} + \frac{0.16\phi}{F_T\,K} \qquad (1)$$

where,

β_o = initial gain,

β_ϕ = post neutron gain,

ϕ = exposure level (n/cm^2),

K = silicon damage constant of 1.6×10^6 n sec/cm ,

F = initial bandwidth product (MHz).

The relation indicates that most modern, high F_T (> 500MHz), small signal transistors and IC transistors have little trouble at levels up to 10^{13}n/cm^2 particularly when conservative design is employed. For low F_T power transistors and the pnp section of many silicon controlled rectifier SCR devices the problem becomes severe. Many hardening efforts revolve around a search for suitable high power devices.

Ionizing Radiation (31)

A large amount of gamma and hard X- radiation is produced by a nuclear explosion. This radiation produces a "photocurrent" in a p-n junction and apparently produces a change in the insulation oxide of MOS devices. The photocurrent can be sufficiently large to cause excess current flow beyond the device design limits and/or produce a latch-up condition. The threshold voltage of a MOS device can be affected for a long period (or permanently) after exposure. The unit of dose is the Rad, this is the radiation absorbed dose equivalent to 100 ergs per gram of material (silicon).

Damage can be divided into three areas for convenience of analysis:

(a) burnout failure,
(b) upset (transient) effects,
(c) accumulated dose failure.

The burnout occurs due to unrestricted current flow. When a bipolar transistor has a high value of base-emitter impedance, the base-collector photocurrent is amplified by the current gain to produce a large secondary current. In these circumstances systems can be extremely (unnecessarily) sensitive to ionizing radiations. A relationship which may be used to predict the photocurrent is:

$$I_p = \rho\dot{\gamma} \tag{2}$$

where,

I = primary photocurrent,

$\dot{\gamma}$ = dose rate (Rad/sec),

ρ = a constant for a particular device (mA/Rad/sec).

Typical small signal devices have ρ values of 1-2 x 10^{-9} mA/Rad/sec. With careful attention to current limiting series resistance, use of low values of base-emitter and gate-cathode resistors, it is possible to eliminate burn-out failure and upset (transient) false outputs. Careful examination of many IC equivalent circuits will reveal direct p-n junction paths between supply lines. Clearly, small series resistances are needed or a method of instantaneously closing down power supplies are needed. Latch-up is a problem for both CMOS and non gold doped, TTL. It is difficult to avoid the need to reset and protect such devices.

The threshold voltage change of MOS LSI devices present a serious problem. Each device requires a test prior to use in a hard system. Considerable difficulty has been experienced with recent N-channel LSI components which have poor design tolerances.

Heating Effects

For the purpose of this discussion the heating effect of the fireball and the soft X-rays will be considered together. The heating pulse can be of only one second or less duration but produces up to 50 calories per square cm. This can burn the insulation of all parts which are externally mounted but the effect on metal cases and internal electronics is minor.

Blast

Based on a 'typical' specification of 5 lbs. per square inch, a 400 square inch panel would exert a pressure of 2000 lb. Obviously, equipment needs protection against being struck by loose objects and requires careful mounting.

Internal EMP

The IEMP problem is only significant where long cables are enclosed in a metal container. Charged particles are ejected from the inside surface and produce fields which couple energy to internal conductors.

Most hardening efforts revolve around the selection of devices and protection circuits to provide cost effective and reliable equipment. Since this technique has not had wide application it is usual to provide specialist help to the design engineer. A large body of literature exists, however, partially due to the classified nature of the field and partially due to the rapid introduction of new logic families and LSI devices there are no single sources of data or expertise. This area presents a prime challenge to the reliability sciences.

OTHER FACTORS

Equipment is designed to be used. Human interaction is therefore an inevitable part of the hazard budget, to be allowed for during design. Unless equipment is designed to be convenient to use and maintain it will be subject to abuse. A design which takes account of the 'human factor' will inevitably show higher reliability even when it has no intrinsic superiority that can be quantified. Designs, which subject the user to annoyances and provide difficulties during maintenance, will not gain rapport and will suffer as a result. A body of literature and specification on human factors exists which should be made a source of reference in the early design phase and during design reviews.

The complexity of contemporary designs has dictated that special efforts are directed toward maintenance, handbooks and test equipment. Large systems now are produced with attendant special purpose test equipment. By this means a major unknown has been removed from the maintenance routines. Alternatively, large established test system complexes such as the US Navy's VAST system, provide known sequences and levels of stimulus during the test and maintenance, procedures. In the absence of such standardization large variations could exist between the desired and actual, test and maintenance procedures. It is quite possible for unstructured test procedures to damage equipment. It is rarely possible to ensure that a high level of understanding of the equipment is available at the test and maintenance level, particularly during the learning phase. More damage has been done to equipment than has ever been recorded. It is important, therefore, to standardize and simplify test and maintenance routines. Where conventional test equipment is used the procedures should be written and illustrated with the utmost care and clarity. A section of this book on handbook design (Chapter 26) describes the various forms currently required. Where the luxury of special purpose test equipment is possible many of the risks are removed.

Dormancy effects are rarely considered as separate factors from the environmental effect during operation. The environmental factors described can have quite different effects during long term dormancy, particularly if the equipment is totally unpowered. Obvious items such as batteries, electrolytic capacitors, switch and relay contacts, slip rings and desicants are major risk items. Less obvious is the tendency for such equipment to be used as a spare source for operating systems. For both hazards the solution is periodic checks and operations, together with careful storage and packaging. The latter element has become, quite properly, a major speciality and the subject of several military specifications. Commercial carriers also offer standardized packaging requirements. Equipment should be shipped experimentally over the likely distance and routes to be employed and then tested extensively.

CONCLUSION

In this chapter a broad discussion of environmental problems has been provided. Although for military systems the applicable standard specifications for environmental requirements and testing must be the starting point for the design, a great many factors are ignored in these specifications. The poor correspondence between test results and actual results is in large part due to what is ignored and what, indeed, cannot be quantified.

REFERENCES

1. Meth, M.A., "A DOD Approach to Establishing Weapon System Reliability Requirements", Defence Management Journal, April 1976, p2-11.

2. Parker, R.N., "R and D Emphasis on Reliability", op. cit., p19.

3. Klass, P.J., "Reliability Test Procedure Changes Set", Aviation Week and Space Technology, 19 April 1976, p62-63.

4. Marsh, R.T., "Avionics Equipment Reliability, An Elusive Object", Defence Management Journal, April 1976, p24-29.

5. Duhig, J.J. and Weaver, T.E., "Effects of Temperature on Avionics Reliability", Proceedings of 1977 Annual Reliability and Maintainability Symposium, IEEE, New York, N.Y., p409-413.

6. Burkhard, A.H., "Combined Environmental Reliability Test", Proceedings of 1977 Annual Reliability and Maintainability Symposium, IEEE, New York, N.Y., p460-466.

7. Sontz, C. and Wallace, W.E., "MIL-STD-781", Proceedings 1977 Annual Reliability and Maintainability Symposium, IEEE, New York, N.Y., p448-453.

8. Meeder, E.A., "Random-Vibration Analysis", Machine Design, 10 September 1964, p179-184.

9. Stovall, F.A., "Is MIL-STD-781B a Good Reliability Test Specification? Part II, IEEE Transactions on Reliability, Vol. R-26, No. 2, June 1977, p85-87.

10. Sumerlin, W.T., "High Reliability Design Techniques", Agard, Lecture Series No. 81, Technical Editing and Reproduction Ltd., Hartford House, 7-9 Charlotte St., London, 1976, p3-1, 3-4.

11. Steinberg, D.S., "Taking the Shake out of Circuit Boards" Machine Design, 22 September, 1977, p96-99.

12. Tustin, W., "A Practical Primer on Vibration Testing", Evaulation Engineering, November/December 1969, p21-23/53-54.

13. Harris, C.M., and Crede, C.E. (editors), Shock and Vibration Handbook, 2nd ed., McGraw-Hill, New York N.Y., 1975.

14. Anderson, R.T., Reliability Design Handbook, IIT Research Institute Chicago, Ill., 1975.

15. Markstein, H.W., "Packaging for the Military Environment", Electronic Packaging and Production, February 1977, p36-40.

16. Buckley, J.E., "System Environmental Factors", Computer Design, September 1977, p18-22.

17. Ratelle, W.J., "Seal Selection Beyond Standard Practice", Machine Design, 20 January 1977, p133-137.

18. Agnew, J., "Choose Cleaning Solvents Carefully", Electronic Design, 1 March 1974, p54-57.

19. Nagel, W.B., "Designing with Rubber, Part 4, some Real-life Applications", Machine Design, 11 August 1977, p101-106.

20. Russel, C.A., "How Environmental Pollutants Diminish Contact Reliability", Insulation Circuits, September 1976, p43-46.

21. Peel, M.E., "Environmental Evaluation of Contact Finishes", Insulation Circuits, April 1975, p55-63.

22. Kimble, G., "Hot Time in the Old Factory", Digital Design, February 1977, p36-42.

23. ____________, "Helpful Tips on Coping with Static Electricity", Evaluation Engineering, July/August 1977 ,p42-43, (see also September/ October 1977).

24. Trigonis, A.C., "Electrostatic Discharge in Microcircuits", Proceedings 1976 Reliability and Maintainability Symposium, IEEE, New York, N.Y., p162-169.

25. ____________, "Protect your Machine Control with Voltage Regulations", Product Engineering, November 1974, p 24-25.

26. Jones, J.P., The Causes and Cures of Noise in Digital Systems, Computer Design Publishing Corp., 1964.

27. Rickets, L.W., Fundamentals of Nuclear Hardening of Electronic Equipment, John Wiley and Sons, New York, N.Y.

28. Melville Clark, O., EMP Transient Suppression, General Semiconductor Industries Inc., 1974.

29. ______________, "EMP Protective Systems", TR-61-B, Department of Defense/Office of Civil Defense, Washington, D.C., November 1971.

30. Messenger, G.C., "Hardness Assurance Considerations for the Neutron Environment", IEEE Conference on Nuclear and Space Radiation Effects, July 1975.

31. Cassidy, E.T., "Nuclear Radiation Testing of Diodes, Silicon Controlled Rectifiers, Transistors and Integrated Circuits", Technical Report 4692, Picatinny Arsenal, December 1974.

11 FAULT TREE ANALYSIS

S. D. G. Williams

Fault Tree Analysis (FTA) is a systems engineering technique which provides an organized, illustrative approach to the identification of high risk areas. This technique has evolved a look of its own due to the nature of its development and flexibility in various fields of application. Generally speaking, FTA is applied to systems where safety and/or operational failure modes are of concern. This is not a restriction, however, for its effectiveness stems from providing the analyst with a medium to focus on how things can fail to work. It is not suggested that FTA replace other forms of system analysis but instead be utilized on conjunction with inductive techniques.

FTA does not resolve complex design problems, pinpoint hazardous situations or reveal overstressed parts but provides the analyst with a qualitative system evaluation procedure enabling detection of such things as system failure modes, potential safety hazards, and subsystems with high failure rates.

From managements' point of view the full impact of any reliability analysis is best felt at the projects' front end. So too with FTA. To achieve this, a FTA should be prepared for the preliminary system design review (PDR) and once again during the critical design review (CDR). This allows design changes resulting from the analysis to be incorporated in a cost effective manner prior to the equipment going into service, as shown in Figure 1.

Basically a fault-tree (FT) is an event logic diagram, providing a logical representation of events occurring within a comlex system. Construction of a fault tree usually begins with the definition of the top undesired event (the system failure). The causes are then indicated and connected to the top event by conventional logic gates. The procedure is repeated for each of the causes and the causes of the causes, etc., until all the events have ben considered. Figure 2 illustrates the construction of a simple FT.

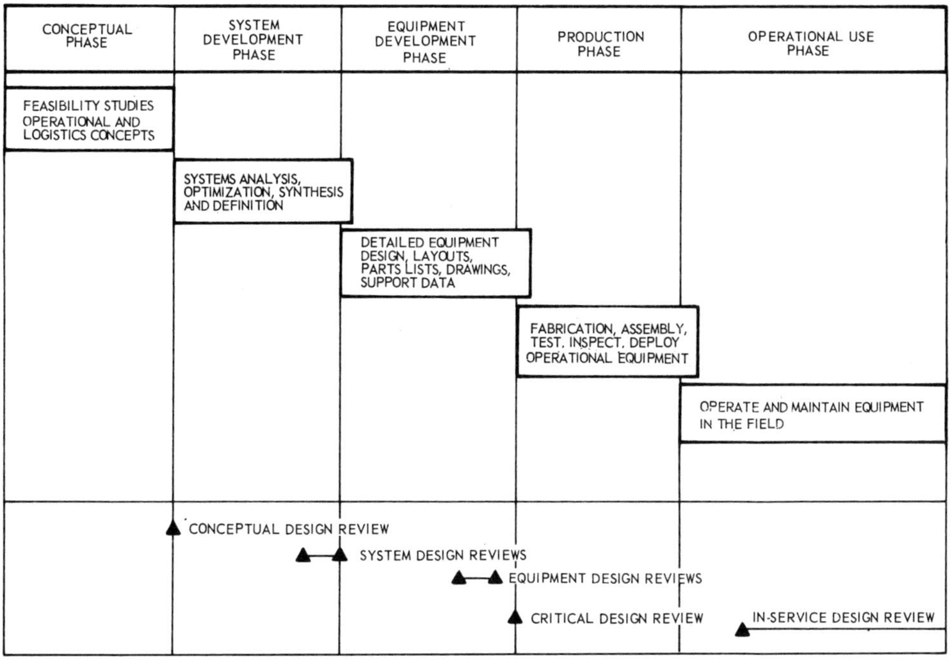

Figure 1. Product Development Chart with Scheduled FTA Inputs

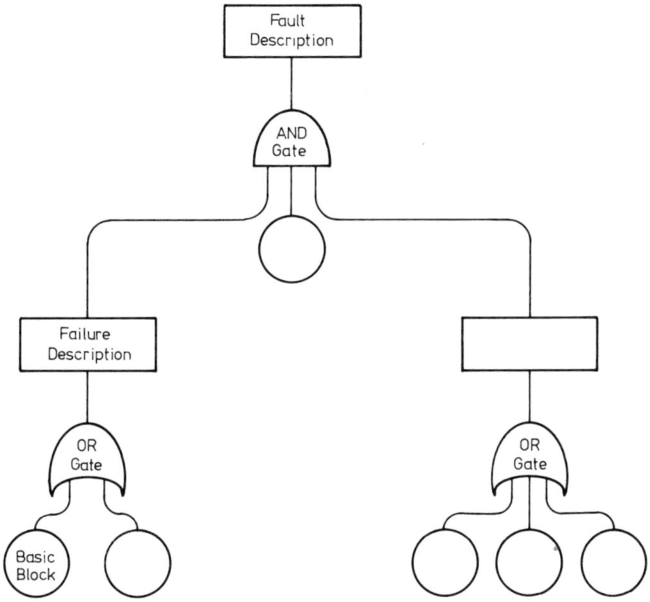

Figure 2. A Simple Fault Tree

The strength of event logic diagramming lies in its straight forward application of simple familiar logic building blocks, most of which have been standardized. Because fault tree construction utilizes a Top-Down approach the system failure becomes the top event with the tree "branches" beneath it, as illustrated in Figure 3. Advantages of the Top-Down technique have been displayed in software engineering, where modules are integrated to form a software program. This permits more than one analyst to work on the same problem.

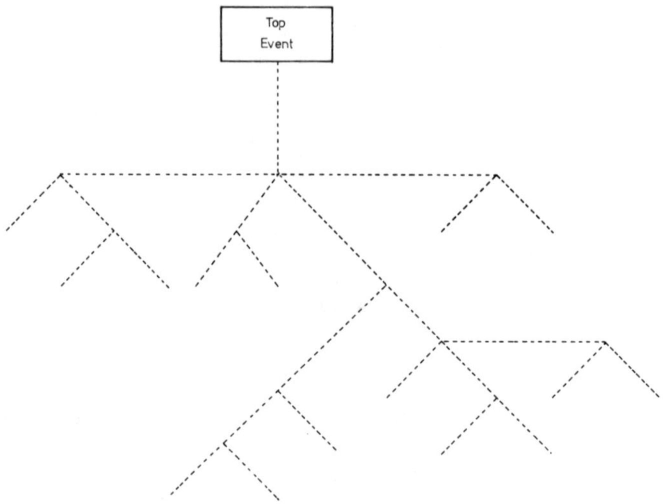

Figure 3. Illustration of the Top Down Approach

The event logic diagram is completely developed only when the causes have been shown for all events except basic input events (independent component failures). Most FTs are compounded using interconnectivity among parts, assemblies, units and finally systems. Usually the subsequent event logic diagram is large and has many basic blocks requiring a computer program to produce anumerical analysis. However, the analyst has options for qualitative or quantitative analysis which lead to, (a) analyzing system failure rates, (b) justifying system changes, (c) performing trade-off studies, (d) analyzing common mode failures, (e) illustrating compliance with safety requirements.

FTA lends itself to systematic consideration of both explicit and implicit functional requirements, illuminates design 'blind spots' and provides an excellent system debugging tool. Designer's tend not to probe beyond the obvious failure modes of their creations because they spend most of their time making the design work. It must be stressed that FTA should begin while hardware concepts materialize; waiting for engineering schematics to stabilize in terms of changes could result in untimely comments. FTA is also useful in the following: (a) quality control; extending analysis down to low levels (resistors, I.C. packages) and complemented with a critical design review, helps to illuminate areas for special attention (i.e., material requirements not ordinarily appreciated), (b) human factor considerations; FTA identifies critical man-machine interfaces, (c) reliability prediction; fault tree format is ideal for quantitative mathematical modelling of a system, simplifying reliability and availability predictions.

Later in this chapter we will discuss FTA and the computer with specific mention of some specialized techniques. This will illustrate the power of FTA in reliability analysis.

FAULT TREE CONSTRUCTION

General

As part of its scope, FTA could include administrative and managerial applications, as well as hardware/product applications discussed here. The steps involved in the construction of any type of FT are shown below in Table 1.

Table 1. Steps In FT Construction

1	Understand the system to be evaluated.
2	Define the undesired event (systems reliability context)/ critical hazard (systems safety context)/loss of an account (commercial business world)/etc.
3	Analyze the system to determine the logical interrelationships of higher and lower functional events which may cause a predefined system fault condition.
4	Apply logical relationships to input fault events which are defined in terms of (a) basic, (b) independent and (c) identifiable faults that may be assigned probability values.
5	Using FT symbols connect this information.
6	Reduce the FT if possible.
7	Eliminate any feedback paths.
8	Check to ensure all FT rules have been followed.

Fault Tree Symbology

The symbols represented in Figure 4 are the fundamental building blocks used in FTA. Although they are not the only blocks used, for most applications they are the only ones that need to be understood. These symbols are defined as follows:

(a) The output of an "OR gate" is an intermediate event which is formed by taking the union of the events which are input to the gate,

(b) The output of an "exclusive OR gate" is an intermediate event which represents the statement that there is no output unless one and only one of the input events is present,

(c) The output of an "AND gate" is an intermediate event which is formed by taking the intersection of the events which are input to the gate,

Note: For the AND gate, the OR gate, or the exclusive OR gate with one input, the output event is set equal to the input event.

(d) The output of an "inhibit gate" is an intermediate event which is identical to the (single) input event if an additional enabling condition is present. This condition is represented by a conditioning event which is drawn to the right of the inhibit symbol. The inhibit gate is logically equivalent to an AND gate which has two input events but is defined separately to allow the plotting of a different symbol,

(e) A primary event represents an event which (for one reason or another) is not developed further in a particular fault tree. The conditioning event and the external event represent the existence of a given condition. The basic event represents the failure of a component of the system which cannot be modelled as due to a combination of other failures. (i.e., the lowest element in the hierarchy),

(f) A description symbol is plotted above each event (primary or intermediate) and encloses a prose description of the event,

(g) The transfer symbols are used to show connections with parts of the tree for which it is not feasible to draw in the connections directly.

Each logic gate relates a set of input conditions to an output event. The input events are considered to be causes of the output event, which become an input to an even higher output event until eventually the top of the tree is reached. An input event which is not the output of any gate is termed a "basic" input event.

Any system failure can be defined in terms of an undesirable event. However, before anticipating all credible causes for an undesirable event, there must be a precise definition of the system under scrutiny in terms of functionality. This can be accomplished with a functional block diagram which is ordinarily part of any system reliability analysis. Essentially the analyst will trace the functions which must occur to "reach" the top event and then further divide these into the respective basic events. This identifies the prerequisites for the undesirable event.

Thus, part of the analyst's task is to produce a functional block diagram at the hardware level, where each block is further reduced to the desired level. It is imperative to perform this task for every conceivable functional requirement of the hardware within the system under consideration. In an analogy to civil engineering, the problem may be stated as follows. A hardware item will be subjected to various stresses in performing some particular function, and it has some characteristic strengths which enable it to resist those stresses. Failure occurs when the stress exceeds the strength. Of particular importance in FTA is the analysis of the failure effects. Providing, of course, that there is failure, one must ask: did the failing hardware item have sufficient built-in reliability, and did the surrounding system produce the excessive stresses? This, in turn, will lead to a focus on other hardware items, or environmental factors, which were responsible for the high stress.

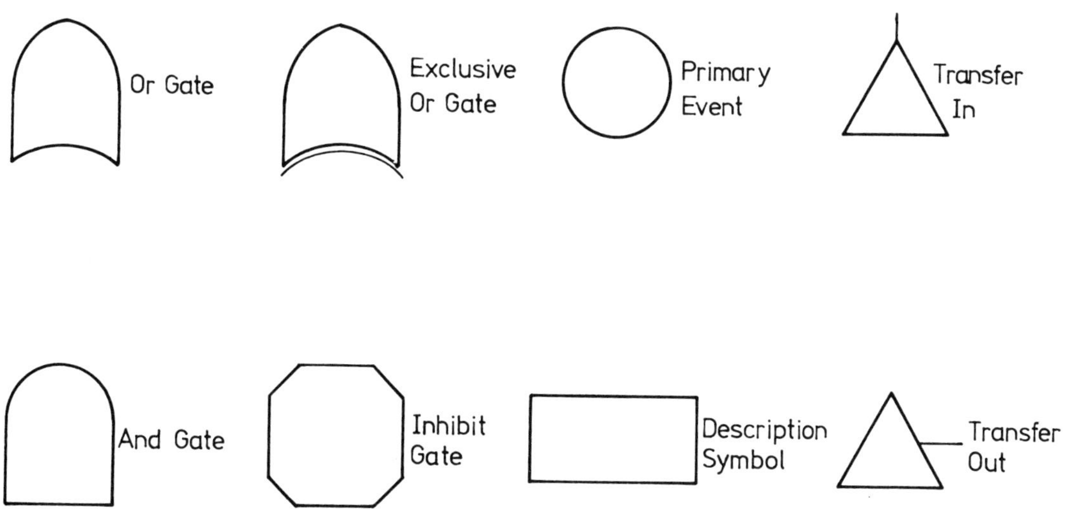

Figure 4. Fundamental FT Symbols

Top-Down/Bottom Up

The 'top-down' approach describes a specific method of FT development. In reference to FTA it identifies the more serious high level failure modes quickly, without waiting for them to be discovered from a laborious failure modes approach which starts at the lower levels (a 'bottom-up' based technique). The top events could have been selected based on one of the following, (a) a 'good decision' based on experience, (b) chemical and physical property data used to define the system hazards, (c) an understanding of the input/ output requirements, if analyzing a sub-system. Predetermined mini-trees illustrating how the hazard (not necessarily the top event) occurs is the events' Boolean description. The more complete the list of top level functional requirements the better the top-down analysis proceeds. Having designated the failure modes approach as a bottom-up technique no further comparison between top-down versus bottom-up need be made.

Mathematics of FTA

Before discussing the quantitative aspects of FTA, a very important subject warrants emphasis. All basic input events for a FT must be independent for reliability calculations. Although it is understood that there is more than one way to analyze a fault tree it is nevertheless true to say that dependencies create inaccuracies. Unfortunately the FT is not always completely free of dependencies, although it does yield a nonredundant form, and special care must be taken during Boolean reduction and substitution techniques. Providing all probabilities are small, significant error does not creep in. Difficulty arises, however, when devising an algorithm to implement on the computer. Thus, if one calculates the probability of an event by considering its immediate inputs independent, then the answer is incorrect if some input appears elsewhere in the FT. Unfortunately, this wastes effort, money, and time for there is no way to correlate the true value to that calculated. Thus the FT must either be drawn so that a basic event appears only once, or a method must be devised that eliminates dependencies during calculations (without introducing errors), i.e., the property of prime numbers. The analyst usually can eliminate dependencies by insuring that a basic event appears only once but it is more difficult to test if a particular intersection is unique and is not a subset of another intersection of events. This concept is illustrated in Figure 5.

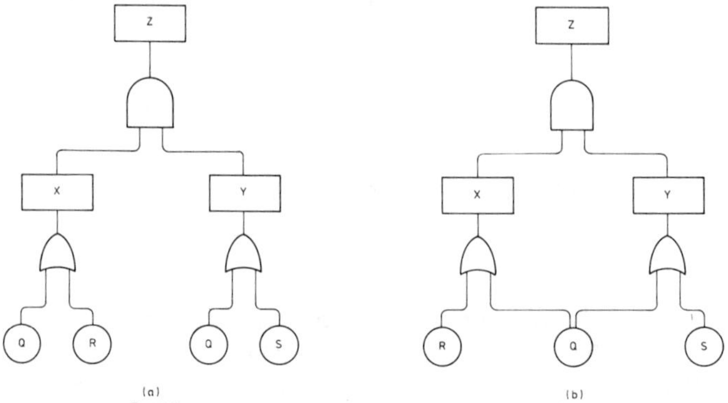

Figure 5. Example Illustrating Dependencies

The events X and Y in both Example 1 and Example 2 of Figure 5 are not independent since Q could cause both. Boolean reduction is applied as follows:

$$Z = X\ Y \tag{1}$$

$$X = Q + R \tag{2}$$

$$Y = Q + S \tag{3}$$

and therefore,

$$Z = (Q + R)(Q + S) \tag{4}$$

which can be simplified to,

$$Z = Q\ Q + Q\ S + Q\ R + R\ S \tag{5}$$

Since,

$$Q\ Q = Q \tag{6}$$

$$Q + Q = Q \tag{7}$$

$$Q + Q\ S = Q \tag{8}$$

then,

$$Z = Q + R\ S \tag{9}$$

Hence, the approximate probability is

$$P_Z \overset{\Delta}{=} P_Q + P_R\ P_S \tag{10}$$

which is a different answer than that obtained if the dependencies were ignored.

Many technqiues have been tried for fault tree probability calculations but most have been discarded because of their inefficiency when analyzing complex systems. Others have too many underlying assumptions making them prohibitive. In the following, time will not enter into the calculations, so that occurrence of the top event depends solely upon which basic event occurs and not when it occurs.

Efficiency of FTA Formats

It would seem that in order to efficiently analyze a fault tree after its construction, two ingredients must exist; (a) standardization of the fault tree art, (b) shortcuts to correct analysis. Generally it is considered that a great deal of time is spent having analysts come to grips with the problem at hand. These analysts must interface with designers and operators and be exposed to the physics and economics of the system. G.J. Powers stated that, "One needs to give the safety analysis group the time to analyze the system. It might be a long time. A recent United States Atomic Energy Commission study of part of the emergency safety systems for a nuclear reactor required over 25 man-years" (1).

Models are needed to describe failure modes at all system levels, i.e., the component, assembly and unit level. The rules for FTA must be rigidly adhered to because, unfortunately, a mistake at a lower FT level propagates to the top event, corrupting the calculations throughout. It seems that the predominant objective of FTA is to simplify system analysis, yet the difficulties encountered in doing so are equally troublesome. The short-comings are cumulative due to the top down structure, so that over-sight, omissions, inapplicable failure data, and poor assumptions can be fatal to the results and hard to detect. The FT can be a very simple or very complex analytical reliability tool depending on (a) the analyst, and (b) the system to be modelled. Either or both can determine the degree of sophistication. Even development of a simple FT in first-time application is costly and time consuming. However, the cost is somewhat offset by future application of the models in system modifications, maintenance scheduling and accident prevention.

The FTA must not be started at too low a level. This could lead to frustration and possible omission, especially if the system is large. Once you have decided on the level, be consistent. Often the analyst will jump ahead two or more logical steps in a chain process and overlook important contributing factors. Ways to avoid this problem are to (a) have two analysts work on the same system independently (b) have one analyst try to produce two fault trees using two different vantage points. This could certainly be done by first applying the top-down techniques and then the bottom-up.

Often a good idea for the novice is to document his procedure, similar to providing a plan before composing an essay. This can be an aid in constructing a fault tree which will be legible to people familiar with the system.

FTA-FMEA

Failure Modes and Effects Analysis (FMEA) is also a design evaluation procedure. It involves examination of the mechanism(s) responsible for particular failure modes and estimation of their probability of occurrence. FMEA determines the consequence of each failure mode's occurrence on the higher events in the system. A detailed FMEA does anticipate possible failures but is very restrictive due to its bulky format, especially for a large system. It has generally been agreed that to identify and quantify potential failure modes economically, a detailed FMA/FMEA should not be considered viable. They simply do not allow simultaneous visibility of contributing failure modes - implying a loss of any deductive reasoning required to anticipate failure mechanisms. The rapid growth of technology has placed restrictions on the FMEA, especially in the development of new materials and their possibly unknown failure modes.

FTA is not necessarily a replacement for other forms of system reliability analysis, but its popularity may be due to it being systematic. An undesired event is selected and its causes are identified - hence the inherent top-down structure. This approach identifies hardware and non-hardware causes, such as human error, revealing specific failure mode combinations. The top-down approach provides an economical means of identifying crucial failure modes at the component level and feasible component/subsystem/module interface failure modes. Table 2 shows a comparison of the techniques.

Table 2. FTA-FMEA Comparison

Characteristic	FTA	FMEA
The best technique for multiple failure description.	X	
The best technique for single failures.		X
Technique which avoids analysis of non-critical failures.	X	
Technique best suited to identify higher level events liable to failure due to lower level events.	X	
Technique which identifies external influences easily.	X	
Technique with broader sphere in observation of failure mode phenomena.		X
Technique with fewer restrictions on intensity of analysis.	X	

FTA AND THE COMPUTER

Overall system reliability is dependent on system structure and on reliabilities of individual components. After the system fault tree has been drawn, failure probabilities are assigned to the basic input events/components/blocks from available data. Probability theory can be used to calculate the failure probability of the system. The 'OR' gates are replaced by 'UNION' operators and the 'AND' connectives by 'INTERSECTION' operators as shown in Figure 6.

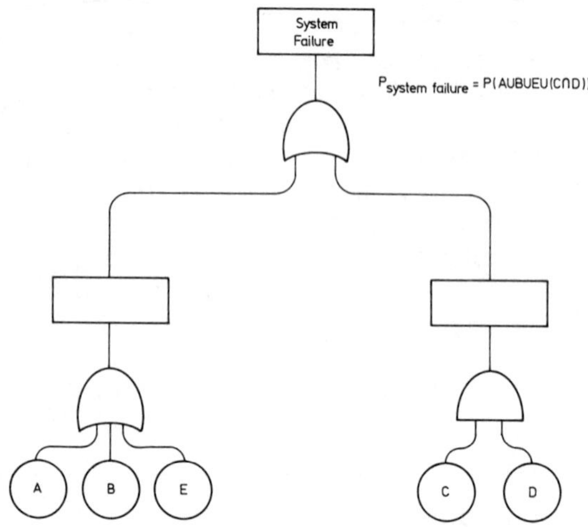

Figure 6. Boolean Reduction Applied to a Simple Fault Tree

Boolean reduction is now applied. From the theory for independent non-mutually exclusive events it can be shown that:

$$P\ (\text{system failure}) = P(A \cup B \cup E \cup (C \cap D)) \quad (11)$$

$$P\ (1 \cup 2 \cup 3 \cup 4) = P(1) + P(2) + P(3) + P(4)$$
$$- [P(1)\ P(2) + P(1)\ P(3) + P(1)\ P(4) + P(2)\ P(3) + P(2)\ P(4) + P(3)\ P(4)]$$
$$+ P(1)\ P(2)\ P(3) + P(2)\ P(3)\ P(4) + P(3)\ P(4)\ P(1) + P(4)\ P(1)\ P(2) \quad (12)$$
$$- P(1)\ P(2)\ P(3)\ P(4)$$

and

$$P(5 \cap 6) = P(5)\ P(6) \quad (13)$$

Substituting equations (12) and (13) into equation (11) yields:

$$P(A \cup B \cup E \cup (C \cap D) = P(A) + P(B) + P(E) + P(C)\ P(D)$$
$$- [P(A)\ P(B) + P(A)\ P(E) + P(A)\ P(C)\ P(D) + P(B)\ P(E) + P(B)\ P(C)\ P(D) + P(E)\ P(C)\ P(D)]$$
$$+ P(A)\ P(B)\ P(E) + P(B)\ P(E)\ P(C)\ P(D) + P(E)\ P(C)\ P(D)\ P(A) + P(C)\ P(D)\ P(A)\ P(B)$$
$$- P(A)\ P(B)\ P(E)\ P(C)\ P(D) \quad (14)$$

This equation, which represents an extremely simple FT, is long and awkward, with its solution prone to error. When the FT is complicated, the accompanying equations become complicated as well. We only wish to allocate our time to the careful development of the fault tree and not to the computation of the numerical results, which can be calculated by computer far more efficiently and accurately.

Event Space Method

There are alternative ways to calculate the combinatorial probabilities of a fault tree. It is up to the individual to select the best one for his purpose. One method (although very inefficient with respect to CPU time) is the Event Space Method, which relies on the definition of a probabilistic universe. This so called universe is the union of all 'mutually exclusive' event states of a system. Each system element has two states; it has either failed or it has not. Hence a system with n components will possess 2 discrete mutually exclusive states. These 2 discrete events are separated into 2 groups; one group containing favorable events and the other containing those that are not. Consider the following:

$$Y = X\ T + S \tag{15}$$

where,

$n = 3$

Thus $2^3 = 8$ discrete mutually exclusive states are possible. Only some of these contribute to the success of Y. Table 3 lists these states, favorable plus unfavorable.

Table 3. Event Universe

Probability Of Occurrence	Favorable	Y	X	T	S
$P(\bar{X}\ \bar{T}\ S)$	no	0	0	0	0
	yes	1	0	0	1
	no	0	0	1	0
$P(\bar{X}\ T\ S)$	yes	1	0	1	1
	no	0	1	0	0
$P(X\ \bar{T}\ S)$	yes	1	1	0	1
$P(X\ T\ \bar{S})$	yes	1	1	1	0
$P(X\ T\ S)$	yes	1	1	1	1

As illustrated above the probability of Y occurring is,

$$P(Y) = P(\bar{X} \cap \bar{T} \cap S) + P(\bar{X} \cap T \cap S) + P(X \cap \bar{T} \cap S) + P(X \cap T \cap \bar{S}) + P(X \cap T \cap S) \quad (16)$$

This technique does not rely strictly on a Boolean equation. What is needed is a means to determine the favorability of each of the mutually exclusive events in the event universe. The fault tree contains this means in its structure and so the event universe is defined by assigning the fault tree basic blocks to a bit in a binary number: a 1 indicates a failed block, a 0 indicates a functioning block. The n bit binary number will produce 2^n mutually exclusive discrete states. We are now able to test the favorability of each of these events by examining the output state for each of the 2^n input states. The failure probability is then given by the sum of the probabilities of occurrence of the events, favorable to the failure of the system. The Event Space Method is extremely accurate but requires an excessive amount of computer time.

Thus it appears that the major problem in the quantitative analysis of fault trees has been the inability to compute event probabilities economically where the trees are complex and the components reliable. "Attempts to utilize digital computer speed have included approximate solutions using Monte Carlo simulation techniques. However, for reasonable accuracy with complex and highly reliable systems such as those used in nuclear power plants, computer run times are excessive" (2).

Attempts at 'sealed simulation' have not been very successful and newer techniques are slow coming. Faster programs have built-in checks against dependencies (such that a basic event can appear only once in a particular intersection, and an intersection of basic events is unique and not a subset of another intersection of events) by utilizing the property of prime numbers.

Monte Carlo Techniques

There is more than one fault tree simulation program developed, at this time, to describe systems and provide quantitative performance results. The development criteria is to closely model the operating modes of the system under discussion. The Monte Carlo technique has the ability to include considerations which would be very difficult to include in analytical calculations. Written in Fortran IV it can assume whatever input basic event, failure distribution, and repair times the programmer wishes. The program views the system represented by the fault tree as a statistical assembly of independent basic input events. The output is a randomly calculated time-to- failure (TTF) for each basic block, based on the assigned mean-time-between-failure (MTBF). The system is tested as each basic input event fails, to detect system failure within the mission time. A time-to-repair (TTR) is predicted, based on the mean-time-to-repair (MTTR) values with detection times and a new TTF value assigned to each failed basic input event to permit failure after repair, as shown in Figure 7.

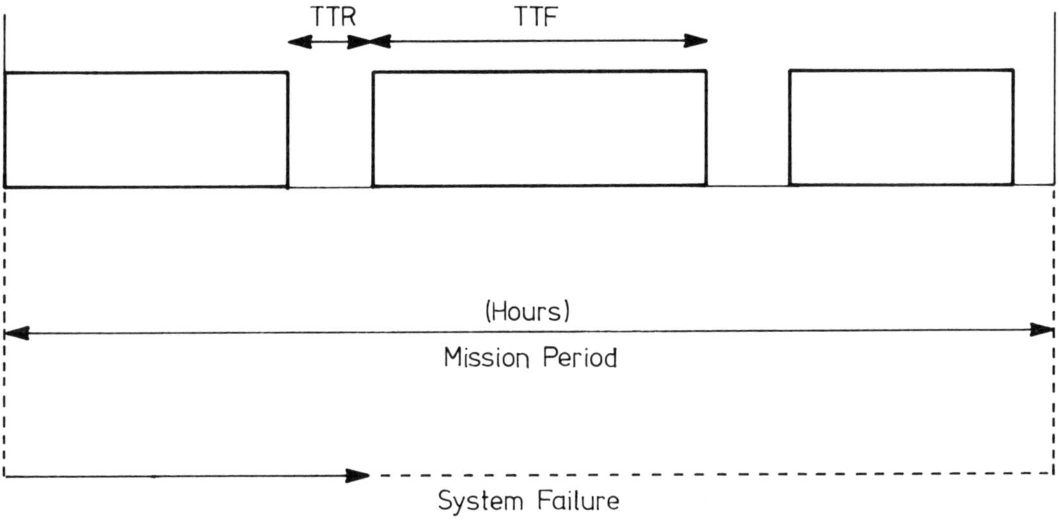

Figure 7. Effect of TTR and TTF on Mission Performance

This process continues until the mission period is reached or the system fails. A new set of randomly selected values are assigned to the basic blocks and the program is rerun. After a significant number of such trials the user obtains:

(a) system probability of failure,
(b) probability of success,
(c) subsystem/component contributions to system failure identification,
(d) subsystem failures are recorded for performance comparisons.

Typical program assumptions are:

(a) mission period is not terminated for a trial when a secondary failure occurs (i.e., subsystem failures),
(b) any basic input failure which results in subsystem failure is not used to reduce system availability and is not included in the mean downtime,
(c) approximately 3,000 trials are required to obtain consistent results for a system probability of failure estimate of 10 and approximately 30,000 trials for a probability estimate of 10.

Some useful outputs which can be obtained are:

(a) probability of system failure/success,
(b) total number of system failures,

(c) average failure mission time,
(d) system availability,
(e) mean and total downtime,
(f) sequential listing of basic input failures which cause system failure for each first mission period system failure,
(g) weighting of all basic block failure paths and of all logical gates which cause system failure directly,
(h) failure path weighting of basic blocks which are in a failed state when system failure occurs,
(j) rank plus listing of basic input failure and availability performance results, including optional weighting or cost effectiveness information if desired,
(k) number of times repair is attempted and found completed for each basic input event,
(m) number of times each basic input and logical gate cause a shutdown.

The Monte Carlo (importance sampling) program outputs a set of equations for critical paths illustrating the items that influence the critical path in time for failure and phase of failure. In accomplishing the foregoing program outputs, Monte Carlo simulates the tree and, using the input data, randomly selects rate data from assigned statistical distribution parameters. Many trials are run until the desired quantitative resolution is achieved. The variety of output obtained yields a very detailed perspective of the system being simulated. One great advantage of Monte Carlo is the technical and engineering alternatives which could be implemented to produce a sensitivity analysis, leading to selection and scheduling of system improvements.

Direct Simulation

Around 1944 Vlam, Fermi, and von Neumann used the Monte Carlo technique to perform neutron-diffusion calculations. Shortly after this,fault tree simulation programs were using straight-forward Monte Carlo sampling techniques called direct simulation. Here events were generated with real frequencies equal to their "true" occurrence frequencies. This type of simulation meets the requirement that the most probable combinations happen most frequently. The importantance of the sampling technique is that it reduces CPU run time as it generates events in a manner which increases the frequency with which the various event combinations occur.

EXAMPLE

General

In this section, an example of a fault tree based on real hardware is given in Figure 8. It is important to explain the steps taken to arrive at the final tree. These are outlined in the following paragraphs.

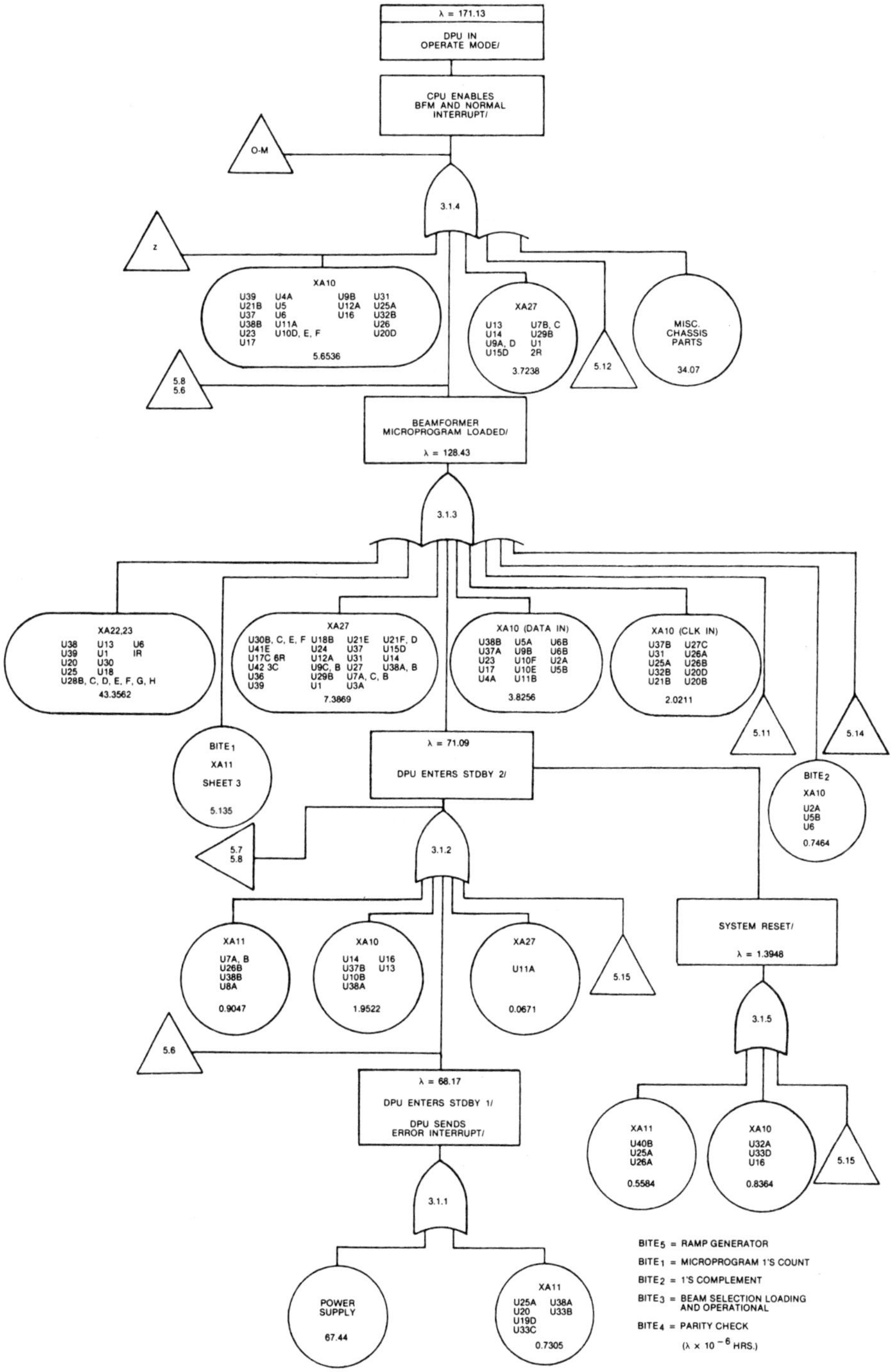

Figure 8. Fault Tree Example

Reference (3) provides a good example of a FT constructed for an analog system. A digital system is used here.

FT Construction

The first step taken was to examine the Procurement Specification (PS), in this case for a special purpose computer unit, in order to identify the various performance modes specified in the PS. Several were identified, such as unit initialization, data processing and data input-output protocol, for example. For each performance mode identified, a FT was prepared and for the purposes of this discussion the FT constructed for the initialization mode is used.

The second step is to identify the hardware and the boundary signals involved in performing the mode. This is accomplished by examining schematics for the various assemblies (circuit cards) and the associated wiring lists. It is now possible to begin the construction of the FT. Only a small portion of the FT construction will be described in detail. The same method applies throughout the entire FT.

Consider the intermediate event, System Reset, shown in the lower right of Figure 8. In the course of analysis it was found that a hardware failure, on either of two assemblies XA11, XA10, or the failure on input signal 5.15, will cause a non-occurrence of the intermediate event. This is shown by OR gate 3.1.5, fed by two primary events on the two assemblies and the input signal 5.15. The primary events indicate the affected parts, e.g., U40B, U25A, etc., and the failure rate for the parts is totalled and shown for each primary event. These in turn are added to give the failure rate for the intermediate event, System Reset. This procedure is followed until the FT is completed up to the top event.

FT Processing

Having assembled the basic data into a FT the next step is to input the data to a FTA program and run it with requests for the output listings of interest.

CONCLUSION

FTA has been recognized as a necessity for thorough system reliability analysis. With the dramatic advances seen in technology today it has become imperative to utilize an accurate and efficient failure analysis tool. It would seem a viable solution for the dilemma posed by complex systems, whether avionic, nuclear, automotive, and many others. Because of its processing capacity, the computer alleviates the routine and formidable task of calculations. However, it is impossible for this technique of system representation to guarantee that all problems will be solved before they occur. However, as with other reliability predictions, it is impossible for this technique of system representation to guarantee that all problems will be solved before they occur.

REFERENCES

1. Power, G.J., "Fault Trees - A State of the Art Discussion", IEEE Transactions on Reliability, Vol. R-23, No. 1, April 1974, p. 51.

2. Semanderes, S.N., "ELRAFT, A Computer Program for the Efficient Logic Reduction Analysis of Fault Trees", Transactions of the American Nuclear Society, (USA), Vol. 13, No. 2, 1970, p78.

3. Human, C.L., "The Graphical FMEA", Proceedings 1975 Annual Reliability and Maintainability Symposium, IEEE Cat. No. 75CHO918-3RQC, IEEE, New York, January 1975, p298-303.

BIBLIOGRAPHY

1. Aggarwal, K.K., "Comments on the Analysis of Fault Trees", IEEE Transactions on Reliability, Vol. R-25, No. 2, June 1976, p126.

2. Baldonado, O.C. and Wee Pierson, J., "Fortran vs GPSS in Reliability Simulations", Proceedings 1973 Annual Reliability and Maintainability Symposium, Vol. 6, No. 1, p550.

3. Eisner, R.L., Fault Tree Analysis to Anticipate Potential Failure, ASME Publication, F2-DE-22, p1.

4. Fussell, J.B., Powers, G.J. and Bennetts, R.G., "Fault Trees - A State of the Art Discussion", IEEE Transactions on Reliability, Vol. R-23, No. 1, April 1974, p51.

5. Schroder, R.J., "Fault Trees for Reliability Analysis", Proceedings of the 1970 Annual Reliability and Maintainability Symposium, Feb 1970, p198.

6. Semanderes, S.N., "ElRAFT, A Computer Program for the Efficient Logic Reduction Analysis of Fault Trees", Transactions of the American Nuclear Society, (USA), Vol. 13, No. 2, 1970, p781.

7. Shooman, M.L., Probabalistic Reliability: an Engineering Approach, McGraw - Hill Book Company, New York, N.Y., 1968, p418.

8. Nieuwhof, G.W.E., "An Introduction to Fault Tree Analysis with Emphasis on Failure Rate Evaluation", Proceedings 1975 Canadian Reliability Symposium, Vol. 14, No. 2, April 1975, p105.

9. Kanda, K., "Computerization of System/Product Fault Trees", AIAA/SAE/ASME 8th Conference on Assurance Engineering and Sciences, 1969, p325.

10. Eagle, K.H., "Fault Tree and Reliability Analysis Comparison", Proceedings 1969 Annual Symposium on Reliability, Vol. 2, No. 1, p12.

11. Jordan, W.E., "Failure Modes, Effects and Criticality Analysis", Proceedings 1972 Annual Symposium on Reliability, Vol. 5, No. 1, p30.

12 SNEAK CIRCUIT ANALYSIS

R. C. Clardy

An overview of the development and methodology of a circuit analysis technique called Sneak Circuit Analysis will be presented in this chapter. The technique is based on the discovery that topological criteria exist which can be used to recognize unplanned operational modes of a circuit. The step by step methodology of performing a Sneak Circuit Analysis will be discussed and the results obtained on a variety of complex systems will be presented as verification of the technique's effectiveness. The development and application of Digital Logic Sneak Circuit Analysis and Software Sneak Analysis, similar techniques applicable to digital and software systems respectively, will also be discussed.

A sneak circuit is a latent path or condition in an electrical system which inhibits a desired condition or initiates an unintended or unwanted action. This condition is not caused by component failures, but has been inadvertently designed into the electrical system. Sneak circuits often exist because subsystem designers lack the overall system visibility required to electrically interface all subsystems properly. When design modifications are implemented, sneak circuits frequently occur because changes are rarely submitted to the rigorous testing that the original design undergoes. Some sneak circuits are evidenced as "glitches" or spurious operational modes and can be manifested in mature, thoroughly tested systems after long use. Sometimes sneaks are the real cause of problems thought to be the result of electromagnetic interference or grounding "bugs".

The unpredictable nature of sneak circuit conditions prompted the National Aeronautics and Space Administration (NASA) to fund an investigation to determine a method of identifying sneak circuit conditions before their occurrence could pose any possible threat to the safety of Apollo, Skylab or Apollo-Soyuz Test Project crew members. In November of 1967, the investigation was initiated with a detailed review of historical incidents of sneak circuit conditions in various electrical systems. A sneak circuit was defined for this study as " . . . a designed-in signal or current path which causes an unwanted function to occur or which inhibits a wanted function". The definition excluded component failures and electrostatic, electromagnetic, or leakage paths as causative factors. The definition also excluded improper system performance because of marginal parametric factors or slightly out-of-tolerance conditions.

The 1967 historical incident investigation resulted in two significant findings, i.e.:

(a) sneak circuits are universal in complex electrical systems and their analogs; and,
(b) topological criteria exist which enable recognition of all planned or unplanned operational modes within a system.

A computer-aided analysis technique was developed based on the 1967 findings which established the analytical goals illustrated in Figure 1. The following paragraphs will discuss each of these tasks in greater detail.

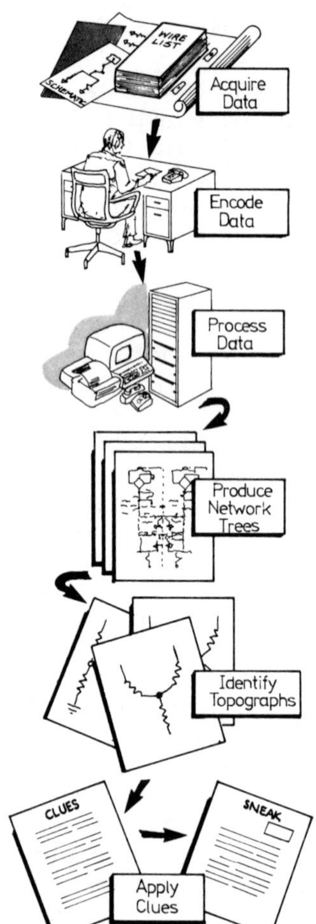

Figure 1. Sneak Circuit Analysis Procedure

SNEAK CIRCUIT METHODOLOGY

Data Acquisition

The first major consideration that must be satisfied in order to identify sneak circuit conditions is to ensure that the data being used for the analysis represents as closely as possible the "as built" circuitry of the system. Functional schematics, integrated schematics, and system level schematics do not always accurately represent the constructed hardware. Detailed manufacturing and installation schematics must be used because these drawings specify exactly what is built, contingent upon quality control checks, tests, and inspections. The data requirements necessitate performance of a Sneak Circuit Analysis after the detail level circuitry design is available, typically after a program's preliminary design review. If the analysis is performed earlier in the program, subsequent design changes can invalidate much of the analysis results.

On the other hand, while Sneak Circuit Analysis is applicable to mature operational systems, the analysis ideally should be applied at very early stages of the system development to minimize integration problems and to permit "paper fixes" instead of more costly hardware modifications. Balancing the project's economic incentives for an early analysis with the analytical data requirements, the most cost effective period in a project's life cycle for performing sneak circuit analysis would be when the detailed circuit design has been completed but before all hardware assembly and testing have been accomplished.

An example of the detailed level circuitry information required for an analysis is presented in Figure 2. While extremely simplified, this example illustrates that manufacturing and installation schematics rarely show complete circuits. The schematics are laid out to facilitate hookup by technicians without regard to circuit or segment function. As a result, analysis from detail schematics is extremely difficult. So many details and hidden continuities exist in these drawings that an analyst becomes entangled and lost in the maze. However, these schematics are the data that must be used if analytical results are to be based on true electrical continuity. The next task of the sneak analyst is, therefore, to convert this detailed, accurate information into a form usable for analytical work. The magnitude of data manipulation required for this conversion necessitates the use of computer automation which requires the encoding of all continuity information, of whatever form, into a single input format for processing.

Data Encoding

Automation has been used in Sneak Circuit Analysis since 1970 as the basic method of network tree production from manufacturing detail data. Computer programs have been developed to allow the simple encoding of continuities in discrete from-to segments from detail schematics and wire lists. An analyst must first study the system intensively and

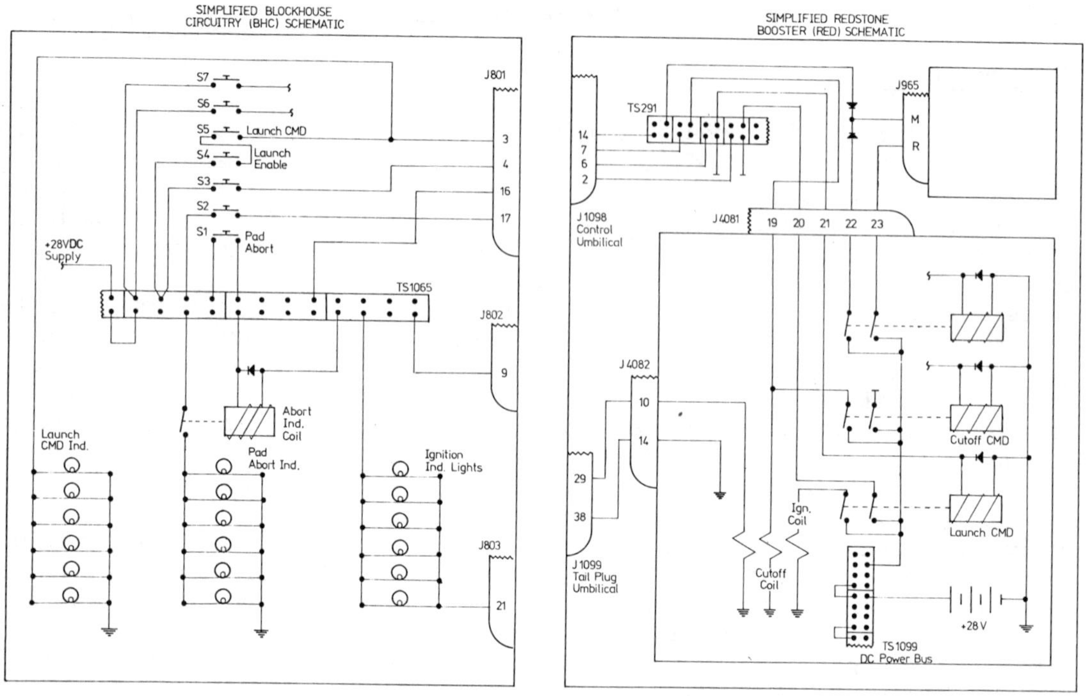

Figure 2. Redstone Data

partition it at key points, such as power and ground busses, to avoid large and unnecessarily cumbersome network trees. Once partitioned, the rigorous encoding rules for both schematic and wire list data are used to ensure that an accurate representation of the circuitry is maintained. Figure 3 illustrates the encoding format that would be used on the accompanying segment of Redstone circuitry. The same rules are applied on all continuity data regardless of source, format, or encoding analyst. The single resultant format for all continuity data ensures that the computer can tie all connected nodes together accurately to produce nodal sets.

Computer Processing

The Automated Sneak Path Analysis Program recognizes each reference designator/item/pin as a single point or node. This node is tied to other nodes as specified by the input records. The computer connects associated nodes into paths and collects associated paths into nodal sets. The nodal sets represent the interconnected nodes that make up each circuit.

Each node is categorized by the computer as "special" or "non-special". Special nodes include all active circuit elements such as switches, loads, relays, and transistors, while non-special nodes are merely interconnecting circuit elements such as connectors, terminal boards, and tie points. Before a nodal set can be output for analysis, the set is simplified by the elimination of the non-special nodes. This simplification removes all non-active circuit elements from the nodal sets while leaving the circuit functionally intact. This simplified nodal set is then output from the computer. Computer sketched plots of each nodal set are first generated. Other output reports that the computer generates include: (a) Path Reports wherein every element of each path is listed in case the analyst needs to trace a given path node by node, (b) an Output Data Index which lists every item and pin code, and provides the nodal set and path numbers in which each appears and (c) Matrix Reports which list each node of a nodal set, lists all labels associated with active circuit elements of that nodal set, and provides cross-references to other related circuitry (e.g., a relay coil in one nodal set will be cross-referenced to its contacts which may appear in another nodal set). Once these subsidiary reports are generated, the reports and the nodal set plots are turned over to sneak circuit analysts for the next stage of the analysis: network tree production.

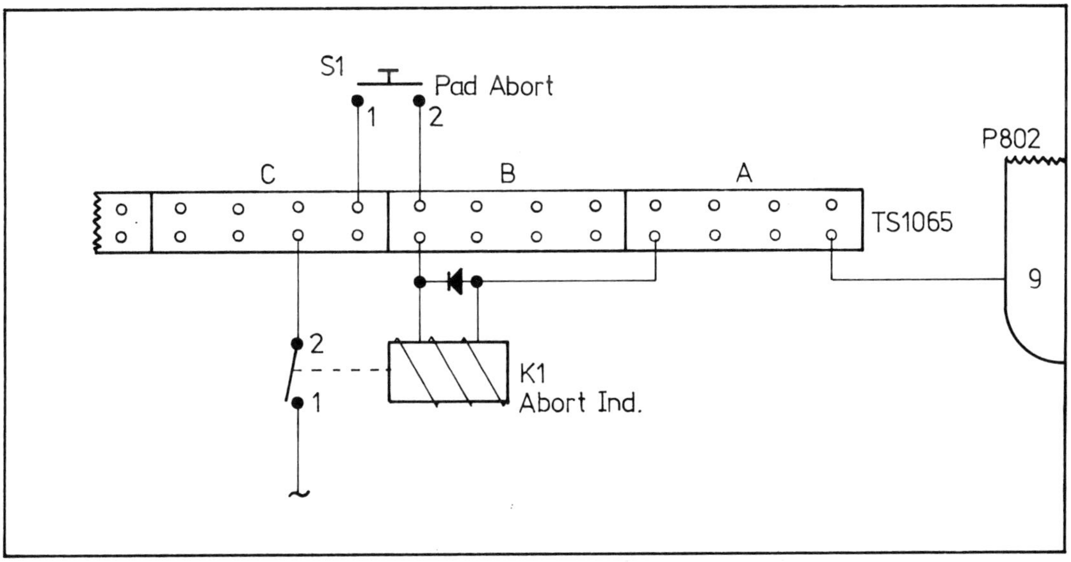

Figure 3. Redstone Encoding Sample

Network Tree Production

Network trees are the end result of all data manipulation activities undertaken preparatory to the actual analysis. The network trees characterize two major goals of the Sneak Circuit Analysis technique. First, the network trees represent a simplified version of the system circuitry by employing selective deletion of extraneous circuitry detail to reduce system complexity, while retaining all circuit elements pertinent to an understanding of all system operational modes. Secondly, network trees are characterized by their topologically oriented layout. All power sources are drawn at the top of each network tree with grounds appearing at the bottom. The circuit is oriented on the page such that current flow would be directed down the page. Figure 4 illustrates a network tree and the Redstone continuity data which were used to construct the tree. Figure 4a presents an integrated version of the data shown earlier in Figure 2. The highlighted circuitry in the integrated schematic of Figure 4a has been simplified to the tree pictured in Figure 4b. As can be readily seen, circuit function can be grasped more easily from a network tree than it can be from even a simplified, integrated schematic, much less the detailed manufacturing schematics used for a Sneak Circuit Analysis.

To produce completed network trees, the analyst begins with the computer nodal set plots. Occasionally these must be redrawn to ensure "down-the-page" current flow, then thoroughly labeled and cross-referenced with the aid of matrix reports and the other computer output reports. If these simple guidelines are followed in the production of network trees, identification of the basic topographs is greatly simplified.

Topological Pattern Identification

The analyst must next identify the basic topological patterns that appear in each network tree. Five basic patterns exist: (a) the single line (no node) topograph, (b) the ground dome, (c) the power dome, (d) the combination dome, and (e) the "H" pattern. These patterns are illustrated in Figure 5. One of these patterns or several together will characterize the circuitry of any given network tree. Although at first glance a given circuit may appear more complex than these basic patterns, closer inspection reveals that the circuit is composed of these basic patterns in combination. As the sneak circuit analyst examines each intersect node in the network tree, he must identify the pattern or patterns which contain that node and apply the basic clues that have been found to typify sneak circuits involving that particular pattern. Once every intersect node in the tree has been examined, all sneak circuit conditions within the tree will have been uncovered.

The Redstone network tree in Figure 4b, for instance, has three intersect nodes: A, B, and C. Figure 4c illustrates several of the topographical patterns which contain each of these nodes.

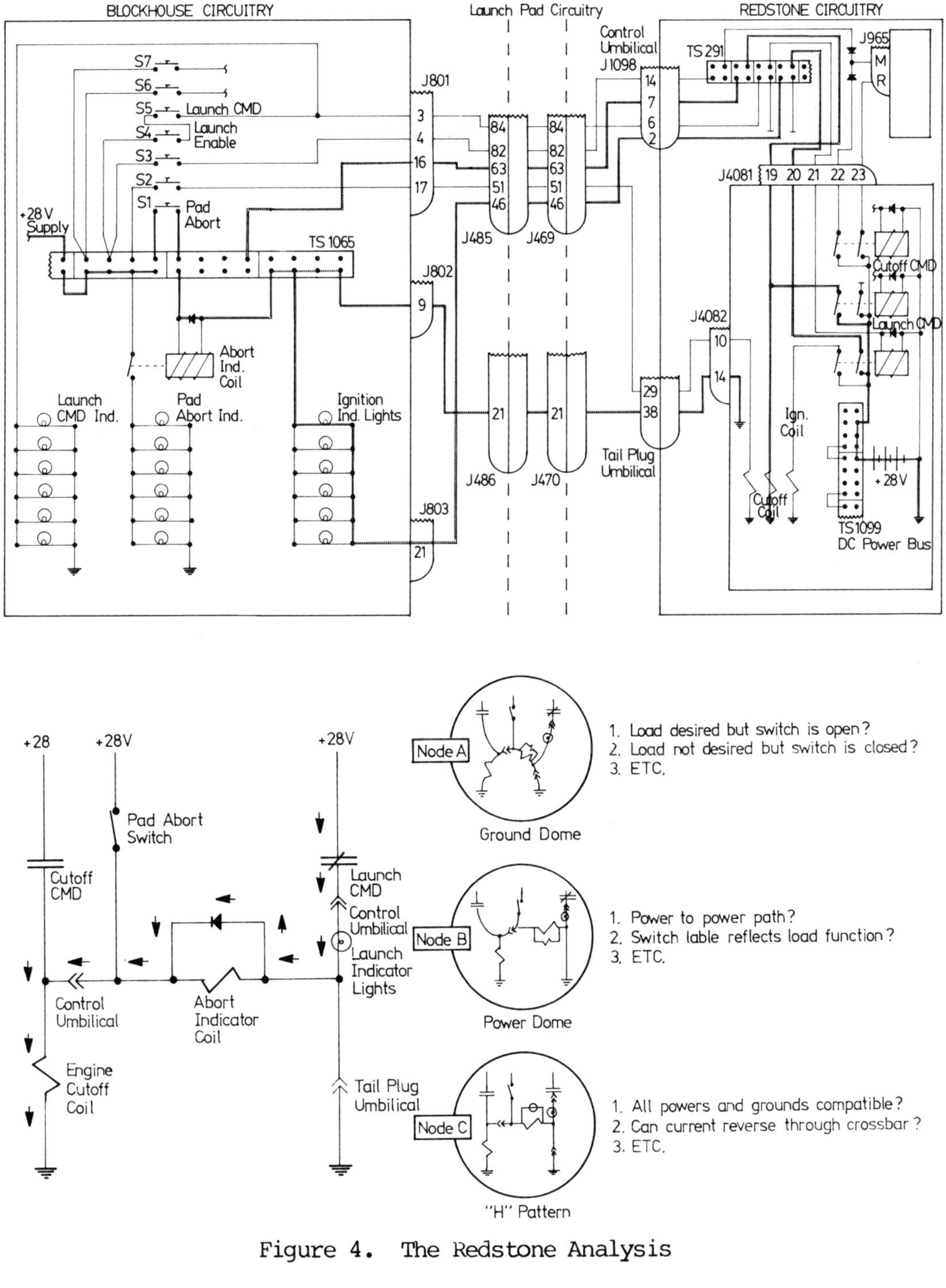

Figure 4. The Redstone Analysis

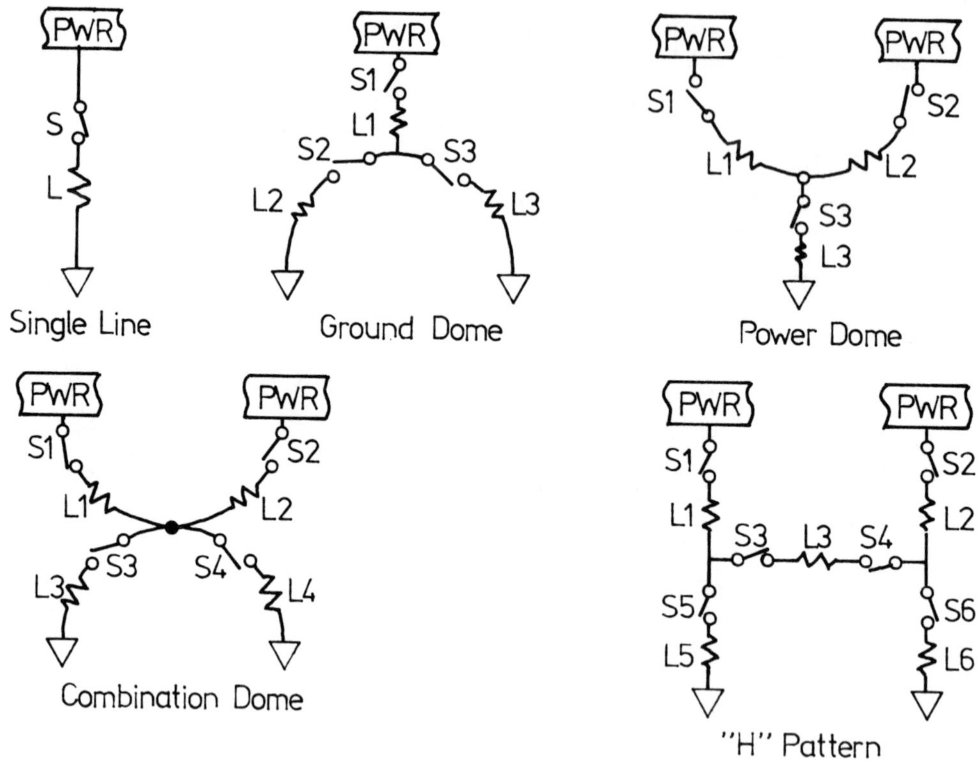

Figure 5. Basic Topographs

Clue Application

Associated with each pattern is a list of clues to help the analyst identify sneak circuit conditions. These lists were first generated during the original study of historical sneak circuits. The lists were updated and revised during the first several years of applied sneak circuit analysis. Now, the clue lists provide a guide to all possible design flaws that can occur in a circuit containing one or more of the five basic topological configurations, subject to the addition of new clues associated with new technological developments. The lists consist of a series of questions that the analyst must answer about the circuit to ensure that it is sneak free. As an example, the single line topograph (Figure 5a) would have clues such as:

(a) Is switch S1 open when load L1 is desired?,
(b) Is switch S1 closed when load L1 is not desired?,
(c) Does the label S1 reflect the true function of L1.

Sneak circuits are rarely encountered in this topograph because of its simplicity. Of course, this is an elementary example given primarily as the default case which covers circuitry not included by the other topographs.

With each successive topograph, the clue list becomes longer and more complicated. The clue list for the "H" pattern includes over 100 clues. This pattern, because of its complexity, is associated with more sneak circuits than any other pattern. Almost half of the critical sneak circuits identified to date can be attributed to the "H" pattern. Such a design configuration should be avoided whenever possible. The possibility of current reversal through the "H" crossbar is the most commonly used clue associated with "H" pattern sneak circuits. This pattern and clue revealed the sneak circuit in the Redstone circuitry. The clue would be stated as: "Is it possible for current to flow in both directions through the "H" cross bar"? If the answer to this question is yes, a sneak circuit is not automatically indicated as this may have been the design intent. The analyst must determine whether or not this is a desired condition within all foreseeable circumstances. In the case of the Redstone circuit, the condition definitely was not desired. The design intent was that the Cutoff Command relay contacts or the Pad Abort switch should energize the Engine Cutoff coil and the Abort Indicator coil (refer to Figure 4b). When the Launch Command contacts close, they should turn on the Launch indicator lights only. The possible reverse current can only exist when the ground below the indicator lights is lost. This would occur if the Tail Plug umbilical opened before the Control umbilical separated. If this happened, current would flow through the Launch Command relay contacts, the Launch indicator lights, through the suppression diode of the Abort Indicator coil and finally through the Engine Cutoff coil to ground. By this means, the Launch Command can activate the Engine Cutoff coil if the Tail Plug umbilical separates before the Control umbilical.

Unlikely as this event may appear, it did occur on November 21, 1960. After more than 50 sequentially successful Redstone booster launches, a Redstone booster with a Mercury capsule on it was to be launched. When the Launch Command was given, the booster fired. After lifting several inches from the pad, the engine inexplicably cut off. The booster settled back on the pad. The Mercury capsule separated and deployed its parachutes. Damage was slight, but a highly explosive rocket with no means of control sat on the pad, the dangling parachutes threatening to topple it. No one dared to approach the Redstone booster for 28 hours until its batteries drained down and the liquid oxygen evaporated. Later investigation revealed that the Tail Plug umbilical had separated 29 milliseconds prior to Control umbilical disconnect. This was enough time for the sneak circuit illustrated in Figure 4 to occur and abort the mission.

While this historical sneak circuit occurred before the advent of Sneak Circuit Analysis, the present day technique would have discovered the potential for this problem. After all pertinent clues have been applied to each topograph of each network tree, all sneak circuits inherent in a system will have been uncovered and reported.

Sneak Circuit Analysis Reports Produced

Sneak Circuit Analysis of a system produces the following four general categories of outputs: (a) Drawing Error Reports, (b) Design Concern Reports, (c) Sneak Circuit Reports and (d) Network Trees and Supplementary Computer Output Reports.

Drawing Error Reports disclose document discrepancies identified primarily during the data encoding phase of the Sneak Circuit Analysis effort. Design Concern Reports describe circuit conditions that are unnecessary or undesirable but which are not actual sneak circuits. These would include single failure points, unsuppressed inductive loads, unnecessary components, and inadequate redundancy provisions. A number of such conditions usually are identified whenever an analyst examines a circuit at the level of detail required for a formal Sneak Circuit Analysis. Sneak Circuit Reports delineate the sneak conditions identified during the analysis. These reports fall into four broad categories, outlined as follows:

Sneak Path. A sneak path is one which allows current or energy to flow along an unsuspected path or in an unintended direction. The example in Figure 6 illustrates an automotive sneak path (notice the "H" pattern). Design intent was for the radio to be powered only through the ignition switch and for the tail lights to be powered from the ignition through the brake switch or via the hazard flasher switch. A reverse current condition occurs when the hazard switch is closed and brake switch closes. The radio blinks on and off even though the ignition switch is off.

Sneak Timing. A sneak timing condition is one which causes functions to be inhibited or to occur at an unexpected or undesired time. The example in Figure 7 illustrates a sneak that occurred in the digital control circuitry of a mine. The enable logic for U4 and U5 allows them, briefly, to be enabled simultaneously. Being CMOS devices in a "wired or" configuration, this allows a potential power to ground short through the two devices, damaging or destroying them during operation.

Sneak Label. A label on a switch or control device which could cause incorrect actions to be taken by operators. The example in Figure 8, taken from an aircraft radar system, involves a circuit breaker which provides power to two disparate systems, only one of which is reflected in its label. An operator attempting to remove power from the liquid coolant pump would inadvertently deactivate the entire radar.

Sneak Indication. An indication which causes ambiguous or incorrect displays. Figure 9 illustrates a sneak indication which occurred in a sonar power supply system. The MOP (Motor Operated Potentiometer) OFF and ON indicators do not, in fact, monitor the status of the MOP motor. Switch S3 could be in the position shown, providing a MOP ON indication even

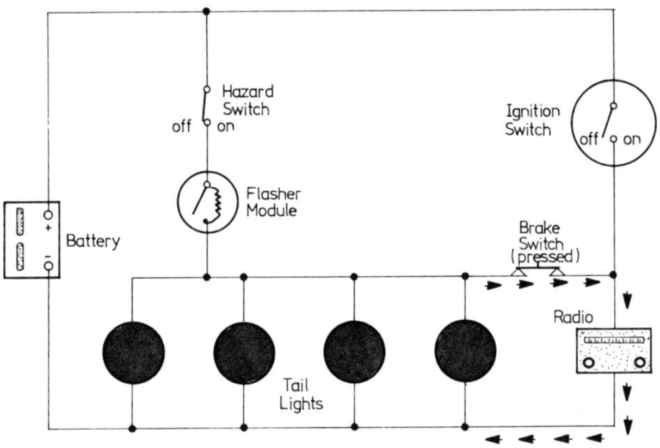

Figure 6. Sneak Path

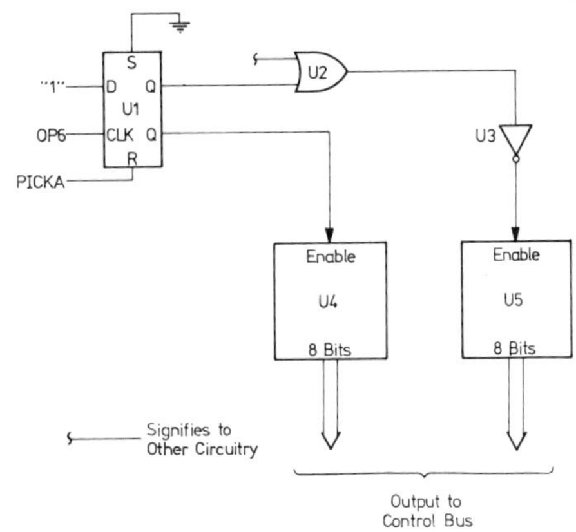

Figure 7. Sneak Timing

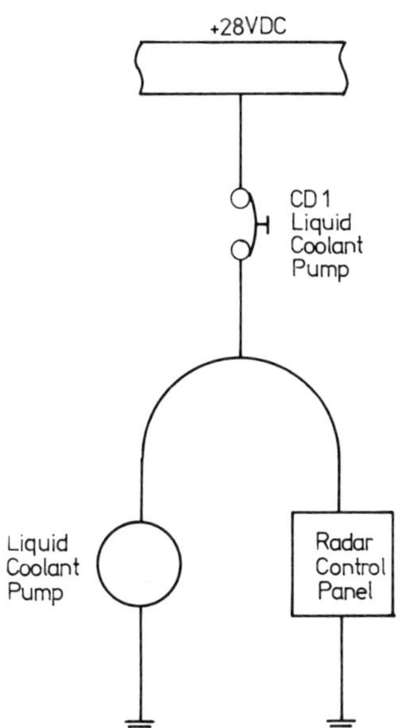

Figure 8. Sneak Label

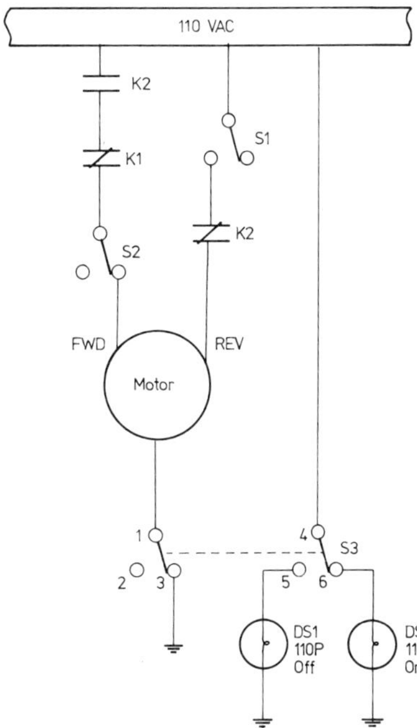

Figure 9. Sneak Indicator

though switches S1 or S2 or relay contacts K1 or K2 could be open, inhibiting the motor.

Network trees and the subsidiary computer output reports are also provided for the customer's use. The data index provides a means of locating quickly the network tree in which any given circuit element appears. The tree itself then provides a simplified picture of the circuitry associated with that element. The trees are, therefore, particularly useful in evaluating effects of design modifications, studying effects of component failures, troubleshooting, and providing mission operational support. The network trees can be used also to facilitate other circuit analyses such as Fault Tree Analysis, Failure Modes and Effects Analysis, Hazard Analysis, etc. In short, the network trees provide another form of circuitry drawings, which in many cases are easier to use than more conventional schematic systems.

Sneak Analysis of Other Flow Oriented Systems

The success of Sneak Circuit Analysis in standard electrical/ electronic systems prompted an investigation of the possibilities of applying similar techniques to other complex systems. Similar techniques should be applicable to digital logic systems, software systems, hydraulic systems, pneumatic systems, mechanical systems, and other flow oriented systems which have properties analogous to electrical systems. To date, analagous techniques have been developed and applied to digital logic and software systems.

DIGITAL LOGIC SNEAK CIRCUIT ANALYSIS

History

The increasing use of integrated circuitry in recently developed systems created several new problems in the performance of Sneak Circuit Analyses. In digital logic circuitry, inputs from many separate circuits are frequently tied together, creating one or two very large nodal sets instead of many small simple ones. Furthermore, actual current flow begins at the output of one device or gate and ends at the input of the next. A path from one gate to the next is too short to be of any analytical value. Logic flow is, therefore, more important than current flow in digital circuitry. Finally, the types of problems commonly occurring in digital circuitry are frequently different from those present in conventional circuitry. Additional clues are therefore required to identify digital sneak circuits.

Many of the methodological adjustments that must be made to perform a Sneak Circuit Analysis on a system containing digital circuitry were developed during the analysis of several of the Skylab Workshops experiments in 1972. The methodology has been improved upon during subsequent digital system analysis until, in May of 1975, an effort to define the

Digital Logic Sneak Circuit Analysis methodology formally was implemented. The remainder of this section will outline the principal differences between Sneak Circuit Analysis and Digital Logic Sneak Circuit Analysis as specified by the 1975 study.

Network Tree Production

The primary advantages of using computer automation in the production of network trees are the automatic selective loss of extraneous circuitry detail and the aid that the computer provides in extracting a circuit from multiple schematics and wire lists and representing it on a single page. Digital circuitry, however, frequently does not lend itself to simplification by automation. In many cases, all or most of a given digital circuit already appears on a single schematic with few extraneous circuit elements present. In these cases, network trees can be produced directly and more cost effectively without the aid of computer processing.

In producing network trees manually, the analyst must perform extensive partitioning of the digital circuits to obtain manageable network trees while maintaining all important circuit interactions. The analyst would also simplify the digital circuitry to a limited degree by functionally grouping discrete logic elements. Several flip-flops, which together form a counter circuit, would be combined into a single element to reduce unnecessary circuit complexity. For many systems containing digital electronics, therefore, the network trees of the digital circuitry are produced manually while all conventional circuitry, conventional digital interface circuitry and digital subsystem interface circuitry are processed in much the same manner as for totally non-digital systems.

The manually produced digital network trees are not oriented with current flow down the page as are conventional circuitry network trees. Logic flow, not current flow, is the more important characteristic of digital circuitry, so digital network trees are oriented with circuit inputs on the left, circuit outputs on the right, and logic flow across the page from left to right. Manual cross referencing between the manually produced and computer produced network trees must be performed next. Network tree analysis proceeds as in non-digital circuitry with topological pattern recognition and clue application. The latter task involves the second major difference between a digital and a non-digital system analysis.

Clue Application

The digital Logic Sneak Circuit Analysis study performed in 1975 included a historical search of the occurrence of sneak circuits in digital systems. The study concluded that the clues used in non-digital system analyses are also applicable to analyses of digital electronics. There are, however, a number of additional categories of problems which occur only in digital circuitry. An analysis of systems containing digital circuitry therefore uses a clue list augmented by more than 50 clues, applicable only to digital circuitry. These clues help identify a broad range of problems characteristic only of digital electronics. Using the network trees as a data base, a timing and race analysis may also be performed using auxillary software routines to identify several other potential digital electronic problems.

Digital Logic Sneak Circuit Analysis has been applied to a variety of complex digital systems and has successfully identified a variety of design deficiencies. Table 1 illustrates some of the major categories of digital circuitry problems that are identified by Digital Logic Sneak Circuit Analysis. Table 1 also lists some typical problems identified in a conventional Sneak Circuit Analysis and by the more recently developed Software Sneak Analysis.

Table 1. Typical Problem Areas

Conventional	Digital Logic	Software Logic
- Relay Race	- Incorrect Signal Polarities	- Unused Paths
- Backfeed	- Load Exceeds Drive Capabilities	- Inaccessible Paths
- Current Reversal	- Incorrect Wiring Interconnections	- Improper Initialization
- Ground Switch	- Improper Voltage Levels	- Lack of Data Storage/Usage Synchronization
- Bus Ties	- Excessive Fanout	- Bypass of Desired Paths
- Power Supply Cross-ties	- Potential Race Conditions	- Improper Branch Sequencing
- Conflicting Commands	- Lack of System Synchronization	- Potential Undesirable Loops
- Ambiguous Labels	- Inadequate System Reset	- Infinite Looping
- False Indicators		- Incorrect Sequencing of Data Processing
- Intermittent or Random Glitches		- Unnecessary (Redundant) Instructions

SOFTWARE SNEAK ANALYSIS

From the foregoing discussion it is apparent that Sneak Circuit Analysis offers a somewhat universal methodology and therefore the question of its applicability to software naturally arose. In 1970, after three years of successful application of the Sneak Circuit Analysis technique, methods of analyzing software systems with analogous techniques were studied seriously. For several years, general approaches to Software Sneak Analysis were investigated and case histories of software problems were studied. In 1975, a feasibility study was performed which resulted in the development of a formal technique involving the use of mathematical graph theory, electrical sneak theory, and computerized search algorithms which are applied to a software package to identify software sneaks. A software sneak is defined as a logic control path which causes an unwanted operation to occur or which bypasses a desired operation, without regard to failures of the hardware system to respond as programmed.

During the feasibility study, the Software Sneak Analysis techniques were applied to several test subroutines in three different assembler languages. The subroutines contained a number of known problem areas which were successfully identified by the Software Sneak Analysis techniques. In addition, 46 software problems not previously identified by other methods were uncovered. The feasibility study concluded that:

(a) Software Sneak Analysis is a viable means of identifying certain classes of software problems,
(b) Software Sneak Analysis works equally well on different software languages,
(c) Software Sneak Analysis does not require execution of the software to detect problems.

Software Sneak Analysis (SSA) was applied successfully only to individual subroutines during the feasibility study. Since multiple subroutine software systems are likely to involve different categories of problems from a single subroutine, it was felt that Software Sneak Analysis should be applied next to a total operating software system. In August, 1976, an analysis of the Terminal Configured Vehicle (TCV) software system was performed for NASA Langley Research Center. During this analysis, The SSA techniques were modified and expanded to include subroutine-to-subroutine interactions, to enable an analysis of a total software system package. The technique was modified to provide sufficient cross referencing to allow tracing of both logic flow and data flow through the software network trees. In this manner, the effect of a hardware input can be traced through the logic to determine its eventual impact on the software. The effect of a given piece of data can also be traced through the trees to evaluate its impact/effect on the total system.

The Software Sneak Analysis technique has evolved along lines very similar to electrical Sneak Circuit Analysis. Topological network trees are used with electrical symbology representing the software commands to allow easy cross analysis between hardware and software trees and to allow the use of a single standardized analysis procedure.

Since topological pattern recognition is the keystone of both Sneak Circuit Analysis and Software Sneak Analysis, the overall methodologies are quite similar. The software package to be analyzed must be encoded, processed, and reduced to a standardized topographical format, the basic topological patterns identified and the appropriate problem clues applied to each pattern. The major differences between a hardware and software analysis will be outlined below.

Network Tree Production

Unlike electrical system elements, software elements occur in a wide variety of formats. High level languages such as FORTRAN, COBOL, PL1 and the lower level languages such as the many assember languages must all be represented by some common symbology to allow an efficient analysis. Fortunately, regardless of the language used, there are only a limited number of discrete operations that any given command can invoke. Arithmetic operations, branches, conditional branches, data manipulation statements, etc. form the building blocks of all other commands. Since a single symbology was to be chosen, more or less arbitrarily, to represent the basic elements, electrical circuit symbols were chosen to facilitate total system analysis by having all the system's network trees appearing in a single format. A conditional branch, for instance, would be represented as a switch, a register load or store instruction as a relay, and data transfer between subroutines as connectors. Unusual software instructions can have other electrical symbols assigned for a given analysis. Coding techniques, computer processing, and network tree production is performed in much the same manner as in electrical systems utilizing this symbology.

A software network tree appears the same as its electrical counterpart. Figure 10 illustrates a software sneak which occurred in the operating software of a military aircraft. Figure 10a illustrates the design intent of the section of software with the sneak. When the actual code was produced, however, the two tests were inadvertently interchanged. The network tree of the actual software code (see Figure 10b) makes the sneak readily apparent. This historical problem was uncovered only during the software system integrated testing when it was found that the instructions represented by LOAD 1 could never be executed. Generally, it would be much more cost effective to identify as many potential software problems as possible before the software is run.

Topological Pattern Recognition

In identifying the basic topological patterns in software nodal sets, the inapplicability of the "H" pattern (see Figure 5) to software becomes apparent. The primary characteristic of the "H" pattern is that current can flow in both directions in the cross bar. Since the node-to-node branches in software always have a directed flow of logic, there is nothing analogous to a reverse current flow. In software, therefore, the "H" pattern breaks up into the other basic topographs. In all other respects, however, topographical pattern recognition is the same as in electrical Sneak Circuit Analysis.

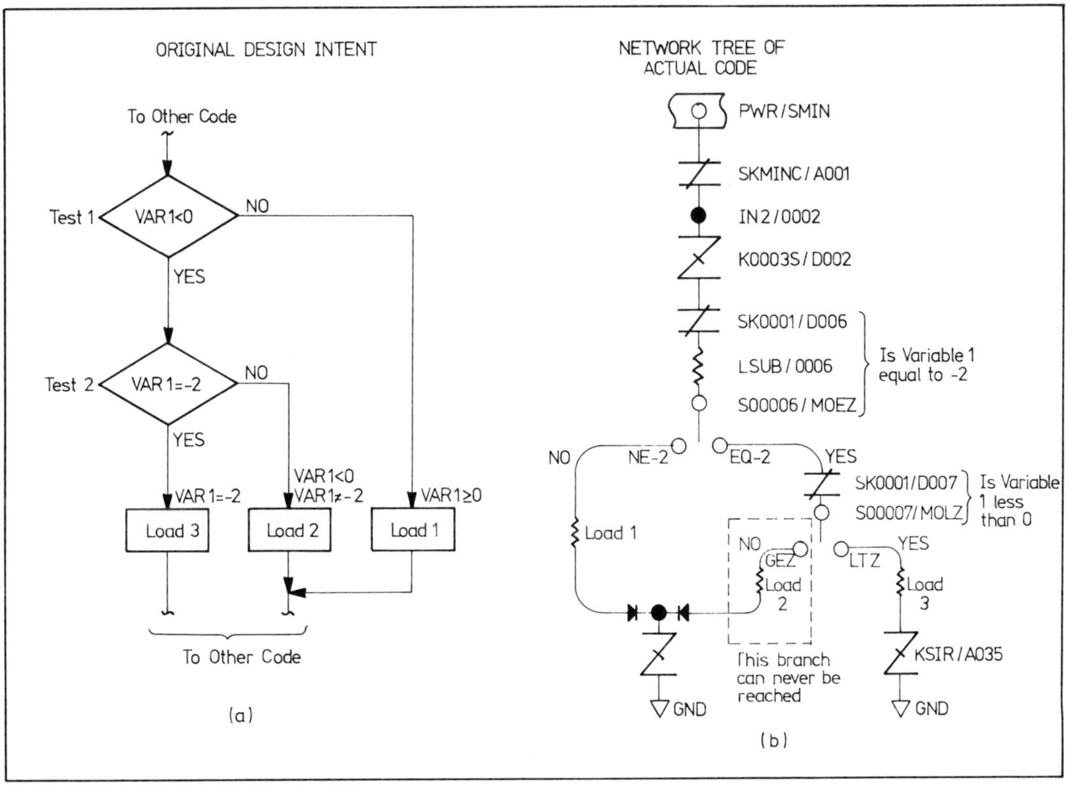

Figure 10. Software Sneak Example

Clue Application

Once the basic topographs have been identified, the clue lists compiled during the historical case studies are applied, and software sneaks are identified. The basic clues are generally the same as in Sneak Circuit Analysis but the sneaks identified are quite different. Table 1 identifies some of the categories of problems found in software as compared with analog and digital circuitry. Results obtained on several selected Software Sneak Analyses appear in Table 2.

Table 2. Sneak Circuit and Software Sneak Results

	Sneaks	Design Concern	Document Error
Electrical Systems			
Skylab (Complete)	259	91	307
N Reactor Safety Subsystems	18	13	22
Bay Area Rapid Transit Door Control	25	21	24
F-8 Digital-Fly-By-Wire (FCS)	19	7	28
E-3 AWACS Power System	57	2	194
Software Systems			
Terminal Configured Vehicle (FCS)	20	1	3
Incipient Failure Detector	4	1	1
Sneak Circuit Subroutines	7	—	---
Bl Avionics Subroutines	35	—	---

INTEGRATION OF HARDWARE/SOFTWARE ANALYSIS

After a Sneak Circuit Analysis and a Software Sneak Analysis have been performed on a system, the interactions of the hardware with the system software can be readily determined. The analyst has, at his disposal, diagramatic representations of these two elements of the system in a single standardized format. The effect of a control operation initiated by some hardware element can be traced through the hardware trees until it impacts the system software. The logic flow can then be traced through the software trees to determine its ultimate impact on the system. Similarly, the logic sequence of a software-initiated action can be followed through the software and electrical network trees until its eventual total system impact can be assessed.

CONCLUSION

The joint analysis of a system's software and electrical circuitry described above is termed simply Sneak Analysis. This system analysis tool helps provide visibility of the interactions of a system's hardware and software and hence will help reduce the difficulties involved in the proper integration of two such diverse, complex system designs. As hardware and software systems increase in complexity, the use of interface bridging analysis tools such as Sneak Analysis, becomes imperative to help provide assurance of total system reliability and maintainability.

BIBLIOGRAPHY

1. Clardy, R. C., "Sneak Circuit Analysis Development and Application", 1976 Region V IEEE Conference Digest, 1976, pp112-116.

2. Rankin, J. P., "Sneak Circuit Analysis", Nuclear Safety, Vol. 14, No. 5, 1973, September-October, 1973, pp461-487.

3. Hill, E.J. and Bose, L.J., "Sneak Circuit Analysis of Military Systems", Proceedings of the 2nd International System Safety Conference, July, 1975, pp351-372.

4. ________________, Software Sneak Analysis, CDRL No. A005 of Contract F2901-75-C-0069, October 1975.

5. ________________, Application of Software Sneak Analysis to the Incipient Failure Detection System, Boeing Aerospace Company, Houston, Texas, September 1976.

6. ________________, Application of the Software Sneak Analysis to the Terminal Configured Vehicle System, Boeing Aerospace Company, Document No. D2-118594-1, Houston, Texas, August 1976.

7. ________________, Sneak Circuit Analysis of N Reactor (Prepared for Atomic Energy Commission, Richland Operations Office), Report D2-118542-1, Boeing Aerospace Company, Houston, Texas, 31 July 1974.

8. ________________, Skylab Saturn Workshop Sneak Circuit Analysis, Report D2-118461-1, Boeing Aerospace Company, Houston, Texas, May 1973.

9. ________________, Sneak Circuit Analysis of the AWACS Electrical Power System, Boeing Aerospace Company, Document No. D2-118547-1, Houston, Texas, October 1974.

10. ________________, Sneak Circuit Analysis of Bay Area Rapid Transit Door Control System, Boeing Aerospace Company, Document No. D2-118611-1, Houston, Texas, March 1977.

11. ________________, "First Mercury-Redstone Flight Test Fails On Pad", Aviation Week, 28 November 1960.

12. ________________, "Second MR-1 Test Planned in Two Weeks", Aviation Week, 5 December 1960.

13 INFORMATION INTEGRITY

A. H. K. Ling and D. C. Honkanen

Reliability has been classically defined as the probability that an item (system) will perform satisfactorally for a specified period of time under a stated set of use conditions. This definition is broad and the first efforts in improving reliability were directed initially at system hardware and more recently towards system software. In both areas, methods for improving reliability have been developed and applied, results measured and progress recorded. However, in larger systems problems still remain and one of them relates to the rate at which information is presented to, processed by, and outputted from, the system. If information is not processed fast enough by a system, even on an occasional basis, it must be considered to be unreliable. A branch of information theory known as queuing theory attempts to make these problems manageable. If the correctness of information cannot be maintained during processing for example, then obviously this also represents an unreliable state of affairs. Another branch of information theory, i.e., error detection and correction, deals with this problem.

This chapter first deals with queuing theory and then provides some practical examples. This is followed by the basic theory behind error detection and correction codes and some applications.

QUEUING THEORY - GENERAL

At one time or another, we must all have been standing in a line waiting to be served. This line of waiting customers is the elementary queue. The basic model of a queuing system consists of a processor, which accepts inputs and produces an output. The processor (or server) provides a service to the arrivals (customers) who require it for a period of time, the service time. The output consists of the processed input.

Figure 1 shows an elementary queuing model. It can represent the flow of mail in the post office, passengers waiting for the arrival of subway trains, flow of information to central processors, communications networks, customers in the bank waiting to be served, etc.

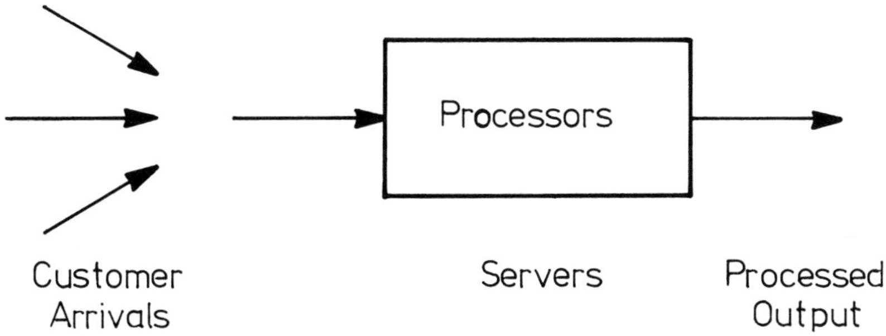

Figure 1. An Elementary Queuing Model

To analyze a queuing system, the following system characteristics must be known or assumed:

(a) The manner in which the inputs arrive and enter the queue. The arrival can be:
   (i) separated by unequal but definite known intervals of time;
   (ii) separated by equal intervals of time,
   (iii) separated by unequal intervals of time whose probability of distribution is known, and
   (iv) in groups or batches,

(b) The number of service centers available to process the arrivals and the amount of service that is allowed in the system, is the queue length or size,

(c) The queue discipline, i.e., the order of service. There are basically three distinct service disciplines, namely:
   (i) first-come-first serve, FCFS,
   (ii) last-in-first-out, LIFO,
   (iii) round-robin, RR,

(d) The service time pattern, called the service time. The service time may be:
   (i) constant,
   (ii) variable but known, and
   (iii) random with known probability distribution.

The characteristics (b) and (c) are the design criteria of the system. They are usually defined to satisfy the characteristics (a) and (d). By knowing the four characteristics of the queuing system, measures of performance can be obtained (1-3, 7, 9, 10).

DEFINITIONS, NOTATION, AND MEASURES OF PERFORMANCE

The following are the general definitions and notations commonly used in queuing analysis:

$P_n(t)$ = probability of n arrivals in system at time t.

$L(t) = \sum_{n=0}^{\infty} nP_n(t)$ = expected number of customers in the system at time t.

$L_q(t) = \sum_{i=C}^{\infty} (i-C)\, P_i(t)$ = expected queue length in a system with C servers at time t.

$C(t) = \sum_{n=0}^{C-1} nP_n(t) + C\sum_{n=C}^{\infty} P_n(t)$ = expected number of busy servers at time t.

$\overline{C}(t) = C - C(t)$ = expected number of idle servers at time t.

$W(t)$ = average waiting time (or average delay).

$W_q(t)$ = average waiting time in the queue.

$1/\lambda$ = average time between arrivals.

Observe that:

$$L(t) = L_q(t) + C(t)$$

Applying Little's theorem (8), we have for the steady state case,

$$W = L/\lambda$$

$$W_q = L_q/\lambda$$

where W, W_q, L and L_q are the steady state representation of their counterparts.

If T_s = average service time, then:

$$W = W_q + T_s.$$

By using the definitions given above and the following discussion the reader will be sufficiently armed to handle other more complex problems.

SPECIFICATION OF QUEUING SYSTEMS

To facilitate the treatment of Poisson queues which are very common, the Poisson distributed arrivals and exponentially distributed service times are discussed briefly. If λ is the expected number of arrivals per unit time, then the probability of n arrivals in the interval (0, t) is given as

$$P_n(t) = \exp(-\lambda t) \frac{(\lambda t)^n}{n!}, \quad n = 0,1,2, \ldots \infty \tag{1}$$

and the probability density function for the time between consecutive arrivals is given as:

$$f(t) = \exp(-\lambda t)\,\lambda, \quad t > 0 \tag{2}$$

If μ is the inverse of the expected service time, or the service rate, then the probability function for the number of departures (i.e., services) n in time t, is given as

$$P_n(t) = \exp(-\mu t) \frac{(\mu t)^n}{n!}, \quad n = 0,1,2, \ldots \infty \tag{3}$$

and the probability density function for service time is

$$f(t) = \exp(-\mu t)\,\mu, \quad t > 0 \tag{4}$$

Equations (1) and (3) can be recognised as a Poisson distribution. A plot of the Poisson distribution is shown in Figure 2, for different number of arrivals and their corresponding probabilities.

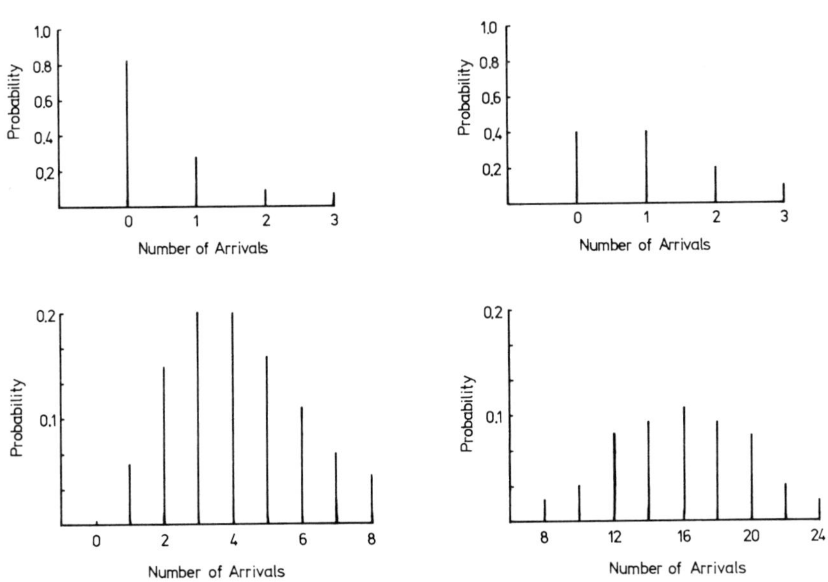

Figure 2. Poisson Distributed Processes

The simplest queuing system is the single server system as shown in Figure 1. The system can serve one arrival at a time. A multiserver system has n parallel identical servers and can serve up to as many as n customers simultaneously. The number of servers in a system depends on the arrival rate, service rate and the importance of the input to the system.

The service discipline is the rule for selecting the next customers for service. The most common queue disciplines are:

(a) First-in-first out, FIFO (or the common first-come-first serve FCFS);
(b) Last-in-first-out, LIFO (or last-come-first serve, LCFS);
(c) Round-robin, RR; where there exist time sharing pre-emptive resume priority, i.e., service is provided for time t in a cyclic form.

An important factor that describes a queuing system is the server utilization factor, ρ . ρ is the probability that any given server is busy and thus is for the single server model the fraction of time the server is busy.

GENERAL BIRTH AND DEATH PROCESS

Let us now develop the general birth and death process formulation to describe a queuing system. Let λ_n and μ_n denote the arrival and service rates, respectively, when n customers are in the system. For a general birth and death process, with a sufficiently small time increment, Δt, the probability of arrival in the system is:

$\lambda_n \Delta t$ = probability of one arrival during Δt,

$1-\lambda_n \Delta t$ = probability of no arrivals during Δt,

$\mu_n \Delta t$ = probability of one service during Δt,

$1-\mu_n \Delta t$ = probability of no services during Δt

The probability of n customers in the system at $t+\Delta t$ can be derived as the sum of the probabilities of the following mutually exclusive events

(a) n customers in the system at t and no arrivals or service during Δt,
(b) n-1 customers in the system at t, one arrival and no service during Δt, and
(c) n+1 customers in the system at t, no arrival and one service during Δt.

Thus, for a single server system:

$$P_n(t + \Delta t) = P_n(t)\,(1-\lambda_n\Delta t)\,(1-\mu_n\Delta t) + P_{n-1}(t)\lambda_{n-1}\Delta t(1-\mu_{n-1}\Delta t) + P_{n+1}(t)(1-\lambda_{n+1}\Delta t)\mu_{n+1}\,\Delta t \qquad (5)$$

where P_n are the state probabilities. As $\Delta t \to 0$, equation (5) can be simplified to the following differential difference equations:

$$\frac{dP_n(t)}{dt} = -P_n(t) \quad (\lambda_n + \mu_n) + P_{n-1}(t)\lambda_{n-1} + P_{n+1}(t)\mu_{n+1}, \; n \geq 1 \tag{6}$$

$$\frac{dP_n(t)}{dt} = -\lambda_o P_o \quad (t) + \mu_1 \, P_1(t), \; n = 0 \tag{7}$$

The above two equations are generally too complex for manual solution. Of interest is the steady state case, when $dP_n(t)/dt = 0$. Thus:

$$0 = -P_n \, (\lambda_n + \mu_n) + P_{n-1}\lambda_{n-1} + P_{n+1}\mu_{n+1} \quad , n \geq 1 \tag{8}$$

$$0 = -\lambda_o P_o + \mu_1 \, P_1. \tag{9}$$

Equations (8) and (9) are the general steady state equations for the M/M/C queue system, where M/M/C is exponential arrival/exponential service/C server. Equations (8) and (9) can be also derived using the steady state approach which is conveniently applicable, where the concept of state is used to represent the queue size. Let us define state i as the probability of having i customers in the system. The state diagram is given in Figure 3.

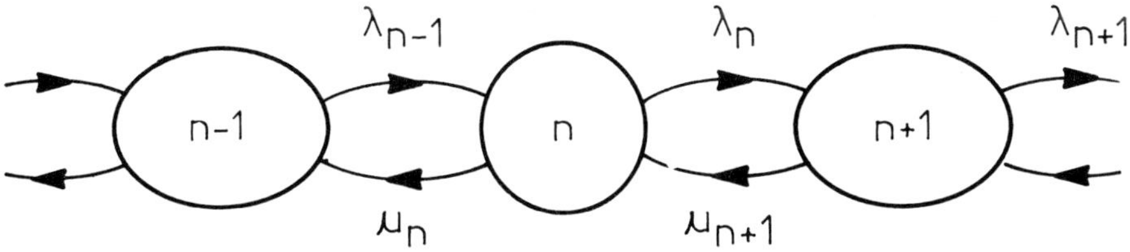

Figure 3. State space diagram

By using the steady state approach we can obtain equation (8) and by substitution, equation (9).

From equation (9),

$$P_1 = \frac{\lambda_0}{\mu_1} P_o \tag{10}$$

For n = 1,

$$P_2 = \frac{\lambda_1+\mu_1}{\mu_2} P_1 - \frac{\lambda_1}{\mu_2} P_o \tag{11}$$

Substituting equation (10) into (11) we get

$$P_2 = \frac{\lambda_1 \lambda_2}{\mu_1 \mu_2} P_o \tag{12}$$

By induction, we get the general equation for P_n.

$$P_n = \prod_{k=1}^{n} \frac{\lambda_{k-1}}{\mu_k} P_o, \quad 1 \leq n \leq N, \; N = \text{number of allowable customers.} \tag{13}$$

To get P_o, we use the probabilistic axiom that

$$\sum_{n=0}^{N} P_n = 1 . \tag{14}$$

Equation (14) can be reduced to give

$$P_o = 1 - \sum_{n=1}^{N} P_n \tag{15}$$

Substituting equation (13) into equation (15) gives

$$P_o \left(1 + \sum_{n=1}^{N} \prod_{k=1}^{n} \frac{\lambda_{k-1}}{\mu_k}\right) = 1 \text{ or}$$

$$P_o = \left(1 + \sum_{n=1}^{N} \prod_{k=1}^{n} \frac{\lambda_{k-1}}{\mu_k}\right)^{-1} . \tag{16}$$

Substituting equation (16) into equation (13) gives

$$P_n = \left[\prod_{k=1}^{n} \frac{\lambda_{k-1}}{\mu_k}\right] \Big/ \left[1 + \sum_{n=1}^{N} \prod_{k=1}^{n} \frac{\lambda_{k-1}}{\mu_k}\right] \tag{17}$$

Equations (13) to (17) can now be used to obtain the operating characteristics of a queuing system where arrival and services are Poisson distributed.

QUEUING MODELS

The four most commonly occurring queuing models are:

(a) M/M/1 - Exponential arrival/exponential service/single server.
(b) M/M/C - Exponential arrival/exponential service/C server.
(c) M/G/1 - Exponential arrival/general service/single server.
(d) M/E/1 - Exponential arrival/Erlangian service/single server.

M/M/1 Model

In this case, there is only one server.

Therefore $\lambda_k = \lambda$, $\mu_k = \mu$.

$$P_n = (\lambda/\mu)^n P_o \tag{18}$$

$P_o = 1 - \lambda/\mu$ or

$$P_o = 1-\rho \text{ where } \rho=\lambda/\mu \text{ is the utilization factor.} \tag{19}$$

The utilization factor determines the percentage of time the system is idle or busy. Substitution of equation (19) into (18) gives,

$$P_n = (1-\rho)\rho^n, \; n = 1,2,\ldots\infty \tag{20}$$

The expected number of customers in the system (waiting and in service), L, is obtained as,

$$L = \sum_{n=0}^{N} nP_n$$

$$= (1-\rho) \sum_{n=0}^{N} n\rho^n \tag{21}$$

which can be written as,

$$L = (1-\rho)\rho \frac{\partial}{\partial\rho} \sum_{n=0}^{N} \rho^n$$

$$= (1-\rho)\rho \frac{\partial}{\partial\rho} \frac{1}{1-\rho} .$$

Therefore,

$$L = \frac{\rho}{1-\rho} . \tag{22}$$

Applying Little's theorem, the expected waiting time in the system is given as:

$$W = \frac{\rho}{\lambda(1-\rho)} = \frac{1}{\mu-\lambda} = \frac{1}{\mu(1-\rho)} \tag{23}$$

and $W_q = W - T_s$

$$= \frac{\rho}{\mu-\lambda} = \frac{\rho^2}{\lambda(1-\rho)} = \frac{\lambda}{\mu(\mu-\lambda)} . \tag{24}$$

Thus,

$$L_q = \frac{\rho^2}{1-\rho} = \frac{\lambda^2}{\mu(\mu-\lambda)} . \tag{25}$$

The probability of having at least k customers in the system is,

$$P(\geq K \text{ in system}) = \sum_{n=K}^{N} P_n$$

$$= \sum_{n=K}^{N} (1-\rho)\rho^n = \rho^k . \tag{26}$$

The system measure of performance for the cases with finite storage space and infinite storage space are given in Table 1.

Table 1. System Measure Of Performance For M/M/1

Allowable Queue Size	N	∞
P_n	$\rho^n P_o \quad n \leq N$	$(1-\rho)\rho^n$
P_o	$\frac{1-\rho}{1-\rho^{N+1}}$	$1-\rho$
L_q	$\rho^2 \frac{1 - N\rho^{N-1} + (N-1)\rho^N}{(1-\rho)(1-\rho^{N+1})}$	$\frac{\rho^2}{1-\rho}$

Table 1 (Cont'd)

W_q	L_q/λ	$\frac{\rho^2}{\lambda(1-\rho)}$
L	$\rho\,\frac{(1-(N+1)\rho^N + N\rho^{N+1})}{(1-\rho)(1-\rho^{N+1})}$	$\frac{\rho}{1-\rho}$
W	L/λ	$\frac{\rho}{\lambda(1-\rho)}$

M/M/C Model

Here there are C servers. The state space diagram is given in Figure 4.

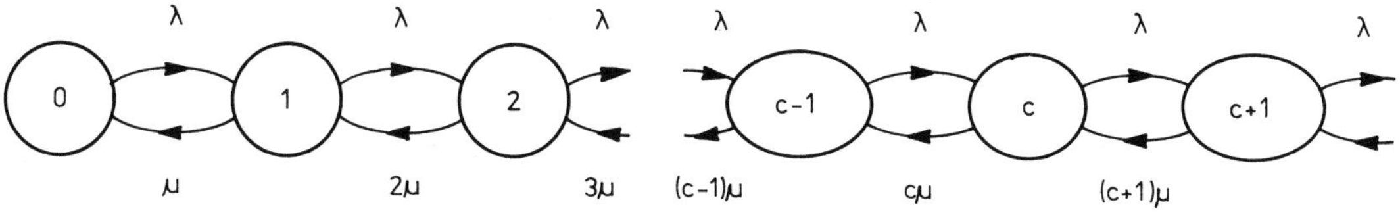

Figure 4. State Space Diagram For M/M/C

Now, $\lambda_n = \lambda$, $n = 1,2,\ldots\infty$

$$\mu_n \begin{cases} = n\mu, 0 \le n \le C \\ = C\mu, C \le n \end{cases}$$

Using equation (13), we obtain,

$$P_n \left.\begin{aligned} &= P_o \prod_{k=0}^{n-1} \frac{\lambda}{(k+1)\mu} \\ &= P_o \frac{1}{n!}\left(\frac{\lambda}{\mu}\right)^n \end{aligned}\right\} 0 \le n \le C \qquad (27)$$

and,

$$P_n = \begin{cases} = P_o \prod_{k=0}^{C-1} \frac{\lambda}{(k+1)\mu} \prod_{k=C}^{n-1} \frac{\lambda}{C\mu} \\ = P_o \frac{\lambda}{\mu} \frac{1}{C!\, C^{n-c}} \end{cases} \quad C \leq n \tag{28}$$

Equations (27) and (28) can be simplified to give,

$$P_n = \begin{cases} = P_o \frac{(C\rho)^n}{n!} & 0 \leq n \leq C \\ = P_o \frac{\rho^n C^c}{C!} & n \geq C \end{cases} \tag{29}$$

where $\rho = \frac{\lambda}{C\mu} < 1.$ (30)

Again using equation (16), we have,

$$P_o = \left[1 + \sum_{n=1}^{C-1} \frac{(C\rho)^n}{n!} + \sum_{n=C}^{\infty} \frac{(C\rho)^n}{C!} \frac{1}{C^{n-c}} \right]^{-1}$$

which can be futher simplified to give,

$$P_o = \left[\sum_{n=0}^{C-1} \frac{(C\rho)^n}{n!} + \frac{(C\rho)^c}{C!} \left(\frac{1}{1-\rho} \right) \right]^{-1} \tag{31}$$

The probability that all servers are busy when a customer arrives is given by

$$P\,(\text{queue}) = \sum_{n=C}^{\infty} P_n = \sum_{n=C}^{\infty} P_o \frac{(C\rho)^n}{C!} \frac{1}{C^{n-c}} \tag{32}$$

By substituting equation (31) into (32), we have:

$$P\,(\text{queue}) = \frac{\frac{(C\rho)^c}{C!} \frac{1}{1-\rho}}{\sum_{n=0}^{C-1} \frac{(C\rho)^n}{n!} + \frac{(C\rho)^c}{C!} \frac{1}{1-\rho}} \tag{33}$$

Using the above derived characteristics the operating measures of performance are obtained. The results are tabulated in Table 2.

Table 2. Measure Of Performance For M/M/C

Allowable Queue Size	N	∞
P_n	$\frac{C\rho^n}{n!} P_o$ if $0 \le n \le C$	$\frac{\rho^n}{n!} P_o$ if $0 \le n \le C$
	$\frac{C^c}{n!} \rho^n P_o$ if $n \ge C$	$\frac{\rho^n}{C!\, C^{n-c}} P_o$ if $n > C$
P_o	$\left[\sum_{n=0}^{C-1} \frac{(C\rho)^n}{n!} + \frac{(C\rho)^c}{C!} \frac{1}{1-\rho}\right]^{-1}$	$\left[\sum_{n=0}^{C-1} \frac{\rho^n}{n!} + \frac{\rho^c}{C!} \frac{C\mu}{C\mu-\lambda}\right]^{-1}$
L_q	$\rho P_o \frac{C^c}{C!}\left[\frac{\rho^c(C+1-(C+2)\rho) - \rho^N(N+1)-(N+2)\rho}{(1-\rho)^2}\right]$ $-P_o \frac{C^{N+1}}{C!}\left[\frac{\rho^{c+1} - \rho^{N+1}}{(1-\rho)}\right]$	$\frac{\lambda\mu\rho^c}{(C-1)!(C\mu-\lambda)^2} P_o$
W_q	L_q/λ	$\frac{\mu\rho^c P_o}{(C-1)!(C\mu-\lambda)^2}$
L	$\lambda(W_q + 1/\mu)$	$\frac{\lambda\mu\rho^c}{(C-1)!(C\mu-\lambda)^2} P_o + \rho$
W	$W_q + 1/\mu$	$\frac{\mu\rho^c}{(C-1)!(C\mu-\lambda)} P_o + 1/\mu$

M/G/1 Model

To a great extent, arrivals at a service facility follow a Poisson distribution. It is not as common to find exponentially distributed service times. In this situation, the formulation from the above two cases are not applicable.

Let us consider two random variables A_n and B_n to represent the number of customers after n customers depart and the number of arrivals during the service of n customers respectively. Therefore, we can represent,

$$A_{n+1} = A_n - 1 + B_{n+1} \qquad A_n > 0$$
$$= B_{n+1} \qquad A_n = 0 \quad \text{for all } n. \tag{34}$$

We assume that A_n and B_{n+1} are statistically independent. If service time is identical and independently distributed, then B_1, B_2, ..., are identical and independently distributed random variables.

Suppose,

$$U(x) \begin{cases} = 1 & \text{if } x > 0 \\ = 0 & \text{if } x \leq 0 \end{cases}$$

Substitution of the above into equation (34) gives,

$$A_{n+1} = A_n - U(A_n) + B_{n+1} \tag{35}$$

Squaring of equation (35) gives,

$$A^2_{n+1} = A^2_n + B^2_{n+1} + 2A_n B_{n+1} + U^2(A_n) - 2A_n U(A_n) - 2B_{n+1} U(A_n) \tag{36}$$

The expected value of equation (36) is

$$E(A^2_{n+1}) = E(A^2_n) + E(B^2_{n+1}) + E(2A_n B_{n+1}) + E(U^2(A_n)) \tag{37}$$

$$- E(2A_n U(A_n)) - E(2B_{n+1} U(A_n))$$

By using equation (35) and probability theory, equation (37) can be simplified to give,

$$E(A^2_{n+1}) = E(A^2_n) + E(B^2_{n+1}) + 2E(A_n) E(B_{n+1}) + E(U(A_n)) \tag{38}$$

$$- 2E(A_n) - 2E(B_{n+1}) E(U(A_n))$$

In the steady state case $E(A^2_{n+1}) = E(A_n)$ and it can be proven that $E(U(A_n)) = \lambda / \mu = \rho$. Therefore, equation (38) can now be simplified to,

$$E(A_n) = \frac{\rho + E(B^2_{n+1}) - 2\rho E(B_{n+1})}{2(1-E(B_{n+1}))} \tag{39}$$

Reference (11) gives

$$E(B^2_{n+1}) = E(B_{n+1}) = \text{Var}(B_n) + E(B_n)^2 . \tag{40}$$

Similarly,

$$E(B_n) = \rho \tag{41}$$

and,

$$\text{Var}(B_n) = \lambda^2 \text{Var}(t) + \rho \tag{42}$$

where Var(t) is the variance of the service time distribution. Substituting equations (40) and (42) into equation (39) we get,

$$E(A_n) = \rho + \frac{\lambda^2 \, Var(t) + \rho^2}{2\,(1-\rho)} \tag{43}$$

where,

$E(A_n)$ is the expected number of customers in the system.

Thus,

$$L = \rho + \frac{\lambda^2 Var(t) + \rho^2}{2\,(1-\rho)} \,. \tag{44}$$

Applying Little's theorem (8), $W=L/\lambda$, the expected time a customer spends in the system is given as,

$$W = 1/\mu + \frac{\lambda(Var(t) + 1/\mu^2)}{2(1-\rho)} \,. \tag{45}$$

Since,

$$W_q = W + T_s \tag{46}$$

where T_s= service time or $1/\mu$.

Thus,

$$W_q = \frac{\lambda(Var(t) + 1/\mu^2)}{2(1-\rho)} \tag{47}$$

which is the Pollaczek - Khintchine formula, and

$$L_q = \frac{\lambda^2(Var(t) + 1/\mu^2)}{2(1-\rho)} \,. \tag{48}$$

If $\alpha^2 = \mu^2 \, Var(t)$, substitution of α^2 into equations (44) and (48), yields,

$$L_q = \frac{\rho^2(\alpha^2 + 1)}{2(1-\rho)} \tag{49}$$

$$L = \rho + \frac{\rho^2(\alpha^2 + 1)}{2(1-\rho)} . \tag{50}$$

The characteristics of L_q and L for equations (49) and (50) respectively are plotted on Figures 5 and 6 for various ρ and α^2. Table 3 gives a summary of the performance characteristics of the M/G/1 queuing model.

Table 3. Measure Of Performance For M/G/1

Measure of Performance	Equation
L	$\rho + \frac{\lambda^2 \, Var(t) + \rho^2}{2(1-\rho)}$
W	$\frac{1}{\mu} + \frac{\lambda \, (Var(t) + 1/\mu^2)}{2(1-\rho)}$
L_q	$\frac{\lambda^2 \, (Var(t) + 1/\mu^2)}{2(1-\rho)}$
W_q	$\frac{\lambda \, (Var(t) + 1/\mu^2)}{2(1-\rho)}$

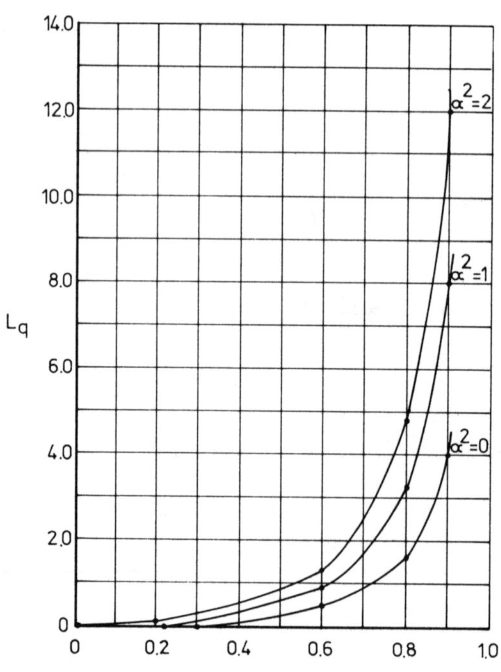

Figure 5. Plot of L_q for M/G/1

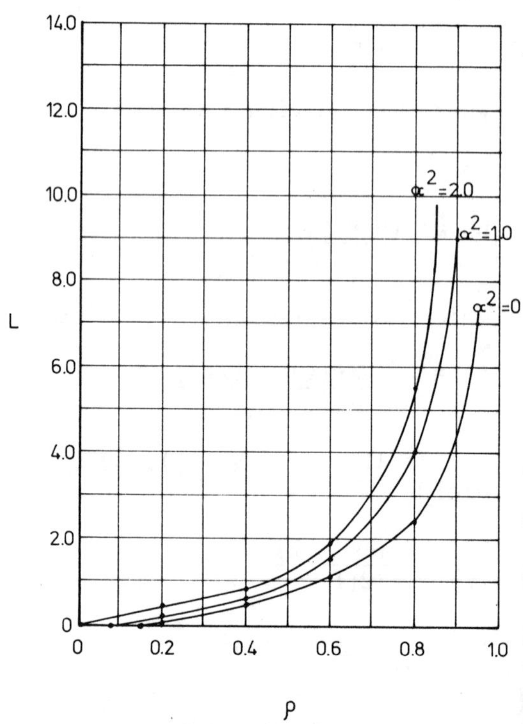

Figure 6. Plot of L for M/G/1

M/E/1 Model

Very often, a situation arises where services are provided in stages. That is, a customer who enters a channel or system must pass through successive stages of processing or services. It can be assumed that the time required to pass through each stage is exponentially distributed. For a k-stage system, the service time is $k\mu$, assuming that services are all the same for simplicity. Figure 7 shows a single channel or service, with k-stages.

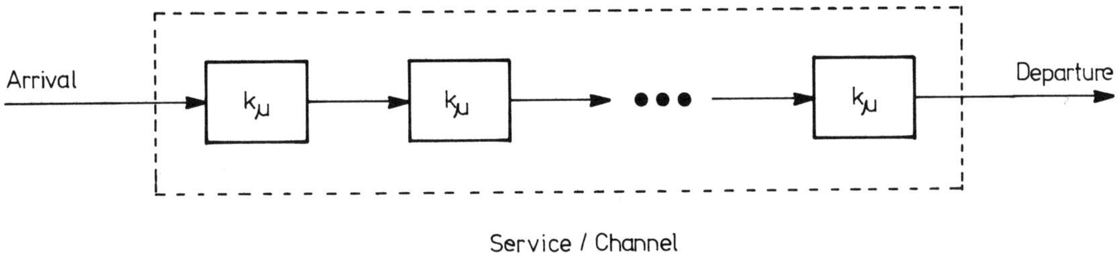

Figure 7. k-stage Erlangian Model

The state space diagram for the M/E /1 system is given in Figure 8.

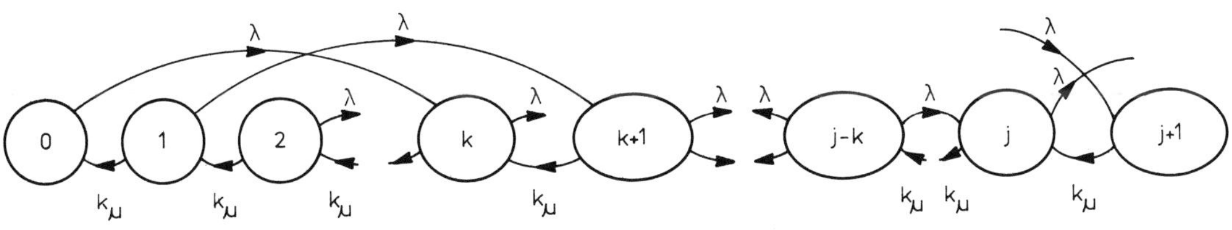

Figure 8. State Space Diagram For M/E /1

Taking the balance equation, we have,

$$\lambda P_0 = k\mu P_1 \tag{51}$$

$$(\lambda + \lambda\mu)\ P_j = \lambda P_{j-k} + k\mu P_{j+1}\ , j = 1, 2, \ldots \infty . \tag{52}$$

Since the state probabilities with negative subscripts are zero (11), i.e.,

$$P_j = 0,\ j < 0 . \tag{53}$$

We can multiply equation (52) by Z^j and sum over $j \geq 1$ to obtain,

$$(\lambda + k\mu) \sum_{j=1}^{\infty} P_j Z^j = \lambda \sum_{j=1}^{\infty} P_{j-k} Z^j + k\mu \sum_{j=1}^{\infty} P_{j+1} Z^j . \qquad (54)$$

Equation (54) can be rearranged to give,

$$(\lambda + k\mu)\left\{\sum_{j=0}^{\infty} P_j Z^j - P_o\right\} = \lambda Z^j \sum_{j=1}^{\infty} P_{j-k} Z^{j-k} + \frac{k\mu}{Z} \sum_{j=1}^{\infty} P_{j+1} Z^{j+1} . \qquad (55)$$

Let us now define the Z-transformation generating function for G(Z) as,

$$G(Z) = \sum_{j=0}^{\infty} Z^j P_j . \qquad (56)$$

Substitution of equation (56) into equation (55) yields,

$$(\lambda + k\mu) (G(Z)-P_o) = \frac{k\mu}{Z} (G(Z) - P_o - ZP_1) + \lambda Z^k G(Z) . \qquad (57)$$

Equation (51) can be simplified to $P_1 = P_o \lambda/k\mu$. Therefore, equation (57) simplifies to,

$$G(Z) = \frac{k\mu(1-Z) P_o}{k\mu (1-Z) - Z\lambda (1-Z^k)} . \qquad (58)$$

Dividing both denominator and numerator by $(1-Z)$ and letting $k\mu = \mu_1$, simplifies G(Z) to,

$$G(Z) = \frac{\mu_1 P_o}{\mu_1 - Z\lambda ((1-Z^k)/(1-Z))} = \frac{\mu_1 P_o}{\mu_1 - Z\lambda (\sum_{j=0}^{k-1} Z^j)} \qquad (59)$$

From Z-transformation theory, the generating function $G(Z)|_{Z=1}$, we get,

$$G(Z) = 1 = \frac{\mu_1 P_o}{\mu_1 + k\lambda} . \qquad (60)$$

From which, $P_o = 1 - \dfrac{k\lambda}{\mu_1}$. (61)

Thus,

$$G(Z) = \frac{\mu_1 (1 - \frac{k\lambda}{\mu_1})}{\mu_1 - Z\lambda \sum_{j=0}^{k-1} Z^j} = \frac{\mu_1 (1-\rho)}{\mu_1 - Z\lambda \sum_{j=0}^{k-1} Z^j} \text{ since } \rho = k\lambda/\mu_1 . \tag{62}$$

This is the general solution for P_j by Z-transformation G(Z). By taking the first derivative of G(Z) with respect to Z, and letting $Z \to 1$, the expected value of j, E(j) gives,

$$E(j) = \frac{k+1}{2} \frac{\lambda}{\mu - 1} . \tag{63}$$

The average waiting time is given as

$$W_q = \frac{k+1}{2k\mu} \frac{\lambda}{\mu - \lambda} . \tag{64}$$

From which,

$$W = \frac{k+1}{2k\mu} \frac{\lambda}{\mu-\lambda} + \frac{1}{\mu} \tag{65}$$

$$L_q = \frac{k+1}{2k\mu} \frac{\lambda^2}{\mu-\lambda} \tag{66}$$

$$L = \frac{k+1}{2k\mu} \frac{\lambda^2}{\mu-\lambda} + \frac{\lambda}{\mu} . \tag{67}$$

QUEUING THEORY - EXAMPLES

In this section, examples are given to show the applicability of the equations obtained.

Example 1

A single traffic intersection is to be controlled by a pre-timed signal having a cycle of 60 seconds. From the approaches there is a left-turning movement amounting to 180 vehicles per hour. The layout of the intersection is such that 2 left-turning vehicles per cycle can be handled without difficulty, whereas three or more left-turning vehicles per cycle will cause congestion in that lane. In what percent of the cycles would such delays occur?

$$\text{Average left turns per cycle} = \frac{60 \times 180}{3600} = 3.0$$

It has been shown that traffic flow can be approximated by a Poisson distribution. Thus, if m = number of vehicles,

$$P(m \geq 3) = 1 - \sum_{m=0}^{2} \frac{3^m}{m!} \exp(-3)$$

$$= 1 - \exp(-3)(1 + 3 + 4.5) = 1 - \exp(-3)(8.5)$$

$$= 0.5768.$$

Therefore, 57.68% would experience delay in the lane.

Example 2

A data processor receives data at a constant rate, λ bits/sec. The input data is processed one bit at a time, on a first-come-first-serve discipline according to an exponentially distributed rate of $1/\mu$ seconds. The buffer can store N bits. When the buffer is full, the incoming bit will be rejected. What shall the smallest storage capacity be so that the probability of non-saturation is at least α where $0 < \alpha < 1$ in a steady state situation?

Data arrival rate = λ bits/sec.

Processor service rate = μ bits/sec.

Probability of saturation = $1-\alpha$.

From Table 1,

$$P_N = \rho^N \frac{1-\rho}{1-\rho^{N+1}} = 1-\alpha$$

$$\rho^N(1-\rho) = 1-\rho^{N+1} - \alpha + \alpha\rho^{N+1}$$

$$\rho^N = \frac{1-\alpha}{1-\alpha\rho}$$

$$N \log \rho = \log \frac{1-\alpha}{1-\alpha\rho}$$

$$N = \frac{\log \frac{1-\alpha}{1-\alpha\rho}}{\log \rho}.$$

Example 3

The data arrival rate to the data processor was found to be 2.771 msg/sec. The message was processed at a rate of 5.38 msg/sec. The study showed that both the arrival and service rate of data are approximately exponentially distributed. The storage capacity of the buffer is infinite. The messages are processed according to the FIFO discipline. Determine the performance characteristics for the cases, (a) single processor and, (b) 2 processors.

Arrival rate of message, $\lambda = 2.771$ msg/sec.

Service rate of processor, $\mu = 5.38$ msg/sec.

Therefore, $\rho = \lambda/\mu = 0.5151$.

Case (a). From Table 1

$$P_n = 0.4850 \times 0.5151^n$$

P_o = probability of no message in system = 0.4850

$$L_q = \frac{\rho^2}{1-\rho} = \frac{(0.5151)^2}{0.4850} = 0.547 \text{ msg. in queue.}$$

$$W_q = \frac{0.547}{2.771} = 0.1974 \text{ second waiting in queue}$$

$$L = \frac{\rho}{1-\rho} = \frac{0.5151}{0.4850} = 1.062 \text{ msgs in system.}$$

$$W = L/\lambda = \frac{1.062}{2.771} = 0.3833 \text{ second waiting in system}$$

Processing time $= 1/\mu = 0.1859$ second.

Note that $W = W_q + 1/\mu$.

Case (b). $\rho = \dfrac{\lambda}{2\mu} = 0.2576$ and from Table 2, C=2

$$P_o = \left[\sum_{n=0}^{1} \frac{\rho^n}{n!} + \frac{\rho^2}{2!}\,\frac{2\mu}{2\mu-\lambda}\right]^{-1}$$

$$= 0.7424$$

$$L_q = \frac{\lambda\mu\rho^c}{(C-1)!(C\mu-\lambda)^2} P_o = \frac{2.771 \times 5.38 \times 0.2576 \times .7424}{(2x5.38-2.771)^2}$$

$$= \frac{0.7344}{63.82} = 0.01150 \text{ msg. in queue .}$$

$$W_q = \frac{\mu\rho^c P_o}{(C-1)!(C\mu-\lambda)^2} = 0.00415 \text{ Sec. waiting in queue .}$$

$$L = \frac{\lambda\mu\rho^c}{(C-1)!(C\mu-\lambda)} P_o + \rho = 0.01150 + 0.2576$$

$$= 0.26910 \text{ msg. in system}$$

$$W = W_q + 1/\mu = 0.1900 \text{ sec. waiting in system.}$$

Example 4

It has been found that arrivals of customers to a system follow a Poisson distribution with an average time between arrivals of 30 sec. There is one server to handle demand. Time studies show that the service per customer is 0.20 minute and the variance for service 0.02 minute. Based on the study, determine the average time a customer spends in the system? What is the average waiting time in the queue and the number of customers in the queue?

From Table 3, the following characteristics are found,

$\lambda = 2$ customers/minute

$\mu = 5$ customers/minute, $1/\mu = 0.2$ minutes

$\rho = \lambda/\mu = 2/5 = 0.4$

$\text{Var}(t) = 0.02 \text{ min}^2$

$$W = \frac{\lambda(\text{Var}(t) + 1/\mu^2)}{2(1-\rho)} + \frac{1}{\mu^2}$$

$$= \frac{2 \times (0.02 + 0.04)}{2 \times 0.6} + 0.2 = 0.3 \text{ minutes}$$

The expected waiting time in the system = 0.3 minutes.

$$W_q = \frac{\lambda(\text{Var}(t) + 1/\mu^2)}{2(1-\rho)} = 0.1 \text{ minute}$$

$$L_q = \frac{\lambda\,(\text{Var}(t) + 1/\mu^2)}{2(1-\rho)} = \frac{2^2(0.02 + 0.04)}{2 \times 0.6} = 0.2 \text{ customer}$$

$L = 0.6$ customers

The expected waiting time in queue = 0.1 minutes. The queue length = 0.2 customers, and the number of customers in the system = 0.6 rounded to the whole number = 1.

Example 5

Studies were conducted on the arrival and service rate at a processor. It was found that arrivals are Poisson distributed with a mean of 100 arrivals/hour and that service is Erlangian distributed with a mean of 2 minutes. Determine the performance of measure characteristic of the system for k=2.

λ = 10 customers/hour

μ = 30 customers/hour

k = 2

From equations (64) to (67):

$$W_q = \frac{k+1}{2k\mu} \frac{\lambda}{\mu-\lambda} = \frac{3}{120} \frac{10}{20} = 0.0125$$

Expected waiting time in queue = 0.0125 hr.

$W = W_q + 1/\mu = 0.0458$ hr

Expected waiting time in system = 0.0458 hr.

$L_q = 0.125$ customers

Expected number of customers in queue = 0.125.

Expected number of customers in system = 0.458.

ERROR DETECTION AND CORRECTION - THEORY

Techniques for error detection and correction have developed along with the growth of digital communications. The information to be transmitted is coded so that when received, it will be possible to detect and, in some cases, correct any errors due to transmission. Coding involves adding redundant bits to the message to be transmitted.

One of the simplest and most common error detection techniques is the parity bit method (12,15). An extra bit is added to the message information bits so that there will always be an even number of ones in each word (even parity) or an odd number of ones (odd parity). This method detects an odd number of bit errors in a message. However, for double or even count errors the number of ones will still provide the required even or odd number parity. The method does not provide enough information to assure reliable error detection for multi-bit errors or to correct the code when an error is detected. For greater protection, that is to increase the number of detectable errors and to introduce correctability, more check bits must then be added.

It is now necessary to introduce some coding terminology. In general, the coded message will have a total of n bits, of which there are the k original information bits and $n-k=r$ check bits. This structure is identified as an (n,k) algebraic code. For example, a (7,4) code has a total of 7 bits of which there are 4 information and 3 check bits.

As the number of check bits increase, so do the number of detectable and correctable errors. The parameter used to indicate the degree of error control is the "minimum Hamming Distance d," named after R.W. Hamming who in the 1950s devised the first complete error detecting and error correcting procedure (25). The distance between two code words is the number of bit positions in which they differ. For example $X_1 = 011, X_2 = 010$. The two words in the above example differ only in the last position, hence the distance "d" is one . With distance d = 1, no error can be detected, since a code word with one error may result in a valid code word. If we then devise a set of code words with minimum distance of two it will be possible to detect single bit errors. The error, however, cannot be corrected since there are two valid code words, which are both distance one from the invalid code word. Codes with minimum distance of three allow one or two errors to be detected and allow one error to be corrected. With a minimum distance, d, equal to three, a code word with one error lies two units away from the first valid code word, and one unit away from the second valid code word. This allows a correction scheme to be devised to find the correct code word. In general, the "minimum Hamming Distance d" guarantees that for a set of code words of minimum distance d, that in principle d-1 or fewer errors can be detected and in principle that (d-1)/2 errors can be corrected where (d-1)/2 is the largest integer $\leq$ (d-1)/2.

Code Word Representation

It is sometimes useful to represent a codeword of n bits by a polynomial:

$$1101001 = 1X^6 + 1X^5 + 0X^4 + 1X^3 + 0X^2 + 0X^1 + 1X^0$$

$$= X^6 + X^5 + X^3 + 1$$

The data bits are interpreted as the coefficients of a n-1 degree polynomial. In the above example a 7 bit codeword becomes a 6th order polynomial.

Modulo-2 Arithmetic

The arithmetic will be consistent with the modulo-2 system; this is arithmetic in which there are no carries or borrows. Subtraction therefore yields the same result as addition, e.g.,

1 + 0 = 1, 1 - 0 = 1, 1 - 1 = 0, 0 - 1 = 1 etc.

This leaves only three arithmetic operations to consider addition, multiplication and division. The addition of two codewords is straight forward, e.g.,

$$\begin{array}{rcl}
X^4 + X^3 + X^2 + 0 + 1 & = & 011101 \\
\underline{X^5 + X^4 + 0 + X^2 + X^1 + 1} & = & \underline{110111} \\
X^5 + 0 + X^3 + 0 + X^1 + 0 & & 101010
\end{array}$$

Multiplication is left to the reader to investigate, (17).
Polynomial division is best shown by an example, consider two polynomials $G(X) = X^3 + X + 1$ and $M(X) = X^6 + X^4$. Then $M(X)/G(X)$ is:

$$\begin{array}{r|l}
 & X^3 + X^0 \\
X^3 + X + 1 & X^6 + X^4 \\
 & \underline{X^6 + X^4 + X^3} \\
 & \qquad\qquad X^3 \\
 & \qquad\qquad \underline{X^3 + X + 1} \\
 & \qquad\qquad\quad X + 1
\end{array}
\qquad\qquad
\begin{array}{r|l}
 & 1001 \\
1011 & 1010000 \\
 & \underline{1011} \\
 & \;\;\,1000 \\
 & \;\;\,\underline{1011} \\
 & \;\;\,\;011
\end{array}$$

Hamming Codes

The Hamming codes are a class of single error correcting codes named after R.W. Hamming (13, 15). The check bits are linear combinations of the message bits. For illustration consider a (7,4) Hamming code defined by the following set of equations:

$$M_1 + M_3 + M_4 = C_1 \tag{68}$$

$$M_1 + M_2 + M_3 = C_2 \tag{69}$$

$$M_2 + M_3 + M_4 = C_3 \tag{70}$$

M_1, M_2, M_3, M_4 are the information bits and C_1, C_2, C_3 are the check bits.

Assume that the received codeword $V_r = M_{1r}, M_{2r}, M_{3r}, M_{4r}, C_{1r}, C_{2r}, C_{3r}$ where subscript r indicates received bits.

After transmission of a codeword, errors may have been introduced. The check bits are computed from the received message bits and we obtain:

$$M_{1r} + M_{3r} + M_{4r} = \hat{C}_{1r}$$

$$M_{1r} + M_{2r} + M_{3r} = \hat{C}_{2r}$$

$$M_{2r} + M_{3r} + M_{4r} = \hat{C}_{3r}$$

We then calculate the sum of the check bits calculated, as above and the check bits received.

$\hat{C}_{1r} + C_{1r} = S_0$

$\hat{C}_{2r} + C_{2r} = S_1$

$\hat{C}_{3r} + C_{3r} = S_2$

The pattern of S_0 S_1 S_2 is known as the syndrome and from this, a transmitted codeword can be found assuming at most a single bit error. Consider that S_0 ,S_1 ,S_2 equal 0,0,0 , then no errors occurred. However if S_0 ,S_1 ,S_2 , were to equal 1,0,1 then an error occurred. Since S_0 and S_2 equal 1 and S_1 equals 0 then the error is common to S_0 and S_2 and not to S_1 . Assuming a single bit error it can be deduced that M_4 is the bit in error.

BCH Codes

Another useful set of linear codes are the BCH codes (after the founders, Base - Chaudhuri-Hocquenghem), (12, 13, 15, 16, 17, 23). These codes extend the single error correcting Hamming codes to multiple error detection and correction. The distance attainable with these codes enhances their error protection power. BCH codes allow detection of up to twice as many errors (2E) as it was designed to correct (E) . The factor E, the number of correctable errors is determined from the following set of conditions:

The Codeword length $n = 2^a - 1$, a=2,3... (71)

Number of check bits, $r \leq aE$ (72)

Minimum distance, $d \leq 2E + 1 \leq n$. (73)

From Equation (71), only codewords of length 2^n-1 (i.e., 3, 7, 15, 31, etc.) are permissible. Equations (72) and (73) allow a trade off to be made between the number of check bits, r, and E, the number of correctable bit errors.

The basic encoding involves dividing the information polynomial (say M(X) by a fixed polynomial G(X). The fixed polynomial G(X) is the generating polynomial of the code. The information polynomial is modified prior to this by adding zeros (equal to the number of check bits) to the end of M(X). For example, consider $M(X) = X^3 + X^1$ then for the (7,4) BCH code , M(X) is modified to $X^6 + X^4$.

1010 000 = modified $M(X) = X^6 + X^4$ = 1010000

M(X), r=3

The remainder then forms the check bits which are appended to the information bits. Referring back to the example on polynomial division where this is calculated, the remainder is 011 and the coded message becomes:

1010 + 011 = 1010011

M(X) Remainder

A simplistic hardware configuration for generating this (7,4) BCH Code generated by the polynomial $X^3 + X + 1$ is shown in Figure 9 (17, 23 24). It involves a 4 Stage Shift Register with Modulo-2 Adders (Exclusive-OR) forming the feedback circuit. The original four information bits are shifted into the shift register, then the feedback path is closed, and then seven successive shifts produce seven encoded bits at the output.

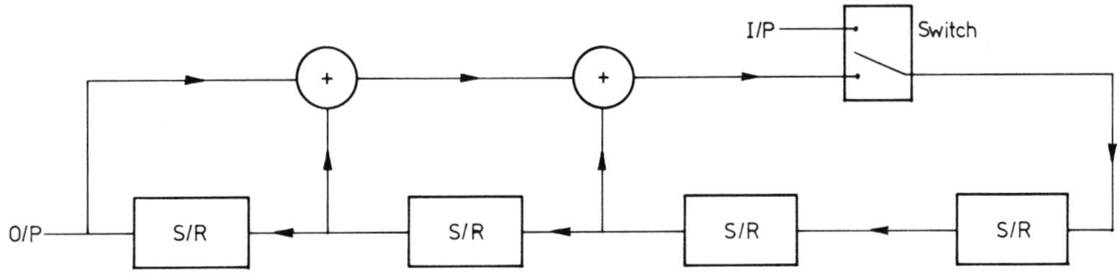

Figure 9. BCH Code Cenerator

Table 1, lists code words of length n, number of information bits k , and minimum distance d , for some shorter BCH Codes, (12, 16).

Table 1. Some Useful BCH Codes

n	k	d	E	Polynomial
7	4	3	1	$X^3 + X + 1$
15	11	3	1	$X^4 + X + 1$
15	7	5	2	$X^8 + X^7 + X^6 + X^5 + 1$
31	26	3	1	$X^5 + X^2 + 1$
31	21	5	2	$X^{10} + X^9 + X^8 + X^6 + X^5 + X^3 + X + 1$
31	16	7	3	$X^{15} + X^{11} + X^{10} + X^9 + X^8 + X^7 + X^5 + + X^3 + X^2 + X + 1$

The choice of the generator polynomial is not arbitrary and fortunately an extensive list was derived by Stenbit (14).

At the receiver the code word is divided by G(X) and if the remainder now is zero then no detectable errors occurred in transmission. A non-zero remainder means that error(s) occurred. The procedure to correct them is more complex and goes beyond the scope of this introduction and the interested reader is referred to (13, 16).

BCH codes can be extended by the addition of a partiy check bit (12). This extended BCH code will now provide detection for odd count errors along with error patterns having up to 2E errors. The error correction capability is unchanged. In choosing a code for an application the knowledge of how errors are likely to occur is important. Errors are either random, due to White noise or to equipment faults, or the errors occur in bursts or strings as caused by fading radio transmission. The codes discussed so far are excellent for a random or independent error environment. It has not been shown here but BCH Codes provide burst detection for burst lengths up to r bits.

In a burst error environment, Fire Codes (12) provide double burst detection and single burst correction. The generator polynomials for these codes can be modified BCH polynomials. In general the number of check bits become quite large and therefore these codes find application only for long blocks of data.

The following section on applications will develop these properties further. The material presented here is a brief exposure to the complex theory of coding. This introduction, should allow the reader to pursue the subject further with the extensive references provided.

ERROR DETECTION AND CORRECTION - APPLICATIONS

Memory Systems

Error correction and detection have been used to improve the reliability of semiconductor memory storage (18, 19, 20, 21, 22). An example is shown in Figure 10.

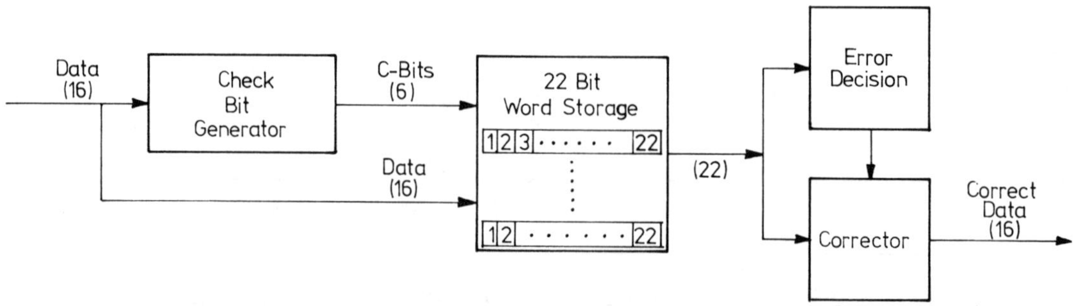

Figure 10. Memory System with Error Detection and Correction

The example illustrates 16 bits of information (k=16) with 6 check bits (n-k=6) generated by a Hamming type code. When the stored data word, equal to 22 bits is read from memory the error decision logic regenerates the check bits from the 16 information bits and then a comparison is made with the c-bits read from memory. From this syndrome the location of errors is determined and this application corrected in the corrector stage. The degree of error protection required is single bit correction and double bit detection.

For a single bit error correcting code to be effective, the predominant failure mode of the memory system has to be single bit. The memory is comprised of ROM/RAM components which have single bit, row, column or total device failure modes. Arranging the devices so that each device stores one bit of the data word, prevents any device failure from causing more than one faulty bit in any data word (21).

Data Transmission

There are two control strategies, Automatic Repeat Request (ARQ) and Forward Error Correction (FEC) (24). Using FEC, an attempt is made to correct the error messages at the receiving end. The check bits are used for both detection and correction. With ARQ, repeat transmission for all erroneous messages is requested by the receiving end. The check bits are used solely for detection. The advantage of FEC is that it allows continuous non-interrupted tranmission, with the penalty that each message requires a large number of check bits and so reduces the information data rate. A FEC system also places a demand for more complex error correction hardware at the receiving end than the ARQ technique. However, ARQ requires storage at the transmission end since any word may have to be retransmitted. Cyclic codes, especially the Bose-Chaudhuri-Hocquenghem (BCH) set of codes are applicable in random and burst noise fields and are extensively used in ARQ and FEC systems.

CONCLUSION

The mathematical presentation and examples given here is by no means complete. The interested reader is urged to refer to the references. By following and understanding the foregoing, the reader will have a full taste of the problems in information integrity and will be prepared to study much more difficult and complicated situations.

REFERENCES

1. Allen, A.B., "Elements of Queuing Theory for Systems Design", IBM System Journal, Vol. 14, No. 2, 1975, p161-187.

2. Coffman, E.G. and Denning, P.J., Operating System Theory, Prentice-Hall, Englewood Cliffs, New Jersey, 1973.

3. Cox, D.R. and Smith, W.L., Queues, Methuens & Co. Ltd., London, 1961.

4. Genlough, D.L. et al, Poisson and Other Distributions in Traffic, Eno Foundation for Transportation, Inc.

5. Kleinrock, L.,Communication Nets, Stochastic Message Flows and Delay, Dover Publication, Inc., New York, N.Y., 1963.

6. Kleinrock, L., "A Continuum of Time Sharing Scheduling Algorithms", Spring Joint Computer Conference, 1970, p453-458.

7. Kleinrock, L., Queuing Systems, Vol. I and II. J. Wiley & Sons, Inc. New York N.Y., 1975 and 1976 respectively.

8. Little, J.D.C., "A Proof for the Queuing Formula L = w", Operations Research Journal, 1961, p383-387.

9. Martin, J., System Analysis for Data Transmission, Prentice-Hall, Englewood Cliffs, New Jersey, 1972.

10. Rau, J., Optimization and Probability in System Engineering, Van Nostrand Reinhold, New York, N.Y., 1970.

11. Beckman, P., Elements of Applied Probability Theory, Harcourt, Brace & World, New York, N.Y., Inc.

12. Martin, J.D. "Using Polynomial Codes". Electronic Engineering, July 1976, p46-49.

13. Peterson W.W. and Weldon. E.J. Error-Correcting Codes, MIT Press, Cambridge, Mass, 1972, p224-285.

14. Stenbit, "Table Generators for Base-Chaudhuri Codes". IEEE Transactions ON Information Theory IT-10, 1964, p390-391.

15. Owens A., and Harknett, M.R., "Introduction to Coding Techniques". Electronic Engineering, Aug. 1975, p28-31.

16. Owens A., and Harknett, M.R., "Basic and Extended Cyclic Hamming Codes". Electronic Engineering, Dec. 1975 and Jan. 1976, p34-37, 45-47.

17. Ralfapalli. K., "CRC Error Detection Schemes Ensure Data Accuracy". EDN, Sept. 1978, p119-123.

18. Hewlett-Packard, "Innovations in Memory Technology", Datamation, June. 1976, p67.

19. Levine, L. and Meyers, W. "Semiconductor Memory Reliability with Error Detecting and Correcting Codes", Computer, Oct. 1976, p43-50.

20. Richard, B. "Automatic Error Correction in Memory Systems", Computer Design, May 1976, p179-182.

21. Application Note 40, Memory System Relability with Intersil 4K RAM's, Intersil, Sunnyvale, CA, April 1977.

22. Sanyal, S. and Venkataramen, K.N. "Single Error Correcting Code Maximizes Memory System Efficiency", Computer Design, May 1978, p175-183.

23. Cavell, P. "Implementation of Cyclic Redundancy Check Circuits". Electronic Engineering, Feb 1977, p51-55.

24. Sindhu, P.S. "Retransmission Error Control with Memory". IEEE Transactions on Communications, COM-25, May 1977, p473-479.

25. Hamming, R.W., "Error Detecting and Error Correcting Codes", Bell System Technical Journal, Vol. 26, No. 2, April 1950, p147-160.

ERROR DETECTION AND CORRECTION - BIBLIOGRAPHY

1. Toschi, E.A. and Watanabe, T. "An All-Semiconductor Memory with Fault Detection, Corection, and Logging", Hewlett-Packard Journal, Aug 1976, p8-13.

2. Constable, T. "ID GCR a Tape Drive Revolution". Peripheral Review, Pertec Computer Corp. Vol. VI No. 2, Winter 1978.

3. Schweber, W. and Pearce, L. "Software Signature Analysis Identifies and Checks PROMs". EDN, Nov 5, 1978, p79-81.

4. Gordon, G. and Nadig, H. "Signature Analysis Spots Faulty Bit Streams", Canadian Electronics Engineering, April 1977, p26-30.

5. Lee, D. and Bellotti, O. "A Random Error Generator for PCM System Testing". Canadian Electronics Engineering, Sept. 1976, p21-22.

6. Montgomery, R. "Simple Hardware Approach to Error Detection and Correction". Computer Design, Nov. 1978, p108-117.

7. Fortune, P. "Two-Step Procedure Improves CRC Mechanism". Computer Design, Nov. 1977, p116-129.

14 APPLICATION ENGINEERING

J. A. Roberts and C. B. Chabot

There are several reasons why a part/component or a group of associated components may fail prematurely or malfunction in service:

(a) poor component design,
(b) inadequate component manufacturing quality control and testing,
(c) deficient purchase specifications,
(d) damage in shipping, handling and in test equipment,
(e) poor component application leading to performance overstress, testability or maintainability problems.

Here, the latter factor, which is in effect a design error, will be discussed together with some methods of prevention. The emphasis will be on ensuring:

(a) correct part usage (application) to provide adequate performance,
(b) the use of conservative stress levels (derating) to maintain performance for long time periods.

COMPONENT APPLICATION

Most electronic equipment manufacturers at the present time are involved in assembling large numbers of 'bought-out' complex components in a system or configuration designed by 'in-house' design groups. Often the 'in-house' manufactured elements of a system constitute less than 10 per cent of the manufacturing cost of a system. Clearly any communication problems between the component manufacturer and the systems manufacturer will show up in the form of application errors. In order to avoid such problems, application engineering groups are maintained by the major component suppliers. A wide variety of publications are produced by component suppliers. However, in practice few designers have time to do any more than try to understand the component manufacturers data sheet. Often, detailed applications data can be obtained only from direct discussions with manufacturers' engineers. These avenues take time to search out and the great majority of components have no supporting applications data. Also it is a rare applications report that discusses the weaknesses of a component.

The solution adopted by many companies is to appoint a component specialist, who is able to gather detailed specification and qualification data on each of the components that the design engineer plans to use. However, more than this level of involvement is required. The component specialist should provide a service to the designer that reduces the risk of design errors. A collection of component-related design information should be the responsibility of the component specialist. The designer clearly has the responsibility for meeting the electrical functional requirements. Usually the creative excitement or difficulty of doing this precludes a deeper examination of component limitations. Methods of building a knowledge of component problems and providing a service to a design group within a large company will be described.

Sources of Information

It is interesting to observe the process by whch engineers gain experience and knowledge of design methods and application errors. There are several major sources of such knowledge:

(a) manufacturer and competitor application reports,
(b) house journals,
(c) personal design experience and the experience of colleagues,
(d) company production and Q.A. test reports,
(e) company field failure data,
(f) component engineer recommendations,
(g) military and NASA applications data,
(h) customer equipment purchase specifications
(j) specialized textbooks,
(k) Government Industry Data Exchange Program (GIDEP) Alerts,
(m) technical seminar and journals,
(n) design reviews.

Since the information is multi-sourced, the designer does not usually have the time to monitor all of those listed. Given a large scale design center, the task of monitoring component-related design information and its subsequent dissemination can be assigned to the component specialist.

While the acquisition of detailed component information is clearly within the function of a component specialist, equally important is the effective use of methods of reporting design hazards. In a general way, checklists offer a compendium of experience and are in effect a report of known design errors. Checklists should be in the hands of all designers. It is still necessary to carry out an independent component application check, however, since checklists are never complete.

Product Questionnaire

A method to increase component knowledge which several companies are using very effectively involves a mailed questionnaire. Directed to gain specific knowledge of a particular component, the questionnaire measures component and vendor viability in several ways. The questionnaire shown in Figure 1 has been used to gain information on LSI (Large Scale Integrated) semiconductor components.

Product Questionaire
Device Type Number__________________
Description__________________________

1. Does your company manufacture the above product? pin/pin________________, similar_________________
Please attach data sheets.

2. Do you have reliability data on this particular product?___________or a similar product_________, type
__________Please attach reliability reports or state reliability figure.

3. Do you have application reports on this or similar products?__________________
__________Please attach application reports.

4. Please outline any possible application problems or handling problems we may encounter.

5. Do you consider that this device is the most suitable for use in a new design.

6. Please state your approximate monthly production rate and total amount of this product that has been produced by your company.
________________per month
________________total

7. Do you anticipate that a military version of this component will become available?_____
______ If none is likely please state the reasons.

8. Please provide the names of applications engineers and customer service engineers who we may call if we encounter problems.

9. Please outline the recommended test and inspection procedures that we should specify and that will be convenient for you to carry out.

10. Do you anticipate that radiation hardened version of this device will become available?

11. Are there other sources of supply of this product besides your company?
Please list:

If none, suggest possible alternate sources who have the capability of producing the device:

Please return to:

Figure 1. Product Questionnaire

Design and Component File

Another method which has proven useful for providing easy access to recently published design and component information is the Design and Component File. Given that a future design program may require an identifiable range of advanced components or techniques, a file of all published information is maintained by the component specialist. An example concerned a design program involving wide temperature range liquid crystal displays and large CMOS (Complementary Metal Oxide Semiconductor) memory components. During a six-month period it was possible to collect more than a dozen published articles from the commercial technical periodicals, which were of direct value to the program. This practice avoids errors of omission such as overlooking advanced components which could ease design problems.

Component Application Checklist

Component knowledge is vital to correct application. A checklist for each major class of components provides a rapid means of assessing the viability of a new design. A component checklist is essentially a collection of design experience. A good deal of the merit of a checklist is lost if it becomes dated. Unfortunately, a checklist dealing with components is particularly prone to becoming dated and needs regular updating. Another factor which is overlooked with respect to checklists is the necessity for the original sources of information to be given as references. As an example, extracts from a forty question capacitor checklist are shown below.

(a) Are metallized capacitors used? If insufficient energy is available for clearing the transient shorts, the circuit will cease to function. Can the transient voltage drops be tolerated during the clearing operation? Energy required is 10 microwatt-seconds typically.
(b) Are dc rated capacitors used in the ac or pulsed dc applications? This practice often leads to exceptionally large failure rates since the dissipation factor and series resistance can be very large.
(c) Are there any capacitors shown which have polychlorinated biphenyl? Because of the extreme toxicity of this compound, such capacitors are on the prohibited import list of many countries.

Preferred Parts List

Preferred Parts Lists (PPL) offer the opportunity, to the components specialist, of controlling the use of components to those of known quality. Exactly what appears in a PPL depends on the product category being manufactured. Most companies producing conventional military grade equipment will be constrained to write a PPL around assessed reliability components and MS components. However, it is not uncommon to find that the contractual requirements of a particular program dictate the quality level and types of components to be used. It is vital that a clear statement of the components to be used reaches the designers at an early stage. Non-Standard Part submissions are very costly, particularly when qualification data has to be supplied.

In any organization, even those engaged in purely commercial manufacturing a PPL has a place. An ideal PPL should include an outline of the relative costs of components.

Reporting Design Errors

A design check stage as envisaged here is a relatively informal but documented review at an early phase of the design. Often, under the pressure of deadlines, a designer's first thoughts could be carried toward final practice. A feasible design is not necessarily a reliable or economic one. It is desirable that the comments made at the early design check stage be limited to communications directly between the individuals immediately concerned. However, the comments are recorded and are available to the customer of the equipment. The form shown below in Figure 2 is one of a three-copy set that is used to report errors to the designer.

Project: .... Drawing: ....

Drawing Number: .... Revision: ....

COMPONENT	COMMENT

FORM 186-NEW2/74

Date: ____________ Signed: ____________

Figure 2. Part Applications Review

The Nature of Design Errors

Here the term 'design error' includes deficiencies of the design which cause performance, overstress, testability or maintainability problems. Clearly, bench testing or even environmental testing will reveal only a fraction of the problems. Long-term production, system operation and maintenance will certainly reveal all of those problems. Early critical examination of a proposed design, together with advisory programs, is the only reasonable course of action to reduce design errors.

(a) TTL (Transistor Transistor Logic) system design problems are few since a sound system of standardization has been imposed by the industry leader. The general lack of problems has, however, led to carelessness; errors vary from simple oversights (excessive fanout) and ignoring design rules (no isolation resistors) to subtle test or system difficulties. As an example of a test problem, the flip-flop shown in Figure 3 has both its set and reset inputs tied to a common isolation resistor. With a single resistor the element cannot be set into a defined state and the logic card canot be initialized easily prior to test. Often, a significant part of the test time can be consumed by applying a homing sequence to initialize the logic card under test (1,2). An example of the subtle type of overstress problem is the requirement of open-collector drivers (shown in Figure 4) to have V_{CC} (Nominal, of 5V) applied if the output transistors are to retain their rated maximum breakdown voltage. (3). Unless Vcc is applied before the 30V stress, the output transistor breaks down at its $V_{(BR)CEO}$ which is much lower than its rated $V_{(BR)CER}$. Supply sequence sensitive circuits pose severe test maintenance and design problems. A lack of appreciation of the limitations of integrated circuits is all too common and is usually due to the communications gap between vendors and users.

(b) MOS (Metal Oxide Semiconductor) LSI products are growing in complexity and playing a much greater part in current designs. Unfortunately, several interface, handling and application problems have become apparent. Some of these difficulties are due to a poor appreciation by the user of the true internal structure of the device.

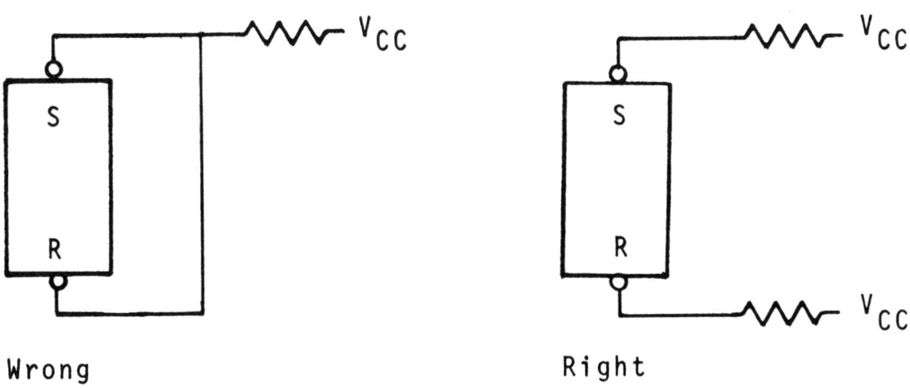

Figure 3. Testability Hazard

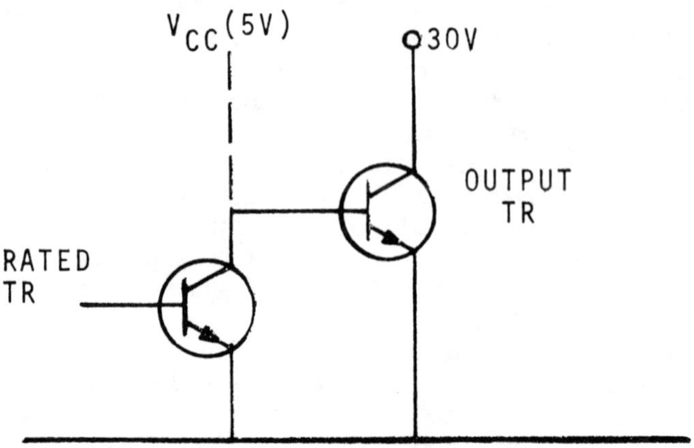

Figure 4. Output structure of a TTL Decoder/Driver

The charge injection problems of LSI MOS dynamic shift registers are typical of interfacing problems encountered. The example shown in Figure 5 involves a P-channel MOS shift register which requires a high level clock driver.

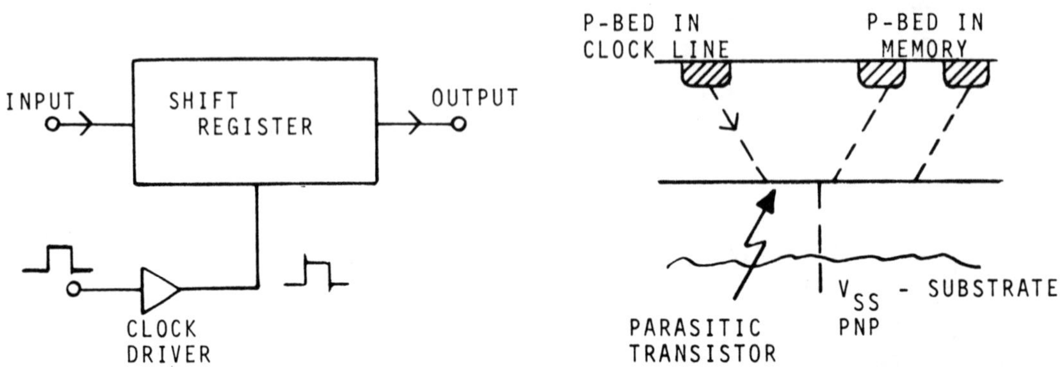

Figure 5. Clock Spike Problems in P-channel Shift Registers

Positive going clock spikes result in stored data loss through parasitic PNP transistor action. Such spikes therefore require special attention which involve adding clamp diodes to the circuit(4). More recently, microprocessors have taken a significant place in circuit designs. One of the most popular of these requires a 5V and a -9V supply. Unfortunately, the component is supply sequence sensitive such that, unless the 5V line rises before the -9V line the internal reset circuitry does not operate and a long software initializing sequence is required prior to use(5).

(c) Discrete component circuit design seems to be a dying art. Consequently, it is an area in which a wide range of mistakes are made. Judging by field failure reports, most of the component failures occur in circuits which have high (or hidden) transient stresses. The examples following have been chosen to illustrate this latter factor. However, errors of tolerancing, failure to account for component aging and electrical noise problems are quite common.

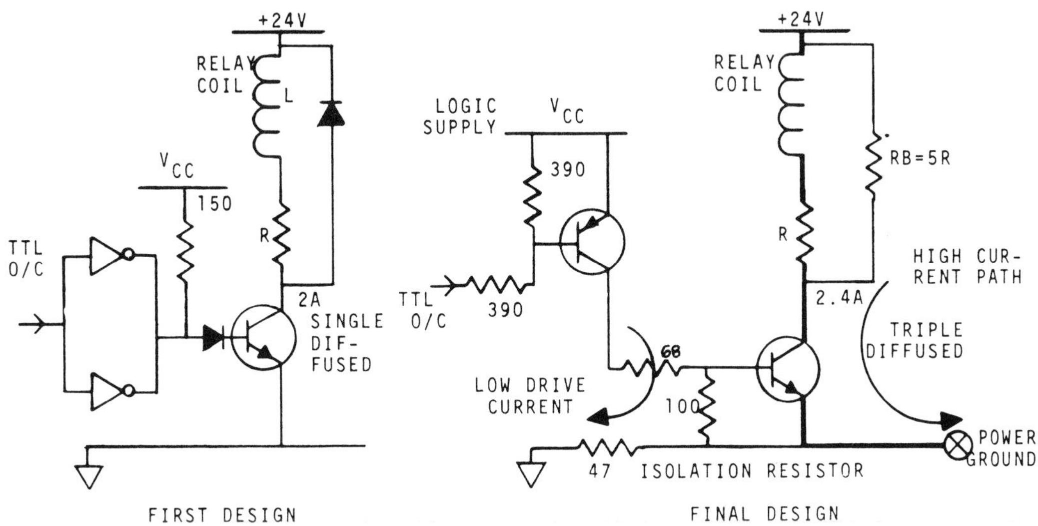

Figure 6. Relay Drivers

The first relay drive circuit shown in Figure 6 uses a single diffused transistor, no base emitter resistor, a catching diode and common ground for both the logic supply and the power ground. The diode can have several unwanted effects, the worst of which is a serious reduction in relay contact life (6) due to long relay release times and contact bounce. The lack of base emitter resistor lowers the breakdown voltage of the transistor, increases the switch-off time of the transistor, and thereby increases the dissipation of the transistor which is single diffused and, therefore, least qualified to handle it. The common ground of the first circuit is a special hazard to TTL circuitry, since TTL is sensitive to ground-borne noise. The new design attempts to solve the problems by using a damping resistor (7) which allows the relay to drop out fairly quickly. A base-emitter resistor preserves the high breakdown voltage of the triple diffused device, and a separate ground is used for the power circuits (8). It is important to note that in a simple bench test the differences between the circuits is not apparent. This latter point is also true of the circuit shown in Figure 7.

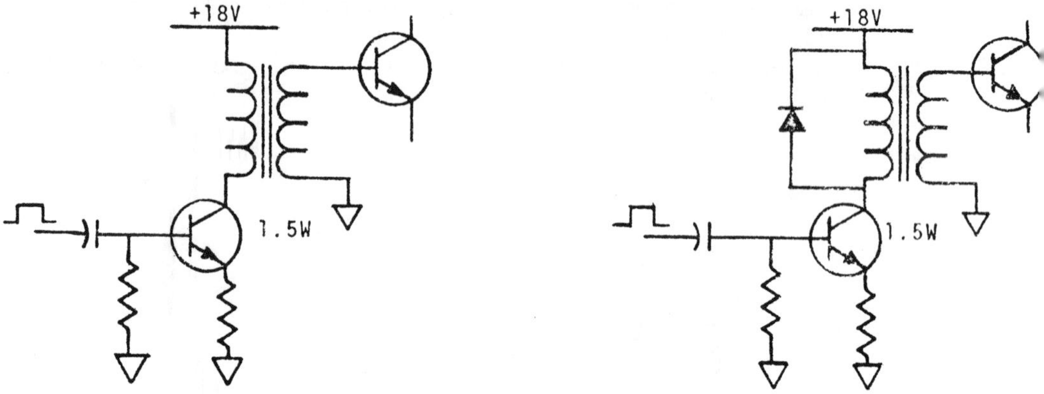

Figure 7. Catching diode reduces transient stress

The left circuit in Figure 7 has a higher voltage stress due to the absence of the catching diode. In terms of computed Mean Time To Failure, the first circuit offers 530,000 hours and the second almost 1,000,000 hours, (9).

DERATING

It is generally recognized in accordance with the laws of chemical and physical degradation, that increasing the electrical, thermal and mechanical stresses on electronic parts will decrease either the time to failure or the time required to accumulate a given amount of degradation. Conversely, decreasing these stresses will reduce the rate of degradation, reduce the probability of catastrophic failure, and thus improve reliability. Derating is defined as the practice of limiting these stresses on electronic parts to levels well within their specified or proven capabilities, in order to enhance reliability. The benefit of derating parts in application is clearly established. Even the best parts, when operated at maximum rated stress levels, do not have low enough failure rates for economical operation and maintenance of complex systems. A major contributing factor to the success of many reliability programs has been a conservative design approach incorporating substantial derating of parts.

In some respects, derating practices are subjective, in that either the manufacturers' ratings or the users' procurement specification ratings are used as the basis from which to derate. The published ratings always contain margins of safety. These margins are within the limiting values of stress conditions which would result in permanent impairment of the serviceability of individual parts. The safety margin of a part for a given stress is a function of the manufacturer's design, as well as the uniformity and repeatability of his production. It is expected, therefore, that these margins will not only vary considerably between part numbers of a given part type, but will vary between interchangeable parts supplied by different manufacturers. In general, these margins, related to the ultimate capabilities of the parts, are unknown.

Tables 1-14 give basic information for the derating of parts. The specified derating ratios and applicable notes will assist the designer in obtaining reliable operation of parts. It must be emphasized that the designer should evaluate all parts to the requirements of his applications, since he is responsible for the implementation of adequate deratings. The recommended derating factors are based on the best information currently available (10-14). The derating factors indicate the maximum recommended stress values and do not preclude further derating. When derating, the designer must determine the difference between the part specification environmental and operating conditions and part actual environmental and operating conditions of the application, derate, if applicable, and then apply the recommended derating factor(s) contained herein. Parts not appearing in these guidelines are lacking in empirical data and failure history. Since the operating characteristics for such parts cannot be guaranteed, it is a good policy to derate generously to provide an additional margin of safety. For new devices it is advisable to obtain vendor life test data for analysis which may lead to suggested derating figures. This is particularly true of step-stress data where distinct breakpoints may be identifible such that reducing stress beyond the breakpoint will not significantly affect device failure rate.

Definition Of Stress Ratios

The stress ratios that apply herein are defined as follows:

Voltage Stress (S_V) - ratio of the actual operating voltage to the maximum rated voltage of the part.

$$S_V = \frac{V\ \text{applied}}{V\ \text{rated}}$$

Current stress (S_I) - ratio of actual operating current to the maximum rated current of the part.

$$S_I = \frac{I\ \text{applied}}{I\ \text{rated}}$$

Power Stress (S_P) - ratio of the actual operating power to the maximum rated power of the part.

$$S_P = \frac{P\ \text{applied}}{P\ \text{rated}}$$

Table 1. Resistors, Fixed

Part & Type	Parameter	Maximum Stress Ratio	Ref.	Remarks
Carbon composition per MIL-R-39008, RCR; MIL-R-11, RC	Power	0.50	13,14	
Film per MIL-F-39017, RLR; MIL-R-22684, RL	Power	0.50	13,14	
Film per MIL-R-10509, RN; MIL-R-55182, PNR	Power	0.50	13,14	
Power film per MIL-F-11804, RD	Power	0.70	10	
Wire wound per MIL-R-39005, PBR; MIL-R-93, RB	Power	0.40	13,14	
Power wire wound Per MIL-R-39007, PWR; MIL-R-26, RW	Power	0.50	13,14	
Power wire wound per MIL-R-39009, RER; MIL-F-18546, RE	Power	0.50	13,14	

Table 2. Resistors, Variable

Part & Type	Parameter	Maximum Stress Ratio	Ref.	Remarks
Wirewound lead screw per MIL-R-39015, RTR; MIL-R-27208, RD	Power	0.70	13,14	
Potentiometer per MIL-R-12934, RR	Power	0.20	10	

'able 2 (Cont'd)

W semi-precision er MIL-R-19, RA; IL-R-39002, RK	Power	0.50	10,13	
W power per IL-R-22, RP	Power	0.60	10	
on WW trimmers per IL-R-22097, RJ	Power	0.20	10	
omposition pot per IL-F-94, RV	Power	0.70	10	

Table 3. Capacitors, Fixed

Part & Type	Parameter	Maximum Stress Ratio	Ref.	Remarks
aper or plastic per IL-C-14157, CPV; IL-C-39022, CHR; IL-C-19978, CQ & CQR	Voltage	0.50	10,13, 14	The ripple current in all capacitors should be limited to values which do not bring the temperature above the derated rating. The peak voltage (including surges and transient) should be limited to 0.5 of the manufacturer's rating.
ca (molded) per L-C-5, CM; L-C-39001, CMR.	Voltage	0.50	10,13, 14	
tton mica per L-C-10950, CB	Voltage	0.50	10	
ass per MIL-C-23269, R	Voltage	0.40	10,13, 14	

Table 3 (Cont'd)

Ceramic per MIL-C-11015, CK; MIL-C39014, CKR	Voltage	0.50	14	
Ceramic per MIL-C-20, CC	Voltage	0.30	10	
Tantalum, solid electrolytic per MIL-C-39003, CSR	Voltage	0.50	14	
Tantalum non-solid electrolytic per MIL-C-39006, CLR; MIL-C-3965, CL	Voltage	0.50	13,14	
Aluminum per MIL-C-39018, CU; MIL-C-62, CE	Voltage	1.0	use at 0.80 min to prevent gas formation	

Table 4. Capacitors, Variable

Part & Type	Parameter	Maximum Stress Ratio	Ref.	Remarks
Ceramic per MIL-C-81, CV	Voltage	0.50	5	
Glass per MIL-C-14409, PC	Voltage	0.50	14	

Table 5. Microelectronic Devices

Part & Type	Parameter	Maximum Stress Ratio	Ref.	Remarks
Digital	Fan-out	Derate one load or	11,14	The maximum jun tion temperatu

able 5 Cont'd

		10% whichever is greater.		shall not exceed: a) 90 degrees C for CMOS, b) 110 degrees C for TTL.
	Operating frequency	0.75		
inear & hybrid	Current	0.65	11,14	The maximum junction temperature shall not exceed 100 degrees C.
oltage regulator	Current Power	0.60 0.50	11,14	The maximum junction temperature shall not exceed 100 degrees C.

Table 6. Transistors

Part & Type	Parameter	Maximum Stress Ratio	Ref.	Remarks
neral purpose	Power Current Voltage	0.30 0.50 0.60	14	The maximum junction temperature for all types of transistors shall not exceed 100 degrees C.
wer	Power Current Voltage	0.30 0.50 0.60	14	
itching	Power Current Voltage	0.50 0.50 0.60	14	
r	Power Current Voltage	0.20 0.50 0.60	14	

Table 6 (Cont'd)

Unijunction	Power Current Voltage	0.30 0.50 0.60	10,14	

Table 7. Diodes

Part & Type	Parameter	Maximum Stress Ratio	Ref.	Remarks
General purpose, switching and SCR	Power PIV Surge current Forward current	0.30 0.50 0.30 0.40	10,14	The maximum per missible junctic temperature shal not exceed 1(degrees C.
Zener, reference	Power Forward current	0.30 0.40	10,14	Zener/reference current shall b limited to no mor than $I_Z = 0.5(I_{Zmax} + I_{Znom})$ Where, I_{Zmax} = specifi maximum operati reverse current I_{Znom} = nominal specifi operati reverse current
Varactor	Power Breakdown voltage, Forward current	0.30 0.75 0.60	10,14	

Table 8. Inductive Devices

ᵓart & Type	Parameter	Maximum Stress Ratio	Ref.	Remarks
ansformers & induc- :s (audio, power and -power pulse) per :-T-27			10	All inductive devices must be isolated or shielded to eliminate effect of radiation. Hot
ils, radio frequency r MI-T15305			10	spot temperature shall not exceed rated insulation temperature. This is estimated by:
ansformer pulse, low wer per MIL-T-21038			10	$T_{HS} = T_A + 1.1x\Delta T$ where, T_{HS} = hot spot temperature (degrees C) T_A = ambient temperature (degrees C) ΔT = temperature rise (degrees C)

Table 9. Relays

Part & Type	Parameter	Maximum Stress Ratio	Ref.	Remarks
·lays per .L-R-5757; .L-R-6106; .L-R-19523; .L-R-39016; :L-R-83725; :L-R-83726;	Resistive current, Inductive current, Filament current,	0.75 0.40 0.10	10,11, 13,14	The relay switching current derating shall be in accordance with the specified load.

Table 9 (Cont'd)

MIL-R-19648	Capacitive current,	0.75		
	Motor current	0.20		

Table 10. Circuit Breakers

Part & Type	Parameter	Maximum Stress Ratio	Ref.	Remarks
Circuit breakers per MIL-C-5809; MIL-C-39019	Current	0.80	14	

Table 11. Switches

Part & Type	Parameters	Maximum Stress Ratio	Ref	Remarks
Push-button & sensitive (limit) per MIL-S-8805	Resistive current	0.75	10,14	All switches sh be current dera according to load indicated.
Lighted push-button per MIL-S-22885	Inductive current	0.40		
Rotary per MIL-S-3786	Filament current	0.10		
Toggle per MIL-S-3950	Capacitive current	0.75		
	Motor current	0.20		

Table 12. Indicators

Part & Type	Parameter	Maximum Stress Ratio	Ref.	Remarks
ncandescent	Voltage	0.80	12	
eon	Voltage	0.80	13	
ED	Power	0.80	13	

Table 13. Connectors

Part & Type	Ref.	Remarks						
		Number Contacts	Contact Size	Wire Size 20	22	24	26	28
nnectors	11	1 - 4	20	6.0	4.5	3.3	-	-
		1 - 4	22	-	4.0	3.0	2.2	1.5
		5 - 14	20	5.0	3.5	2.7	-	-
		5 - 14	22	-	3.0	2.3	1.8	1.2
		15 & UP	20	3.7	2.5	2.0	-	-
		15 & UP	22	-	2.2	1.8	1.5	1.0

Maximum full time current (amps) per contact at 25 degrees C ambient.

The maximum current may be carried by only 20% of the contacts or by one contact, whichever is greater, at one time. The other contacts shall be limited to 100ma.

Table 14. Crystals

Part & Type	Parameter	Maximum Stress Ratio	Ref.	Remarks

Table 14 (Cont'd)

Crystal	Power	0.50	11	Power refers t drive level.

CLASSIFYING DESIGN ERRORS

While the range of possible design errors is almost limitless, it i possible to classify errors into specific groups. These classificatio headings provide a degree of awareness of just what form errors can take While a checklist is a list of specific past experience, a classificatio list provides a means of avoiding future errors.

1.	Class EXCESS	The design has an excessive number o components, some having no purpose or value.
2.	Class COST	The components used are of excessive cos relative to the function of the design.
3.	Class TEST	The design is impossible or difficult to tes with conventional or automatic test equipment Also the circuit lacks properly chosen tes points.
4.	Class DEGRADE	The performance of the design will degrad slowly as the components age.
5.	Class NUISANCE	The design is such as to cause safety shutdow circuits to operate, when there is no safet risk.
6.	Class EMI	Circuit chosen will produce excessiv Electromagnetic Interference (EMI) or i excessively sensitive to EMI and will requir expensive line and signal filters.
7.	Class MECHANICAL	The physical nature of the components chose prevents good packing practice or makes th design susceptible to vibration and shock.
8.	Class SOURCE	The components chosen are single source and/or difficult to obtain. Alternativel they are not listed on the preferred part list, or are from a poor vendor or have r reliability history.

9.	Class DOMINO	A failure of one component will lead not only to system failure but directly to the subsequent destruction of a large number of other components or an expensive component.
10.	Class SPECIFICATION	The design exercises the components outside the limits set by the manufacturer or subjects them to a stress which degrades the MTBF below that specified.
11.	Class SAFETY	The circuit fails or has an operating condition which makes the system or the circuit itself dangerous to operators or test personnel.
12.	Class NOISE	The design is highly sensitive to any small, random noise signal.
13.	Class CHOICE	The choice of circuit configuration and type leads to excessive complexity and/or makes the circuit undesignable.
14.	Class TOLERANCE	A build-up of the end limit tolerance of several components will result in system failure.
15.	Class THERMAL	The design has thermal problems, such as insufficient heatsinking or the placement of high thermal emitters close to heat-sensitive components.
16.	Class DRAW	The circuit is drawn in such a manner as to make the circuit difficult to recognize, understand, test or fault find. Also uses non-standard symbols.
17.	Class DANGER	The components chosen contain dangerous materials or store energy so that handling is dangerous.
18.	Class SHELF	The storage life of the components used is such that the equipment cannot be put into service without extensive re-work.
19.	Class SEQUENCE	The switch on/switch off sequence of signal sources, power supplies or relays is a critical factor in preventing component damage or initializing difficulties.

20.	Class NARROW	The circuit operates only over a narrow temperature and/or voltage range.
21.	Class CHANGED	The components listed for the production unit are significantly different than those used to build prototypes.
23.	Class SELECTED	The circuit requires specially selected components or (worse) components selected on final test.
24.	Class PROTECTION	The circuit lacks essential protection circuitry against internal transients or external factors such as static electricity and high input signals.
25.	Class INTERACT	The circuit requires several interacting adjustments to be made.
26.	Class PARASITIC	The circuit is prone to parasitic oscillation at specific voltage levels or under certain load conditions.
27.	Class CONTAMINATION	Fungus, moisture or dust will cause the components chosen to degrade or fail.
28.	Class FRAGILITY	A single component having very poor reliability dominates the failure rate of the equipment and can be easily replaced by a reliable item.
29.	Class MISSING	The circuit has components missing which are essential to its long-term reliable operation.
30.	Class DISSIPATES	The circuit has a design error which causes components to dissipate an unnecessary amount of heat.
31.	Class FIRE	The components chosen are known to have significant fire risk.
32.	Class OBSOLETE	The design uses obsolete components which can be replaced with superior new types.
33.	Class LIST	The component numbers listed are in error or insufficiently complete to be certain of purchasing the component quality needed.

34. Class RULES — The circuit does not reflect component manufacturers design rules or published application limitations.

35. Class PRESET — The circuit uses preset components which can easily be eliminated by redesign.

36. Class RACE — The logic output of the circuit is critically dependent on minor timing delays between logic elements.

37. Class SNEAK — The circuit or system has a sneak path which bypasses the intended circuit.

CONCLUSION

The cost savings and competitive advantages of getting a design 'right first time' should not be in doubt. Time saved alone should meet the cost of mounting the program described. Nothing shown here, however, should be considered as taking the place of a formal review meeting. Essentially, what has been described is a back room activity. The operations of a component specialist, although based within a Reliability and Maintainability department, can traverse departmental boundaries providing a service of value is seen to be available. According to organizational psychologists, professional employees regard a growth in job competence as a prime element in work satisfaction and motivation. This factor has been evident and a high level of support for the activities described has been obtained.

REFERENCES

1. Boswell, F.R., "Designing Testability into Complex Logic Boards", Electronics, 14 August 1972, p116-119.

2. Henckels, L. and Schneider, D., "Testability of Logic Circuit Boards", Communitronics, November/December 1974, p11-14.

3. Morris, R.L. and Miller, J.R., (Editors), Designing with TTL Integrated Circuits, McGraw-Hill, New York, N.Y., 1971.

4. Wilnai, A., "Eliminating Clock Waveform Imperfections in MOS Dynamic Circuits", Computer Design, September 1973, p94 and 95.

5. Buchanan, G., Private Communication.

6. Thomas, E.U., "The $2.5 Billion Wayward Diode", Evaluation Engineering, September/October, 1973, p22 and 23.

7. Wetzel, K., Uses of Triple-Diffused Transistors, Siemens Application Report B11/1173.101.

8. Boaen, V., "Designing Logic Circuits for High Noise Immunity", IEEE Spectrum, January 1973, p53-59.

9. Deger, E. and Jobe, T.C., "A Design Factor in Reliability", Electronics, August 30 1973 p83-89.

10. ____________, MIL-HDBK-217B, Reliability Prediction of Electronic Equipment, U.S. Department of Defense, Washington, D.C., 20 September 1974.

11. ____________, C-4 Program derating Policy for Electronic/Electromechanical Parts, MSD Parts Reliability, Bulletin No. 49, 27 March 1974.

12. Curran, W., "Pick the Proper Subminiature Lamps", Evaluation Engineering, January/February 1974, p4-10.

13. ____________, Design Data NoteBook, Reliability Design Guidelines, Technical Memorandum, TM-1037/003/00, Computing Devices Company, Ottawa, Canada, February 1976.

14. ____________, Reliability Design Guidelines, Avionic Equipment, Report No. MDC A3807, McDonnell Aircraft Company, St. Louis, Missouri, 17 February 1976.

15 THE SELECTION AND SPECIFICATION OF PARTS

J. A. Roberts

Parts must be selected to meet the performance and contractual requirements of a program. A weak parts selection and control activity is the first and most obvious defective element found in any critical evaluation of a contractors product. Poor parts choice is instantly perceived in Design Reviews. When specific part selection criteria have not been written into specifications and contracts, part selection errors have produced chaos. Even when the eventual customer has not been specific, the prime contractor should develop guidance documentation for the subcontractors and in-house designers. At the very minimum, repairable subassemblies should be developed using multisourced parts which are chosen to meet the environment to which the equipment is exposed. These criteria, together with a contractually binding Mean Time Between Failures (MTBF) specification, based on (1) are absolute minimum requirements in any system purchase. Military production programs are often at the opposite extreme with part selection documents such as (2) or an agency approved prime contractor Preferred Parts List (PPL).

In this chapter the factors involved in choosing parts, controlling the choice of parts and specifying parts will be discussed. The reasons for a strong parts control program may not, however, be obvious to those without direct exposure. Therefore a discussion of the need for part selection criteria and control will be presented.

THE NEED FOR CONTROL AND GUIDELINES

Perhaps the most solid evidence of widespread poor parts control is that provided by Military Parts Control Advisory Group (MPCAG) of the U.S. Defense Electronics Supply Center (DESC). They have reviewed the part nominations of some 60 contractors and found that 59% of the non-standard parts could be replaced with preferred standard types (3). The cost of each new non-standard part involves a drawing (specification) costing from \$500 to \$8000, qualification tests costing between \$5000 and \$25,000 and a government life cycle cost (10 years) of \$1200. Since Integrated Circuit (I.C.) qualification is one of the most costly items, the problem is likely to increase as I.C. usage grows.

Several examples of not using parts control have been published (3,4). Quite apart from the costs involved in the use of non-standard parts, lack of parts control can lead directly to reliability problems. In a study by MPCAG of Test Sets which showed 60% failures after short term storage, 69 of the 118 non-standard parts were inadequately described to assure reliable performance. Fourteen parts were known to part specialists to be trouble causing parts, of which all but one could be replaced with high reliability military parts. Three of the parts had been the subject of Government-Industry Data Exchange Program (GIDEP) warnings.

A study of a particular prototype system carried out by ARINC Research Corporation (4) showed the usual mixture of strictly commercial parts, non-hermetic parts, unreliable parts, unqualifiable parts and of course, single sourced parts. Almost inevitably design changes are required when standard parts have to be incorporated. Therefore, the need to use standard parts from 'day one' of any design effort which could reach production status cannot be over emphasized. Means by which parts control can be achieved and criteria for the selection of parts will be described in the following sections.

DESIGN FACTORS IN PARTS SELECTION

The prime object of any system is that it must meet the basic operational performance. High reliability, good maintainability, electro-magnetic compatibility and other desirable goals are of course important, but they are secondary factors. The struggle between the basic performance requirements and the Reliability and Maintainability requirements are often reflected in the part selection problems. Rapid technical advances have occurred in part performance, particularly in I.C.s. Valuable performance and sometimes reliability gains can be made by choosing from the latest part range available. Where no military part range has been established, great care is needed to ensure the following:

(a) the new part (range) is truly superior,
(b) the new part will become a 'defacto' standard and thereby multi-sourced,
(c) the new part is qualifiable to a grade equal to standard military parts of roughly similar function.

There are always tradeoffs in the choice of any part. A typical example which often arises is the choice of logic family. An investigation of the relative merits of, for example, emitter coupled logic (ECL) versus standard transistor-transistor logic (STTL) in a system can be complex and expensive. What is required is a detailed technical study which provides a numerical basis for a design decision. Good examples of such studies are provided in the literature (5,6). In any trade study which is to be presented to a procurement agency, very detailed presentations together with clear comparisons are required and expected.

EXAMINATION OF CUSTOMER DOCUMENTS

The responsible part specialist must regard the customer requirements as the prime guidance documents used in parts selection. The nature and complexity of the contract, will largely determine the extent of customer control. Considerable variation now exists between different contracts even when let by the same agency of government. A source of this variation is the experimentation that has been implemented to achieve Design to Cost (DTC), Life Cycle Cost (LCC), Minimum Procurement Cost (MPC) or other prime overall customer costing aim. Considerable care is therefore needed when examining the customer bid set. In preparing a proposal based on the customer bid set, the parts specialist must have an understanding of the exact overall nature of the contract. If the aim is specified for example as a DTC, the part specialist must not only illustrate compliance to the directed part quality, but must also ensure that the parts chosen enable the overall cost to be met.

The prime contractors must ensure that the aims of the final customer are properly reflected in their communications with subcontractors. This means that procurement documents such as Specification Control Drawings (SCD), Statements of Work (SOW) and other control documents fully reflect the customers requirements. Where a subassembly or part has to be specially developed for a particular contract the rights to the design may become the subject of some negotiation and particular care is required in such areas. Familiar documents such as (7) defining SCDs are inadequate to deal with specialized procurement. A complete reading of the customer documents could reveal more precise requirements such as reference to (8).

When an agency produces a design specification, it is common practice for Parts, Materials and Processes to be referenced in a single paragraph which simply provides imposition of that agency's general specification for electronic equipment. If for example, the U.S. Army references (2), over 300 significant part specifications are thereby applicable. In addition, such a document could impose the non-standard part reporting requirements of (9). This allows the agency to monitor the degree of control being applied to parts selection.

It is sometimes valuable to break down the agency's general specification into a simple list of prime part headings. Such a list enables others to gain a rapid understanding of the requirements of a program. Usually (10) is the core document of the procurement. Everyone involved in the design should be given a clear understanding of the requirements. Clearly, thereafter, a policing operation is required to ensure that the ground rules are understood and applied.

PARTS CONTROL AND STANDARDIZATION

Parts selection must not be left to chance. A design engineer, presented with a creative and possibly difficult task, requires direct and simple aids to ensure correct part selection. One of the best ways of assisting a designer is through a PPL. This must be drawn up or reviewed with the exact program requirements in sharp focus. In general, the parts used for military applications will be chosen from the military parts lists. The existence of the part in a military specification is a first step in its possible selection (11). It is necessary that the part appears in the Qualified Parts List (QPL) and that there are qualified vendors. Certain parts have limited qualification and may not be approved by all services. The order of precedence for selecting parts is specified in (12).

When a Non-Standard Part (NSP) has to be selected, it is often necessary to initiate an NSP request to the customer (11). If the part is approved, an SCD or related document has to be prepared. In most contracts, a qualification procedure, vendor audit and extensive part screening procedures will have to be imposed. The introduction of a new NSP is a serious undertaking and is not viewed lightly by most customers (13). A very serious attempt must be made in Military Systems to maximize the use of Standard Parts such as Established Reliability (ER), passive components, TX (Test-Extra) semiconductors and microcircuits (14). While the cost of such parts exceeds the initial cost of Non-Standard Parts the system Life Cycle Cost and the cost of initial support elements such as the SCD will be lower.

PURCHASE SPECIFICATIONS AND DRAWINGS

Currently in military systems it is usually not possible to reduce Non-Standard Part usage to much below 10% or 20% of the total part count. Digital systems tend toward the larger figure while analog systems can often reach lower due to the wide availability of ER passive parts. Inevitably, purchase specifications are required for a rather large number of parts due to the large overall parts count of a modern system. Usually, even for the best controlled system design of any complexity, somewhere between 50 and 200 NSP purchase specifications are required.

Part purchase specifications are usually written to conform with (7). The types of drawings defined by (7) are;

(a) Envelope Drawings which define input/output parameters, space envelope, environmental requirements and any other characteristics (i.e., Reliability, Maintainability, Electromagnetic Compatibility, Human Factors, etc.) of an item to be developed. When the development is completed the Envelope Drawings evolve to Specification or Source Control Drawings,

(b) Specification Control Drawings which depict existing off-the-shelf items that are used without alterations, selections, or limitations,

(c) Source Control Drawings which define existing off-the-shelf items that, for reasons identified in the drawing (i.e. Maintainability, Reliability, etc), may be procured only from a specified source.

(d) Altered Item Drawings which define existing off-the-shelf items (commercial or Military parts) with complete detail of alterations required,

(e) Selected Item Drawings which define existing off-the-shelf items (commercial or Military) with details of additional selections for form, fit, tolerance, performance or reliability.

When complex electronic parts are to be purchased it is usual to adopt the book form part drawing which uses multiple 8 1/2 by 11 inch sheets. The drawing then takes on a format akin to a specification and it is common practice to use the six part specification style described by (15). In general, section headings included are;

(a) Scope,
(b) Applicable documents,
(c) Requirements,
(d) Quality Assurance Provisions,
(e) Preparation for delivery,
(f) Notes.

The paragraphs and precise tests called for in a part specification are usually based on the nearest applicable Military Part Specification or the Vendors published data describing their standard procedures. Exactly how close a purchase part drawing conforms to Military Drawing Standards is controlled by the agency procurement specifications. Reference (8) defines a classification system which controls the Form and Category of Drawings. This specification describes the extent to which (7) shall apply in a particular development. In a similar manner Specifications are controlled by reference to the Types and Forms described by (15), as determined by an appropriate reference to the definitions provided by (16). In general, system and subassembly specifications are developed more closely under the guidance of the Military Specification Control documents while Part Drawings conform to the Military Drawing Control documents.

The cost of producing adequate purchase specification depends on the complexity of the part that is to be defined and the screening/qualification requirements specified. The overall complexity of a purchase can vary considerably depending on whether the item is a standard production part which is one of a similar series or a custom development of a new complex assembly (17, 18). Factors involved in procurement may include:

Pre-Purchase

(a) vendor data sheet study, comparison and recompilation,
(b) vendor representative and vendor engineer interview,
(c) visit to vendor facilities for audit,
(d) qualification and test of sample lots,
(e) government Industry Data Exchange Program (GIDEP) search for data,
(f) questions to other users of the vendor parts,
(g) production of SCD to accommodate vendor and purchaser requirements,

(h) vendor review of SCD, agreed changes implemented and special part numbers assigned (where required),
(i) review of vendors internal production specifications and special part numbers based on SCD,
(j) Statement of Work (SOW) produced by purchasers,

Post-Purchase

(k) incoming inspection of vendor's lot qualification data and compliance statement,
(l) incoming inspection of vendor parts to SCD,
(m) monitoring of vendor Engineering Change Orders (ECO) and configuration control documents,
(n) periodic requalification,
(o) feedback to vendor of failure data.

If most of the above actions are implemented the cost of a procurement can be extremely high for each part. The most valuable of the above is the creation of an SCD which describes a multisourced part. Even so the cost of producing an SCD is quite large, as shown in Figure 1, (19). On the other hand the possible losses which can be incurred through not having SCD's can be startling (17, 20, 21). Without feedback to vendors of failure data based on the customer's tests of a part, bought and in-house tested to the SCD, the SCD can be become meaningless. Therefore, vigilance is the key item in part procurement.

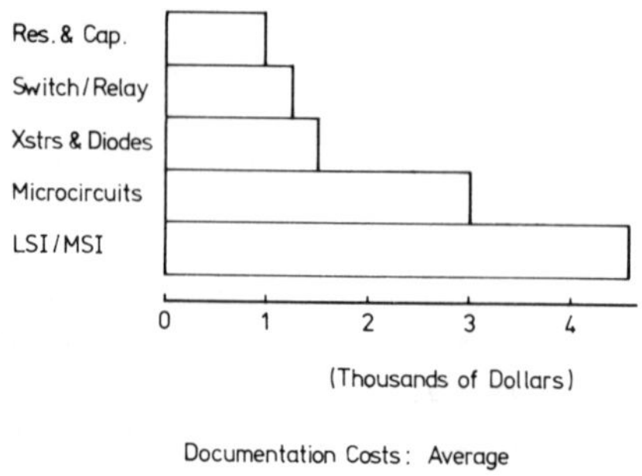

Figure 1. Cost of Producing an SCD.

QUALIFICATION AND QUALITY CONFORMANCE

It would be reasonable to expect that a vendor would have internal quality assurance (Q.A.) procedures which ensured that parts were subject to initial Qualification to its published data and would be subject to requalification at defined intervals. Not all vendors are even this careful. Qualification implies more than a simple test to verify published characteristic data. Perhaps the relevant paragraph of (15) best describes the intent;

Qualifications: Qualification, as used in this Standard, refers to the verification or validation of item performance in a specific application. This qualification results from design review, test data review, and configuration audits. Where performance qualification of a design or an end item (including its components) is required, either on a one-time basis or a periodic basis, to achieve design approval, proof of producibility, assessment of production or other reason, provisions for such qualification testing shall be stated in this paragraph. Requirements shall be included which state the conditions for testing, the time (program phase) of testing, period of testing, number of units to be tested, and other requirements relating to qualification or requalification.

Qualification activities should be included in the requirements section of a purchase specification. The Q.A. section should include mention of the particular tests required in initial and periodic qualification. Quality conformance inspections are listed in order to verify that all requirements are achieved. These inspections include tests and checks of performance and reliability, measurement of physical characteristics, verification of workmanship and tests for environmental performance.

The qualification costs of NSPs is significant, as shown in Figure 2 (19). However, the cost of using parts that are not qualified is highly damaging.

Figure 2. The Qualification Costs of NSPs

CONCLUSION

Parts selection and specification errors can impact cost, reliability and delivery dates. Efforts made in the area of parts selection can repay the cost incurred many times over. The attitude of many part vendors is one of almost total irresponsibility (22). Control is vital. In addition to control, an effective aid to good design should be provided. In this way, enthusiasm for parts control can be achieved; this is the vital factor in a successful effort (23).

REFERENCES

1. ______________, MIL-HDBK-217, Reliability Prediction of Electronic Equipment, U.S. Department of Defense, Washington, D.C.

2. ______________, MIL-P-11268, Parts, Materials and Processes Used in Electronic Equipment, U.S. Department of Defense, Washington, D.C.

3. Schade, R.W., "Military Standardization of Electronic Parts", IPC Technical Review, October 1976.

4. ______________, Investigation and Selection of Qualified Components Forcasts, GIDEP E14-2879, ARINC Research Corp., Santa Ana, CA, Pub. W2-D14-TN01, June 1972.

5. Alfke, P., "Improved ECL Opens Application", Electronic Engineering Times, November 1974.

6. Atterbury, G., "The Tortoise and the Hare: A Study in Technology Blending", Digital Design, May 1976.

7. ______________, MIL-STD-100, Engineering Drawing Practices, U.S. Department of Defense, Washington, D.C.

8. ______________, MIL-D-1000(5), Drawings Procurement (Identical Items), for Electronics Command Equipment, U.S. Department of Defense, Washington, D.C.

9. ______________, MIL-STD-749, Preparation and Submission of Data for Approval of Nonstandard Parts, U.S. Department of Defense, Washington, D.C.

10. ______________, MIL-STD-454, Standard General Requirements for Electronic Equipment, U.S. Department of Defense, Washington, D.C.

11. Simonton, D.P., "Standard or Non-Standard Parts?" Machine Design, August 1962.

12. ______________, MIL-STD-143, Standards and Specifications, Order of Precedence for the Selection of, U.S. Department of Defense, Washington, D.C.

13. Klass, P.J., "Standard Military Components Use Sought", Aviation Week, March 3, 1975.

14. ______________, MIL-M-38510, Microcircuits, General Specifications for, U.S. Department of Defense, Washington, D.C.

15. ________________, MIL-STD-490, Specification Practices, U.S. Department of Defense, Washington, D.C.

16. ________________, MIL-S-83490, Specifications, Types And Forms, U.S. Department of Defense, Washington, D.C.

17. Hamblen, D., "Technical Factors in Selecting Components for Mixed Systems", Computer Design, March 1976.

18. Caldwell, G., Tichnell, G., "Selecting a Custom Microelectronic Hybrid and Vendor", Evaluation Engineering, January/February 1977.

19. Williams, J., "The Military System Manufacturers View on Microcircuit Standardization", Electronic Packaging and Production, April 1977.

20. Henderson, J., "I.C. Screening, Reliability of Rip-off", Proceedings of IEEE 1976, Reliability and Maintainability Symposium, 1976.

21. Rowen, G., (Ed), Speaking of Standards, (Part 5/6), Cahners Books, Boston, MA, 1972.

22. Ames, F.H., "Clowns, Clowns, Clowns", IEEE Engineering Management Society Newsletter, July/August 1975.

23. Zizzi, N.M., "A New Look at Standards Engineering", Evaluation Engineering, January/February 1964.

16 SCREENING

J. E. Arsenault

One of the most powerful techniques developed to achieve reliability in electronic systems is known as screening. The technique is implemented by stressing items in a prescribed manner, usually to specification limits, to induce the early failure of as many marginal items as possible before they find their way into operational use. Screening may be conducted at any combination of equipment levels, i.e., Part, Assembly, Unit and System. Unit and System screening may be jointly considered because of similarity.

Screening tests are designed around historical data patterns based on actual performance and projections related to failure mechanisms, i.e., the fundamental reasons for failure. The driving forces behind screening are reliability and cost, i.e., the desire to achieve reliability at an acceptable cost.

It was recognized early that reliability could be improved by applying screening at two principal equipment levels, i.e., Part and Unit/System, and this approach is embodied today in documents primarily concerned with screening at these levels, for example, see (1) for Parts and (2) for Units/Systems. Screening at the Assembly level is a less frequently employed technique. In the remainder of this chapter, we will discuss the subject of screening in some detail for the Part, Assembly, and Unit/System levels after a brief look at the economics of screening. Figure 1 illustrates a scheme where a screening program has been introduced at all equipment levels.

It is generally agreed that the removal of defects from systems, starting at the lowest equipment level, is an economical approach to controlling production costs in addition to field costs,which may in fact, represent warranty costs early in the life of a program. This is illustrated in Table 1 adapted from (3).

Thus, for example, if we consider the Industrial sector it costs approximately $6.00 to catch a defective part at incoming inspection as opposed to catching the part at the assembly level at a cost of $37.50. A rather complete discussion in (4) recommends the use of 100% inspection at incoming and provides an economic analysis of how such a procedure can be justified.

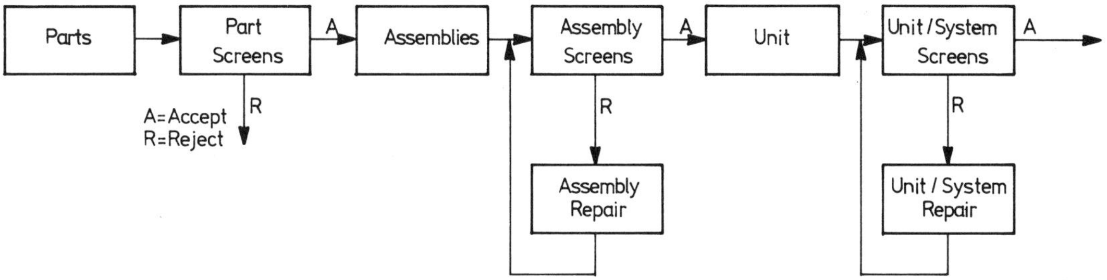

Figure 1. Screening At All Equipment Levels.

Table 1. A Comparison Of Costs At Four Stages In Four Markets For Remaining Defective Semiconductor Devices (Dollars)

	Incoming Piece Part	Board Mount Removal	System Test	Field Use
Consumer	3.00	7.50	7.50	75.00
Industrial	6.00	37.50	67.50	322.00
Military	10.50	75.00	180.00	1500.00
Space	22.50	112.50	450.00	300M

Note: Prices adjusted to 1978 costs in U.S. Dollars.

PART SCREEENING

General

It is at the part level that the largest body of screening information has been developed for general use and most of that body is concerned with the failure mechanisms peculiar to semiconductors and in particular microcircuits. This has occurred because this segment of the industry is subject to rapid changes, with ever more new and complex devices being introduced in a highly competitive marketplace. As a result, new failure mechanisms are being discovered continually and new screens devised to uncover and control them. This body of knowledge is essential as most of today's electronic systems are comprised largely of microcircuits. In the more mature areas of the parts industry screening also plays an important role in delivering parts with consistent quality.

Screened parts are readily available for use in programs where their use is mandatory or essential and this is illustrated in Table 2 which shows the range of screening designations used in the military, space, and industrial fields to denote different degrees of screening. Definitions for the screening designators will be found in (1, 5, 6, 7).

Table 2. Part Type Vs. Screening Designator

Part Type	Screening Designators
Microelectronics	A, B, C
Discrete Semiconductors (transistors, diodes, etc.)	JANTXV, JANTX, JAN
Capacitors, Established Reliability (ER)	L, M, P, R, S
Resistors, Established Reliability (ER)	M, P, R, S

The question of part screening designator selection can be a critical one when reliability and cost are in competition. The higher the screening level, the higher the cost. Generally, a trade-off can be performed between reliability (MTBF) and cost until requirements are met or until a balance is struck (see Chapter 8). In some applications the screening designations or minimums are specified.

As an illustration of the effect on reliability of the screening designators consider that a transistor screened to level JANTX could be expected to fail, on average, about 10 times less often than a device screened to the JAN level (8). However, some recent data (9) would call into question this order of magnitude gain through screening.

As has been mentioned previously, systems today contain a very large percentage of microelectronic parts compared to the total and therefore microcircuits represent most of the system failure rate. Therefore we will start the detailed part screening discussion with microelectronics. It should be noted that these discussions refer to hermetic microcircuits, plastic encapsulated microcircuits are discussed in a separate section.

Microelectronic Screening

As with any screening program, it is important to understand the underlying processes which are involved in the manufacture of the item. With the manufacture of microelectronics especially, the more complex types such as microprocessors, random access memories and read only memories involve highly sophisticated processing steps each of which introduces additional possibilities for the development of failure mechanisms. Figure 2 is an example of the different, generalized steps involved in the manufacture of microelectronic devices. Table 3 (10) indicates the failure mechanisms unique to each process step. A failure mechanism will be exhibited as failure mode, i.e., a failure of a device to meet its specified performance characteristics. Common microcircuit failures can be classified with respect to failure as follows:

(a) degradation of performance characteristics,
(b) shorts,
(c) opens,
(d) intermittents.

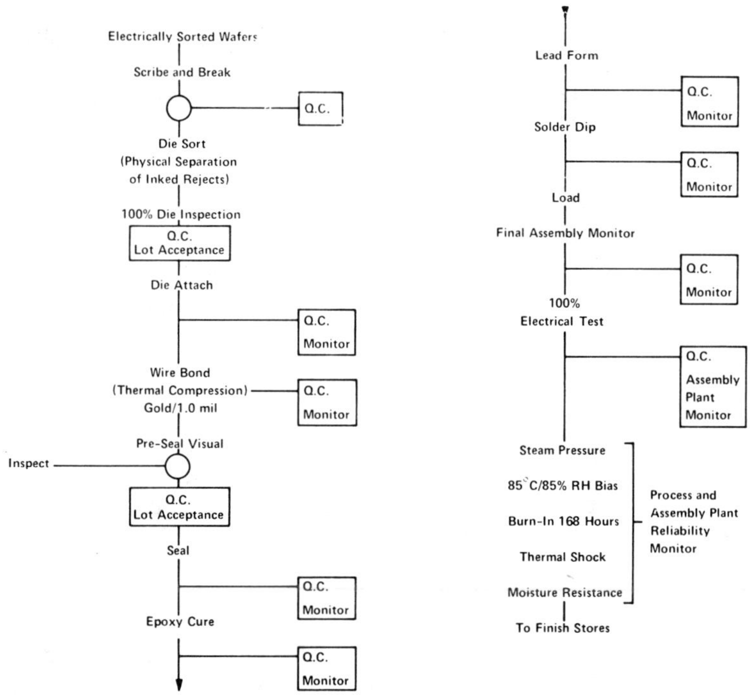

Figure 2. Generalized Microelectronic Manufacuring Processes

Table 3. Process Step Vs. Failure Mechanism

Process Step	Failure Mechanism Introduced As A Reliability Influencing Variable
Slice Preparation	Dislocations and stacking faults. Nonuniform resistivity. Irregular surfaces. Cracks, chips, scratches (general handling damage). Contamination.

Table 3. (Cont'd)

Process Step	Failure Mechanism Introduced As A Reliability Influencing Variable
Passivation	Cracks and pin holes Nonuniform thickness.
Masking	Scratches, nicks, blemishes in the photomask. Misalignment. Irregularities in photoresist patterns (line widths, spaces, pinholes).
Etching	Improper removal of oxide. Undercutting. Spotting (etch splash). Contamination (photoresist, chemical residue).
Diffusions	Improper control of doping profiles.
Final Seal	Poor hermetic seal. Incorrect atmosphere sealed in package. Broken or bent external leads. Cracks, voids in kovar-to-glass seals. Electrolytic growth of metals or metallic compounds across glass seals between leads and metal case. Loose conducting particles in packages. Improper marking.
Metallization	Scratched or smeared metallization (handling damage). Thin metallization due to insufficient deposition or oxide steps. Oxide contamination, material incompatibility. Corrosion (chemical residue). Misalignment and contaminated contact areas. Improper alloying temperature or time.
Die Separation	Improper die separation resulting in cracked or chipped dice.
Die Bonding	Voids between header and die. Overspreading and/or loose particles of eutectic solder. Poor die-to-header bond. Material mismatch.
Wire Bonding	Overbonding and underbonding. Material incompatibility or contaminated bonding pad.

Table 3. (Cont'd)

Process Step	Failure Mechanism Introduced As A Reliability Influencing Variable
	Plague formation. Insufficient bonding pad area or spacings. Improper bonding procedure or control. Improper bond alignment. Cracked or chipped die. Excessive loops, sags, or lead length. Nicks, cuts and abrasions on leads. Unremoved pigtails.

Not all failure mechanisms appear in failed devices in equal measure as shown in Table 4 (11) for two microcircuit technologies used in manufacturing memories.

Table 4. Non-Package Malfunction Summary For Bipolar And MOS Memory Device Technologies

Failure Classification	Percentage	
	Bipolar	MOS
Surface	29.00	45.09
Oxide	14.00	24.86
Diffusion	1.00	9.83
Metallization	21.00	1.74
Bond	5.00	4.05
Interconnection	29.00	4.05
Die(Mechanical)	-	1.74
Degraded Input - Circuitry	1.00	8.64
	100.00	100.00

Now that the failure mechanisms and modes of microcircuit devices have been examined, we will next look at the screens that are generally applied and their effectiveness in exposing failure mechanisms as shown in Table 5 (10).

Table 5. Screen Vs. Failure Mechanism

Screen	Defects
Internal visual inspection	Lead dress, Metallization, Oxide, Particle, Die bond, Wire bond, Contamination, Corrosion, Substrate.
X-ray	Die bond, Lead dress (gold), Particle, Manufacturing (gross errors), Seal, Package, Contamination.
High temperature storage	Electrical (stability), Metallization, Bulk silicon, Corrosion.
Temperature cycling	Package, Seal, Die bond, Wire bond, Cracked substrate, Thermal mismatch.
Thermal shock	Package, Seal, Die bond, Wire bond, Cracked substrate, Thermal mismatch.
Constant acceleration	Lead dress, Die bond, Wire bond, Cracked substrate.
Shock (unmonitored)	Lead dress.
Shock (monitored)	Particles, Intermittent short, Intermittent open.
Vibration fatigue	Lead dress, Package, Die bond, Wire bond, Cracked Substrate.
Vibration variable frequency (unmonitored)	Package, Die bond, Wire bond, Substrate.
Vibration variable frequency (monitored)	Particles, Lead dress, Intermittent open.
Random vibration (unmonitored)	Package, Die bond, Wire bond, Substrate.
Random vibration (monitored)	Particles, Lead dress, Intermittent open.
Helium leak test	Package, Seals.
Radiflo leak test	Package, Seals.
Nitrogen leak test	Package, Seals.

Table 5. (Cont'd)

Screen	Defects
Gross-leak test	Package, Seals.
High-voltage test	Oxide.
Isolation resistance	Lead dress, Metallization, Contamination.
Intermittent operation life	Metallization, Bulk silicon, Oxide, Inversion channeling, Design, Parameter drift, Contamination.
AC operating life	Metallization, Bulk silicon, Oxide, Inversion channeling, Design, Parameter, Contamination.
DC operating life	Metallization, Bulk silicon, Oxide, Inversion channeling, Design, Parameter Contamination.
High-temperature AC operating life	Same as AC operating life.
High-temperature	Inversion channeling.

As may be seen, not all screens are effective with all failure mechanisms and the problem is one of selecting the best screens for the application. Although not a simple task, there are some standard screening sequences which are available, such as offered by Method 5004 of (12). These provide for mandatory screening as a function of quality designator, as summarized in Table 6.

Table 6. Standard Screening

Screen	Quality Designator		
	A	B	C
Internal Visual	Condition A	Condition B	Condition B
Stabilization Bake	24 h	24 h	24 h
Thermal Shock	15 cycles and	15 cycles or	15 cycles or
Temperature Cycle	10 cycles	10 cycles	10 cycles
Mechanical Shock	20,000 g	No	No
Centrifuge	30,000 g	30,000 g	30,000 g
Hermeticity	Yes	Yes	Yes

Table 6 (Cont'd)

Screen	Quality Designator A	B	C
Critical Electrical Parameters	Yes	Yes	No
Burn-in	168 + 72 h	168 h	No
Final Electrical	Yes	Yes	Yes
X-ray Radiograph	Yes	No	No
External Visual	Yes	Yes	Yes

It is quite possible that the application does not demand the use of devices screened to the higher levels just discussed because of (a) cost, (b) environment, but never the less screened parts are required. In these cases it is possible to obtain parts screened by manufacturers employing screening programs, similar to Method 5004 while omitting certain requirements. Nearly all large part manufacturers have available parts screened in this manner. For example, the screens shown in Figure 3 represent a manufacturer's "vendor equivalent" to Method 5004 screening designation C.

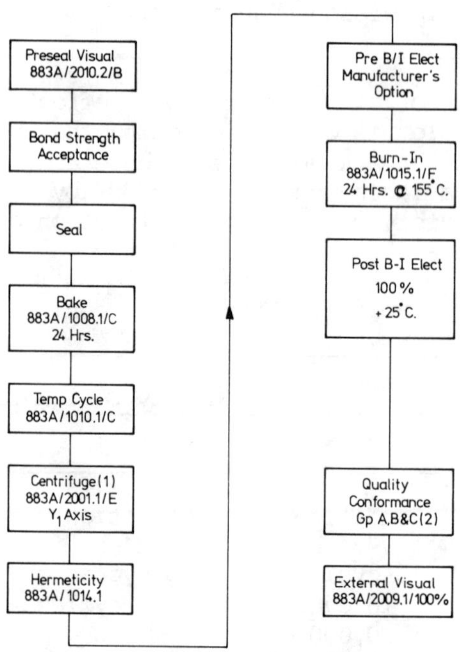

Figure 3. Vendor Equivalent - Screening Designation B.

If it is necessary to use custom screened parts or parts screened to test lower levels than those already discussed then the use of outside facilities or setting up in-house facilities should be considered. Figure 4 shows the standard test sequence offered by an outside test house which could be considered adequate for commercial applications or applications such as military test equipment.

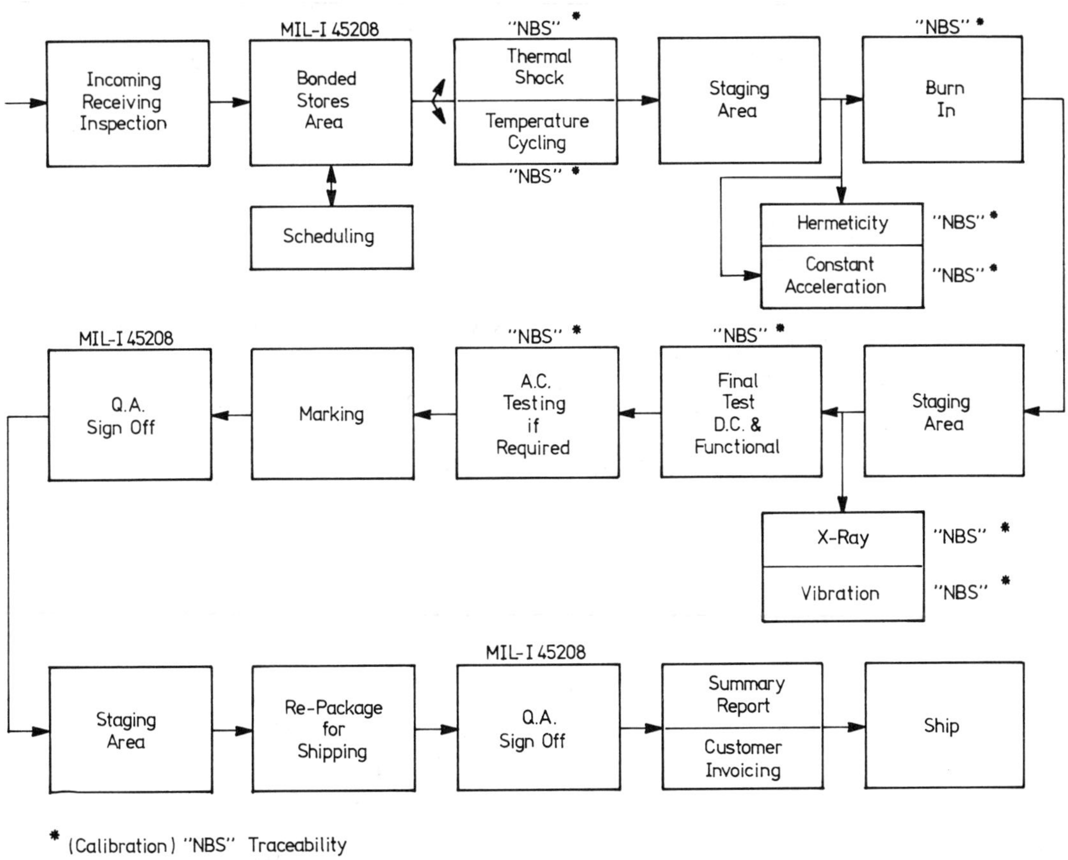

Figure 4. Screening Tests - Outside Test House

An economic study of in-house vs. external screening trade-off is given in (13).

Recent Developments

Some newer trends in the screening area are clearly evident and these are discussed briefly below.

Accelerated Testing. As we have seen in Table 4, one of the major causes of microcircuit failure is the lack of adequate process control during manufacture, which results in surface defects in some devices. The screen used to control this defect effectively is burn-in, in which a device is operated at a specified temperature for a specified time, for example, 125°C for 168 hours. Extensive research (14) has shown that by increasing the temperature to say 200°C, the test time required to achieve the same screening effect is reduced to just a few hours. In fact, "accelerated testing" of this type has demonstrated a bi-modal characteristic for devices where a "freak" population is mixed with the "main" population, each demonstrating a characteristic life as shown in Figure 5.

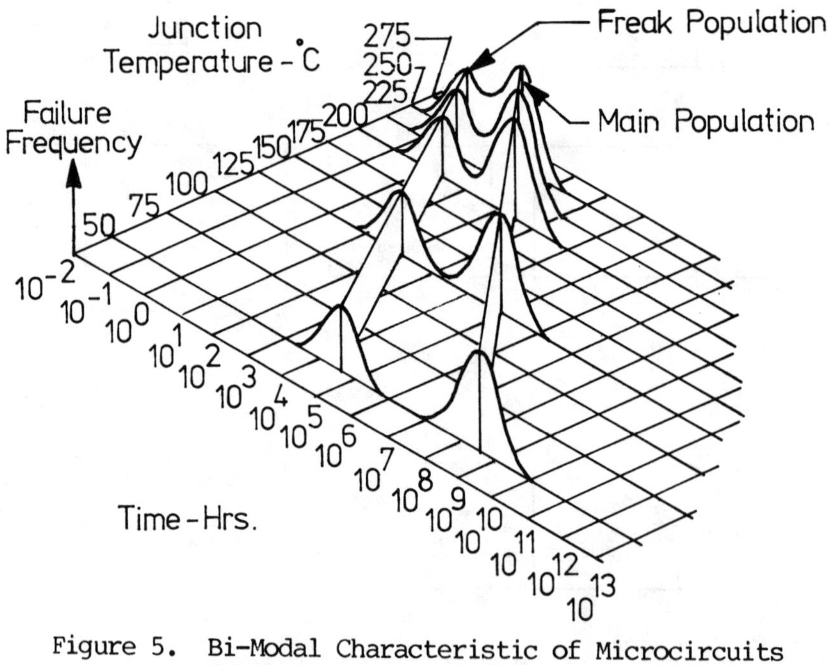

Figure 5. Bi-Modal Characteristic of Microcircuits Under Accelerated Test Conditions

Physical examination of failed devices showed that acceleration testing did not degrade significantly the reliability of the main population devices. The tests further showed, that close control of test bias as well as temperature is important because of voltage effects at higher temperatures.

The time temperature relationship is the well known Arrhenius equation which holds well at lower temperatures.

$$t\,(50\%) = A \exp\left[\frac{Ea}{k\,Temp}\right] \tag{1}$$

where,

t (50%) = median life at temperature,

A = a constant,

Ea = apparent activation energy,

k = Boltzman's constant,

Temp = absolute junction temperature.

At higher temperatures, however, the bias voltage must be considered and equation (1) must be adjusted using the proposed Erying modification:

$$t\,(50\%) = \frac{G}{Temp} \exp\left[\frac{ETA}{k\,Temp} - V\left(C + \frac{D}{k\,Temp}\right)\right] \tag{2}$$

where,

G, C and D = positive constants,

ETA = activation energy,

V = bias voltage.

The Erying modification indicates that for a fixed temperature, median life increases as bias voltage decreases, which agrees with observation.

Inspection of Table 6 shows that the burn-in screen takes 168 hours and that this screen takes the longest to perform, by far. An accelerated testing approach to burn-in could reduce the time to a few hours and thus remove about 1 week from the delivery time of screened devices.

Plastic Encapsulated Devices. Microcircuit packages in solid encapsulant packages rather than hermetic packages, are widely used in applications where environmental variations are minimal, and achieve acceptable reliability performance. However, their use in applications where environmental extremes prevail is not fully proven or even recommended, if reliability performance is critical. This is due to additional potential failure mechanisms inherent in the packaging system, viz., (a) moisture penetration, (b) material imcompatibility and (c) thermal instability vs. hermetic devices.

Early work produced packaging systems composed of silicones and epoxies, neither of which were found to be satisfactory over the complete environmental range. In 1971 an epoxy known as "epoxy B" was widely introduced and represented a material that attempted to combine the desirable environmental characteristics of both silicone and epoxy. A definite improvement was observed but generally devices were unable to qualify to screens and qualification criteria of Joint Electronic Device Engineerng Council (JEDEC) for example, JC-13.1 (15), and as evidenced by (16, 17). Recent developments (18) indicate that plastic encapsulated devices are capable of passing JC-13.1; however, it will likely be some time before such products will find widespread application in severe environments.

In general, plastic encapsulated device reliability performance will be optimal where, (a) relative humidity is low, (b) temperature is fairly constant and (c) power is normally applied.

Transistors and Diodes

This part group has a set of failure mechanisms and modes similar to those discussed for microcircuits. It is, therefore, natural to expect the use of similar screening techniques and such is the case. Details can be found in (5, 19), for example, and the approach developed for microcircuits with respect to screening by the part manufacturer, a test house, or the user also applies.

Capacitors

Capacitors may be purchased off-the-shelf as Established Reliability (ER) devices for most parts. They are available to various screening designators, established on the basis of life test criteria. As an example, ceramic capacitors may be screened at twice rated voltage for a specified time period, at maximum rated temperature. Details may be found in (6).

Resistors

Comments similar to those made for capacitors apply to resistors and details may be found in (7).

ASSEMBLY SCREENING

Assembly screening is used infrequently and is usually applied to circuit cards for use as spares, supporting screened systems to maintain reliability performance. Other applications include assembly screening to increase reliability performance where lower screening designator parts have been used, i.e., the parts are screened up while assembled on a card. If the proper screening tests are used microcircuits with a C screening designator can be raised close to a B, while mounted on a card. This is particularly useful in the case of memory cards where the parts are similar and fault detection and isolation is rapid.

When assembly screening is employed the results are substantial and up to 70% of failures that would occur at the system level, can be eliminated (20). A typical assembly screening profile is shown in Figure 6.

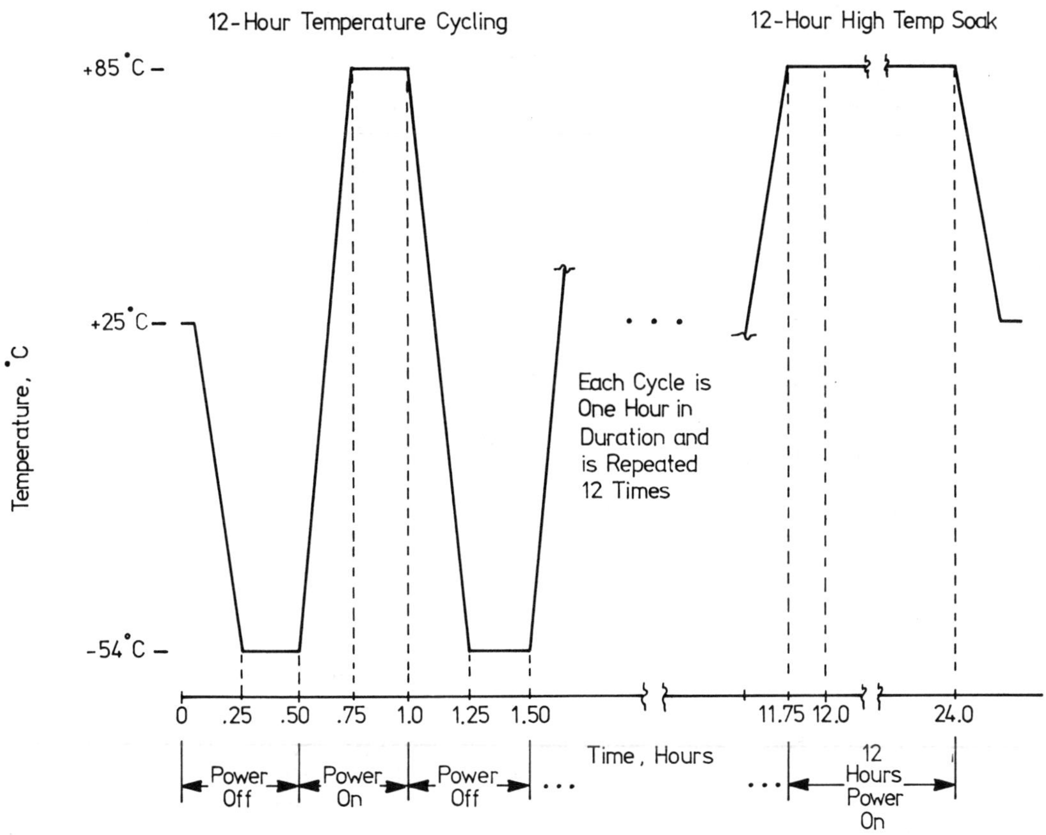

Figure 6. Typical Assembly Screening Profile

UNIT/SYSTEM SCREENING

At this equipment level we will discuss screening with respect to units composed of assemblies, e.g., circuit card assemblies. Units may in turn be integrated with other units to form systems. This type of screening has proven to be highly effective in military programs and is gaining acceptance in the commercial segment of the industry. In general, a system is subjected to the environmental extremes specified as expected in operational use, while operating in normal modes.

Military production programs rely on a form of Test Plan 29 for example, (see Equipment Screening Tests of (2) with suitable test conditions chosen from Table 7). In addition, the test time or number of test cycles is specified. More detailed information with respect to test plans, conditions and equipment will be found in Chapter 18.

Table 7. Summary of Test Levels

Test Level	Temperature	Temperature Cycling	Vibration	Equipment On/Off Cycling
A	+20 to +30	None	Yes	Yes
A-1	+20 to +30	"	None	None
B	+35 to +45	"	Yes	Yes
C	+50 to +55	"	"	"
D	+60 to +70	"	"	"
E	-54 to +55	Yes	"	"
F	-54 to +71	"	"	"
G	-54 to +95	"	"	"
H	-65 to +71	"	"	"
J	-54 to +125	"	"	"

Apart from the above, manufacturers usually devise their own test plans to suit their particular products. A television set manufacturer might screen his sets for 72 hours at room ambient with a 4:1: power on: power off proportionality. A computer manufacturer might employ test conditions similar to those in Table 7 but not use elaborate environmental test equipment.

Results typical of such a program based on thermal cycling are given in Table 8 as given in (21).

Table 8. Temperature Cycling Failures

Failure Classification	Percentage (%)
Design	5
Fabrication Workmanship	33
Parts	62
Total	100

Table 8 (Cont'd)

Note: In immature hardware, a greater incidence of design failure can be expected. In programs using very thoroughly screened parts a lower incidence of part failures would be expected.

CONCLUSION

Screening is a key element in the production of electronic equipment as it not only leads to reliable equipment, but also to reduced production and warranty costs. In the long term the effect of screening will result in lower life cycle costs (LCC) and help improve the cost ratio of research, development, test and engineering to maintenance.

REFERENCES

1. ___________, Microcircuits, General Specification for MIL-M-38510, U.S. Department of Defense, Washington, D.C.

2. ___________, MIL-STD-781, Reliability Tests, Exponential Distribution, U.S. Department of Defense, Washington, D.C.

3. Danner, F. and Lombardi, J.J., "Setting up a Cost-Effective Screening Program for ICs", Electronics, Vol.44, 30 August 1971, p44-47.

4. Foster, R.C., "Why Consider Screening, Burn-In, and 100-Percent Testing for Commercial Devices?", IEEE Transactions On Manufacturing Technology, Vol. MFT-5, No. 3, IEEE, New York, September 1976, p55-58.

5. ___________, MIL-S-19500, Semiconductor Device, General Specification for, U.S. Department of Defense, Washington, D.C.

6. ___________, Capacitor, Selection and Use of, MIL-STD-198, Supplemental Information, U.S. Department of Defense, Washington, D.C. 7.

7. ___________, Resistor, Selection and Use Of, MIL-STD-199, Supplemental Information, U.S. Department of Defense, Washington D.C. 7.

8. ___________, MIL-HDBK-217, Reliability Stress and Failure Rate Data for Electronic Equipment, U.S. Department of Defense, Washington, D.C.

9. Hnatek, E.R. "High-Reliability Semiconductors: Paying more Doesn't Always Pay Off", Electronics, Vol. 50, No. 3, February 1977, p101-105.

10. ___________, MIL-HDBK-175, Microelectronic Device Data Handbook, U.S. Department of Defense, Washington, D.C.

11. __________, Microcircuit Device Reliability Memory/LSI Data, Rome Air Development Center, Griffiss Air Force Base, New York, 1975-76, p24.

12. __________, MIL-STD-883, Test Methods and Procedures for Microelectronics U.S. Department of Defence, Washington, D.C.

13. Hnatek, E.R., "The Economics of In-House Versus Outside Testing", Electronic Packaging and Production, August 1975, p T29.

14. Johnson, G.M., Evaluation of Microcircuits Accelerated Test Techniques, RADC-TR-76-218, Rome Air Development Centre, Griffiss Air Force Base, New York, 13441.

15. Reiche, B. and Hakin, E.B., "Can Plastic Semiconductor Devices and Microcircuits be Used in Military Equipment?", 1974 Annual Reliability and Maintainability Symposium, IEEE, New York, p396-402.

16. Fox, M.J., "A Comparison of the Performance of Plastic and Ceramic Encapsulations Based on Evaluation of CMOS Integrated Circuits", Microelectronics and Reliability, Vol. 16, Pergamon Press, London, 1977, p251-254.

17. Bailey, C.M., "Effects of Burn-In and Temperature Cycling on the Corrosion Resistance of Plastic Encapsulated Integrated Cirucits", Reliability Physics 1977, IEEE No. 77CH1195-7PHY, IEEE, New York, p120-124.

18. __________, "Information about Semiconductor Grade Moulding Compounds", Dow Corning Corp., Midland, Michigan, 48640.

19. __________, MIL-STD-750, Test Methods for Semiconductor Devices, U.S. Department of Defense, Washington, D.C.

20. Rue, H.D., "System Burn-In for Reliability Enhancement", Proceedings 1976 Annual Reliability and Maintainability Symposium, IEEE, New York, 1976, p336.

21. Burrows, R.W., "Role Of Temperature in the Environmental Acceptance Testing of Electronic Equipment", 1973 Annual Reliability and Maintainability Symposium, IEEE, New York, p42.

BIBLIOGRAPHY

__________, Reliability of Semiconductor Devices - Special Issue On, Proceedings of the IEEE, IEEE, New York, February, 1974.

17 FAILURE REPORTING AND CORRECTIVE ACTION SYSTEMS

W. L. Brown

In general, failure reporting and analysis is a necessary operation to ensure that a product's reliability and maintainability will be achieved and sustained. It is used to form the data base pointing the way to corrective action when warranted. Figure 1 illustrates the closed loop nature of the activity throughout a product's life cycle (1).

Several important data sources usually are available as inputs to the data base, i.e.,

(a) in-house or plant testing,
(b) user or field data,
(c) major sub-contractors.

The most useful of the three is the in-plant failure reporting, analysis and corrective action, from which procedures, policies and results can readily be applied. Other sources are unreliable since they are, in most cases, incomplete, have little substantiating back up or information file for reference and cover only a small percentage of such failures. However, there are exceptions.

Providing an agreed policy and procedure has been set up, major sub-contractors can provide as much data as in-plant users and this source should not be overlooked. A useful tutorial paper covering most aspects of data collection techniques has been provided by (2) and this should be consulted for details on methods.

The reasons for electronic equipment operational malfunctions can be classified into five major areas:

(a) design,
(b) the quality of the parts and materials used in the construction,
(c) the standard of workmanship,
(d) the intended equipment application,
(e) the amount of proper maintenance carried out.

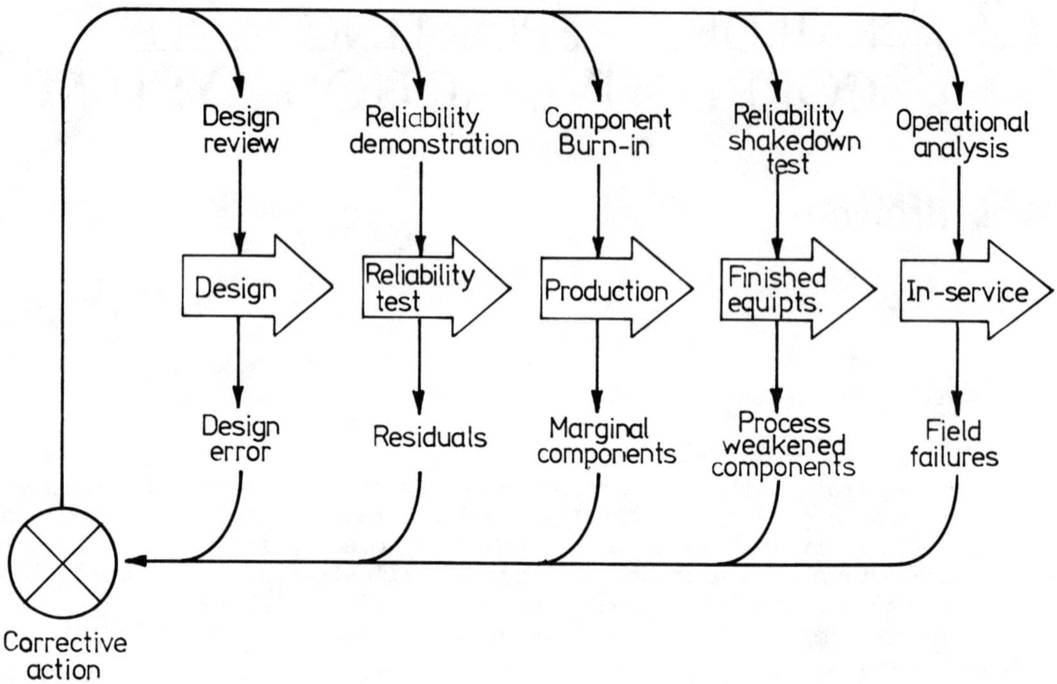

Figure 1. Corrective Action Closed Loop System

Failure reporting resulting from any of the above categories must result in a corrective action feedback loop. This corrective action must be implemented in a timely fashion and consequently needs internal policies and guidelines to define areas of responsibility.

A failure is the inability of the equipment to meet minimum performance requirements. Where applicable, operator control adjustment is allowed to the extent stated in the specification, in order to maintain performance. Where such adjustment can be made to bring the equipment back to its permissible performance requirements, it will not be counted as a failure.

FAILURE REPORTING SYSTEMS

Responsibilities

In order to establish an operable failure analysis system, areas of responsibility must be defined and adhered to. This is normally accomplished by means of departmental instructions or directives, outlining the responsibilities of the various groups and individuals in the chain. For example, Quality Assurance (Q.A.) engineering and inspection personnel are responsible for issuing a failure report whenever an out of tolerance condition is observed during the Q.A. tests (or measurements). A Q.A. Failure Analysis Section should be responsible for analyzing the failed parts and describing the failure mode. A Quality Assurance Planning and Analysis (QAPA) Section should be responsible for recording failures in such a manner that the information may be readily retrieved when required for analysis. A Q.A. Engineering Division should review all failures reported from the operations or manufacturing group and from the field, and make technical proposals for correction of such failures in conjunction with the Reliability and Maintainability Departments.

Applicability

Failure reports should be raised for failures occurring on all fabricated items, which in the case of military operations would be all those models defined in (3) including:

(a) Advanced Developmental Models,
(b) Service Test Models,
(c) Pre-production Models,
(d) Production Models,
(e) Contracted Models.

Authorization

Generally speaking, all failures should be reported, reviewed and where necessary an analysis made on the failed part, in order to maintain the quality and reliability of the end product. However, in practice this may not be possible from a financial point of view or realistic if the failure is a batch problem (i.e., vendor sent in bad material). There are cases where a great deal of effort is required on failed parts and no further production can progress until the customer has approved the failure analysis and corrective action. In these instances the failure philosophy is defined in the contract for that specific job order. It is recommended that each project have the extent of the failure program defined by the contract, in order that the effort of endeavours become known to all groups involved.

Field Failures

The field service representative should compile a failure report whenever he finds an equipment failure and return both the failed part and the report to contracts for disposition. Contracts should then prepare a project work statement which will direct the work to be performed on the failure such as: analysis of part, investigation of failure, search for trends or previous failures of a like nature, response required detailing corrective action or response to customer and, finally, the disposition of the part after rework; i.e., quarantine, scrap, rework, etc.

Failure Analysis

Failure analysis of the part is normally carried out when the failure mode of the failed part cannot be established during the trouble-shooting phase with complete accuracy. It is well known that many good parts are removed from a circuit due to poor trouble-shooting procedure or due to inexperience of the troubleshooter in that product line. As the familiarity with the product increases the 'failed' parts that are good usually decrease. If they do not, indications are that a training course should be re-introduced to make the personnel more familiar with the system.

Acquisition

The gathering of failure data should be made at the same time that the failure occurs. All requested data should be provided and the report, along with the failed part (if applicable), be sent to the decision body or group for continued processing.

Routing

It is essential that the system of handling data and monitoring corrective action is followed consistently. A management procedure keyed to the particular organization concerned must be created. A useful guide to such a system has been provided by (4). Figure 2 outlines the decision and action flow requirements of a typical system.

Failure Report Form

The data analysis originates from information on the failure reporting form and is heavily reliant upon complete data. If vague or sloppy reporting is provided the data file base will be inadequate and of little use in determining trends.

The system begins with confirmation of a failure at which point Part 1 of the Failure Report Form, shown in Figure 3, is completed. In other words, there was an inability of material to perform its required function within previously established limits and the part number of the failed item is entered.

The Failure Data section of the form, Part 2, should be filled out while the information is still fresh, as this is the symptom or failure mode. Proper attention to detail here could prevent unnecessary costly analysis of a failure later on in the process. The statements used here should be brief and factual but unambiguous since they may be used by an analyst days, months or even years later.

Part 3 of the form pertains to the activities carried out in the process of identifying which is the failed part or parts of many available on the sub-assembly. The value of this section is in establishing whether the diagnostic routine was adequate or resulted in difficult diagostic decisions having to be made. This would indicate the need for more sophisticated test routines or additional test parameters being added to the test procedure.

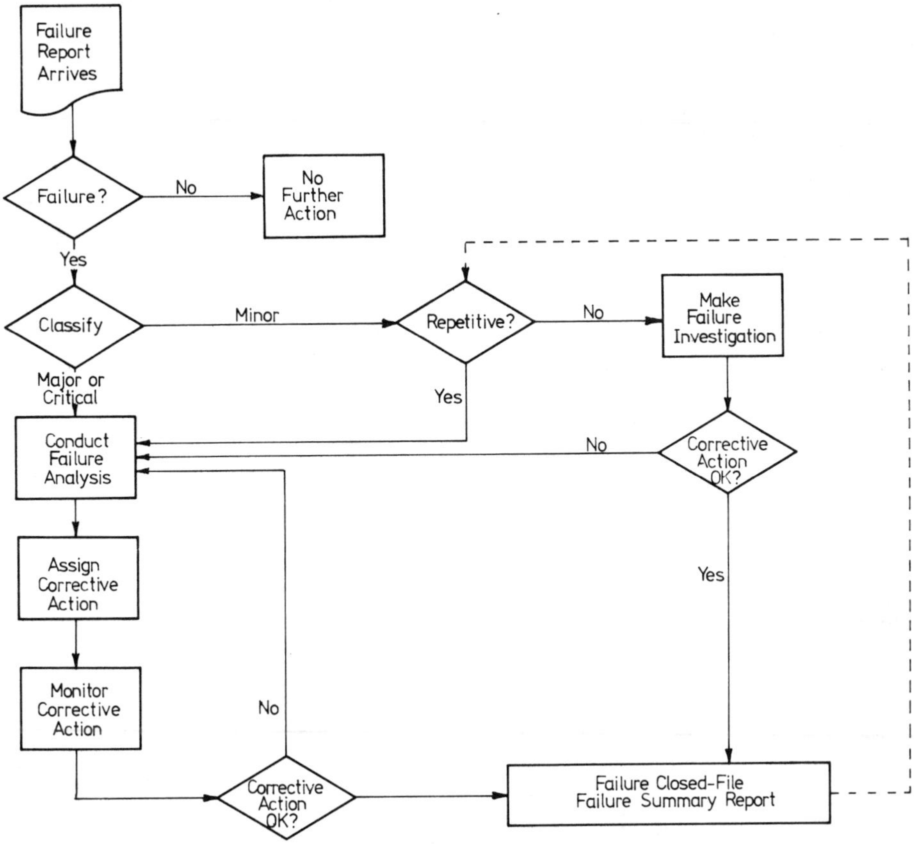

Figure 2. Basic Failure Analysis Decision and Action Flow

Part 4 indicates what parts were removed as being the suspected failure items and contains a data code requirement so that vendor discrepancies can be identified readily. It also specifies the time required to fit a new part. This is a good indication of the practical aspects of unit or system maintainability.

Failed Part Analysis Form

When a part is sent for Analysis, a form such as shown in Figure 4 or an equivalent format could be utilized. The form caters to several failed parts which would be all the parts removed from the suspected unit. The analysis would then show which was the primary failure and which were secondary failures or non-failures.

1 FAILED ITEM CONFIG.

FAILED ITEM PART NO. 1	FAILED ITEM NAME 2		REF. DES. 3	SER. NO. 4	OPER. HRS 5
NEXT HIGHERASSY/BLACK BOX PART NO. 6	NEXT HIGHER ASSY/BLACK BOX NAME 7		REF. DES. 8	SER. NO. 9	OPER. HRS 10
UNIT OR SYSTEM 11	SERIAL NO. 12	OPERATING HRS 13	PROJECT NO. 14	W/O 15	16

2 FAILURE DATA

DESCRIBE SYMPTOMS OF FAILURE, EFFECT ON SYSTEM, DEGREE OF MALFUNCTION 17

WHO, WHERE DISCOVERED –

- ☐ ENGINEERING ☐ TEST AND CALIBRATION ☐ QA
- ☐ CUSTOMER ☐ FIELD SERVICE ☐ R & O
- ☐ BENCH ☐ ☐ IN AIRCRAFT
- ☐ PRE-FLIGHT ☐ IN FLIGHT ☐ OTHER DETAILS IN WRITE-UP

DURING –

- ☐ DEVELOPMENT ☐ RECEIPT – STORES ☐ MECHANICAL INSPECTION ☐ CALIBRATION
- ☐ PRODUCTION ☐ RELIABILITY ☐ ACCEPTANCE ☐ TEST
- ☐ FINAL ☐ EQUIPMENT TURNON ☐ AFTER PROLONGED OPERATION
- ☐ ROOM TEMPERATURE ☐ HOT ☐ COLD ☐ OTHER DETAILS WRITE-UP 19

HUMAN ERROR OR TEST ACCIDENT – ☐ YES ☐ NO 20	RECURRING FAILURE – ☐ YES ☐ NO 21	FAILURE DATE – DAY MONTH YEAR 22	ORIGINATORS NAME 23	SITE/LOCATION 24

3 TROUBLESHOOTING ACTION

ITEM NO.	TROUBLESHOOTING PROCEDURE USED, DESCRIBE CONDITION FAILURE MODE OF DEFECTIVE PARTS, REPAIR INSTRUCTIONS	LOCATE HOURS MINS	TROUBLESHOOTERS NAME
25	26	27	28
	BASIC CAUSE OF FAILURE 29		

4 REPAIR ACTION PARTS REMOVED

	PART NUMBER	PART NAME OR REFERENCE DESIGNATION	DATE OF MANUFACTURE	MANUFACTURER	SERIAL NUMBER	REPAIRED BY	Q/A INSPECT	HOURS/MINUTES REPAIR	HOURS/MINUTES TEST	REPAIR CODE
30	31	32	33	34	35	36	37	38	39	40

REPAIR CODES – (A) REPLACED PART (B) ADJUSTED RE/CAL (C) RESOLDERED (D) REWIRED (E) ECO INC. (F) ENTER OTHER 41

QA INSPECTORS NAME AND DATE OF ACCEPTANCE – DAY MONTH YEAR 42	FAILED PARTS ANALYSIS REQUESTED – DAY MONTH YEAR 43	SUPPLEMENTS RAISED (✓) 44: 1 2 3 4 5 6 7 8

FR SERIAL NO. № 14510 56

Figure 3. Typical Failure Report Form

REQUEST FOR ANALYSIS

TO	DEPARTMENT	DATE REQUESTED (DAY / MONTH / YEAR)	PRIORITY: URGENT ☐ ROUTINE ☐
FROM	DEPARTMENT	PROJECT NUMBER	WORK ORDER

ITEM	PART NUMBER	GENERIC DESCRIPTION	NAME/REFERENCE DESIGNATION	SERIAL NO.	DATE CODE	MANUFACTURER
1						
2						
3						
4						

ACTION REQUESTED: CONFIRM FAILURE MODE ☐ DETERMINE MECHANISM OF FAILURE ☐ DISASSEMBLY PERMITTED: YES ☐ NO ☐ MODE REPORTED: OPEN ☐ SHORT ☐ DRIFT ☐ OTHER ☐

ANALYSIS

INVESTIGATION: DETAILS, ANALYSIS METHODS, EQUIPMENT USED

PHOTO YES ☐ NO ☐
X-RAY YES ☐ NO ☐

ITEM	FAILURE MODE, MECHANISM OF FAILURE, BASIC CAUSE:

REMARKS:

PAI CODE	ITEM	FAILURE REPORT NO.
	1	
	2	
	3	
	4	

INSPECTION REPORT NUMBER	ANALYST	DATE (DAY / MONTH / YEAR)

Figure 4. Typical Failed Parts Analysis Form

The form is divided into two major sections, i.e., Request for Analysis, and Analysis for Documenting Investigation Details. The Request for Analysis section is used to log the basic information before any failed parts analysis may begin. In most cases it is necessary to expand and make more complete the part information supplied. For example, the manufacturer, manufacturer's part number, data code, etc., should all be entered on the form. The details of analysis required will normally be given in general terms only. The Analysis section is used by the analyst to detail the methods of investigation and apparatus used. A subsection is provided to describe failure modes and mechanisms observed. Here considerable detail is required. For example, if overstressing is observed as in the case of a microcircuit, it is necessary to state the pins that are involved; e.g., pins #7 and #8 overstressed. The remarks section is used for summing up and classifying observations. For example, primary or secondary failure, and in the case of semiconductor material state whether a surface, bulk, or mechanical defect was observed and other remarks as required.

FAILED PARTS ANALYSIS

Purpose

Failure Analysis is a post-mortem examination of failed devices employing, as required, electrical measurements and analytical techniques of Physics, Metallurgy, and Chemistry in order to verify the reported failure and identify the mode or mechanism of failure as applicable. The Failure Analysis procedure should be sufficient to yield adequate conclusions, for determination of cause or relevancy of failure for initiation of corrective action in device application, production processing, device design, device testing and to eliminate the cause or prevent recurrence of the failure mode or mechanism reported. Specific terms are used in failure analysis and the definitions that follow are frequently used with respect to semiconductors.

Reduction and Final Classification of Observations

The final reduction of observations made during failed parts analysis, in nearly all cases, is sufficient to classify the part failure as primary or secondary. In the case of semiconductor material it should, in most cases, be possible to classify failure mechanism as to whether surface, mechanical or bulk.

(a) Failure Mode - the electrical or mechanical manifestation of a failure; it is the particular manner in which the failure occurred. Examples of failure modes are open metallization or shorted lead wires,

(b) Failure Mechanism - the fundamental physical or chemical process responsible for a particular failure mode. It is an explanation of the physical and chemical changes leading to a failure. The failure mechanism associated with open metallization might be a scratch,

(c) Primary Failure - occurs when it has been determined that a part itself is the intrinsic cause for the failure when operating within specification. In many cases, a number of parts will have failed and it will be necessary to determine the part (if any) causing the prime failure,

(d) Secondary Failure - occurs when it has been determined that the part failure has been caused by some out of tolerance stimulus. A transistor with a vaporized emitter lead wire would normally indicate a secondary failure caused by excessive current,

(e) Bulk Defect - an inherent imperfection in the silicon or a defect introduced during processing which may cause a device to fail electrical testing. Bulk Defects are temperature and time dependent. Bulk Defects are the least significant factors affecting reliability in well designed devices, but they are the hardest to substantiate through failed parts analysis,

(f) Surface Defect - this is one of the major failure mechanisms in semiconductors but it has been eliminated to a great extent by use of passivating layers,

(g) Mechanical Defect - those caused by centrifuge testing, shock tests, fatigue caused by thermal cycling, etc. and mainly confined to weaknesses in assembly operations. An example is a poor thermal compression bond attaching the fine electrode wire to an integrated circuit.

Apparatus

The Apparatus required for Failure Analysis includes electrical test equipment capable of complete electrical characterization of the device types being analyzed, photographic equipment, micro-manipulators capable of point-to-point probing on the surface of device dies or substrates, as required, and microscopes capable of making the observations at adequate magnifications. In addition, special analytical equipment for bright field, dark field and phase contrast microscopy, apparatus for x-ray radiography, hermeticity test and other specific test methods may be needed. Cleaning agents should be available as required.

Failure Analysis Techniques

The following failure analysis techniques are provided as guidelines for active semiconductor devices, which includes most types of transistors and integrated circuits:

(a) Electrical Verification Procedures - This test condition calls for the measurement of all important electrical parameters in the applicable procurement document. In addition to the minimum electrical test as per procurement document, this section provides for curve tracer pin to pin measurements and other nonstandard measurements which allow electrical characterization of significant physical properties,

(1) Threshold test, determine the forward characteristic obtained for each pin to substrate and compare to the device schematic and structure. Excessive forward voltage drop may indicate an open or abnormally high resistance in the current path,

(2) Case Isolation (for metal packages or those with metal lids or headers only), applying a voltage between the package and the external leads. Current flow determines the presence of shorts-to-case,

(3) Pin-to-pin two and three terminal electrical measurements, (utilizing a transistor curve tracer, picoammeter, capacitance bridge, and oscilloscope as required), may be performed and results recorded for lead combinations involving the defective portion of a microcircuit. Gain, transfer, input vs. output, forward and reverse junction characteristics, should be observed and interpreted. Resulting characteristics may be compared to those obtained from a good unit, and differences interpreted for their relationship to the device failure,

(b) X-Ray Radiography - a film record is required of the failed device taken normal to the top surface of the device; where applicable, additional views shall be recorded. This may be performed when open or shorted leads, or the presence of foreign material inside the device package are indicated from electrical verification of failure, or when there is evidence of excessive temperature involved in the device failures,

(c) Fine and Gross Seal Testing - this should be performed when warranted,

(d) Internal Examination - the lid of the failed device should be removed carefully and an optical examination made of the internal device construction, at a minimum magnification of 30X. A color photograph, at suitable magnification to show sufficient detail, should be taken of any anomalous regions which may be related to the device failure. Where there is evidence of foreign material inside the device package, it may be removed using a stream of dry compressed inert gas or appropriate solvents. The relationship of the foreign material to device failure (if any) should be noted and if possible, the nature of the material should be determined. A multipoint probe should be used where necessary to probe active regions of the device to further localize the cause of failure. A curve tracer should be used to measure resistors, the presence of localized shorts and opens, breakdown voltages, and transistor gain parameters. A picoammeter should be used for measuring leakage currents, and where applicable, a capacitance bridge should be employed for the determination of other junction properties. It may be necessary to open metallization strips to isolate components,

(e) Electron Microscopy - an examination at extremely high magnification, of the structure of failure metallization and bulk materials, is best accomplished using electron microscopy.

Information Obtainable

The following is a partial list of failure modes and mechanisms which may be identified using the above analysis techniques:

(a) corroded metals within package,
(b) cracked doe of substrate,
(c) degradation of lead at lead frame,
(d) degradation of time response or frequency dependent parameters,
(e) excessive leakage currents indicating degraded junctions,
(f) further definition of failed device region,
(g) hermeticity problems,
(h) intermetallic formation,
(i) mask misregistration,
(j) migration of metal
(k) missing or peeling metals,
(l) open and shorted leads and/or metallization land areas,
(m) overstress conditions resulting from device abuse, transients or inadequate power supply regulation,
(n) oxide contamination, discoloration,
(o) poor bond placement and lead dress,
(p) quality of junctions, diffusions and elements,
(q) radiographically determined defects such as poor wire dress, loose bonds, open bonds, voids in die or substrate mount, presence of foreign materials,
(r) reactions at metal/semiconductor contact areas,
(s) resistance changes,
(t) shorts through the oxide or dielectric,
(u) stability of surface parameters,
(v) undercut metals.

Case History

During acceptance testing a fault was found which was traced to a circuit card and then upon further troubleshooting identified as part number 905141-002 with a data code 7249. This integrated circuit was found to have pins 2 and 7 short circuited.

A visual examination of the chip (Figure 5) shows the problem:

(a) top of photo - poor wire bond,
(b) right hand side - below pin 2 a voltage scar

This was a secondary failure resulting in a surface defect due to over-stressing.

Figure 6 shows a manufacturing defect. This is a dual clocked JK flip flop and the failure modes identified were:

(a) no clock input,
(b) clock pulse output one way only,
(c) pin 5 high, pin 6 low, pin 4 low reverses input.

A visual examination showed a pin hole under metallization allowing the track to go to the substrate. There was poor marking in one area and chemical contamination at the collector of the transistor to pin 6 with metallization removed. There were numerous pin holes on the chip surface.

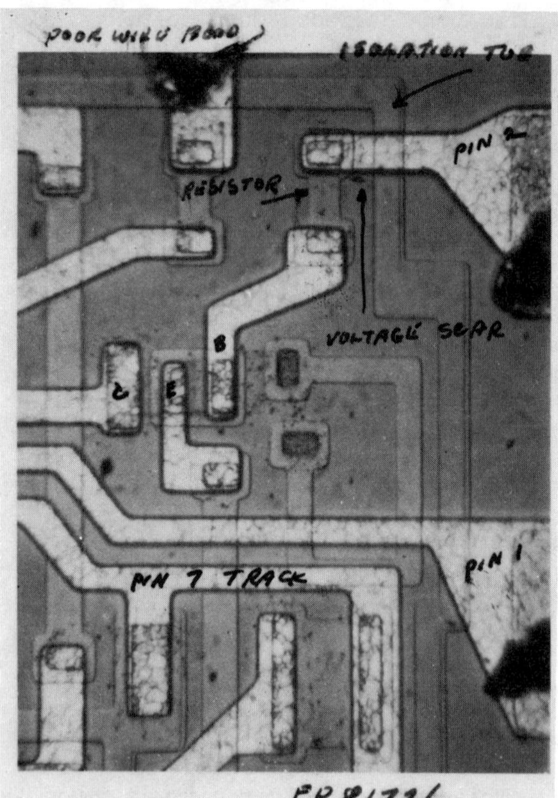

Figure 5. Photo Showing Poor Wire Bond and Voltage Scar

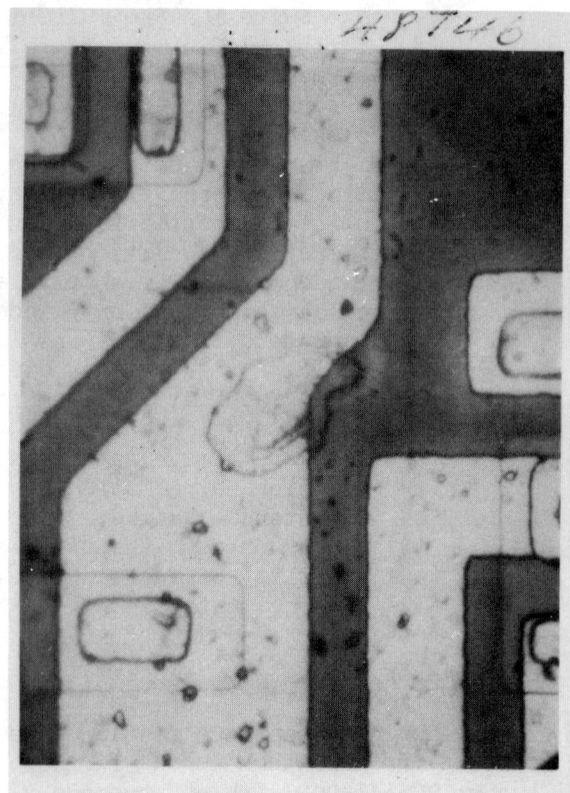

Figure 6. Photo Showing Manufacturing Defect

FAILURE DATA HANDLING

In any large company data accumulation requires the use of computer based systems. A system in which the data appears on a conversational display terminal, as well as being printed out (on request) can be completely flexible in that faults can be grouped by project number or by part number. The total of all of these failures on one unit or over a period of time can be ascertained. A listing of failures by type can be requested by keyboard interrogation of the computer. An example of this is shown in Figure 7. The failure analysis data handling personnel scan the failure sheets and produce a weekly summary for engineering and management, indicating the failures, trends, bad batches of devices and effectivity of corrective action, thus providing a valuable service with timely corrective action techniques and real time information on any discrepancy. Consider the value of this service in its entirety; without "Failure Analysis" a company could spend a huge amount of money on re-fitting or reworking systems that have been shipped to customers with defects and perhaps there also would be loss of repeat business due to animosity between the customer and supplier.

CONCLUSION

Table 1 shows the costs of failures based on practical experience. A fault found on an assembled circuit card is many times more expensive than a fault found on the part at receiving or at burn-in. A part defect found at the unit level is many times more expensive to correct than that found on a circuit card and a field failure could prove to be the most costly problem of all.

Table 1. Failure Costs

	Loss/1000 after burn-in	Circuit Cards	Burn-in Cycling	System Test	Field Failure
Non-burned-in, bought-out parts	48	25.0%	30.0%	5	12
Vendor or plant burned-in parts	98	15.8%	5.9%	1	1
Costs	$0.10 (1)	$20.(2)	$150.(2)	$60.(2)	$200.(2)

Notes: (1) Part cost
(2) Direct labor plus part cost

Unit	UPN	USN	C	SPN	SSN	PART NUMBER	COUNT	REF DESIG	MFGR	DOM	SOU	ENV	ANAL	PAI	MECH	FAILREP	OF	DATE
MGU1 2		00	A	H785	17	906349-001		A3U24	SIGNE	7745	37	09	N		997	218592	3	790313
MGU2 2		00	A	H785	27	906349-001		A4U24	SIGNE	7739	36	09	N		997	223134	7	790403
							2											
MGU2 2		00	A	H765	06	906350-001		A3U23	SIGNE	7715	37		T	NPF	995	190752	4	780417
MGU2 2		00	A	H765	06	906350-001		A3U40	TEXAS	7617	37		T	NPF	995	190754	4	780417
							2											
MGU1 2	H765	19	U	H765	39	906354-001		A14U32	TEXAS	7727	73	08	Y	PCU	425	210823	3	790129
MGU1 2		00	A	H765	48	906354-001		A14U38	TEXAS	7836	37	09	N		997	233585	5	790627
MGU1 2	H765	08	U		23	906354-001		A15U32	TEXAS	7710	74	09	Y	PCU	400	192814	4	780524
							3											
MGU1 2		00	A	H785	05	906355-001		A10U26	TEXAS	7740	37	09	T	NPF	995	195901	1	780821
MGU1 2		00	A	H765	19	906355-001		A10U26	TEXAS	7736	37	09	T	NPF	995	195931	2	780822
MGU1 2		0	A	R854	05	906355-001		A10U26	FAIRC	7816	37	09	T	NPF	995	196472	2	780907
MGU1 2	H765	21	U	R854	04	906355-001		A10U26	FAIRC	7816	38	09	T	NPF	995	206763	3	781212
MGU1 2	H765	15	U	R854	01	906355-001		A11U39	FAIRC	7816	38	09	Y	PMW	400	198361	1	780928
MGU1 2		00	A	H765	03	906355-001		A3U25	TEXAS	7801	37	09	T	NPF	995	193833	4	780628
MGU1 2		00	A	H785	03	906355-001		A3U25	FAIRC	7810	37	09	T	NPF	995	193834	4	780628
MGU1 2		00	A	H765	10	906355-001		A3U7		7630	37	09	T	NPF	995	173301	1	771223
MGU1 2		00	A	H785	17	906355-001		A3U7	TEXAS	7827	37	09	T	NPF	995	218591	3	790313
MGU2 2		00	A	H785	23	906355-001		A10U26	FAIRC	7816	37	09	T	NPF	995	208711	2	781213
MGU2 2		00	A	H785	27	906355-011		A4U7	FAIRC	7816	36	09	T	NPF	995	223133	7	790403
							11											
MGU1 2		00	A	H765	03	906362-001		A12U1	TEXAS	7711	37	09	T	NPF	995	188941	1	780330
MGU1 2		00	A	H765	08	906362-001		A12U1	TEXAS	7711	37	09	T	NPF	995	188951	2	780330
MGU1 2		00	A	H765	22	906362-001		A13U38	ADVAN	7742	37	09	T	NPF	995	190614	4	780504
MGU1 2		00	A	H785	18	906362-001		A13U38	RAYTH	7751	37	09	T	NPF	995	209432	2	790129
							4											
MGU1 2		00	A	H765	36	906388-001		A14U10	MOTOR	7741	37		T	NPF	995	190692	5	780511
MGU1 2		00	A	H765	36	906388-001		A14U11	MOTOR	7741	37		T	NPF	995	190693	5	780511
MGU1 2	H765	19	U	H765	39	906388-001		A14U11	MOTOR	7741	73	17	Y	PMW	425	210821	3	790129
MGU1 2		00	A	R854	06	906388-001		A14U8	FAIRC	7815	37	09	T	NPF	995	198822	3	781103
MGU1 2		00	A	H785	21	906388-001		A5U37	FAIRC	7823	37	09	N		997	234832	4	790713
MGU1 2		00	A	H785	06	906388-001		A5U40	MOTOR	7741	37	09	T	NPF	995	210631	1	790119
							6											
MGU1 2		00	A	H765	02	906391-001		A9U31	ADVAN	7409	36	09	N		997	221151	2	790328
							1											
MGU1 2	H785	14	U	H785	16	906392-001		A10U6	FAIRC	7819	73	09	N		997	220051	1	790315
							1											
MGU1 2	H765	07	U		04	906394-001		A12U20	MOTOR	7707	73	17	T	NPF	995	189951	1	780508
MGU1 2		00	A	H765	48	906394-001		A14U17	FAIRC	7809	37	09	N		997	233584	5	790627
MGU1 2		00	A	H785	16	906394-001		A9U18	FAIRC	7820	37	09	T	NPF	995	20880G	R	781218
MGU1 2	H767	11	U	H765	05	906394-001		A9U20	FAIRC	7708	74	09	T	NPF	995	189991	6	780814
MGU1 2		00	A	H785	16	906394-001		A9U20	FAIRC	7820	37	09	T	NPF	995	20880C	R	781218
MGU1 2	H767	11	U	H765	05	906394-001		A9U30	MOTOR	7707	74	09	T	NPF	995	189995	6	780814
MGU8 2		00	A	H765	03	906394-001		A14U17	MOTOR	7707	37		T	NPF	995	171223	3	771128
							7											
MGU1 2		00	A	R854	06	906474-001		A14U27	MOTOR	7706	37	09	N		997	198823	3	781103
							1											

Figure 7. Failure Data Listing

REFERENCES

1. Boardman, K.W., "The A-7 Head-up Display Story", Proceedings of a Symposium Organized by The Royal Aeronautical Society, Luton and Stevenage Branch, England, February 1976, p26-30.

2. Hahn, R.F, "Data Collection Techniques", Proceedings of IEEE Reliability and Maintainability Symposium, 1972, p38-43.

3. ______________, MIL-STD-243, Types and Definitions of Models for Communications Electronic Equipment, U.S. Department of Defense, Washington, D.C., U.S.A.

4. Maloney, W.J. and Crumley, L.M.,"GE/RESD Failure Analysis System", Proceedings of the 3rd. Annual Seminar of Failure Analysis, IEEE Philadelphia Section and Reliability Section, May 1970, p11-21/22.

18 RELIABILITY TESTING

A. H. K. Ling and J. E. Arsenault

Reliability testing has become a very prominent part of a comprehensive reliability program. The importance of a well planned, detailed reliability test cannot be overemphasized, for not only will it prove indicative of the operational reliability of a product, it may also provide a corrective action mechanism for product improvement.

Reliability testing is a tool which attempts to prove that theoretical reliability calculations, or the expectations associated with a design are viable. Although it may not always be possible to duplicate the operational environment exactly for test purposes, most of the time it will be possible to simulate the operational environment to a satisfactory level.

The data obtained from testing becomes a vital part of product history and it would not be wrong to say that a reliability test is different for each product depending upon the type of use each has to undergo. Sometimes products are tested under higher stresses than those which will be encountered during operation but this is not typical for electronic equipment.

Although it is possible to visualize testing at all equipment levels, i.e., part, assembly, unit and system, most testing is, in general, concerned with the part and system levels. Normally the type of testing suitable for parts will also be suitable for assemblies and a similar relationship exists between systems and units. The main thrust of this discussion relates to system testing, although some remarks relating to parts are made in the next section.

In general, reliability testing has four underlying principles which have changed very little since first synthesized in (1), i.e.,

(a) Statistically efficient tests are chosen to minimize cost and time to an accept/reject decision. After each failure or moment of time the test data are examined and a decision is made either to accept, reject or continue testing,

(b) Test data is carefully recorded, is available for review, and there is redundancy in the forms used so that data may be cross-checked,

(c) All failures occurring during testing are to be thoroughly analyzed to determine the failure mechanism, i.e., the fundamental cause of failure traced to its chemical, physical, design or workmanship origin,

(d) Corrective action is taken in the event that the test reveals evidence of systematic failures and the required changes are reflected in the final design.

In brief, reliability testing is a technique which demonstrates that the reliability characteristic(s) of interest will not be less than a certain specified minimum acceptable value(s) to some level of confidence.

PART RELIABILITY TESTING

Part reliability testing generally involves a longevity or life test which is designed to determine the capability of the part to meet performance requirements over an extended period. Reference (2) provides guidance for the testing of microcircuits. However, the testing approach can be applied to most electronic parts where wearout mechanisms are not important. Usually the Lot Tolerance Percent Defective (LTPD) method, outlined below, is used.

LTPD Method

Before discussing the application of the method a few definitions are necessary as follows:

(a) LTPD Series - the following decreasing series of LTPD or Lambda (λ) values: 50, 30, 20, 15, 10, 7, 5, 3, 2, 1.5, 1, 0.7, 0.5, 0.3, 0.2, 0.15, 0.1,

(b) Acceptance Number (c) - an integral number associated with the selected sample size which determines the maximum number of defectives permitted for that sample size,

(c) Rejection number (r) - one plus the acceptance number,

(d) Lambda (λ) - LTPD for 1000 hours.

The variables defined above are shown in Table 1. In performing life testing, Lambda (λ) is specified, and the life-test time is 1000 hours elapsed time. Samples are drawn randomly from the lot to equal the sample size specified under LTPD in Table 1 and the acceptance number associated with the particular sample size is chosen. With respect to test conditions, (3) may be used as a guide. It should be noted that part manufacturers often use life testing for the generation of failure rates. Usually the testing is performed on parts screened to a certain quality level in a laboratory environment and care must be taken in translating these results to the actual parts used in a particular use environment (see Chapter 8 for a discussion of part quality and environmental factors). Further guidance applicable to part testing is given in (4, 5, 6).

Table 1. LTPD Sampling Plans[1,2]

Minimum size of sample to be tested to assure, with a 90 percent confidence, that a lot having percent-defective equal to the specified LTPD will not be accepted (single sample).

Max. Percent Defective (LTPD) or λ	50	30	20	15	10	7	5	3	2	1.5	1	0.7	0.5	0.3	0.2	0.15	0.1
Acceptance Number (c) (r = c + 1)	Minimum Sample Sizes (For device-hours required for life test, multiply by 1000)																
0	5 (1.03)	8 (0.64)	11 (0.46)	15 (0.34)	22 (0.23)	32 (0.16)	45 (0.11)	76 (0.07)	116 (0.04)	153 (0.03)	231 (0.02)	328 (0.02)	461 (0.01)	767 (0.007)	1152 (0.005)	1534 (0.003)	2303 (0.002)
1	8 (4.4)	13 (2.7)	18 (2.0)	25 (1.4)	38 (0.94)	55 (0.65)	77 (0.46)	129 (0.28)	195 (0.18)	258 (0.14)	390 (0.09)	555 (0.06)	778 (0.045)	1296 (0.027)	1946 (0.018)	2592 (0.013)	3891 (0.009)
2	11 (7.4)	18 (4.5)	25 (3.4)	34 (2.24)	52 (1.6)	75 (1.1)	105 (0.78)	176 (0.47)	266 (0.31)	354 (0.23)	533 (0.15)	759 (0.11)	1065 (0.080)	1773 (0.045)	2662 (0.031)	3547 (0.022)	5323 (0.015)
3	13 (10.5)	22 (6.2)	32 (4.4)	43 (3.2)	65 (2.1)	94 (1.5)	132 (1.0)	221 (0.62)	333 (0.41)	444 (0.31)	668 (0.20)	953 (0.14)	1337 (0.10)	2226 (0.062)	3341 (0.041)	4452 (0.031)	6681 (0.018)
4	16 (12.3)	27 (7.3)	38 (5.3)	52 (3.9)	78 (2.6)	113 (1.8)	158 (1.3)	265 (0.75)	398 (0.50)	531 (0.37)	798 (0.25)	1140 (0.17)	1599 (0.12)	2663 (0.074)	3997 (0.049)	5327 (0.037)	7994 (0.025)
5	19 (13.8)	31 (8.4)	45 (6.0)	60 (4.4)	91 (2.9)	131 (2.0)	184 (1.4)	308 (0.85)	462 (0.57)	617 (0.42)	927 (0.28)	1323 (0.20)	1855 (0.14)	3090 (0.085)	4638 (0.056)	6181 (0.042)	9275 (0.028)
6	21 (15.6)	35 (9.4)	51 (6.6)	68 (4.9)	104 (3.2)	149 (2.2)	209 (1.6)	349 (0.94)	528 (0.62)	700 (0.47)	1054 (0.31)	1503 (0.22)	2107 (0.155)	3509 (0.093)	5267 (0.062)	7019 (0.047)	10533 (0.031)
7	24 (16.6)	39 (10.2)	57 (7.2)	77 (5.3)	116 (3.5)	166 (2.4)	234 (1.7)	390 (1.0)	589 (0.67)	783 (0.51)	1178 (0.34)	1680 (0.24)	2355 (0.17)	3922 (0.101)	5886 (0.067)	7845 (0.051)	11771 (0.034)
8	26 (18.1)	43 (10.9)	63 (7.7)	85 (5.6)	128 (3.7)	184 (2.6)	258 (1.8)	431 (1.1)	648 (0.72)	864 (0.54)	1300 (0.36)	1854 (0.25)	2599 (0.18)	4329 (0.108)	6498 (0.072)	8660 (0.054)	12995 (0.036)
9	28 (19.4)	47 (11.5)	69 (8.1)	93 (6.0)	140 (3.9)	201 (2.7)	282 (1.9)	471 (1.2)	709 (0.77)	945 (0.58)	1421 (0.38)	2027 (0.27)	2842 (0.19)	4733 (0.114)	7103 (0.077)	9468 (0.057)	14206 (0.038)
10	31 (19.9)	51 (12.1)	75 (8.4)	100 (6.3)	152 (4.1)	218 (2.9)	306 (2.0)	511 (1.2)	770 (0.80)	1025 (0.60)	1541 (0.40)	2199 (0.28)	3082 (0.20)	5133 (0.120)	7704 (0.080)	10268 (0.060)	15407 (0.040)
11	33 (21.0)	54 (12.8)	83 (8.3)	111 (6.2)	166 (4.2)	238 (2.9)	332 (2.1)	555 (1.2)	832 (0.83)	1109 (0.62)	1664 (0.42)	2378 (0.29)	3323 (0.21)	5546 (0.12)	8319 (0.083)	11092 (0.062)	16638 (0.042)
12	36 (21.4)	59 (13.0)	89 (8.6)	119 (6.5)	178 (4.3)	254 (3.0)	356 (2.2)	594 (1.3)	890 (0.86)	1187 (0.65)	1781 (0.43)	2544 (0.3)	3562 (0.22)	5936 (0.13)	8904 (0.086)	11872 (0.065)	17808 (0.043)
13	38 (22.3)	63 (13.4)	95 (8.9)	126 (6.7)	190 (4.5)	271 (3.1)	379 (2.26)	632 (1.3)	948 (0.89)	1264 (0.67)	1896 (0.44)	2709 (0.31)	3793 (0.22)	6321 (0.134)	9482 (0.089)	12643 (0.067)	18964 (0.045)
14	40 (23.1)	67 (13.8)	101 (9.2)	134 (6.9)	201 (4.6)	288 (3.2)	403 (2.3)	672 (1.4)	1007 (0.92)	1343 (0.69)	2015 (0.46)	2878 (0.32)	4029 (0.23)	6716 (0.138)	10073 (0.092)	13431 (0.069)	20146 (0.046)
15	43 (23.3)	71 (14.1)	107 (9.4)	142 (7.1)	213 (4.7)	305 (3.3)	426 (2.36)	711 (1.41)	1066 (0.94)	1422 (0.71)	2133 (0.47)	3046 (0.33)	4265 (0.235)	7108 (0.141)	10662 (0.094)	14216 (0.070)	21324 (0.047)
16	45 (24.1)	74 (14.6)	112 (9.7)	150 (7.2)	225 (4.8)	321 (3.37)	450 (2.41)	750 (1.44)	1124 (0.96)	1499 (0.72)	2249 (0.48)	3212 (0.337)	4497 (0.241)	7496 (0.144)	11244 (0.096)	14992 (0.072)	22487 (0.048)
17	47 (24.7)	79 (14.7)	118 (9.86)	158 (7.36)	236 (4.93)	338 (3.44)	473 (2.46)	788 (1.48)	1182 (0.98)	1576 (0.74)	2364 (0.49)	3377 (0.344)	4728 (0.246)	7880 (0.148)	11819 (0.098)	15759 (0.074)	23639 (0.049)
18	50 (24.9)	83 (15.0)	124 (10.0)	165 (7.54)	248 (5.02)	354 (3.51)	496 (2.51)	826 (1.51)	1239 (1.0)	1652 (0.75)	2478 (0.50)	3540 (0.351)	4956 (0.251)	8260 (0.151)	12390 (0.100)	16520 (0.075)	24780 (0.050)
19	52 (25.5)	86 (15.4)	130 (10.2)	173 (7.76)	259 (5.12)	370 (3.58)	518 (2.56)	864 (1.53)	1296 (1.02)	1728 (0.77)	2591 (0.52)	3702 (0.358)	5183 (0.256)	8638 (0.153)	12957 (0.102)	17276 (0.077)	25914 (0.051)
20	54 (26.1)	90 (15.6)	135 (10.4)	180 (7.82)	271 (5.19)	386 (3.65)	541 (2.60)	902 (1.56)	1353 (1.04)	1803 (0.78)	2705 (0.52)	3864 (0.364)	5410 (0.260)	9017 (0.156)	13526 (0.104)	18034 (0.078)	27051 (0.052)
25	65 (27.0)	109 (16.1)	163 (10.8)	217 (8.08)	326 (5.38)	466 (3.76)	652 (2.69)	1086 (1.61)	1629 (1.08)	2173 (0.807)	3259 (0.538)	4656 (0.376)	6518 (0.269)	10863 (0.161)	16295 (0.108)	21726 (0.081)	32589 (0.054)

[1] Sample sizes are based upon the Poisson exponential binomial limit.
[2] The minimum quality (approximate AQL) required to accept (on the average) 19 of 20 lots is shown in parenthesis for information only.

SYSTEM RELIABILITY TESTING

Electronic system designers are mainly concerned with the demonstration of their systems and will usually not be required to be familiar with part testing. Normally proper specification of parts to be procured will ensure that adequate part reliability will be achieved (see Chapter 15).

Reliability testing is based on a well developed body of test theory. This makes possible the establishement of efficient and affordable testing. A good overview of the scope of reliability testing will be found in (7) and (8). Before proceeding further it is necessary to define the following essential terms:

(a) Specified MTBF (θ_0) - the MTBF for the equipment as stated in the requirements specification,
(b) Minimum Acceptable MTBF (θ_1) - the MTBF that is obtained by dividing θ_0 by the Discrimination Ratio (θ_0/θ_1),
(c) Discrimination Ratio (θ_0/θ_1) - the ratio of the specified MTBF (θ_0) to the minimum acceptable MTBF (θ_1),
(d) Consumer's Decision Risk (β) - the probability of accepting equipment with a true MTBF equal to the minimum acceptable MTBF (θ_1),
(e) Producer's Decision Risk (α) - the probability of rejecting equipment with a true MTBF equal to the specified MTBF (θ_0).

In general, the types of testing used in conjunction with electronic equipment are defined as:

(a) Standard Probability Ratio Sequential Tests (PRST),
(b) Short Run High Risk PRST,
(c) Fixed Length Tests,
(d) Longevity Tests,
(e) All Equipment Screening Tests.

Standard PRST plans are used when a sequential test plan with normal (10 percent to 20 percent) producer and consumer risks are desired. Short-run high risk PRST plans may be used when a sequential test plan is desired, but circumstances require the use of a short test with a low discrimination ratio and both the producer and the consumer are willing to accept relatively high decision risks. Fixed length test plans may be used when the exact length and costs of tests must be known beforehand. PRST plans with similar risks and discrimination ratio will be more efficient and economical than the fixed length test plans. Longevity test plans are designed to determine the capability of the equipment to meet performance requirements over a specified, extended period of time. All equipment screening test is designed for use when defects due to inadequate quality control, new production lines, design changes, or any incipient defects must be eliminated at the producers facilities. This test is really a screening test and is discussed in Chapter 16.

The most commonly used tests, the PRST and the Fixed Length Tests, are discussed in the succeeding paragraphs.

SEQUENTIAL TESTING

Sequential testing involves the use of efficient tests where three decisions are possible, i.e. (a) accept, (b) reject or (c) continue testing. It usually involves truncation so that the test cannot continue indefinitely, i.e. an (a) or (b) decision is forced. This approach forms the basis for (9) which provides a wide range of choices for sequential testing as shown in Table 2 for $\alpha = \beta$. As with parts testing, test articles are chosen at random, according to Table 3, and placed on life test.

Table 2. Sequential Test Plan with Truncation

Test Plan	Decision Risks	Discrimination Ratio
I	10%	1.5
II	20%	1.5
III	10%	2.0
IV	20%	2.0
lVa	20%	3.0
V	10%	3.0
VI	10%	5.0
VII	30%	1.5
VII	30%	2.0

Single equipment 'on' time should be not less than one half of the average operating time of all equipments on test. When failures occur the failed article is removed from testing while the test continues on the remaining test equipments. When test equipments are repaired they are returned to the testing.

Table 3. Recommended Quantity of Test Articles

Lot Size	Recommended Sample Size	Maximum Sample Size
1 - 3	all	all
4 - 16	3	9
17 - 52	5	15
53 - 96	8	19
96 - 200	13	21
over - 200	20	22

Practical tests for use in reliability demonstrations have their origin in (10) and for the Poisson distribution the continue test region can be described by:

$$\frac{\ln(\beta/(1-\alpha)) + r\ln(\theta_1/\theta_0)}{1/\theta_0 - 1/\theta_1} > T > \frac{\ln((1-\beta)/\alpha) + r\ln(\theta_1/\theta_0)}{1/\theta_0 - 1/\theta_1} \qquad (1)$$

where,

r = number of failures,

T = total test time.

For example, consider Test Plan III, where $\alpha = \beta = 0.10$, i.e., 10 percent and $\theta_0/\theta_1 = 2.0$. Assume that $\theta_0 =$ 1000 hours, by definition $\theta_1 = 500$ hours.

$\ln((\beta/(1-\alpha)) = \ln 0.1/0.9 = -2.197225$

$\ln((1-\beta)/\alpha) = \ln 9 = 2.197225$

$\ln(\theta_1/\theta_0) = \ln 0.5 = -0.693147$

$1/\theta_0 - 1/\theta_1 = 1/1000 - 1/500 = -0.001000$

$$2197.2 + r\ 693.1 > T > -2197.2 + r\ 693.1 \tag{2}$$

Thus equation (2) defines the continue test region for this test plan and is shown in Figure 1.

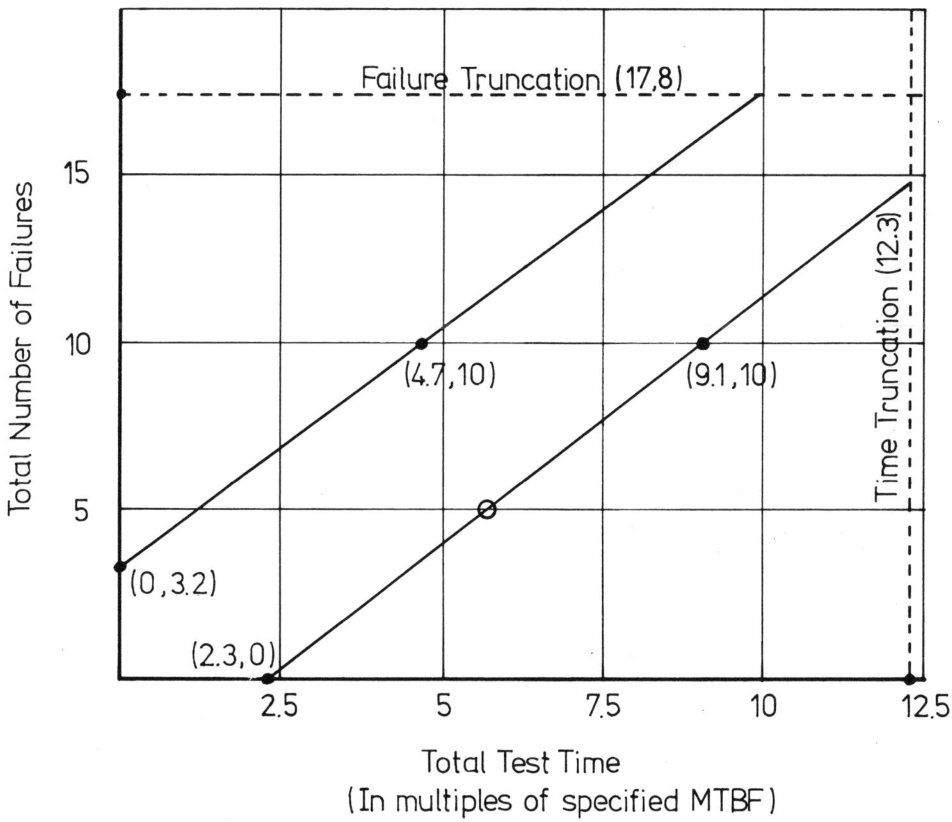

Figure 1. Sequential Truncated Test

The expected number of failures to a decision if the true MTBF = r is given by:

$$\frac{(1-\alpha)\ln((\beta/(1-\alpha)) + \alpha\ln((1-\beta)/\alpha)}{\ln(\theta_0/\theta_1) - (\theta_0/\theta_1 - 1)} \quad (3)$$

where,

$\ln(\theta_0/\theta_1) = \ln 2 = 0.693147$

and,

$r_0 = 5.73$ failures.

From equation (2) r_0 would occur at $T_{ro} = E_T$ (see Table 4), i.e.

2197.2 + 3971.5 = 6168.7 hours

The method used for specifying truncation is arbitrary and can be stated in terms of the maximum number of test failures or test hours allowable. For example, if truncation time, T_T, was specified at $2T_{ro}$ then in the above case truncation would occur at 12337.4 hours. A line parallel to the accept/reject lines is drawn through the origin to represent the truncation line. If the test is to be truncated at T_T when no prior decision has been made then accept, if the number of failures, r, is as follows:

$$r \leq \frac{1/\theta_1 - 1/\theta_0}{\ln(\theta_0/\theta_1)} T_T \quad (4)$$

and for this case is 17.80.

Truncation affects the values of α and β for a test plan. However, it has been found that the effect is on the safe side, e.g., r is actually lower than that computed from (3) and should be adjusted. The adjusted form of Test Plan III taken from (9) is shown in Figure 2 and details for Test Plan III given therein are reproduced below in Table 4.

FIXED LENGTH TESTS

Fixed length tests are of two main types, viz, those which, (a) employ reliability growth principles and (b) those that do not. Typical of (b) are test plans 10 through 27 of (9). The application of the principles of reliability growth to fixed length tests can be used to provide insight into the performance of an equipment. Then, necessary design changes can be incorporated before commitment to production and/or formal demonstration testing.

Table 4. Sequential Test Plan (with truncation) Accept/Reject Criteria

Test Plan III $\alpha = \beta = 10\%$; $\theta_0/\theta_1 = 2.0$; $E_T = 5.1$

No. of Failures	Reject (Equal or less)	Accept (Equal or more)	No. of Failures	Reject (Equal or less)	Accept (Equal or more)
0	N/A	2.20	9	4.51	8.44
1	N/A	2.89	10	5.20	9.13
2	N/A	3.59	11	5.90	9.83
3	.35	4.28	12	6.59	10.30
4	1.04	4.97	13	7.28	10.30
5	1.74	5.67	14	7.97	10.30
6	2.43	6.36	15	8.67	10.30
7	3.12	7.05	16	10.30	N/A
8	3.82	7.75			

Note 1. Reject and Accept figures are multiples of θ_0 and are the total combined hours of individual equipment 'on' times.

Note 2. Expected decision point for MTBF = θ_0 is expressed by E_T as a multiple of θ_0.

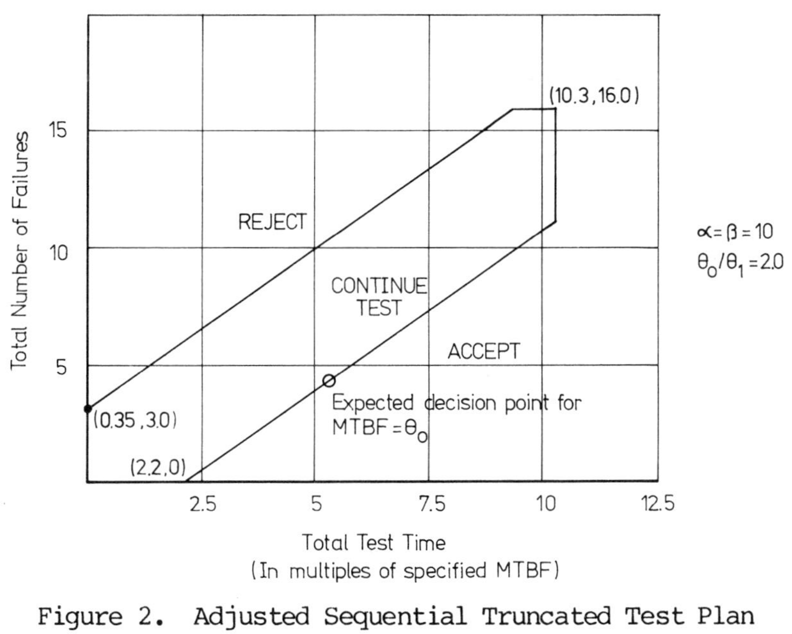

Figure 2. Adjusted Sequential Truncated Test Plan

Growth Testing

To achieve a high level of system reliability acceptable to the user, development programs for sophisticated systems require considerable resources, such as time, manpower and money. The reliability requirements for many systems, ranging from consumers to military and aerospace applications, are high. To achieve such stringent reliability goals, it is a common practice to subject the system to a test-fix-test-fix process. In this process, the total system or sampled subsystems are tested to failure. Failure data, causes, modes and criticality are determined and analyzed. Whenever necessary, design and/or engineering changes are made. Any deficiency will eventually (hopefully) be eliminated. With competent design/engineering modification the system, reliability will grow as development progresses (11, 15, 17, 19).

In the development phase of system design, the system configuration is continuously changing and availability of test data is generally limited or extremely restricted. Consequently, to have a certain degree of confidence in the estimate, a mathematical analysis of the available test data is necessary.

The mathematics required has grown from the statistical method of least squares and regression theory. The parameters will be discussed and examples given to show the applicability of the mathematics discussed.

Linear Regression Modelling (18)

Suppose that we made an observation consisting of n pairs of values (X_1,Y_1), (X_2,Y_2),...(X_n, Y_n) and let these points be plotted as shown in Figure 3. Assume further that from the physical nature of the relation between X and Y, we deduce that the relationship is linear. We can, therefore, postulate that,

$$Y = \alpha + \beta X \tag{5}$$

where α and β are constant. The problem now is to determine the best line of fix and the values of α and β. These are best determined by the least squares method.

The least squares method states that if Y is a linear function of an independent variable X, the most probable positions of a line is such that the sum of the squares of deviation of all points (X_i,Y_i) from the line is a minimum. These deviations are measured in the direction of the Y-axis. The measured quantity, either X or Y, is not error free. This is the ε which is to be included in equation (5) to give,

$$Y = \alpha + \beta X + \varepsilon \tag{6}$$

where,

ε is a random variable.

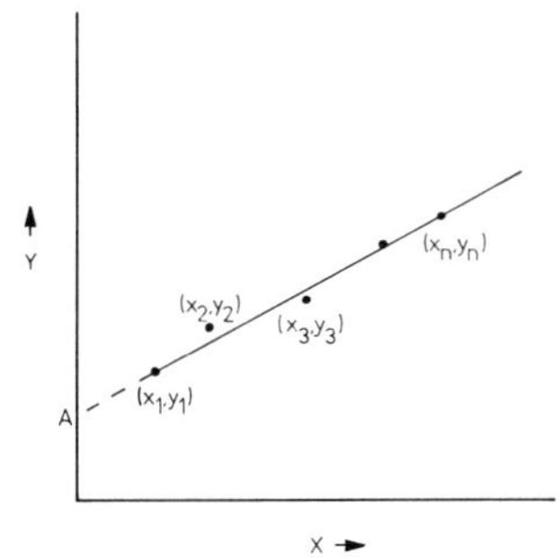

Figure 3. Linear Regression Plot

With n observations we have n simultaneous equations:

$$
\begin{aligned}
Y_1 &= \alpha + \beta X_1 + \varepsilon_1 \\
Y_2 &= \alpha + \beta X_2 + \varepsilon_2 \\
Y_n &= \alpha + \beta X_n + \varepsilon_n
\end{aligned}
\qquad (7)
$$

Statistical theory gives:

$$P = \sum_{i=1}^{n} \varepsilon_i^2 \qquad (8)$$

or,

$$P = \sum_{i=1}^{n} (Y_i - (\alpha + \beta X_i))^2 \qquad (9)$$

To determine the minimum sum of squares of residuals P, we differentiate P and set it equal to zero. That is,

$$\frac{\partial P}{\partial \alpha} = 0 \quad . \qquad (10)$$

$$\frac{\partial P}{\partial \beta} = 0 \quad . \qquad (11)$$

Thus, we have for equations (10) and (11) from equation (9),

$$\sum_{i=1}^{n} (Y_i - (\beta X_i + \alpha)) = 0 \qquad (12)$$

and,

$$\sum_{i=1}^{n} X_i(Y_i - (\alpha + \beta X_i)) = 0 \tag{13}$$

respectively.
Equation (12) can now be rewritten as,

$$\sum_{i=1}^{n} Y_i - \sum_{i=1}^{n} \alpha - \beta \sum_{i=1}^{n} X_i = 0 \quad . \tag{14}$$

Since α is a constant,

$$\sum_{i=1}^{n} Y_i - n\alpha - \beta \sum_{i=1}^{n} X_i = 0 \quad . \tag{15}$$

Equation (15) is simplified to give,

$$\frac{\sum_{i=1}^{n} Y_i}{n} = \alpha + \beta \frac{\sum_{i=1}^{n} X_i}{n} \tag{16}$$

or,

$$\overline{Y} = \alpha + \beta \overline{X} \tag{17}$$

where $\overline{X}$ and $\overline{Y}$ are the mean of the X's and Y's respectively. Equation (17) states that the line will pass through the point $(\overline{X}, \overline{Y})$. Similarly, like equation (12), equation (13) can be simplified to,

$$\sum_{i=1}^{n} X_i Y_i = \alpha \sum_{i=1}^{n} X_i + \beta \sum_{i=1}^{n} X_i^2 \tag{18}$$

Equations (15) and (18) are the normal equations. Solving these two equations we get,

$$\alpha = \frac{\sum_{i=1}^{n} X_i^2 \sum_{i=1}^{n} Y_i - \sum_{i=1}^{n} X_i \sum_{i=1}^{n} X_i Y_i}{n \sum_{i=1}^{n} X_i^2 - (\sum_{i=1}^{n} X_i)^2} \tag{19}$$

and,

$$\beta = \frac{n\sum_{i=1}^{n} X_i Y_i - \sum_{i=1}^{n} X_i \sum_{i=1}^{n} Y_i}{n\sum_{i=1}^{n} X_i^2 - (\sum_{i=1}^{n} X_i)^2} \quad . \tag{20}$$

Duane's Model (13, 16, 19)

The observation was made by J.T. Duane that a plot of cumulative failure rate versus cumulative operating hours on log-log paper, for complex electromechanical devices, gives approximately a straight line. Mathematically, if T is the cumulative operating hours, then the failure rate can be represented by,

$$\lambda_\Sigma = KT^{-\alpha} \tag{21}$$

where,

λ_Σ = cumulative failure rate at time t,

K = constant,

T = total operating time,

α = a measure of reliability growth ("growth rate").

Taking the common logarithm of equation (21) gives,

$$\log \lambda_\Sigma = \log K - \alpha \log T \quad . \tag{22}$$

Equation (22) is a representative of Duane's observation for a linear plot with a negative slope α, the growth rate, on log-log paper. At T = 0, the cumulative failure rate is K and equation (22) is rewritten to give,

$$\lambda_\Sigma = \frac{F}{T} = KT^{-\alpha} \tag{23}$$

where,

F = total number of failures observed in time t.

Therefore, from equation (23):

$$F = KT^{1-\alpha} \quad . \tag{24}$$

The instantaneous failure rate, $\lambda_i = \frac{dF}{dT}$. Thus,

$$\lambda_i = \frac{dF}{dT} = (1 - \alpha)\,KT^{-\alpha} \quad . \tag{25}$$

Substitution of equation (21) into equation (25) gives,

$$\lambda_i = (1 - \alpha)\lambda_\Sigma \quad . \tag{26}$$

From equation (26) and Chapter 7,

$$MTBF_i = \frac{1}{\lambda_i} = \frac{MTBF_\Sigma}{1 - \alpha} \tag{27}$$

where,

$MTBF_i$ = instantaneous MTBF,

$MTBF_\Sigma$ = cumulative MTBF.

The Duane parameters K and α can be determined from the regression analysis of the failure data observed, using equations (19) and (20), with slight modification. That is,

$$\alpha = \frac{\sum_{i=1}^{n} (\log X_i \log Y_i) - (\sum_{i=1}^{n} \log X_i \sum_{i=1}^{n} \log Y_i)/n}{\sum_{i=1}^{n} (\log X_i)^2 - (\sum_{i=1}^{n} \log X_i)^2/n} \tag{28}$$

and,

$$\log K = \frac{(\sum_{i=1}^{n} \log Y_i)}{n} - \frac{\alpha(\sum_{i=1}^{n} \log X_i)}{n} \tag{29}$$

where,

X_i = time to ith failure,

Y_i = cumulative mean time between failure at time X_i

n = cumulative number of failures encountered during test.

In general, K and α are design and system assurance program dependent. Prediction of these two factors can be rather sophisticated and time consuming. Typical values for α have been found to lie between 0.15 to 0.7.

Weibull Model

The Weibull Model is basically generated from Duane's growth model previously discussed. By the substitution of $\beta = 1-\alpha$ in equation (25) we have,

$$\lambda_i = \frac{dF}{dt} = \beta K T^{\beta-1} \tag{30}$$

Equation (30) can be observed to be the instantaneous Weibull failure rate function. From this equation we can obtain the instantaneous mean-time-between failure of the system under test.

$$MTBF_i = \frac{1}{\lambda_i} = \frac{1}{\beta K} T^{1-\beta} \tag{31}$$

The unknown factors β and K can be derived using the method given in reference (15). That is, the Maximum Likelihood (ML) of $\beta = 1 - \alpha$ is,

$$ML(\beta) = \frac{n}{\sum_{i=1}^{n-1} \log \frac{X_n}{X_i}} \tag{32}$$

and the Maximum Likelihood of K is,

$$ML(K) = \frac{n}{X_n^{ML(\beta)}} \tag{33}$$

where,

n = number of failures during test,

X_i = cumulative time at the ith failure,

X_n = cumulative time at the nth failure.

Thus, in the estimation of $\beta = \overline{\beta}$ and $K = \overline{K}$ the instantaneous failure rate function of equation (30) becomes,

$$\overline{\lambda}_i = \overline{\beta}\,\overline{K}\,T^{\overline{\beta}-1} \tag{34}$$

and,

$$MTBF_i = \frac{1}{\overline{\lambda}_i} . \tag{35}$$

For which by simplification, the mean-time-between-failure is estimated by,

$$\overline{MTBF}\ (X_n) = \frac{X_n}{n\overline{\beta}} \tag{36}$$

which is the projected reliability growth of the system.

A discussion of mathematical modelling is incomplete without any examples to show the applicability of the models developed. Two examples will be given here.

Example 1

Suppose in an experiment, four values of Y were obtained as shown in Table 5.

Table 5. Results Of An Experiment

X	1	2	3	4
Observed Y_i	0.5	6.5	21.3	48.6
Most probable Y	α	$\alpha+\beta$	$\alpha + 4\beta$	$\alpha + 9\beta$
Deviation $(Y_i - Y)$	$0.5-\alpha$	$6.5-(\alpha + \beta)$	$21.3-(\alpha+4\beta)$	$48.6-(\alpha+9\beta)$

By physical deduction, it was observed that the general relationship between X and Y is of the form $Y = \alpha+\beta X^2$. Deduce α and β.

The sum of squares of the deviation is,

$$P = (0.5 - \alpha)^2 + (6.5 - \alpha - \beta)^2 + (21.3 - \alpha - 4\beta)^2 + (48.6 - \alpha - 9\beta)^2$$

To determine the minimum P, we must satisfy the conditions:

$$\frac{\partial P}{\partial \alpha} = 0 \quad \text{and} \quad \frac{\partial P}{\partial \beta} = 0 \text{ i.e.,}$$

$$(0.5 - \alpha) + (6.5 - \alpha - \beta) + (21.3 - \alpha - 4\beta) + (48.6 - \alpha - 9\beta) = 0 \text{ and}$$

$$(6.5 - \alpha - \beta) + 4(21.3 - \alpha - 4\beta) + 9(48.6 - \alpha - 9\beta) = 0 .$$

The last two equations reduce to,

$$4\alpha + 14\beta = 76.9$$

$$14\alpha + 98\beta = 592.1 .$$

Hence $\alpha = 0.219$ and $\beta = 5.430$. From these, the line of fix of the form $Y = \alpha + \beta X^2$ for this observation is,

$$Y = 0.219 + 5.430X^2.$$

Example 2

During a test to determine the potential reliability growth of the system, the failure time was recorded for each failure. The failure time and the cumulative failure rate were tabulated as shown in Table 6.

Table 6. Reliability Growth Test Results

Number of System failed	Time to Failure T	Cumulative failure rate $\lambda(T)$
1	0.7	1.428
3	13.2	0.227
5	54.5	0.092
10	163.0	0.061
15	486.3	0.031
17	513.3	0.033
20	688.0	0.029
24	1029.0	0.023
25	1035.0	0.024
30	1430.0	0.021
32	1630.0	0.020
33	1702.0	0.019
35	2073.0	0.017
37	2930.0	0.013
40	3255.0	0.012

Using equation (32) $ML(\beta) = 1 - \alpha = 0.490$, and from equation (33), $ML(K) = 0.761$. From these, we can deduce the reliability growth function of

$$\lambda_i(T) = 0.490 \times 0.761 T^{0.49-1} \text{ or } \lambda_i = 0.37T^{-0.51}$$

which is the Duane growth model. As postulated earlier, for this example $0.15 \leq \alpha \leq 0.7$. Further, the plot of T versus $\lambda(T)$ on log-log scale gives approximately a straight line as shown in Figure 4.

Growth Test Summary

As the concept of reliability growth has become firmly established as a more accurate method of approaching reliability for systems being developed, a controlled method of demonstrating growth has arisen. Recently an approach was formalized (20) which employs a fixed length test. The system development effort consists of two phases. In Phase I a prototype system is fully tested in all respects against the requirements specification including the environmental requirements. The reliability sensitivity of the design is carefully recorded and investigated and the failure data and test time information is used to estimate the reliability prior to Phase II testing. Phase II testing is really a reliability growth test with preproduction systems, conducted along the lines of (9) except that the basis for determining test length is different. This is shown in Table 7.

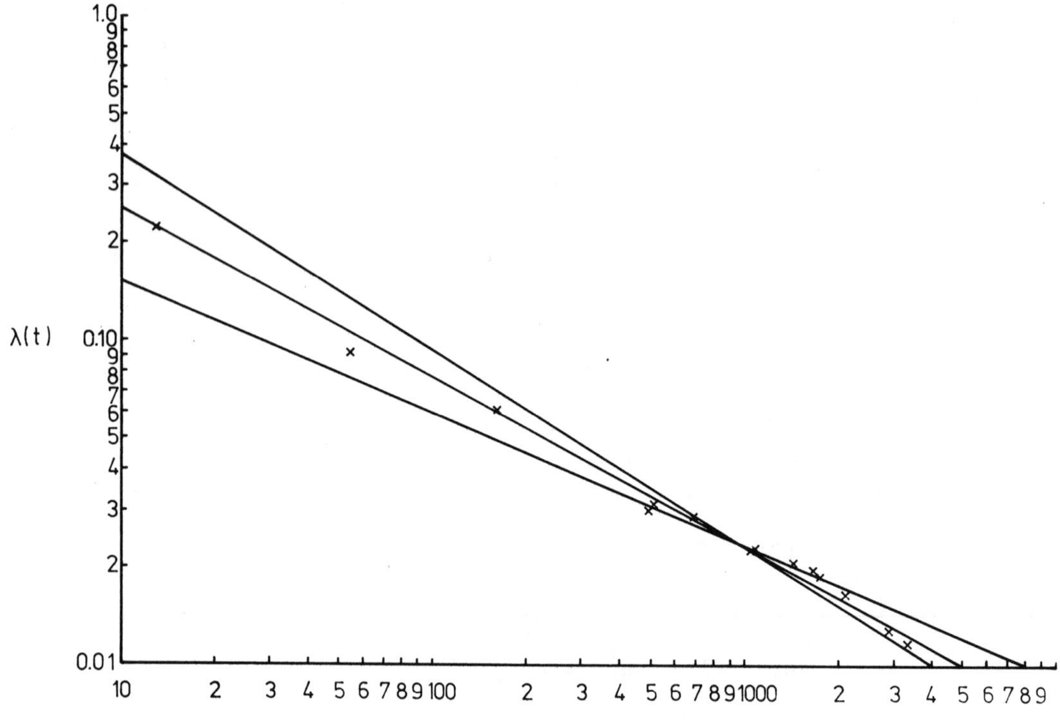

Figure 4. Reliability Growth Plot for Example 2

Table 7. Minimum Test Time per Test Item

Specified MTBF (hours, etc.)	Test Duration (Multiples of θ, Minimum per Item)	Recommended No. of Test Items
1 - 100	$7x\theta$	2
101 - 300	$6x\theta$	2
300 - 500	$5x\theta$	2
501 - 800	$4x\theta$	2
801 - 1000	$3x\theta$	2
1001 - 2000	$2x\theta$	2
2001 - up	$1x\theta$	2

Test time is accumulated based on a mission profile which is representative of operational use and includes only the equipment operating or 'on' time.

The reliability growth of the equipment is monitored throughout the test through a graphic plot of the currect MTBF, starting with the value generated in Phase I testing. The plot is made showing current MTBF versus the cumulative operating time with the specified MTBF (θ_0) shown as a horizontal line. The curve may be adjusted to reflect the level of achieved reliability when it has been proven that acceptable corrective action has resolved a failure. The test is terminated on achievement of the specified MTBF (θ_0).

Test Planning

Reliability testing involves proper plans to be executed at the right time. For the purpose of planning reliability testing the total activity can be regarded as a project and possibly the concept of project management or project engineering should be applied to it. It is suggested that consideration be given to the application of critical path analysis to reliability test planning should the size of test suggest it.

The purpose of the reliability testing should be agreed upon by the various interested parties before anything else is done. It is often useful to write up all or any purpose(s) concerned with testing. The purpose will further determine the organization or the selection of personnel for a particular test. It is often noticed that a fixed organization is responsible for reliability testing or the department most concerned is given full responsibility. This approach is not the best and it is recommended that concerned personnel from the various departments or sections should be invited to participate. Once the personnel have been selected, an agreed schedule can be determined and purpose of the testing will be helpful in deciding when the testing should be done.

Test procedures are usually the responsibility of a reliability specialist. The quality of the test procedure may well determine the end result. Poorly described procedures will result not only in inadequate data but also may not be purpose oriented. The higher the reliability requirements, the more detailed and controlled the procedure should be. Test procedures can be divided basically into two main areas:

(a) Proofing the test equipment to be used,
(b) Test step details.

Proofing the test equipment involves checking and making sure that the equipment which is being used for testing purposes is up to the specified standards. It may involve calibration and other detailed checking of the functioning of the equipment. Test step details involve writing each step to be performed during the test. It is suggested that complete details be given in as clear language as possible. Related elements can be written in sub elements. Once the test procedure is finalized and approved then it should not be changed without proper authority. Depending upon the data required, forms or sheets should be laid out to record the necessary data. Facilities where the testing demonstration will be performed should be arranged, well in advance.

Test Conditions

Before any steps can be taken regarding test equipment, planning or test execution, it is essential to know the test conditions. Test requirements depend on may factors such as contractual requirement or functional requirement, i.e., to demonstrate the functioning of the system or equipment. Strict requirements may be necessary where the products are to be used in critical applications such as weapon systems, etc. Requirements for testing also contribute to the cost of the final system or product.

One of the major contributions to the demonstration field is (9). This standard is written into many military and commercial contracts. The standard describes the test levels to which an equipment may be subjected and included are ten test levels, A to J. Each test level describes temperature range, temperature cycling, vibration and equipment ON/OFF cycle. Table 8 summarizes the test levels.

Temperature

Temperature and temperature range are chosen on the basis of how the equipment will be used operationally. For example if an equipment will be used in an air conditioned environment then test levels A or A-1 may be suitable.

Temperature Cycling

Should the environment be that experienced in a fighter aircraft, then probably there will be rapid and extreme changes in temperature and test level F may be selected, which requires the use of temperature chambers. In order to determine the parameters for thermal cycling, as depicted in Figure 5, a thermal survey is often performed.

Table 8. Summary of Test Levels

Test Level	Temperature °C	Temperature Cycling*	Vibration **	Equipment On/Off Cycle***
A	+20 to +30	None	Yes	Yes
A-1	+20 to +30	None	None	None
B	+35 to +45	None	Yes	Yes
C	+50 to +55	None	Yes	Yes
D	+60 to +70	None	Yes	Yes
E	-54 to + 55	Yes	Yes	Yes
F	-54 to + 71	Yes	Yes	Yes
G	-54 to + 95	Yes	Yes	Yes
H	-65 to + 71	Yes	Yes	Yes
J	-54 to + 125	Yes	Yes	Yes

* Input voltage should be the nominal specified voltage +5 -2%
Input voltage cycling may be applied to any of the Test Plans.
Determined by a thermal survey or as a minimum the heating/cooling period are each 2 hours in duration.

** 2.2G $\pm$ 10% at any frequency between 20 and 60 cps (The vibration level shall be 1 G $\pm$ 10% for equipment designed for use on Navy Ships). If the equipment is designed to meet a vibration requirement less severe than the requirement contained herein, the vibration may be reduced to the level specified as a design requirement.

*** Equipment shall be "on" during the heating period and "off" during the cooling cycle.

Vibration

If vibration is required during the testing, a vibration survey is necessary to search for resonant conditions between 20 and 60 cycles per second, in order that these may be avoided during the reliability tests. Normally vibration is applied for 10 minutes during each 60 minutes of equipment operation.

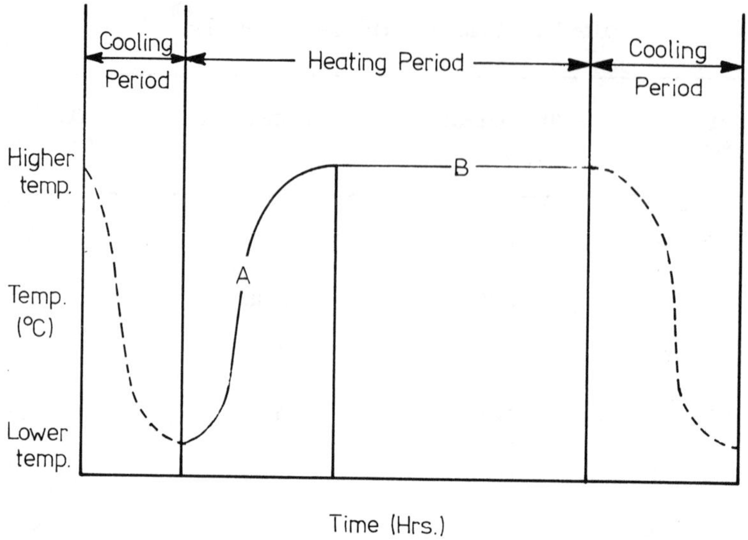

------ Equipment off.

_______ Equipment operated in accordance with duty cycle.

A. Time for Equipment to stabilize at higher temperature.

B. Time of stabilized equipment operations at higher temperature.

Figure 5. Temperature - Time Profile for Temperature Cycling

Duty Cycle

The duty cycle (or performance profile) is the time phase apportionment of modes of operation and functions to be performed by the equipment during the on-time portion of the environmental test cycle. It is intended that the duty cycle be representative of field operation.

Test Equipment

Test equipment for use in demonstration testing can be divided into three areas for consideration:

(a) the equipment which provides the input to the hardware being tested,
(b) the equipment which measures the output,
(c) the equipment which controls the environment to which the hardware is exposed.

It is very important that proper equipment be selected or manufactured for each area, as failure to do so may result in frequent test equipment problems, resulting in test interruptions while the problem source is corrected.

Factors Affecting Test Equipment Selection

Factors which affect the choice of test equipment are many and varied. However, the principle ones are as follows:

(a) Time - the choice of test equipment may depend on time. Some test equipment may be available as a long lead item only and must be designed, built and tested,

(b) Cost - the relative cost of the equipment plays an important role in its selection. Cost also is governed by the complexity of the equipment,

(c) Accuracy - the cost of the equipment is often directly proportional to the accuracy required. It is paramount that the exact accuracy requirement be determined. Another problem typical of higher accuracy equipment is that it sometimes consumes more time for adjustment and calibration than lower accuracy equipment,

(d) General purpose vs. special purpose - the basic difference between the two types is that the first has more applications than the second. The choice should be general purpose equipment but sometimes there are various criteria for using special purpose equipment, such as non-availability, accuracy, test time, delivery time, etc.,

(e) Automated vs.manual - the selection of automated or manual test equipment depends on various factors. Generally, automated equipment is used when the output data results are very repetitive. Another point in favor of automated equipment is that it is usually much easier and faster to diagnose test failures. However, automated equipment generally is more expensive and time is not saved because there is almost no test equipment which is one hundred percent automatic. For some operations such as control of the environment, it is recommended that automatic equipment be used. Another factor in favor of automatic equipment is that it is not always possible to operate controls as precisely as demanded by the test. In contrast, automatic equipment is capable of great test precision.

Whatever type of equipment is chosen, equipment error should be taken into account and analysis of the error will vary according to the testing and type of equipment used.

Test Execution

Test execution involves the carrying out of testing according to the detailed procedures, developed by the equipment producer and approved by the equipment consumer. The equipment to be tested must be representative of production equipment and as such will be received for demonstration testing having passed production acceptance tests normally performed at room temperature. Before testing officially begins the test equipment and the equipment to be tested are integrated and checked out for compatibility. The equipment to be tested is normally burned-in prior to commencement of official testing to eliminate early life failures with the number of burn-in cycles equal to that planned for production units. Some guidelines have been developed to aid in determining the number of cycles required for effective burn-in and are given in Table 9.

Table 9. Burn-In Cycles vs. Complexity

Number of Burn-In Cycles	Item Complexity (Parts)
1	100
3	500
6	2000
10	4000

Note: Last two cycles must be failure free.

Upon official commencement of the test, data recording is of paramount importance as it will form the basis for eventual test acceptance or rejection. Thorough failure analysis is also essential to determine whether a failure is relevant or non-relevant. The categories defined for relevant failures are as follows:

(a) Equipment Manufacturer Design (EMD) - failure in this area places the cause directly upon the design of the equipment, i.e. the design of the equipment caused the part in question to degrade or fail, resulting in an equipment failure, e.g., a circuit design which overstresses a part or other improper application of parts,

(b) Equipment Manufacturer Workmanship (EMW) - these failures are those which are caused by poor workmanship during the equipment construction, testing, or repair prior to start of test. This would include possible overstressing of parts by the assembly process during the construction of equipment,

(c) Part Manufacturer Design (PMD) - This category consists of parts whose failures resulted directly from the inadequate design of the part. This would include such areas as the longevity of the part and its ability to withstand continuous temperature cycling,

(d) Part Manufacturer Workmanship (PMW) - these failures are the result of poor workmanship during assembly of the part, inadequate inspection or testing,

(e) Non-Classified - this type of failure is one which is not immediately determinable as being EMD, PMD, PMW, EMW, or other. It is, therefore, necessary to continue an investigation until a definite conclusion has been reached with sufficient supporting data. This category is not a final category but is considered an intermediate category to be changed to a final category later,

(f) Other - This covers malfunction of the equipment under test which appeared and, before the cause could be determined, the condition corrected itself, i.e., a temporary change from the equipment's normal operating condition due to an unknown cause.

Monthly reports are usually required throughout testing, in order to indicate status and report significant progress. A final report is required summarizing all test results obtained, i.e., failure summary and analysis, a general reliability analysis, reliability design data and correlation of test results. The failure summary and analysis should cover the detailed diagnosis of each failure and include basic corrective action whether indicated or accomplished. Each failure analysis should be cross referenced to the consecutive failure numbers. Full supporting data for all failures classed as 'nonrelevant' or 'test equipment' shall be included.

CONCLUSION

The concept of reliability demonstration was evolved in conjunction with the efforts of the Advisory Group on Reliability of Electronic Equipment (AGREE) which was established in 1952 and resulted, ultimately, in a final report in 1957 (10). The main task of the Group was to make recommendations as to how the poor reliability of electronic systems then apparent could be improved. MIL-H-108 was used extensively as a guide for reliability testing (6) but none of the above documents find much direct use for that purpose. The recommendations of the Group in the area of reliability testing were formalized and embodied in MIL-STD-781 for use in procurement (9).

Major changes reported in (20) are expected to be made to MIL-STD-781 this document will require:

(a) The demonstration of θ_1 and not θ_0 ,

(b) Equipment operation for twice as long as previously specified for a given number of failures,

(c) The use of random vibration over 20-2000 cps vs. the present sinusoidal vibration over 20-60 cps,

(d) Simultaneous exposure to humidity, voltage transients, extreme temperature and vibration,

(e) Closer tailoring of the test environment to the operational use conditions.

The changes stem from the fact that to date it has been observed that passing a factory reliability demonstration does not necessarily mean that acceptable reliability characteristics will be achieved in the field. The suggested changes to MIL-STD-781 should reverse this apparent gap between the factory test environment and the operational environment. In addition to the changes with respect to MIL-STD-781, a newer document (AR-104(21)) will provide guidance for reliability where the demonstration of reliability growth is a requirement.

REFERENCES

1. Beckhart, G.H., The Practical Application of 'AGREE' and Other Advanced Reliability Tests, Materials Management Institute, Publication No. PEN - 16, Boston, Ma., June 1962, p4-5.

2. _______________, MIL-M-39510, Microcircuits - General Specification for, U.S. Department of Defense, Washington, D.C.

3. _______________, MIL-STD-883, Test Methods and Procedures for Microelectronics, U.S. Department of Defense, Washington, D.C.

4. _______________, MIL-HDBK-106, Multilevel Continuous Sampling Procedures and Tables for Inspection by Attributes, U.S. Department of Defense, Washington, D.C., 31 October 1958.

5. _______________, MIL-HDBK-107, Inspection and Quality Control-Single Level Continuous Sampling Procedures and Tables for Inspection by Attributes, U.S. Department of Defense Washington, D.C., 30 April 1959.

6. _______________, MIL-H-108, Sampling Procedures and Tables for Life and Reliability Testing (Based On Exponential Distribution), Office of the Assistant Secretary of Defense, U.S. Department of Defense Washington, D.C., 29 April 1960.

7. _______________, Practical Reliability - Volume III - Testing Research Triangle Institute, Prepared for NASA as Report NASA CR-1128, NTIS Report N68-32779, NTIS Springfield, Va., 22151, August 1968.

8. Ireson, W.G., Editor-in-Chief, Reliability Handbook, McGraw-Hill, New York, N.Y., 1966.

9. _______________, MIL-STD-781B Reliability Tests: Exponential Distribution, U.S. Department of Defense, Washington, D.C. 15 November 1967.

10. _______________, Reliability of Military Electronic Equipment, Advisory Group On Reliability of Electronic Equipment, U.S. Government Printing Office, Washington, D.C., 4 June 1957.

11. Begat, A. et al, "Growth Modelling Improves Reliability Prediction", 1975 IEEE Reliability-Maintainability Symposium Proceedings, p317-322.

12. Bind, G.T. and Hend, G.R., Avionics Reliability Control During Development, AGARD-LS-81 on Avionics Design for Reliability, April 1976.

13. Codier, E.O., "Reliability Growth in Real Life", 1968 IEEE Reliability Maintainability Symposium Proceedings, p458-469.

14. Cox, T.D. and Keely, J., "Reliability Growth Management of Satcom Terminals", 1976 IEEE Reliability Maintainability Symposium Proceedings, p218-223.

15. Crow, L.H., "Tracking Reliability Growth", 1975 IEEE Reliability-Maintainability Symposium Proceedings, p438-443.

16. Duane, J.T., "Learning Curve Approach to Reliability Monitoring", IEEE Transactions on Aerospace, Vol.2, No.2, April 1964, p563-566.

17. Green, J.E., "Reliability Growth Modelling for Avionics", AGARD-LS-81 on Avionics Design for Reliability, April 1976.

18. Kennedy, J.B. and Neville, A.M., Basic Statistical Methods for Engineering and Scientists, A. Dun-Donnelley Publisher, New York, N.Y., 1976.

19. Pollack A.S. et al, "Reliability and Choosing Number of Prototypes", 1974 IEEE Reliability Maintainability Symposium Proceedings, p84-90.

20. Klass, P.J., Reliability Test Procedure Changes Set, Aviation Week and Space Technology, McGraw-Hill, New York N.Y., 19 April 1976, p59-60.

21. _______________, AR-104, Reliability Development Test Program for Avionic Equipment, Naval Air Systems Command, Washington, D.C., 20361, April 1974.

PART THREE

MAINTAINABILITY

19 MATHEMATICAL MODELLING

B. G. Lamarre

Most of the systems with which the engineer deals are subject to maintenance actions at one time or another. There are two distinct types of maintenance action, namely preventive and corrective. Preventive maintenance is performed at regular intervals and can contribute significantly to increased reliability and availability. It must be scheduled carefully, in order that the cost of maintenance will be optimum. Corrective maintenance is performed when the system fails. The occurrence of corrective maintenance action is a random variable that can be predicted. In this chapter, the following subjects will be covered, (a) General Concepts of Maintainability, (b) Maintainability Models, (c) Maintenance Policies, (d) Reliability of Maintained Systems, (e) Availability of Maintained Systems.

GENERAL CONCEPTS OF MAINTAINABILITY

There exist many definitions of maintainability, some emphasize the qualitative aspect and others the quantitative aspect. Reference (1) defines maintainability as follows:

"Maintainability is a characteristic of design and installation which is expressed as the probability that an item will conform to specified conditions within a given period of time when maintenance action is performed in accordance with prescribed procedures and resources."

From this definition, it is clear that time to restore is the most important factor in maintainability. The objective of maintainability is to minimize this time, which is composed mainly of the following, (a) time to diagnose, (b) time to organize technicians, tools, test equipments, etc., (c) time to access faulty item, (d) time to replace or repair and (e) time to inspect, verify, etc.

Numerous techniques to reduce the time to restore are available. Reference (2) contains an exhaustive list of these techniques for different types of equipment.

Mathematically, maintainability may be expressed as $M(t_1)$, where:

$$M(t_1) = \text{probability of completing a maintenance action in time } t_1, \tag{1}$$

or,

$$M(t_1) = P(t \leq t_1). \tag{2}$$

If f(t) is the probability distribution function (p.d.f.) of the time to repair then

$$M(t_1) = \int_0^{t_1} f(t)\, dt. \tag{3}$$

In the case of Figure 1, $M(t_1)$ is represented by the shaded area under the curve.

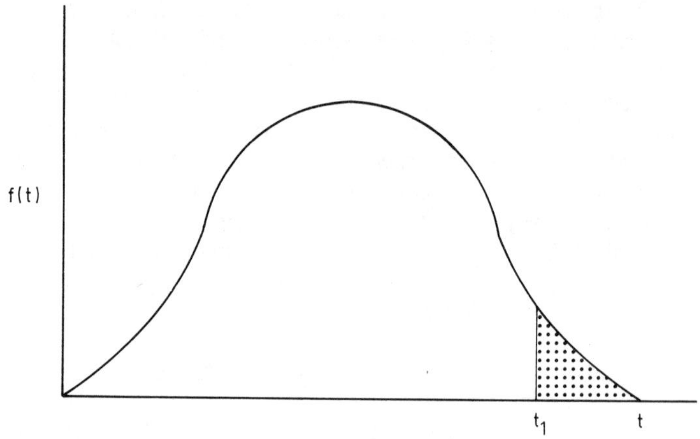

Figure 1. P.D.F. of Time to Repair

It was found that time to repair often follows a Log-Normal or Exponential distribution. However, other distributions like the Weibull and Gamma are also possible. In general, the functions used in maintainability are similar to those used in reliability (Chapter 7). Table 1 illustrates the complementary nature of maintainability and reliability. Algebraic and calculus rules can be applied to many of these fundamental relationships.

MAINTAINABILITY MODELS

In order to determine the maintainability of an item, it is necessary to make a mathematical model. The easiest way to explain how this can be done is by an example. Using fictitious data and common statistical methods, a model will be developed and the necessary information extracted from it.

Table 1. Comparison of Functions Used in Reliability and Maintainability

Reliability		Maintainability	
Time to failure		Time to repair	
$f(t)$	(4)	$f(t)$	(13)
Reliability		Maintainability	
$R(t_1) = \int_{t_1}^{\infty} f(t)\, dt$	(5)	$M(t_1) = \int_{0}^{t_1} F(t)\, dt$	(14)
Unreliability			
$Q(t_1) = \int_{0}^{t_1} f(t)\, dt$	(6)		
Failure rate		Repair rate	
$\lambda(t) = \dfrac{f(t)}{R(t)}$	(7)	$\mu(t) = \dfrac{f(t)}{1-M(t)}$	(15)
Mean Time Between Failure		Mean Time to Repair	
$MTBF = m = 1/\lambda$	(8)	$MTTR = d = 1/\mu$	
$= \int_{-\infty}^{+\infty} t\, f(t)\, dt$	(9)	$= \int_{-\infty}^{+\infty} t\, f(t)\, dt$	(16)
P.D.F. of Time to Failure		P.D.F. of Time to Repair	
$f(t_1) = \lambda(t_1)\; R(t_1)$	(10)	$f(t_1) = \mu(t_1)\; (1-M(t_1))$	(17)
$= \lambda(t_1)\; (1-Q(t_1))$	(11)	$= \mu(t_1) \exp\left(-\int_0^{t_1} \mu(t)dt\right)$	(18)
$= \lambda(t_1) \exp\left(-\int_0^{t_1} \lambda(t)dt\right)$	(12)		

The time to repair data are listed in ascending order and tabulated in column 2 of Table 2. The median, 5% and 95% ranks corresponding to each failure number are tabulated in columns 3, 4 and 5 respectively. For more detail on the concept of ranking and tables, the reader is referred to (3).

Table 2. Time to Repair Data

(1)	(2)	(3)	(4)	(5)	(6)	(7)	(8)	(9)
Rank	Time to Repair(hrs)	Median Rank	5% Rank	95% Rank	M(t) %	|D|	f(t)	μ(t)
1	1.3	3.41	0.26	13.91	3.14	0.0027	0.1134	0.1171
2	1.7	8.31	1.81	21.61	9.68	0.0137	0.2102	0.2327
3	2.0	13.22	4.22	28.26	16.85	0.0363	0.2623	0.3154
4	2.0	18.12	7.14	34.36	16.85	0.0127	0.2623	0.3154
5	2.2	23.02	10.41	40.10	22.36	0.0066	0.2831	0.3646
6	2.5	27.93	13.96	45.56	31.21	0.0328	0.2946	0.4283
7	2.5	32.83	17.73	50.78	31.21	0.0162	0.2946	0.4283
8	2.7	37.74	21.71	55.80	37.07	0.0067	0.2916	0.4634
9	3.0	42.64	25.87	60.64	45.42	0.0278	0.2758	0.5053
10	3.0	47.55	30.19	65.31	45.42	0.0213	0.2758	0.5053
11	3.0	52.45	34.69	69.81	45.42	0.0703	0.2758	0.5053
12	3.6	57.36	39.36	74.13	60.26	0.0290	0.2233	0.5620
13	4.0	62.26	44.19	78.29	68.44	0.0618	0.1851	0.5864
14	4.0	67.17	49.22	82.27	68.44	0.0127	0.1851	0.5864
15	4.5	72.07	54.44	86.04	76.73	0.0466	0.1416	0.6087
16	4.5	76.98	59.89	89.59	76.73	0.0025	0.1416	0.6087
17	5.0	81.88	65.63	92.86	82.89	0.0101	0.1059	0.6192
18	5.5	86.78	71.73	95.78	87.49	0.0071	0.0781	0.6246
19	6.0	91.69	78.39	98.19	90.82	0.0087	0.0572	0.6227
20	8.0	96.59	86.09	99.74	97.32	0.0073	0.0161	0.6002

The time to repair versus median rank is then plotted on log-normal probability paper. Additional information on probability plotting can be found in (4).

A straight line is fitted through the points. This can be done by best eye approximation as in Figure 2 or by linear regression using a linear transformation as discussed in (4). The engineer may be satisfied by the dispersion of the points alongside the line or he may decide to try the same process for another distribution using paper with a different scale.

If the engineer is satisfied with the dispersion of the points, the Kolmogorov-Smirnov test (K-S) can be performed (5) after the moments of the distributions have been estimated. The moments can be estimated graphically from Figure 2, or by using the Maximum Likelihood Estimator (MLE). In the case of log-normal, we shall use:

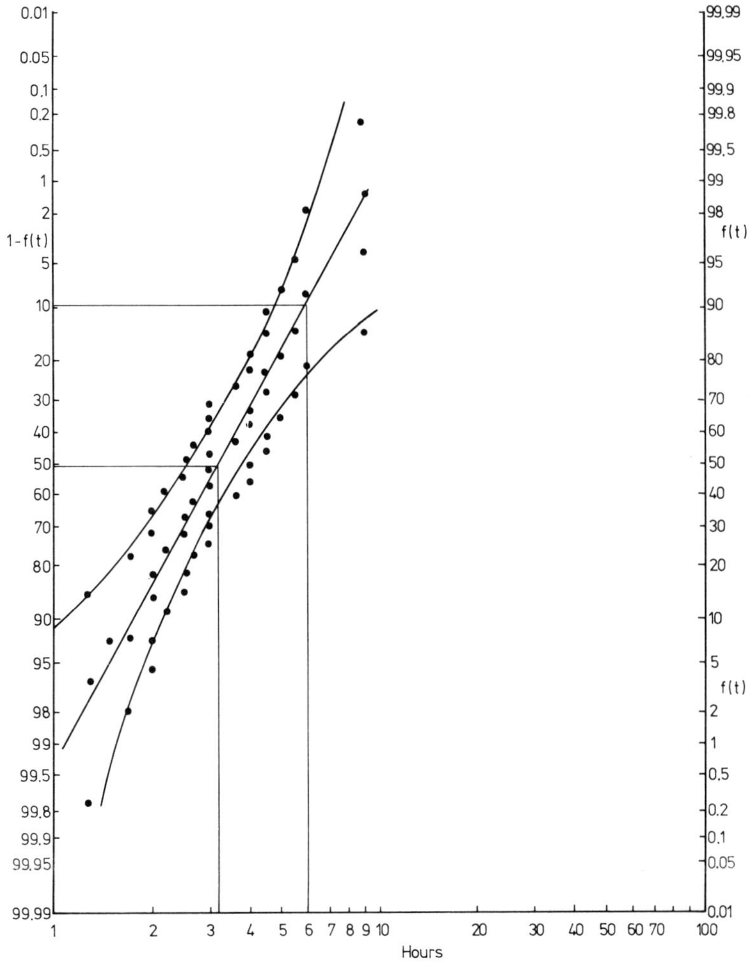

Figure 2. Log-Normal fit to Time To Repair Data

$$t' = \ln t \tag{19}$$

thus,

$$\overline{t}' = \frac{\sum_{i=1}^{n} t_i'}{n} \tag{20}$$

is the MLE of the mean, and

$$S'^2 = \sum_{i=1}^{n} (t_i' - \overline{t}')^2/(n-1) \tag{21}$$

is the MLE of the variance.

The graphical estimates are read directly from the graph of Figure 2.

$$\hat{\bar{t}} = t_{0.5} \tag{22}$$

$$\hat{\bar{t}}' = \ln t_{0.5} \tag{23}$$

$$\hat{\bar{t}}' = \ln (3.17) = 1.15373 \tag{24}$$

$$\hat{S} = 0.4 (\ln (t_{0.9}) - \ln (t_{0.1})) \tag{25}$$

$$\hat{S}' = 0.4 (\ln (5.8) - \ln (1.75)) = 0.4793 \tag{26}$$

The standard form of the log-normal distribution is given by:

$$f (Z') = 1/\sqrt{2\pi} \quad \exp \frac{-Z'^2}{2} \tag{27}$$

where,

$$Z' = \frac{t' - \bar{t}'}{S'}$$

From a table of areas under the standard normal curve, M(t) can be calculated and tabulated as shown in column 6. The absolute difference between columns 6 and 3 is tabulated in column 7. In the present case the maximum value is 0.0703. Since at the 0.2 level of significance the K-S value is 0.231, the log-normal cannot be rejected at this level of significance and

$$\text{Max} |D| = 0.0703 < 0.231. \tag{28}$$

In the log-normal distribution, f(t') is normally distributed. To extract f(t), the following relation has to be used

$$f(t) = \frac{f(t')}{t} = 1/t\,\sigma'\sqrt{2\pi} \, \exp\left[-\frac{1}{2}\left(\frac{t' - \bar{t}'}{\sigma'}\right)^2\right] \tag{29}$$

and $\mu(t)$ can be calculated using equation 15 of Table 1. Figure 3 is a plot of f(t) made using equation (29). Figure 4 is a plot of column 6. Similarly Figure 5 is a plot of μ (t) using the values of column 9 of Table 2. If the specified period of maintenance downtime is 6 hours, it can be seen that M(6 hrs) = 0.90, meaning that 90% of the repair will be completed within 6 hours.

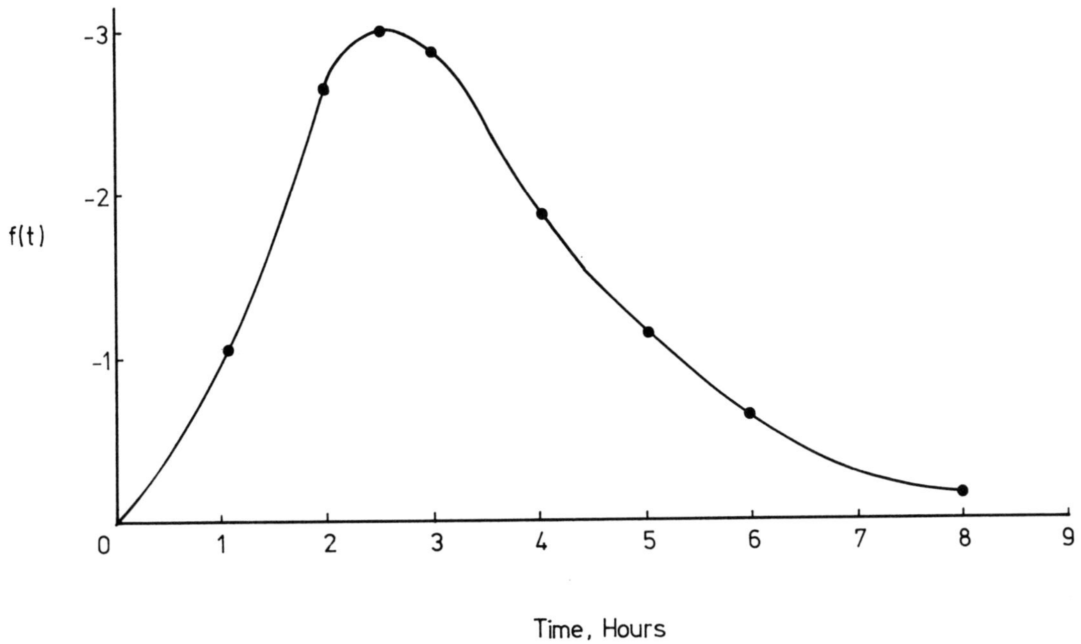

Figure 3. Distribution of Time to Repair

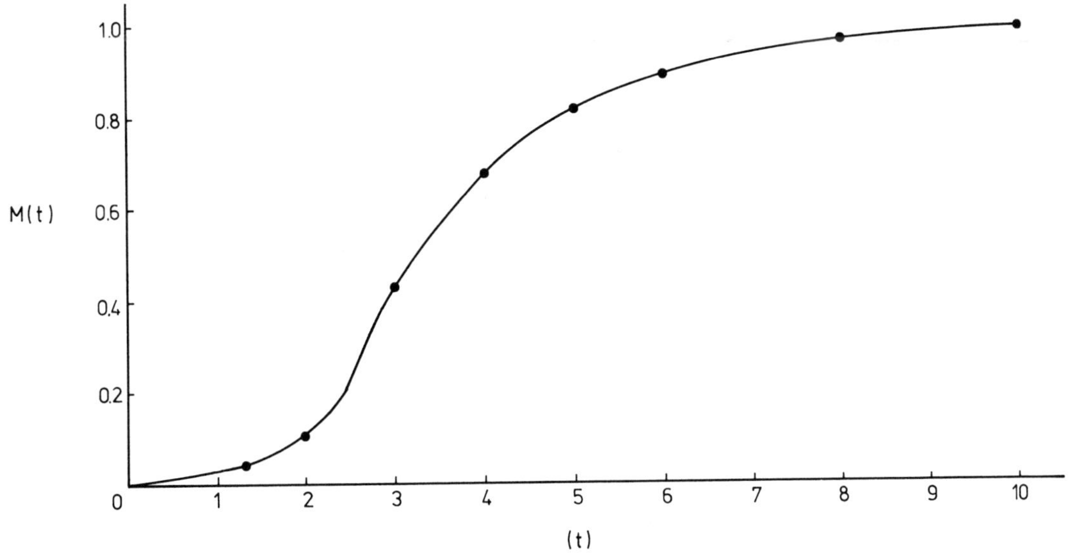

Figure 4. Maintainability Function

One interesting value which can be read directly from the graph, is $M_{0.95}$ which is defined as the time by which 95% of the repair will be completed and for this case is,

$$M_{0.95} = 7 \text{ hrs.} \tag{30}$$

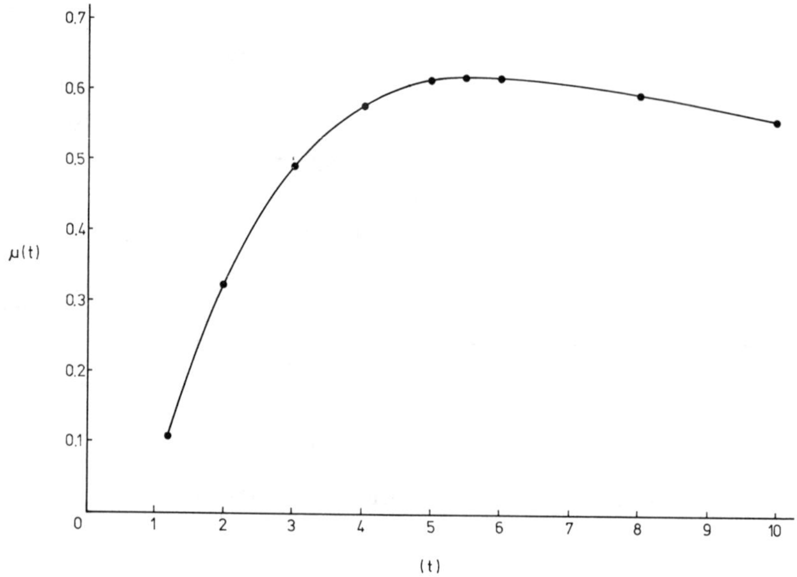

Figure 5. Repair Rate Function

By plotting the 5% and 95% rank on the same graph (Figure 2) one can obtain the 95% confidence limit.

$$M \text{ (6 hrs)} = 0.75 \tag{31}$$

at the lower one sided 0.95 confidence level, and

$$0.75 \leq M(6 \text{ hrs}) \leq 0.975$$

at the two sided 0.90 confidence level.

Mean Time To Repair (MTTR) can also be evaluated as follows:

$$MTTR = \hat{\tilde{T}} = \exp\ (\overline{T} + \sigma^2_{T'}/2) \tag{32}$$

$$MTTR = \exp\ (1.5373 + 0.11486) = 3.55 \text{ hrs.} \tag{33}$$

The time by which 50% of the repairs are completed is the median:

$$\breve{T} = \exp\ (\overline{T}') = \exp\ (1.15373) = 3.169 \text{ hrs.} \tag{34}$$

The most frequently occurring time to repair is the mode:

$$\hat{T} = \exp\,(\overline{T}'-\sigma^2_{T'}) \tag{35}$$

$$\hat{T} = \exp\,(1.15373 - 0.22973) = 2.52 \text{ hrs} \tag{36}$$

The variability of time to repair is σ_T

$$\sigma_T = \overline{T}(\exp(\sigma^2_{T'}) -1)^{1/2} \tag{37}$$

$$\sigma_T = 3.55\ (\exp\,(0.22973) -1)^{1/2} = 1.804 \text{ hrs.} \tag{38}$$

In this section the techniques for obtaining the maintainability parameters of a system have been discussed. In the next section, it will be seen how these parameters can be used.

MAINTENANCE POLICIES

The cost of maintaining an equipment and its reliability are in a large part dependent upon the preventive maintenance policy. The scheduling of preventive maintenance can be done by two different methods, that will be called Policy I and Policy II.

Policy I

In Policy I, the preventive maintenance period is rescheduled after a corrective maintenance action, as shown in Figure 6.

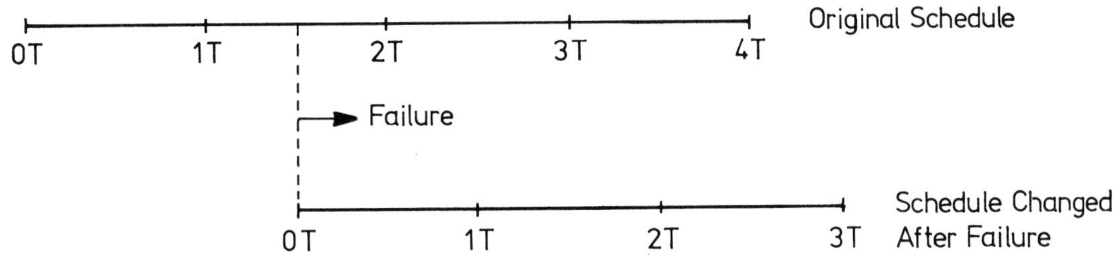

Figure 6. Policy I

This policy assumes that the age of the equipment is noted at a point of corrective maintenance and the replacement period T starts from that point. If corrective maintenance occurs too often between preventive maintenance it indicates that the period T is too long.

We are interested in the MTBF and reliability R (t) where:

$$t = jT + \tau, \ 0 \leq \tau < T \tag{39}$$

then,

$$R_T(t) = (R(T))^j R(\tau) \tag{40}$$

and

$$m_T = \int_0^\infty R_T(t)\, dt. \tag{41}$$

Note that the subscript T indicates that the system is undergoing preventive maintenance every T hours of operation, since the preventive maintenance period, T is discrete. Figure 7 illustrates a particular case of Policy I.

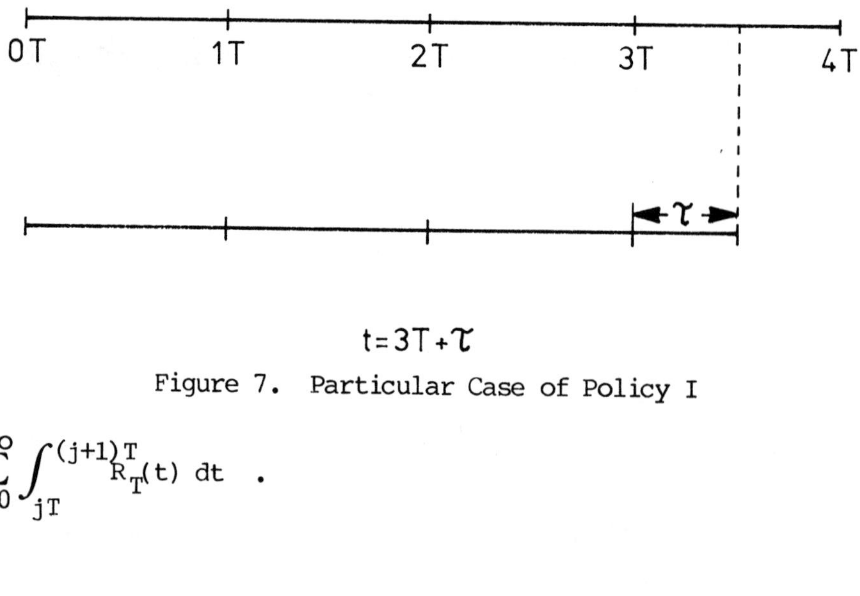

Figure 7. Particular Case of Policy I

$$m_T = \sum_{j=0}^{\infty} \int_{jT}^{(j+1)T} R_T(t)\, dt \quad . \tag{42}$$

Thus,

$$m_T = \sum_{j=0}^{\infty} \int_0^T (R(T))^j R(\tau) d\tau \quad . \tag{43}$$

Rearranging,

$$m = \sum_{j=0}^{\infty} (R(T))^j \int_0^T R(\tau)d\tau \quad . \tag{44}$$

Since,

$$\sum_{j=0}^{\infty} y^j = 1/(1-y) \quad \text{for } y < 1 \tag{45}$$

$$m_T = \frac{\int_0^T R(\tau)d}{1 - R(T)} = \frac{\int_0^T R(\tau)d\tau}{Q(T)} \tag{46}$$

m_T is the corrective MTBF at steady state. At steady state, the reliability is

$$R_T(t) = \exp(-t/m_T) \tag{47}$$

As can be seen from Figure 8 the steady state is never reached completely but is very close to the real reliability.

Since,

$$\int_0^T R(t)\, dT \tag{48}$$

is the mean time between preventive and corrective maintenance action,

$$\frac{1}{\int_0^T R(\tau)dt} = \frac{1 - R(T)}{\int_0^T R(\tau)d\tau} + \frac{R(T)}{\int_0^T R(\tau)d\tau} \tag{49}$$

where the second term of the right side of the equation is the preventive maintenance rate at steady state.

From the above expression, one can determine the mean number of parts required for a mission of duration t.

$$N_{cm}(t) = t \frac{1-R(T)}{\int_0^T R(\tau)d\tau} \tag{50}$$

$$N_{pr}(t) = t \frac{R(T)}{\int_0^T R(\tau)d\tau} \tag{51}$$

where,

$N_{cm}(t)$ = mean number of parts replaced correctively during a mission of length t,

$N_{pr}(t)$ = mean number of parts replaced preventively during a mission of length t.

Knowing the cost per preventive maintenance action, C_P, and the cost per corrective maintenance action C_C, it is possible to find the total maintenance cost per operating hours C_T.

$$C_T = C_C \frac{1-R(T)}{\int_0^T R(\tau)dT} + C_P \frac{R(T)}{\int_0^T R(\tau)dT} \tag{52}$$

It is also possible to optimize the preventive maintenance period, T, by making

$$\frac{d(C_T)}{dT} = 0 \tag{53}$$

However, it has to be checked that the target reliability is met.

Although the analytical solution can be quite laborious, the optimum T can be found numerically as in the following example.

Example

Suppose an item has a Weibull failure distribution with the following parameters; $\gamma = 0$, $\eta = 500$, $\beta = 2.5$. Given that the cost associated with a preventive maintenance action is \$20.00, the cost associated with a corrective maintenance action is \$100.00, and that the reliability target is 0.80 for a mission of 400 hrs, find, (a) C_T for different T, (b) the optimum T that meets the requirements, (c) the total number of parts required for a mission. Also plot, (a) $R_T(t)$ real and stabilized, (b) R(t) without preventive maintenance, (c) $m_T(t)$.

For a Weibull distribution,

$$R(t) = \exp(-t/\eta)^{\beta} \quad . \tag{54}$$

Expression (48) can be evaluated numerically using Simpson's rule for values of T from 100 to 900 in steps of 100. Knowing these values the cost, C_T, can be calculated using equations (52) and (54), and $m_T(t)$ can be calculated using (46).

From column 3 of Table 3, it can be seen that T optimum is 300 hours. The reliability for a 400 hour mission is then, using equation (40) and (54).

$$R(400) = R(300)\, R(100) = 0.75 \tag{55}$$

A preventive maintenance period T of 300 hours, although the most economical, does not meet the reliability target. The period T will then have to be shortened. For T = 200 hours,

$$R(400) = (R(200))^2 = 0.82 \tag{56}$$

Table 3. Optimum Maintenance Cost

(1) T(hrs)	(2) $\int_0^T R(T)\,dt$	(3) C_T	(4) m_T
100	99.49	0.215	5611.55
200	194.38	0.142	2019.71
300	279.76	0.141	1149.62
400	344.39	0.159	790.16
500	382.93	0.184	605.79
600	418.96	0.199	527.99
700	433.75	0.212	481.07
800	437.85	0.220	457.81
900	442.72	0.223	448.53

Table 4. Reliability for a 200 Hour Preventive Period

(1) t	(2) $R_{200}(t)$	(3) R (t)	(4) $R_{200}(t)$ Stabilized
50	0.9968	0.9968	0.9755
100	0.9823	0.9823	0.9517
150	0.9519	0.9519	0.9284
200	0.9038	0.9038	0.9057
250	0.9009	0.8380	0.8836
300	0.8878	0.7566	0.8620
350	0.8603	0.6637	0.8409
400	0.8168	0.5642	0.8203
450	0.8142	0.4637	0.8003
500	0.8023	0.3679	0.7807
550	0.7775	0.2811	0.7616
600	0.7382	0.2065	0.7430
650	0.7359	0.1456	0.7248
700	0.7251	0.0984	0.7071
750	0.7027	0.0631	0.6898
800	0.6672	0.0392	0.6729

A preventive maintenance period T of 200 hours would be recommended and the maintenance cost per operating hour, as calculated by equation (52), is \$0.142 and for a 400 hour mission it would be \$57.00. The total number of parts used for one mission as calculated by equations (50) and (51) is approximately 2.

The benefits of a preventive maintenance schedule can be visualized by the plot of the reliability of the same item with and without preventive maintenance.

Table 4 shows the reliability with a 200 hour preventive period as calculated by equation (40) in column 2. Column 3 shows the reliability calculated by equation (54) without preventive maintenance. Column 4 shows the reliability stabilized, as calculated by equation (47). Figure 8 is the plot of columns 2, 3 and 4. Figure 9 shows how m_T varies with T. It can be seen that at T=700 hours and greater, no real benefit is gained by a preventive maintenance program.

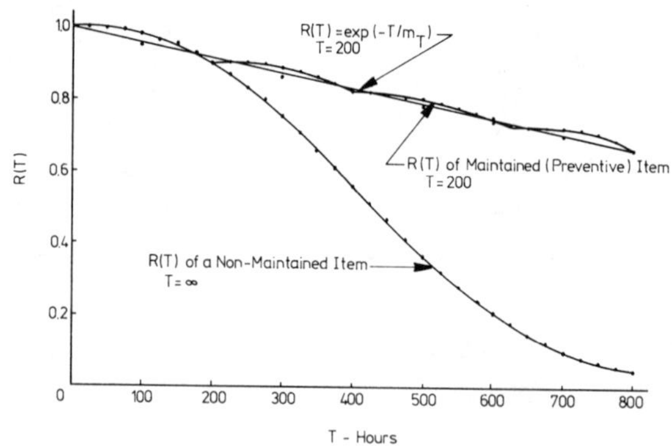

Figure 8. Reliability vs. Time

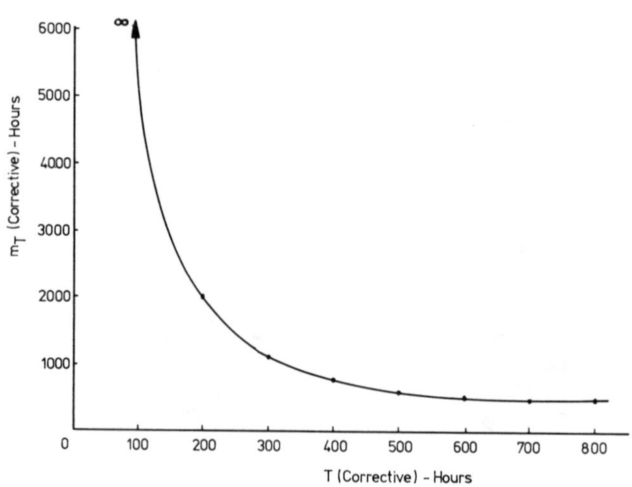

Figure 9. m_T vs. T

Policy II

In Policy II, the preventive maintenance schedule is always adhered to, regardless of the occurrence of a failure. Figure 10 illustrates that if a failure occurs, the item is replaced and it is maintained always at time nT.

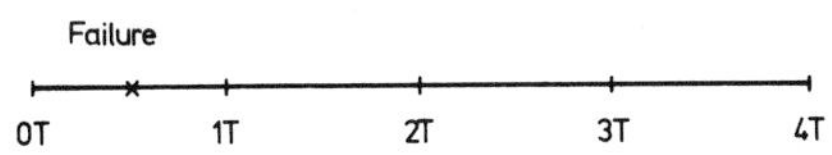

Figure 10. Particular Case of Policy II

The preventive maintenance rate is constant and equal to:

$$\lambda_{pr} = 1/T. \tag{57}$$

This type of policy has applications when an item is overhauled at scheduled time or when it is impracticable to keep a record of when a corrective replacement occurred.

Since,

$$\lambda_{Total} = \lambda_{cm} + \lambda_{pr} \tag{58}$$

and,

$$\lambda_{Total} = \frac{1}{\int_0^T R(t)dt} \tag{59}$$

$$\lambda_{cm} = \frac{1}{\int_0^T R(t)\ dt} - \frac{1}{T}. \tag{60}$$

Thus,

$$\lambda_{cm} = \frac{\int_0^T Q(\tau)d\tau}{T\int_0^T R(\tau)d\tau}. \tag{61}$$

Since reliability is of interest only until the first failure occurs, for a mission starting at any integer multiple of T, equation (40) is the reliability expression. In the general case, the steady state reliability given by:

$$R_T(t) = \exp(-\lambda_{cm}t) \tag{62}$$

is much more meaningful.

The mean number of parts required can be calculated in the same manner:

$$N_{Total}(t) = t(\lambda_{cm} + \lambda_{pr}) = t\,\lambda_{Total} \tag{63}$$

and the total cost C_T is given by:

$$C_T = C_{cm}\,\lambda_{cm} + C_{pr}\,\lambda_{pr}. \tag{64}$$

The period T also can be optimized by taking

$$\frac{d(C_T)}{dT} = 0. \tag{65}$$

It is left to the reader, as an exercise, to do the previous example using policy II. It is interesting to note that at steady state under policy II, the optimum T would be 400 hours and the maintenance cost per operating hour would be \$0.0904 or \$36.16 for a mission. The reliability would be 0.85 and the number of parts required for a mission would be less than 2.

Usually it is not beneficial to have a preventive maintenance schedule if the cost of a preventive maintenance action is close to, or greater than, the cost of corrective maintenance action. However, it might still be required to achieve the reliability goal. When the failure rate is constant, preventive maintenance has no effect, except in the case where there is more than one item with constant failure rate in a parallel or stand-by configuration.

RELIABILITY OF MAINTAINED SYSTEMS

In the case of a maintained system, the repair rate, μ , has an influence as important as the failure rate λ. The approach described below is Markovian and as such the failure and repair rates are constant which implies an exponential distribution of time to failure and time to repair. This restriction is not detrimental to the analysis because it can be proven that for a complex equipment, where the components follow different failure distribution, the failure distribution of the whole equipment can be described by the exponential distribution (6). Furthermore, it has already been shown that a preventively maintained system exhibits a quasi constant failure rate. On the other hand, the exponential distribution is the most conservative to describe time to repair. Before going any further in the analysis, the states of a system and the implications of the Markovian approach will be defined.

State 0: All units have failed. This is an absorbing state in the case of Reliability.

State 1: One unit is operable and the others are being repaired or waiting to be repaired.

State n: All n units of an n unit system are operable or operating, none are waiting to be repaired or are being repaired. This is usually the starting state of the system.

The implications of the Markovian approach are, (a) all states are independent and mutually exclusive of each other, (b) the sum of the probability of being in any state is 1,

$$\sum_{i=0}^{m} P_i(t) = 1 \tag{66}$$

(c) only one failure or one repair is allowed in Δt, (d) the probability of one unit failing in Δt is $\lambda\Delta t$, (e) the probability of one unit not failing in Δt is $1-\lambda\Delta t$, (f) the probability of completing a repair in Δt is $\mu\Delta t$, (g) the probability of not completing a repair in Δt is $1-\mu\Delta t$.

To illustrate the technique, let us consider a simple case, two identical units in parallel, as in Figure 11.

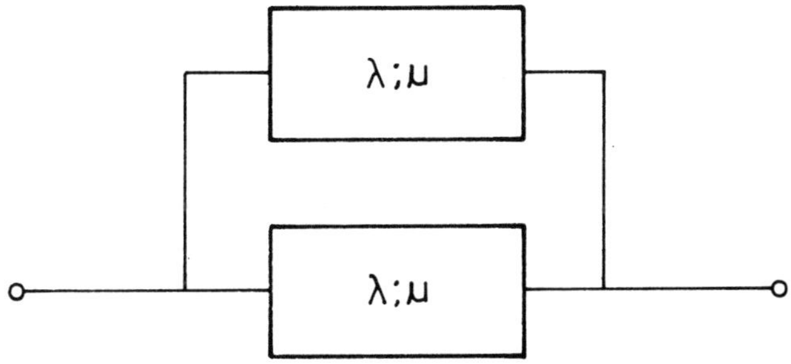

Figure 11. Units In Parallel

The states of the system are:

State 0 : Both units have failed, thus the system has failed.

State 1 : One unit is functioning, the other has failed and is being repaired.

State 2 : Both units are functioning.

If at least one unit is required to assure mission success,

$$R(t) = 1-P_0(t) \tag{67}$$

$$Q(t) = P_0\ (t) \tag{68}$$

The Markovian graph describing the system is shown in Figure 12.

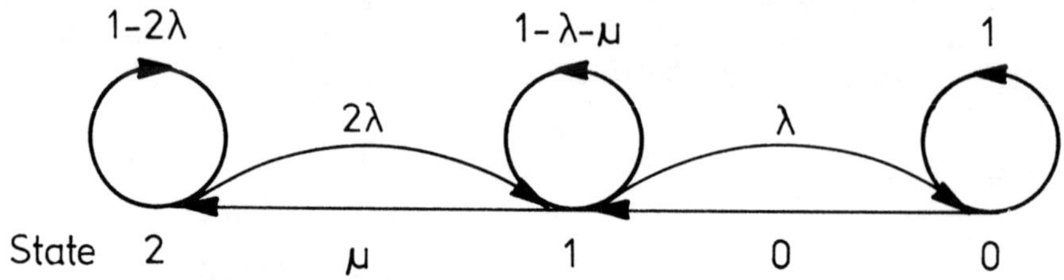

Figure 12. Markovian System Graph

The transition matrix is of the form

$$P = \begin{array}{c} \\ 2 \\ 1 \\ 0 \\ \uparrow \end{array} \begin{array}{c} \begin{array}{ccc} 2 & 1 & 0 \leftarrow \text{Final state at } t+\Delta t \end{array} \\ \begin{bmatrix} 1-2\lambda & 2\lambda & 0 \\ \mu & 1-\lambda-\mu & \lambda \\ 0 & 0 & 1 \end{bmatrix} \end{array} \tag{69}$$

Initial State at t.

It is easier to understand the transition matrix if the columns are formed by analyzing all the events leading to the realization of a certain state after $t+\Delta t$.

$P_2\ (t+\Delta t)$ = Probability of no failure from state 2 in Δt
+ probability of one repair from state 1 in Δt,

thus,

$$P_2(t+\Delta t) = P_2(t)\ (1-\lambda\Delta t)(1-\lambda\Delta t) + P_1(t)\mu\Delta t \tag{70}$$

$$P_2(t+\Delta t) = P_2(t) - 2\lambda\Delta t P_2(t) + P_1(t)\mu\Delta t\ . \tag{71}$$

Since at the limit $(\Delta t)^2 \to 0$

rearranging,

$$\frac{P_2(t+\Delta t) - P_2(t)}{\Delta t} = \mu P_1(t) - 2\lambda P_2(t). \tag{72}$$

By definition the left side of (72) is $P_2'(t)$. Thus:

$$P_2'(t) = \mu P_1(t) - 2\lambda P_2(t). \tag{73}$$

Using the same technique for state 1 and 0 at $t + \Delta t$ leads to,

$$P_1(t + \Delta t) = 2\lambda \Delta t P_2(t) + (1 - \lambda\Delta t - \mu\Delta t)\, P_1(t) \tag{74}$$

$$P_1'(t) = 2\lambda P_2(t) - (\lambda+\mu) P_1(t) \tag{75}$$

and,

$$P_0'(t) = \lambda P_1(t) \quad . \tag{76}$$

Therefore the system is now reduced to,

$$P_2'(t) = \mu P_1(t) - 2\lambda P_2(t) \tag{77}$$

$$P_1'(t) = 2\lambda P_2(t) - (\lambda+\mu) P_1(t) \tag{78}$$

$$P_0'(t) = \lambda P_1(t) \quad . \tag{79}$$

We want to find:

$$R(t) = 1 - P_0(t) \; . \tag{80}$$

The solution is obtained from the three difference equations, (77), (78) and (79). It is to be noted that the three difference equations can be obtained directly from the Markov graph or the transition matrix. One easy way to verify the correctness of the transition matrix is to ensure that each row sums to 1. To solve the difference equations, the use of the Laplace transforms leads to quick results. For a review of Laplace transforms, the reader is referred to (7).

Using the elementary transform,

$$P_n(s) = \frac{P_n(0)}{s} + \frac{1}{s}\int_0^\infty \exp(-st) P_n'(t)\,dt \quad . \tag{81}$$

Thus,

$$P_n(s) = \frac{P_n(0)}{s} + \frac{1}{s} P_n'(s) \tag{82}$$

and,

$$P_n'(s) = sP_n(s) - P_n(0) \; . \tag{83}$$

By applying the transformation of equation (83) to the difference equations of (77, 78, 79) and remembering that,

replacing the values of s_1, and s_2 leads to,

$$MTBF = \frac{3\lambda + \mu}{2\lambda^2} \quad . \tag{105}$$

For the system without maintenance,

$\mu = 0$

and,

$$MTBF = \frac{3}{2\lambda} \tag{106}$$

and taking the ratio of the two MTBF's gives the factor of improvement (FI) due to repair.

$$FI = 1 + \frac{\mu}{3\lambda} \quad . \tag{107}$$

For comparison purposes, let us use the values of the previous examples,

$$\mu = 1/MTTR = 1/3.55 = 0.28 \tag{108}$$

$$\lambda = 1/m_T = (n\Gamma(1/\beta + 1))^{-1} \quad . \tag{109}$$

Thus,

$$\lambda = (500\Gamma(1/2.5 + 1))^{-1} \tag{110}$$

$\lambda = 0.0025 \quad .$

Using equations (90) and (91)

$s_1 = -3.53139 \times 10^{-5}$

$s_2 = -0.28671 \quad .$

The MTBF of the system with repair is,

$$MTBF = 28320 \text{ hours}, \tag{111}$$

and the MTBF of the system without repair is

$$MTBF = 667 \text{ hours} \tag{112}$$

and the factor of improvement is 42.

To show the gain in reliability, especially for a long mission, Table 5 shows the reliability versus time of a system with and without repair.

Table 5. Reliability vs. Time

t	R (with repair)	R (without repair)
100	0.99660	0.9594
500	0.98262	0.54391
1000	0.96542	0.19969
1500	0.94852	0.06727
2000	0.93192	0.02209
2500	0.91561	0.00720
3000	0.89959	0.00234

The same technique can be employed for any reliability configuration. The important trick is to set up the Markov graph and the transition matrix properly; the rest is just mechanical. Reference (9) contains an extensive list of solutions for different reliability configurations. Reference (10) contains more advanced considerations such as the fact that the time to repair may not be exponential.

AVAILABILITY OF MAINTAINED SYSTEMS

For a system composed of a single unit, reliability is not concerned with repair. Once the system has failed the state is absorbing. On the other hand, availability takes into consideration the fact that the system can be repaired. In this sense, availability is a better measure of the performance of the system.

Reference (11) defines availability as operational readiness:

"The probability that at any point in time, the system is either operating satisfactorily or ready to be placed in operation on demand, when used under stated conditions including allowable warning time."

There are different kinds of availability according to how it is calculated. The main ones are (a) Achieved Availability A_A and, (b) Intrinsic or Inherent Availability A_i. Achieved availability A_A is defined as

$$A_A = 1 - \frac{\text{Downtime}}{\text{Total time}} \tag{113}$$

Downtime includes all the repair time, administrative time and logistic time. As can be seen, the achieved availability is strongly dependent on the maintenance systems and procedures. Intrinsic or inherent availability A_i is defined as:

$$A_i = \frac{MTBF}{MTBF + MTTR} \tag{114}$$

at steady state.

To illustrate how (113) is derived, consider a single unit with failure rate λ and repair rate μ as illustrated in Figure 13.

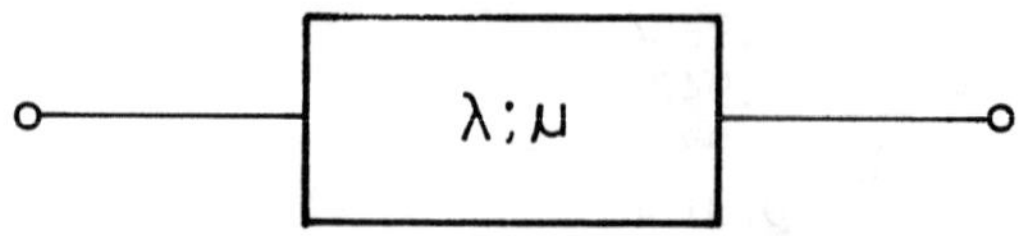

Figure 13. Single Unit

The Markov graph is given in Figure 14.

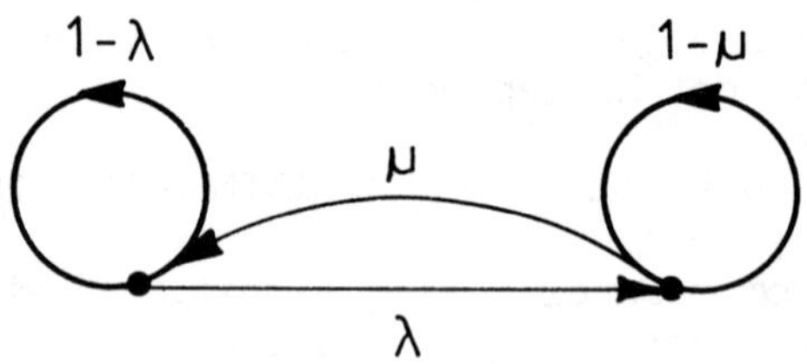

Figure 14. Markov Graph For Single Unit

Note that in this case, state 0 is not an absorbing state.

The transition matrix is,

$$P = \begin{array}{c} \\ 1 \\ 0 \end{array} \begin{array}{c} \begin{array}{cc} 1 & 0 \end{array} \\ \begin{bmatrix} 1-\lambda & \lambda \\ \mu & 1-\mu \end{bmatrix} \end{array} \tag{115}$$

and the difference equations are

$$P_1'(t) = -\lambda P_1(t) + \mu P_0(t) \tag{116}$$

$$P_0'(t) = \lambda P_1(t) - \mu P_0(t) \tag{117}$$

$$A_i(t) = P_1(t) \tag{118}$$

Using the fact that,

$$P_1(t) + P_0(t) = 1 \tag{119}$$

$$P_1'(t) = \mu - (\lambda+\mu)\, P_1(t) \tag{120}$$

and using the Laplace transform, assuming $P_1(0) = 1$

$$sP_1(s) = \mu/s - (\lambda+\mu)\, P_1(s) + 1 \quad . \tag{121}$$

Hence,

$$P_1(s) = \frac{\mu + s}{(s + \lambda + \mu)s} = \frac{\mu + s}{(s - s_1)(s - s_2)} \tag{122}$$

and the roots of the denominator are,

$$s_1 = 0$$

$$s_2 = -(\lambda+\mu) \quad .$$

Taking the inverse transform leads to,

$$P_1(t) = \frac{\mu}{\lambda+\mu} + \frac{\lambda\exp(-(\lambda+\mu)t)}{\lambda + \mu} \quad . \tag{123}$$

Thus the availability is composed of two parts, a constant part which is the steady state availability and a transient term that becomes negligible when:

$$t = \frac{4}{\lambda+\mu} \quad . \tag{124}$$

Remembering that MTBF = $1/\lambda$ and MTTR = $1/\mu$ the constant part of equation (122) is the same as equation (114).

For a mission of $(t_1 - t_2)$ duration, the mission availability is

$$A(t_2 - t_1) = \frac{1}{t_2 - t_1}\int_{t_1}^{t_2} P_1(t)\, dt \tag{125}$$

$$A(t_2 - t_1) = \frac{\mu}{\lambda+\mu} + \frac{\lambda}{(\lambda+\mu)^2} + (\exp(-(\lambda+\mu)t_1 - \exp(-(\lambda+\mu)t_2)) \quad (126)$$

Example

Using the data of the previous problem, i.e., $\lambda = 0.00225$, $\mu = 0.28$ the availability function is tabulated in Table 6.

Table 6. Availability vs. Time

t	A(t)
1	0.99804
2	0.99656
5	0.99397
10	0.99250
15	0.99214
20	0.99205
30	0.99203
50	0.99203

As can be seen, since the constant term is 0.99203, the transient term has influence for only a very short period.

The same technique can be applied to different reliability configurations. As an example, consider a parallel system of two identical units with multiple repair, as shown in Figure 15.

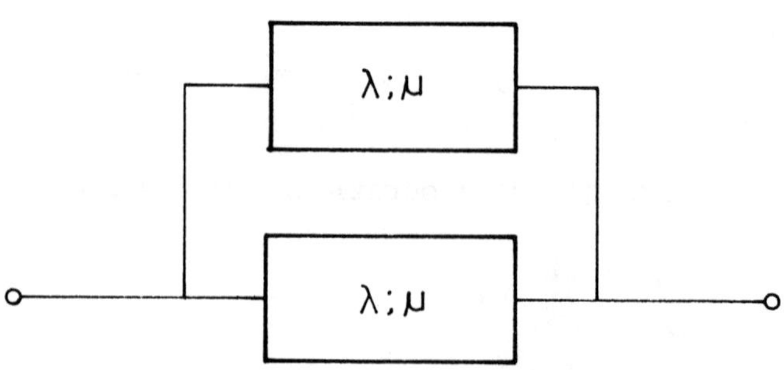

Figure 15. Units In Parallel

The Markov graph is given in Figure 16.

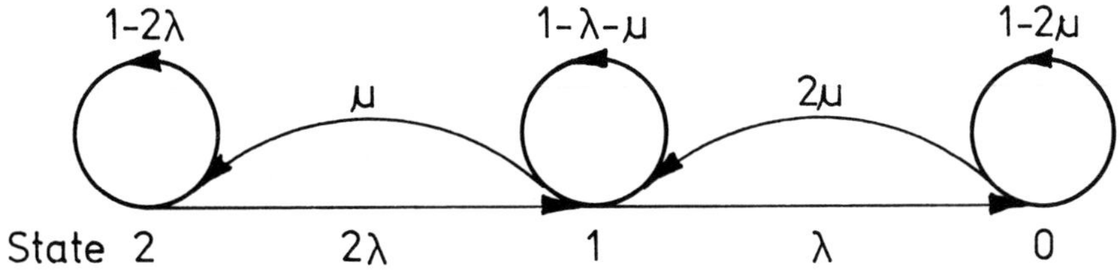

Figure 16. Markov Graph For Parallel Units

The transition matrix is,

$$P = \begin{array}{c} \\ 2 \\ 1 \\ 0 \end{array} \begin{array}{c} \begin{array}{ccc} 2 & 1 & 0 \end{array} \\ \begin{bmatrix} 1-2\lambda & 2\lambda & 0 \\ \mu & 1-(\lambda+\mu) & \lambda \\ 0 & 2\mu & 1-2\mu \end{bmatrix} \end{array} \qquad (127)$$

and the difference equations are,

$$P_0'(t) = \lambda P_1(t) - 2\mu P_0(t)$$

$$P_1'(t) = 2\lambda\, P_2(t) - (\lambda+\mu)P_1(t) + 2\mu P_0(t) \qquad (129)$$

$$P_2'(t) = -2\lambda P_2(t) + \mu P_1(t) \quad . \qquad (130)$$

After using the Laplace Transform, the above equations become:

$$(s + 2\lambda)P_2(s) - \mu P_1(s) = 1 \qquad (131)$$

$$-2\lambda P_2(s) + (s + \lambda + \mu)P_1(s) - 2\mu P_0(s) = 0 \qquad (132)$$

$$-\lambda\, P_1(s) + (s + 2\mu)P_0(s) = 0. \qquad (133)$$

Assuming that at least one unit has to be operational,

$$A(t) = 1-P_0(t) \quad . \qquad (134)$$

We need to find only $P_0(t)$.

After laborious algebraic manipulations we have:

$$P_0(s) = \frac{2\lambda^2}{s\, s_1 s_2} + \frac{2\lambda^2}{s_1(s_1 - s_2)\,(s - s_1)} - \frac{2\lambda^2}{s_2(s_1 - s_2)\,(s - s_2)} \qquad (135)$$

where,

$$s_1 = -2(\lambda+\mu) \qquad (136)$$

$$s_2 = -(\lambda+\mu) \quad . \qquad (137)$$

Taking the inverse Laplace Transform leads to:

$$P_0(t) = \frac{2\lambda^2}{s_1 s_2} + \frac{2\lambda^2}{s_1(s_1 - s_2)} \exp\,(s_1 t)) - \frac{2\lambda^2}{s_2(s_1 - s_2)} \exp\,(s_2 t) \qquad (138)$$

and,

$$A(t) = \frac{\mu^2 + 2\lambda\mu}{\lambda^2 + 2\lambda\mu + \mu^2} - \frac{2\lambda^2}{s_1 s_2 (s_1 - s_2)} (s_2 \exp\,(s_1 t) - s_1 \exp\,(s_2 t)) \quad . \qquad (139)$$

The steady state availability is:

$$A_{ss} = A(\infty) = \frac{\mu^2 + 2\lambda\mu}{\lambda^2 + 2\lambda\mu + \mu^2} \qquad (140)$$

and taking the ratio of A_{ss} for two parallel units to A_{ss} of one unit will give the factor of improvement in the availability.

$$FI = 1 + \frac{\lambda}{\lambda + \mu} \quad . \qquad (141)$$

With this data the improvement is insignificant but in the case where λ is large and μ small, at the limit, the availability can double.

As has been seen, the mathematical manipulations can be quite laborious. Reference (10) contains general formulas for a (m,n) system. Reference (9) contains tables of equations for different reliability configurations. Optimization can make use of the techniques explained above. It might be more cost effective to provide two equipments in parallel with repair, than to expend resources in research and development to improve the performance of one equipment.

CONCLUSION

As we have seen in this chapter, maintainability is an exact science and it is possible to construct mathematical models which can be used to optimize maintainability costs. Considering that in most systems, the cost of maintenance is usually greater than 50% of the total life cycle cost of the equipment, it is very important that a properly developed maintenance plan exist. Often the maintenance costs are concealed because of the time frame in which they occurred. However, maintenance cost accounting systems, like LOMMIS in the Canadian Forces, indicate that the acquisition cost is a small percentage of the maintenance cost. It is to be emphasized that maintainability has to be quantified to be meaningful. Many times the word maintainability is used to mean ease of maintenance. The techniques which ease the maintenance task surely contribute to increase maintainability but maintainability implies much more.

REFERENCES

1. ______________, MIL-HDBK-472, Maintainability Prediction, U.S. Department of Defense, Washington, D.C., May 1966.

2. ______________, CFP 298 (2) Maintainability, Department of National Defence,Ottawa, Canada, July 1971.

3. Lipson, - and Sheth -, Statistical Designs and Analysis of Engineering Experiments, McGraw - Hill, New York, N.Y., 1973, p518.

4. Hahn, R.F. and Shapiro -, Statistical Models in Engineering, John Wiley & Sons, New York, 1967, p353.

5. ______________, AMCP 702-3, Quality Assurance, Reliability Handbook, U.S. Army Material Command, October 1968.

6. Drenick -, The Failure Law for Complex Equipment, Journal of SIAM, Vol. 8, No. 4, p680 - 690.

7. Kreysnig -, Advanced Engineering Mathematics, John Wiley & Sons, New York, 1965, p856.

8. Abramowitz, M.O., and Stegun, I.A. ed., Handbook of Mathematical Functions, Dover, New York, N.Y., 1964, p1046.

9. Meyers, R.H., F.L. Wong, F.L. and Gordy H.M., Reliability Engineering for Electronic Systems, John Wiley & Sons, New York, N.Y., 1964, p360.

10. Rau, J.G., Optimization and Probability in Systems Engineering, Van Nostrand-Reinhold, New York, N.Y., 1970 p403.

11. ______________, CFP 235 Technical Glossary, Department of National Defence, Ottawa, Canada, November 1975.

20 ALLOCATION AND ESTIMATION

J. E. Arsenault

When the operational requirements of a system are being defined, it is essential that the principal maintenance characteristics be specified. The usual fundamental figures given relate to corrective and preventive maintenance. Corrective maintenance time is the time that begins with the observation of a malfunction of an item and ends when the item is restored to a satisfactory operating condition. It may be subdivided into active and non-active maintenance time and does not necessarily contribute to equipment or system downtime in cases of alternate modes of operation or redundancy. Active maintenance time is the time during which preventive and corrective maintenance work is actually being done on the item. Non-active maintenance time is the time during which no maintenance is being accomplished on the item because of either supply or administrative reasons. Preventive maintenance time is the time used in accomplishing preventive maintenance and is made up of time spent in performance measurement, care of mechanical wearout items, front panel adjustment, calibration and alignment, cleaning, etc. Since it is desirable to repair a system in an acceptable time, it is normal to express this as a mean time to repair ($\overline{M}ct$, MTTR) or to limit man-hours spent in system repair by specifying maintenance man-hours per system operating hour (MM/OH).

Generally, the first maintainability design task faced by the system designer is to apportion the specified maintainability requirements to each succeeding lower system level. The allocation process is usually done in the conceptual, proposal or early development stage, when system information is sparse. As the level of system information improves, a point is reached when a detailed maintainability estimate may be accomplished meaningfully.

MAINTAINABILITY APPORTIONMENT

The apportionment of specified system maintainability requirements will usually consist of the allocation of the system $\overline{M}ct$ to lower equipment levels, i.e. sub-system, unit or assembly. The apportionment is carried out to permit the establishment of design goals for equipment against which detailed maintainability estimates may be compared. Maintainability estimates are discussed later in this chapter.

The method used is described in (1) and makes use of a factor K to direct the apportionment of $\overline{M}ct$ in relation to the items considered with respect to the next higher equipment level. The items having a large failure rate will be allocated the least time, and vice versa. This is based on the principle that items most frequently approached for maintenance must consume the least time.

A weighting factor for an item is derived from considerations, for example, of environment, handling, accessibility, packaging, fault isolation technique and complexity for each of it's sub-items, so that for a particular item:

$$K = \frac{K_1 + \ldots + K_n}{n} \qquad (1)$$

where,

K = weighting factor for an item,

$K_1, \ldots, K_n$ = sub-weighting factors for sub-items.

If there is a weighting factor for each sub-item to which an allocation is to be made, then the weighting factor for the item is:

$$K = \frac{\lambda_1 K_1}{\lambda} + \ldots + \frac{\lambda_n K_n}{\lambda} \qquad (2)$$

where,

$\lambda_1, \ldots, \lambda_n$ = failure rate of each sub-item,

$\lambda = \lambda_1 + \ldots + \lambda_m$ = failure rate of each item.

Solving (2) for λ yields

$$\lambda = \frac{\lambda_1 K_1}{K} + \ldots + \frac{\lambda_n K_n}{K} \qquad (3)$$

Consider that

$$\overline{M}ct = \frac{\lambda_1 \overline{M}ct_1}{\lambda} + \ldots + \frac{\lambda_n \overline{M}ct_n}{\lambda} \qquad (4)$$

where,

$\overline{M}ct$ = mean active corrective maintenance time of an item,

$\overline{M}ct_1, \ldots, \overline{M}ct_n$ = active corrective time for each sub-item.

Solving for λ provides,

$$\lambda = \frac{\lambda_1 \overline{M}ct}{\overline{M}ct} + \ldots + \frac{\lambda_n \overline{M}ct}{\overline{M}ct} \qquad (5)$$

and setting (3) equal to (5) gives,

$$\frac{\lambda_1 K_1}{K} + \ldots + \frac{\lambda_n K_n}{K} = \frac{\lambda_1 \overline{M}ct_1}{\overline{M}ct} + \ldots + \frac{\lambda_n \overline{M}ct_n}{\overline{M}ct} \qquad (6)$$

Rearranging gives,

$$\lambda_1\left(\frac{K_1}{K} - \frac{\overline{M}ct_1}{\overline{M}ct}\right) + \ldots + \lambda_n\left(\frac{K_n}{K} - \frac{\overline{M}ct_n}{\overline{M}ct}\right) = 0 \qquad (7)$$

Therefore,

$$\left(\frac{K_1}{K} - \frac{\overline{M}ct_1}{\overline{M}ct}\right) = 0, \text{ etc.} \qquad (8)$$

and

$$\overline{M}ct_1 = \frac{\overline{M}ct\, K_1}{K}, \text{ etc.} \qquad (9)$$

In addition, from (2)

$$K = \sum_{j=1}^{n} \frac{\lambda_j K_j}{\lambda} \qquad (10)$$

Having derived a solution to the maintainability apportionment problem through the use of factor K, it is necessary to provide information that will allow K to be calculated for each item under consideration, given the failure rate for each item. Tables 1 through 8 list individual factors for complexity, fault isolation, packaging, accessibility, handling and environmental factors.

Table 1. Complexity Factors For Units

Type of Unit	Kcu	Considerations
Power supply - simple	2	

Table 1 (Cont'd)

Power supply - complex	3	
Control	3	
Digital computer	2	Including memories
Disk memory	3	Fixed head
Disk memory	4	Moving head
Mag tape	4	Including cassette type
Teletype	6	
Paper Tape	4	Punch or reader
Digital	1	
Analog	2	Electronic - not mechanical
CRT terminal	3	Including keyboard
Keyboard	2.5	Including code/decode and lamp drivers
Printer	6	Impact, serial or line
Printer	4	Non-impact
Card Handler	6	Punch or reader

Table 2. Complexity Factors For Assemblies

Type	Kca	Considerations
Digital	1	CCAs (small)
Digital	1.5	CCAs (large)
Analog - low level	1.5	CCAs (small)
Analog - low level	2	CCAs (large)
Analog - high level	2	CCAs or S/As (small)
Analog - high level	3	CCAs or S/As (large)
D to A (or A to D)	2	CCAs or S/As (small)

Table 2 (Cont'd

D to A (or A to D)	3	CCAs or S/As (large)
Power supply - simple	2	Complete
Power supply - complex	3	Complete
Control	3	Complete
Wiring harness	4	Complete

Note:
1. CCA - Circuit Card Assembly
2. S/A - Sub-assembly

Table 3. Complexity Factors For Parts

Type	Kcp	Considerations
Lamp	1	Any
LED	1.5	Any
Diode	1.5	Any
Resistor	1.5	Fixed
Resistor	2	Variable
Capacitor	1.5	Fixed
Capacitor	2	Variable
Transistor	3	Any low-med power
SCR	3	Any
Relay - mechanical	3	Any
Relay - solid state	4	Any
Hybrid	4	Any semi-conductor
Transformer	3	Any
Inductor	2	Any
Switch	2	Any
Circuit breaker	3	Any
Analog IC	4	Any
Digital IC	5	MSI
Digital IC	7	LSI
PCB	3	Single layer
PCB	5	Multiple layer
Connector	1	Single contact
Connector	5	Multiple contact
Synchro, resolver, MG	4	Any
ETI, counter	3	Any
Fan, blower	3	Any

Table 3 (Cont'd)

Fan	1	Any
RFI filter	2	Any

Table 4. Fault Isolation Technique Factors

Type	Kf	Considerations
Automatic	1	Computer built in test (BIT), i.e., circuits providing auto fault isolation to replaceable item
Semi-automatic	3	BIT circuits controlled manually (includes test point selector switch and meter combination)
Manual	5	Manually making measurements using portable test equipment at circuit test points
Manual	7	As above but no test points, hence other techniques required

Table 5. Packaging Factors

Type		Kp	Considerations
CCA	- quick disconnect	1	No fasteners
CCA	- pigtail connector	2	With screw fasteners
CCA	- screw or clip connection for each wire	3	With screw fasteners
CCA	- solder connections	6	With screw fasteners
Chassis mounted part	- direct plug in	1	No fasteners

Table 5 (Cont'd)

Chassis mounted part - pigtail connector	2	With screw fasteners
Chassis mounted part - screw or clip connection for each wire	3	With screw fasteners
Chassis mounted part - solder connection for each wire	6	With screw fasteners

Table 6. Accessibility Factors

Type	Ka	Considerations
Direct	1	No covers, etc.
Simple	2	Quick release cover fasteners
Difficult	4	Screw cover
Very Difficult	6	Screw cover fasteners plus internal covers, part or wiring displacement

Table 7. Handling Factors

Type	Kh	Considerations
Simple	1	Light weight, one man
Difficult	3	Two men, awkward
Very Difficult	5	Two men plus hoist or dolly

Table 8. Environment Factors

Type	Ke	Considerations
Indoor - mild	1	Normal temp. and rel. hum.
- moderate	2	33 - 50° F and/or > 90% rel. hum. or 90 - 100°F
- severe	3	< 32°F or > 100°F and > 90% rel. hum.
Outdoor - mild	2	50-90°F, < 90% rel. hum. precipitation & wind nil
- moderate	3	33-50°F or 90-100°F and/or > 90% rel. hum. precipitation nil and light wind or mild plus rain and light wind.
- severe	6	Moderate temp. plus rain or snow and heavy wind
- extremely severe	10	< 32°F or > 100°F plus rain or snow, heavy wind

Allocation Example

Consider the design of a program generation system consisting of the items listed in Table 9. The system $\overline{M}ct$ is specified as 1.0 hours and it is necessary to allocate the system $\overline{M}ct$ among the various system units for inclusion in procurement specifications. The failure rate of each unit has been estimated. Since the analysis is for units, Table 1 is used in turn to obtain values for Kc, Kf, Kp, Ka, Kh and Ke and entered in Table 9. The value of K_j is computed from equation (1) and the failure rate λ_j is next entered in Table 9. The products $K_j\lambda_j$ are computed and the sums of λ_j and $K_j\lambda_j$ are next computed. The value of K is computed from equation (10) and the allocated $\overline{M}ct$ are computed from equation (9) and entered in Table 9.

Table 9. System Allocation Example

Item Description	Kc	Kf	Kp	Ka	Kh	Ke	K_j	λ_j	$K_j\lambda_j$	$\overline{M}ct_j$
Digital computer	2	1	1	2	1	1	1.33	500	665	0.658

Table 9 (Cont'd)

Disk memory, fixed	3	3	2	6	3	1	3.00	2500	7500	1.485
Digital	1	1	1	2	1	1	1.67	1000	1670	0.827
CRT terminal	3	3	2	4	1	1	2.33	1500	3495	1.153
Keyboard	2.5	1	2	1	1	1	1.42	8000	11360	0.703
Card handler	6	3	2	2	1	1	2.50	500	1250	1.238
Printer, impact	6	3	2	4	1	1	2.83	2000	5660	1.401
Paper tape, reader	4	3	2	2	1	1	2.17	200	4340	1.074
Magnetic tape unit	4	3	2	4	1	1	2.50	1000	2500	1.238
$\Sigma\lambda_j=19000$								19000	38440	

$\Sigma K_j\lambda_j = 38440$

$K_j = (Kcu + Kf + Kp + Ka + Kh + Ke)/6$

$$K = \frac{\Sigma K_j\lambda_j}{\Sigma\lambda_j} = \frac{38440}{19000} = 2.02$$

$$\overline{M}ct_j = \frac{\overline{M}ct\ K_j}{K} = \frac{1.0\ K_j}{2.02} = 0.495K_j$$

MAINTAINABILITY ESTIMATES

As the system design process progresses to the detailed level more complete design information will become available and consequently the sytem maintainability characteristics may be estimated accurately. Starting with the system operational maintainability characteristics, it is essential that a technique be utilized to estimate its Maintainability characteristics as early as possible in the design stage. This estimate should be updated continuously as the design progresses to provide the visibility necessary to ensure that the specified requirements have a high probability of being met. The estimate of the expected system downtime while it is undergoing maintenance is of vital importance to the user, because of the effect of downtime on mission success. A good maintainability estimate technique will highlight those areas of poor maintainability for the designer, which warrant design changes if specified requirements are to be met. Another useful feature of maintainability prediction is that it permits the user to make an early assessment of whether the predicted downtime, the quality and quantity of personnel, tools and test equipment are adequate and consistent with the needs of system operational requirements.

Although many maintainability estimate techniques are defined (2), all techniques are dependent upon at least two basic parameters, i.e., (a) failure rates of equipment items at the specific system level of interest, i.e., sub-system, unit or assembly, (b) repair time required at the maintenance level involved.

Failure Rate

Many sources record the failure rates, for use in conjunction with electronic systems, as a function of electrical stress, environmental stress, and part quality, (most notable of these being (3), refer to Chapter 8 for more detail). Failure rate is usually expressed as failures per million hours. In a maintainability prediction the failure rate for an item gives the relative frequency of failure and hence the relative frequency of repair.

Repair Times

Repair times are determined from prior experience, simulation of maintenance tasks, or past data secured from similar applications. A maintenance task is any action or actions required to preclude the occurrence of a malfunction or restore an equipment to satisfactory operating condition. The following maintenance tasks are associated with corrective maintenance. Note that these items do not include non-active time, i.e., the defined times refer to active repair time only.

(a) Localization - determining the location of a fault without using accessory test equipment. This task is known as Setup when Automatic Test Equipment (ATE) is used for fault isolation and usually involves connecting the item to the ATE and preparing the ATE for running,
(b) Isolation - determining the location of a failure to the extent possible, by the use of accessory test equipment,
(c) Disassembly - equipment disassembly to the extent necessary, to gain access to the item that is to be replaced,
(d) Interchange - removing the defective item and installing the replacement,
(e) Reassembly - closing and reassembly of the equipment after the replacement has been made,
(f) Alignment - performing any alignment, minimum tests and/or adjustment made necessary by the repair action,
(g) Checkout - performing the minimum checks or tests required to verify that the equipment has been restored to satisfactory performance.

The sum of the individual task times gives the total maintenance task times to repair an item. Table 10 provides a detailed list of maintenance task times associated with manual tasks performed during troubleshooting. Table 11 provides an example format useful in estimating test times associated with troubleshooting using automatic test equipment (ATE). No times are given as the time depends on the particular test system.

Table 10. Corrective Maintenance Task Times

	Task Description	Minutes
A.	INTERCHANGE	
	1. Part handling (part not a plug-in)	
	Pull up part and position in chassis for assembly	0.3
	The following maintenance tasks include handling:	
	2. Plug-in's	
	Modules, pin-type tubes, fuses, parts	0.75
	PCB (Printed Circuit Board)	0.5
	3. Wiring and soldering (Non-PCB)	
	First 2 leads on any part	4.5
	Each 2 additional leads on any part	2.0
	Jumper wire and cable leads (one end)	2.0
	4. PCB wiring and soldering	
	Two terminal part (R,C,L,CR, etc.)	2.0
	Transistors, TO-5 relays	4.0
	Dual transistors, flat packs, dual-in-line package (DIP)	5.0
	5. Hardware (disassembly and assembly)	
	Loose hardware (per screw)	1.5
	Captive hardware (per screw)	0.6
	Dzus type fastener	
	1/4 turn	0.1
	Quadruple thread	0.2
	Connector (quick disconnect)	0.5
B.	MANUAL TROUBLESHOOTING (Includes probe attachments)	
	1. Oscilloscope (includes change of scope settings)	
	Initial measurement	2.0
	Each additional measurement (per channel)	0.75

Table 10 (Cont'd)

2. Volt-ohmmeter	
Each voltage or resistance measurement	0.5
3. Capacitor tester (in-circuit)	
Shorts (with shunt to 6 ohms)	
Opens (down to 7 pf)	
Capacitance (2-450 uf)	1.0
4. Adjustments (multi or single-turn)	0.5

Table 11. Estimated Mean Test Times

Function		Stimulus Application (1)	Measurement Execution (2)
A.	DC VOLTAGE		
	1. 0.1-35 V	X	X
	2. 35-500 V	X	X
	3. 500 V-1 kv	X	X
	4. 1 kv-20 kv	X	X
B.	RESISTANCE		
	1. Below 10 k ohms	X	X
	2. Above 10 k ohms	X	X
C.	DIGITAL SIGNALS		
	1. Transmit 1,000 words to unit under test (n bits)	X	X
	2. Receive 1,000 words from unit under test (n bits)	X	X
	3. Process in ATE computer	X	X

Notes: 1. Stimulus application and measurement: Use the sum of both columns.
2. Measurement only using prior stimulus setup: Use (2) only.

General Estimating Technqiues

Assuming that the failure rate and repair time of each replaceable item is given, some means must be found to compute the mean time to repair and other indicative maintainability indices. Predictions are usually performed assuming that repair times are continuously distributed according to some well known theoretical distribution such as the log-normal and normal distributions. Other distributions have been used such as the exponential and the gamma but are in less use than the normal and especially the log-normal. The following discussion presents the more commonly used maintainability indices and how they are calculated.

Mean (Average) Corrective Maintenance Times - This technique is based on mean (average) maintenance task times (4) assuming a normal distribution.

Mean Corrective Maintenance Time ($\overline{M}ct$) - This is the mean (average) time required to perform corrective maintenance on an item. This is the best known measure of maintainability, most often referred to simply as MTTR, and is calculated as follows:

$$\overline{M}ct = (\sum_{i=0}^{n} \lambda_i \; \overline{M}ct_i) / \sum_{i=0}^{n} \lambda_i \qquad (12)$$

where,

λ_i = failure rate of the ith item to be repaired.

$\overline{M}ct_i$ = total average maintenance task time associated with the corrective action on the ith item.

Mean Preventive Maintenance Time ($\overline{M}pt$) - The mean (average) time required to perform preventive maintenance on an item, expressed as follows.

$$\overline{M}pt = (\sum_{i=1}^{n} fpt_i \overline{M}pt_i) / \sum_{i=1}^{n} fpt_i \qquad (13)$$

where,

fpt_i = frequency of the ith preventive maintenance action,

$\overline{M}pt_i$ = average action, maintenance task time associated with the preventive maintenance of the ith item.

Mean Maintenance Time ($\overline{M}$) - This is the mean (average) of the active time required to perform corrective and preventive maintenance and is expressed by the following equation.

$$\overline{M} = (\lambda \overline{M}ct + fpt\ \overline{M}pt)/(\lambda + fpt) \tag{14}$$

Maintenance Manhours - The frequency indices consider elapsed time and not manhours, which forms a basic input to system cost estimates throughout the useful life of a system. Some additional measures are required. The usual measure used is Maintenance Manhours Per Operating Hour (MM/OH) or a variation thereof.

$$MM/OH = \sum_{i=1}^{n} MTTR_i \lambda_i \tag{15}$$

Median And Maximum Corrective Maintenance Times

In order to obtain median (Mct) and maximum corrective maintenance task time (Mmax), it is necessary to investigate probability distribution functions of interest. The most commonly encountered distributions associated with maintenance task time are the normal and log-normal (5).

Normal Distribution

The normal probability distribution function is

$$f(t) = \frac{1}{\sigma \sqrt{2}\, \pi} \quad \exp\left[-\frac{t-n}{\sqrt{2}\,\sigma}\right]^2 \tag{16}$$

where,

t = variable repair time

n = mean of t

σ = standard deviation of t.

Let the system consist of n replaceable items with failure rates $\lambda_1, \lambda_2, \lambda_3, \ldots, \lambda_n$ and their associated repair times $t_1, t_2, t_3, \ldots, t_n$. It is possible from this data to obtain the parametric estimators of the distribution as follows:

$$\hat{\mu} = \frac{1}{\lambda} \sum_{i=1}^{n} \lambda_i t_i \tag{17}$$

$$\hat{\sigma} = \sqrt{\frac{1}{\lambda} \sum_{i=1}^{n} \lambda_i \quad (t - \mu)^2} \tag{18}$$

where,

$$\lambda = \sum_{i=1}^{n} \lambda_i \tag{19}$$

The system's maintainability characteristics may now be calculated as follows:

$$\overline{M}ct = \tilde{M}ct = \mu \tag{20}$$

where,

$\tilde{M}ct$ = median.

It is now possible also to calculate the maximum repair time for any percentage (x) of repair times, i.e. Mmax (x) which is illustrated below for the two most commonly specified percentages.

$$\text{Mmax } (0.90) = \mu + 1.282\sigma \tag{21}$$

$$\text{Mmax } (0.95) = \mu + 1.645\sigma \tag{22}$$

Log-Normal Distribution

The log normal distribution represents a good fit for the repair time distribution of electronic systems. Some repair times are short (minutes) but never zero, most times cluster about the median while a few repair times are very long (hours). Assuming a log-normal distribution for repair times the probability density function is

$$f(t) = \frac{1}{\sigma t \sqrt{2}} \exp\left[-\frac{\ln t-\mu}{\sqrt{2}\,\sigma}\right]^2 \tag{23}$$

where,

t = variable repair time

μ = mean of ln t,

σ = standard deviation ln t.

Similarly, as for the normal distribution the parametric estimators are:

$$\hat{\mu} = \frac{1}{\lambda} \sum_{i=1}^{n} \lambda_i \ln t_i \tag{24}$$

$$\hat{\sigma} = \sqrt{\frac{1}{\lambda} \sum_{i=1}^{n} \lambda_i \ (\ln t_i - \hat{\mu})^2} \tag{25}$$

where,

$$\lambda = \sum_{i=1}^{n} \lambda_i$$

Again, as for the normal, log normal maintainability characteristics are:

$$\overline{M}ct = \exp\left(\hat{\mu} + \frac{\hat{\sigma}^2}{2}\right) \tag{26}$$

$$\tilde{M}ct = \exp(\hat{\mu}) \tag{27}$$

$$M\max\ (0.90) = \exp(\hat{\mu} + 1.282\ \hat{\sigma}) \tag{28}$$

$$M\max\ (0.95) = \exp(\hat{\mu} + 1.645\ \hat{\sigma}) \tag{29}$$

Maintainability Estimate Example

In this paragraph a practical example of a maintainability estimate is described. Figure 1 shows a portion of the input data to a computer program listed as it appears to the program for processing. The program then computes the required maintainability indices. The listing is self-explanatory except for the columns entitled Ambiguity Group Numbers and Ambiguity Group Sequence which are explained in the next section. Figure 2 is a self-explanatory example of the programmed output.

THE AMBIGUITY PROBLEM

It frequently happens in detailed maintainability analysis that due to practical considerations, it is possible only to isolate a fault down to a group of items. For example, in the case of a test failure at the interface between two or more circuit cards or where feedback is involved. Assuming that a fault is detected by monitoring test points, the question arises as to whether the fault is on the driving card or the driven card. In most cases the fault will be caused by the driver card because it usually has associated with it considerably more circuitry that could cause the fault. The case for the driver card, however, diminishes as the number of driven cards increases. A good strategy in these cases is to replace first, the most likely failed item, followed by the remaining possible items in order of decreasing failure rate.

SYNCHRONIZER AND DEMULTIPLEXER RUN NO. 1 DECEMBER PAGE 1

TEST NO	AM	SQ	DESCRIPTION OF SRA/PART	REFERENCE DESIGNATION	FAILURE RATE	SET UP TIME	ISOLAT TIME	ACCESS TIME	INTCHG TIME	ADJUST TIME	CHECK TIME	TOTAL TIME	MM/OH
10	0	0	BIT SYNC		13.377	10.00	10.00	7.50	.50	N/A	12.00	40.00	535.10
	0	0	BIT SYNC		13.377	10.00	10.00	7.50	.50	N/A	12.00	40.00	535.10
30	0	0	BIT SYNC		13.377	10.00	10.00	7.50	.50	N/A	12.00	40.00	535.10
40	0	0	BIT SYNC		13.377	10.00	10.00	7.50	.50	N/A	12.00	40.00	535.10
50	0	0	B FRAME SYNC		26.747	10.00	10.00	7.50	.50	N/A	12.00	40.00	1069.89
60	0	0	B FRAME SYNC		26.747	10.00	10.00	7.50	.50	N/A	12.00	40.00	1069.89
70	0	0	B FRAME SYNC		26.747	10.00	10.00	7.50	.50	N/A	12.00	40.00	1069.89
80	0	0	B FRAME SYNC		26.747	10.00	10.00	7.50	.50	N/A	12.00	40.00	1069.89
90	0	0	RECEIVER PLATFORM		17.198	10.00	10.00	7.50	.50	N/A	12.00	40.00	687.91
100	0	0	SERIAL INTERFACE		25.523	10.00	10.00	7.50	.50	N/A	12.00	40.00	1020.93
110	0	0	TEST WAVEFORM GENERATOR		39.600	10.00	10.00	7.50	.50	N/A	12.00	40.00	1584.02
120	0	0	COUNT ADDRESS		23.011	10.00	10.00	7.50	.50	N/A	12.00	40.00	920.44
130	0	0	ARITHMETIC		25.784	10.00	10.00	7.50	.50	N/A	12.00	40.00	1031.34
140	0	0	COEFFICIENT		53.391	10.00	10.00	7.50	.50	N/A	12.00	40.00	2135.66
150	0	0	A/D CONVERTER		15.435	10.00	10.00	7.50	.50	N/A	12.00	40.00	617.41
160	0	0	PHASE LOCKED LOOP		23.044	10.00	10.00	7.50	.50	N/A	12.00	40.00	921.75
170	0	0	TRANSMITTER		15.872	10.00	10.00	7.50	.50	N/A	12.00	40.00	634.88
180	0	0	BIPOLAR MEMORY		43.700	10.00	10.00	7.50	.50	N/A	12.00	40.00	1748.00

Figure 1. Maintainability Estimate - Input Data Listing

MAINTAINABILITY ESTIMATE SUMMARY

1.0 MILITARY-STANDARD-472, PROCEDURE 2, CALCULATIONS

1.1 MEAN TIME TO REPAIR = 43.25 MINUTES

2.0 LOG NORMAL DISTRIBUTION CALCULATIONS

2.1 LOG NORMAL PROBABILITY DENSITY FUNCTION PARAMETRIC ESTIMATORS

2.1.1 MEAN ESTIMATOR = 3.72

2.1.2 STANDARD DEVIATION ESTIMATOR = .21983

2.2 LOG NORMAL PROBABILITY DENSITY FUNCTION CALCULATED VALUES

2.2.1 MEAN TIME TO REPAIR = 42.48 MINUTES

2.2.2 MEDIAN TIME TO REPAIR = 41.47 MINUTES

2.2.3 MAXIMUM (0.75) TIME TO REPAIR = 48.10 MINUTES

2.2.4 MAXIMUM (0.80) TIME TO REPAIR = 49.90 MINUTES

2.2.5 MAXIMUM (0.85) TIME TO REPAIR = 52.08 MINUTES

2.2.6 MAXIMUM (0.90) TIME TO REPAIR = 54.97 MINUTES

2.2.7 MAXIMUM (0.95) TIME TO REPAIR = 59.53 MINUTES

Figure 2. Maintainability Estimate - Program Output Listing

Ambiguity Example

The following example is provided to illustrate how data is to be input, where fault ambiguity exists between items using the program demonstrated earlier. The idea is to enter task time data in a consistent manner in order to produce the correct results from the program. It is felt that this approach is flexible and eliminates program customizing. The basic problem then is to input data to give correct total maintenance task times for use in predicting $\overline{M}ct$.

The following assumptions are made, (a) corrective maintenance is carried out at the unit level using automatic test equipment, (b) the unit wiring is intact, (c) maintenance is carried out by one man, (d) the ambiguity group size is three circuit cards, (e) end-to-end test takes 20 minutes to run. The fault causes dropout at 10 minutes. A convenient entry point is found 2 minutes back of dropout or 8 minutes from the beginning of the tape. The checkout time will therefore be 12 minutes. The setup time is 10 minutes, access is 5 minutes, interchange is 1 minute, adjust is 0 minutes.

A flowchart is shown in Figure 3 with the exact sequence of the maintenance tasks and their associated times. The total maintenance task times are also tabulated in Table 12. Maintenance Task Time Analysis Data input is written as in Figure 4. We use an Ambiguity Group Number, which in this case is 3, and an Ambiguity Group Sequence which denotes the order of card replacement as 1st, 2nd and 3rd in decreasing order of failure rate. The data is processed exactly as in the previous section.

Table 12. Detailed Maintenance Task Time Analysis

Assumed Fault	Set Up	Isolate	Access	Inter Change	Adjust	Check Out	Total
1A6	10	10	5	1	0	12	
Total Accum.	10	10	5	1	0	12	38
1A4	0	2	0	2	0	12	
Total Accum.	10	12	5	3	0	12	42
1A24	0	2	0	2	0	12	
Total Accum.	10	14	5	5	0	12	46

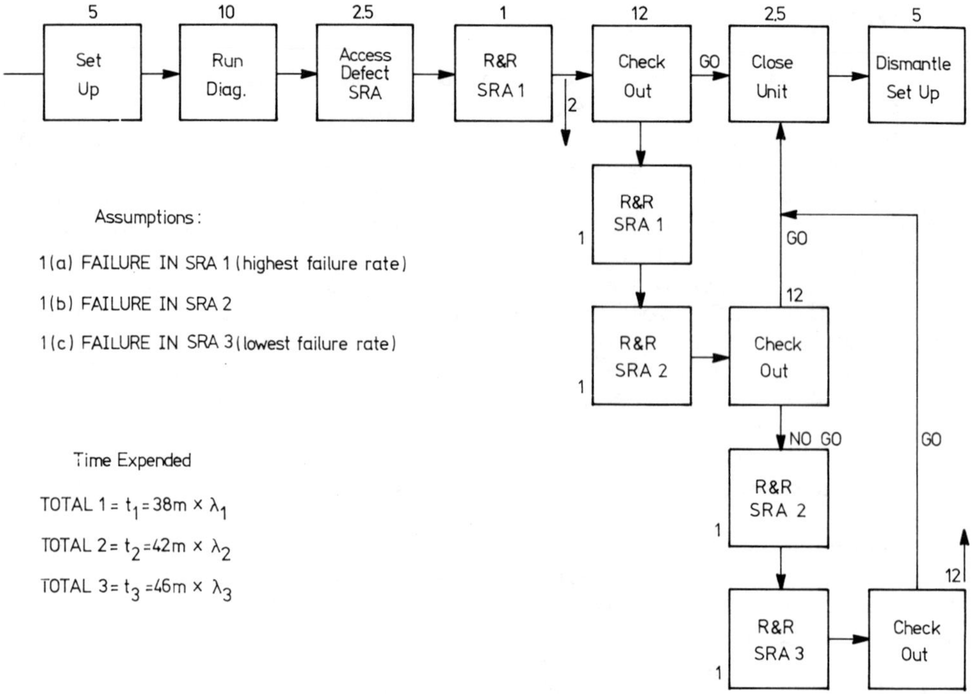

Figure 3. Diagnostic Flowchart

MAINTAINABILITY DATA INPUT FORM

TEST NUMBER	ITEM DESCRIPTION	ITEM NUMBER DESIGNATION	AMBIGUITY GROUP NUMBER	AMBIGUITY GROUP SEQUENCE	FAILURES PER 10^6 HOURS	SET UP TIME	ISOLATE TIME	ACCESS TIME	INTER-CHANGE TIME	ADJUST TIME	CHECK TIME	SPARE	LINE NUMBER
1010	SIN/CØS CCA	1A6	3	1	20.34	10.0	10.0	5.0	1.0	0.0	12.0		
1010	MULTIPLEXER CCA	1A4	3	2	14.62	10.0	12.0	5.0	3.0	0.0	12.0		
1010	MULTIPLIER CCA	1A24	3	3	9.84	10.0	14.0	5.0	5.0	0.0	12.0		

PROJECT .... PREPARED BY .... DATE .... PAGE .... OF ....

FORM 257-NEW5/75

Figure 4. Maintainability Estimate - Data Input

CONCLUSION

Beginning with the specified overall maintainability requirements for a system, the requirements are apportioned to each unit comprising the system and continuing downward to the lowest level of interest. These apportioned values become the goals for realization during design and development. As the detailed design progresses, maintainability estimates can be generated, the accuracy of which are a function of the level of detail available. Tradeoffs can then take place to develop the best solution when corrective action must be applied to comply with system maintainability requirements.

REFERENCES

1. Chipchak, J.S., "A Practical Method of Maintainability Allocation," IEEE Transactions On Aerospace and Electronic Systems, Vol. AES-7,No. 4, July 1971, p585-589.

2. ____________, MIL-HDBK-472, Maintainability Prediction, U.S. Department of Defense Washington D.C., 1966.

3. ____________, MIL-HDBK-217, Reliability Prediction of Electronic Equipment, U.S. Department of Defense, Washington, D.C., September 1974.

4. Blanchard, B.S., Logistics Engineering and Management, Prentice-Hall, Englewood Cliffs, New Jersey, 1974, p38-40.

5. Bazofsky, I., "Distribution of Downtimes," Proceedings 1974, Reliability and Maintainability, IEEE, New York, 1974 p477-489.

21 SYSTEM DESIGN

J. E. Arsenault, P. J. Des Marais and S. D. G. Williams

In an operational system, provided that enough time has passed, failures will occur, even after a large effort has been directed toward preventing their occurence. It is possible to design systems that will mask failures so that a mission may continue uninterrupted, but there is always a need to know that a failure has occurred (detection) and where the failure has occurred (isolation), at all system levels, down to the individual part or line of code. Detection and isolation of failures falls into the general area known as diagnostics. System diagnostics are necessary for complex systems where system repair is accomplished by replacement of a unit or an assembly depending on the Maintenance Concept for the particular system. Careful attention must be attached to the relationship between system hardware and software as it is only through the proper marriage of the two that efficient system diagnostics evolve.

GENERAL CHARACTERISTICS OF DIAGNOSTICS

Diagnostic Strategies

To be efficient, diagnostics must perform fault detection and isolation down to the smallest replaceable element. This is accomplished by a number of strategies, including start small, start big, overlap and marginal checking. The most efficient approach will be determined by the system structure (1,2).

Start Small - this technique starts by checking a small area of circuitry. If the first check passes, then other areas of circuitry are included progressively, and in this manner circuits found to be operating correctly are used to check other circuitry. This process is continued until all circuits that can be checked are checked.

Start Big - the reverse of Start Small wherein a large group of circuits is checked and when a fault is found further tests are performed to locate the fault in the smallest possible area.

Overlap - this technique uses the results from two or more separate tests. Thus the fault can be located at the overlap portion of each test.

Marginal - this technique is used to detect the effect of part drift due to aging. Generally, the changes brought about by aging are gradual and may remain unnoticed until a failure occurs. To accomplish marginal checking, certain operating conditions are varied from nominal values. Two methods of changing operating conditions are, (a) variation of system D.C. voltages, (b) variation of system clock frequency.

Failure Classification

Diagnostics are generally designed to detect and isolate only three types of failures (2):

Catastrophic - this type of failure is permanently present until it is repaired, and for this reason is usually easiest to detect and isolate.

Intermittent - this type of failure is not present permanently because it occurs at random intervals. It is extremely difficult to isolate because it presents an inconsistent set of symptoms.

Machine State - this type of failure occurs only under certain conditions, most likely (a) after a certain sequence of operations, (b) after a specific instruction followed by a delay in time or (c) at a specific clock rate.

THEORY OF SYSTEM TESTING

An ideal system diagnostic design would, when implemented, detect and isolate 100 percent of all system failures. In a large system, e.g., a computer, this is practically impossible because of the almost infinite number of possible machine states. Table 1 shows the possible results of system testing.

Table 1. Possible Results of System Testing

Test Result	Actual System Status	
	Failure	No Failure
Failure Indicated	T_1	F_1
Failure Not Indicated	F_2	T_2

T = True Test Result   F = False Test Result

To assess the validity of test results given to the operator by the system, the magnitude of F_1 and F_2 must be evaluated analytically or measured. The difference between F_1 and F_2 is significant. If the system has not failed but a failure is indicated, the operator will be wrongly informed of the system status but the system will still function (false alarm). If, however, a system failure is undetected, no indication will be made to the operator and he may continue to rely on the failed equipment. The latter case is the more important as the situation could degrade system performance. A measure of F_1 can be obtained from the reliability (failure rate) of the portion of the system dedicated to performance monitoring. F_2 is slightly more involved since it may be due to (a) the portion of the system under test, (b) the portion of the system dedicated to performance monitoring.

F_1 Evaluation - A measure of F_1 as indicated above may be obtained from a knowledge of circuitry dedicated to system performance monitoring. This will be the probability of the monitoring circuitry being failure free, i.e.,

$$F_1(t) = \exp(-\lambda_m t) \qquad (1)$$

where,

λ_m = failure rate of performance monitoring circuitry

t = time.

F_2 Evaluation - Consider a system containing N parts, p of which are untested. The probability of a fault occurring in p parts, i.e., P_1 is given by:

$$P_1(t) = 1-\exp \sum_{i=1}^{p} -\lambda_i t \qquad (2)$$

If n is the number of parts tested then

$$\sum_{i=1}^{p} \lambda_i = \sum_{j=1}^{N} \lambda_j - \sum_{k=1}^{n} \lambda_k \qquad (3)$$

and substituting in (2) above yields,

$$P_1(t) = 1-\exp\left(-\sum_{j=1}^{N} \lambda_j + \sum_{k=1}^{n} \lambda_k\right)t. \qquad (4)$$

Test thoroughness, T, is given by:

$$T = \sum_{k=1}^{n} \lambda_k \sum_{j=1}^{N} \lambda_j \qquad (5)$$

hence by substitution in (4),

$$P_1(t) = 1 - \exp - (1-T)t \sum_{j=1}^{N} \lambda_j \qquad (6)$$

$$P_2(t) = 1 - \exp - (1-T)t\lambda_s \qquad (7)$$

where,

λ_s = system failure rate.

The probability, P_2, of the system being 'undetectable' failure free is $(1-P_1)$ hence

$$P_2(t) = \exp - (1-T)t\lambda_s \qquad (8)$$

and the probability of the system monitoring circuitry being failure free, P_3, is

$$P_3 = \exp(-\lambda_m t) \qquad (9)$$

where,

λ_m = the failure rate of the monitoring circuitry.

The probability of the system performance and monitoring circuitry being undetectable fault free, i.e., F_2 is $P_2 P_3$ hence

$$F_2(t) = \exp - (\lambda_s(1-T) + \lambda_m)t. \qquad (10)$$

The factor $(\lambda_s(1-T)+\lambda_m)$ is a measure of the failure rate of the untested parts. Hence the mean time between undetected faults (MTBUF) is given by:

$$MTBUF = 1/(\lambda_s(1-T)+\lambda_m) \qquad (11)$$

Thus we may note the trade-off situation between the depth of testing, T, and the failure rate of the monitoring circuitry since as the amount of system circuitry tested increases (T increases) so the failure rate of the monitoring circuitry (λ_m) also increases.

False Alarm - When the system indicates a fault, there exists a probability that this could be a false alarm, due to system transients, power supply fluctuations, etc. Therefore, it is recommended that the first generated indication of system failure be ignored and a re-check initiated. This can be proved mathematically by the use of the distribution which assumes only two possible extremes from a trial, i.e., true or false. If the probability of a correct alarm is (1-p), the number of independent trials is n, then the probability of r false alarms in n trials is

$$B(r;n,p) = \binom{n}{r} p \, (1-p)^{n-1} \text{ for } r=0,1,2,3,\ldots n. \quad (12)$$

Application of Graph Theory to System Diagnosis

While classical methods of fault diagnosis remain practical only for small systems graph theory has emerged as an effective and uniform approach to visualizing and analyzing structural characteristics of larger systems at all levels. Various graphical techniques (3-5) have thereby provided much insight into the many-sided problem of fault diagnosis.

One particular problem that often frustrates the test engineer is the lack of testability which should be designed into the system (6) at various levels, if the desired degree of diagnostic resolution is to be attained. As a partial solution, a graphical technique (7) has been developed that optimizes the selection of test points needed to uniquely identify faulty components.

However, in attempting to apply the results to electronic system diagnostics, two changes to the underlying theory are warranted, (a) a redefinition of the edge cover relation which forms the basis of the technique, (b) a more explicit recognition of the diagnostic capabilities already provided at the system's functional (operational) outputs. Fault Isolation Program (FLIP) is the computer program which resulted from these efforts.

Mathematical Basis. This section will describe the graphical representation used to model a system, thereupon defining the context of analysis, and outlining the sequence of matrix computations that lead to the selection of a minimal set of test points. However, let us first establish the following notation:

SEC graph	single-entry single-exit connected graph
n	number of vertices in the graph
m	number of edges in the graph
v_i	i th vertex
e_{ij}	edge directed from vertex v_i to vertex v_j
$e_i = e_{i_1 i_2}$	i th edge directed from v_{i_1} to v_{i_2}
v_1	entry vertex
v_n	exit vertex
e_{in}	output edge directed from v_i to exit vertex v_n
dp_i	i th disconnected pair of vertices (v_{i_1}, v_{i_2})

$\Omega(i \times j)$ measurable set of vertices v_i to v_j
$e_i \subset e_j$ edge e_i covers edge e_j

The technique behind FLIP processing is based on a system representation known as a SEC graph, as illustrated in Figure 1, which corresponds to a directed basis graph whose vertices $v_1,\ldots,v_n$ represent system (or subsystem) components, and whose edges $e_1,\ldots,e_m$ define the connectivity of these components. An entry vertex corresponds to a component that receives an input signal external to the system, while an exit vertex represents a component that generates a system output signal. For simplicity, the graph is assumed to contain a single entry vertex, v_1, that has no incoming edges and whose outgoing edges are connected to the various entry vertices of the system. Similarly assumed is a single exit vertex v_n, having no outgoing edges and whose incoming edges originate from the exit vertices of the system.

A fault is said to have occurred, if the input-output relationship of the system differs from that of its correct operation, a faulty vertex then corresponds to a defective system component. In addition, the following assumptions apply:

(a) faults are of the permanent type, i.e., sustained until repaired,
(b) A single fault at most occurs at any time.

A test point corresponds to a point which, when assigned to edge e_{ij}, enables a signal from v_i to v_j to be monitored and is thereby said to separate v_i and v_j for fault diagnosis. The problem is then to select a minimal set of test points that will collectively separate all pairs of vertices, i.e., achieve 1-distinguishability of the graph.

Computations begin with a translation of the graph into a connectivity matrix $C = (c_{ij})$, where $c_{ij} = 1$, if and only if there is an edge from v_i to v_j and is 0, otherwise. In turn, matrix C is used to generate the reachability matrix $R = (r_{ij})$, where $r_{ij} = 1$, if and only if there is a directed path from v_i to v_j and is 0, otherwise. We define the measurable set from v_i to v_j as $\Omega(i \times j) = (v_k | r_{ik} = r_{kj} = 1)$. We may now formulate the following edge cover relation: $e_{ij} \subset e_{st}$ if and only if $v_s \varepsilon \Omega(1 \times i)$ and $v_t \not\varepsilon \Omega(1 \times i)$. Since, by definition of a SEC graph, $r_{1k} = 1$ $\forall k = 1,\ldots,n$, the set of edges covered by e_{ij} can be easily computed from the reachability matrix as $(e_{st}\ r_{si} = 1$ and $r_{ti} = 0)$.

The problem has thus been reduced to that of computing a minimal cover over all edges of the graph. However, since the above cover relation guarantees the separation of only those pairs of vertices, (v_i, v_j) for which $r_{ij} = 1$ or $r_{ji} = 1$, the set of disconnected pairs of vertices $(v_i, v_j) | r_{ij} = r_{ji} = 0$ must be added to the covering problem in order to fully achieve 1-distinguishability. Similarly to the edge cover relation, e_{ij} covers disconnected pair (v_s, v_t) if and only if $v_s \varepsilon \Omega(1 \times i)$ and $v \not\varepsilon \Omega(1 \times i)$ or vice-versa; again, the relation is easily verified on the basis of the reachability matrix by simply checking if $r_{si} \neq r_{ti}$.

The next step is then to form the final covering matrix Z, where a row of Z is associated with each edge of the graph and a column is associated with each edge and each disconnected pair of vertices. Then $Z = (z_{ij})$ where $z_{ij} = 1$, if $e_i \subset e_j$ or $e_i \subset dp_j$ and is 0, otherwise.
Note, however, that the covering problem is reduced by the following.
Assumption 3: output edges e_{in} are monitored for diagnostic purposes, i.e. test points are assumed a priori on all output edges of the graph. Hence, all rows of Z associated with output edges are excluded and all columns of Z corresponding to edges or disconnected pairs of vertices covered by at least one output edge are also excluded, generally resulting in a significant reduction in the size of Z.

Finally, the minimal set of test points is obtained by a 2-stage minimization procedure on Z. First, Z is reduced using row dominance, column dominance, and row essentiality (8). Under the latter criterion, edges corresponding to essential rows are selected for test points. Repeated application of the above three rules may or may not completely reduce Z. If not, the second and final step is then to compute the minimal cover for the resulting cyclic matrix, using a linear integer programming method (9).

In summary, the procedure for computing the minimal set of test points is stated as follows:

(a) model the system as a SEC graph,
(b) translate the graph into a connectivity matrix C,
(c) compute the reachability matrix R,
(d) form the final covering matrix Z on the basis of the edge cover relation and matrix R,
(e) solve matrix Z as a minimal covering problem, using row dominance, column dominance, row essentiality, and, if necessary, a linear integer programming method to complete minimization.

System Modelling and an Example. FLIP allows the analyst many degrees of freedom. For instance, he may select the level to which fault detection will be achieved and hence the graph complexity. 'Low level' modelling will provide the best point coverage and diagnostics, especially where system fault propagation paths are discrete. Modelling is creative and the rules of graph theory are flexible. However, if the results are to be meaningful, the principles outlined in (7) must be adhered to. Hardware symbolized by vertices may contribute faults within more than one vertex, and vertex communication, symbolized by edges, should be faithfully reproduced. System feedback must be avoided. This may be achieved by using one of FLIP's minimizing sub-routines once the SEC graph is complete, or during the modelling stage which usually requires a thorough understanding of the system.

Figure 1 is a SEC graph of a small digital electronic system. For this particular case the connectivity, reachability and Z matrices are shown in Figures 2, 3 and 4 while Figure 5 shows the minimal set of test points for 1 - distinguishability of the system. The fault signature table shown in Figure 6, which is a program by-product, clearly illustrates how monitoring the 6 suggested edges will uniquely fault isolate the faulty vertex.

REQUIREMENTS DEFINITION

System diagnostic requirements can be generalized to the following statements which will be found to be adequate for most circumstances,

(a) permit end-to-end system tests and confidence tests in all modes of operation and provide the operator with GO/NO GO and DEGRADE indications of system readiness,
(b) isolate failures to individual unit or assembly and provide indicators to locate the failed item,
(c) obviate the need for external test equipment,
(d) indicate what reversionary modes are available.

Equipment reversionary modes aim at managing the effectiveness of a particular system in the event of a failure which cannot be rectified while the system is operating. Reversionary modes do not infer a duplication of hardware although this may be required to meet operational requirements.

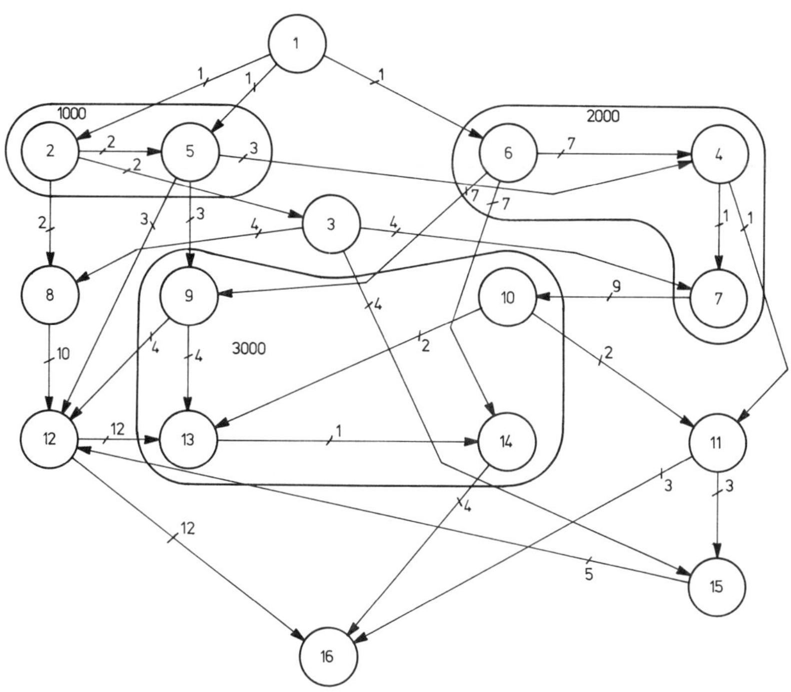

Figure 1. System SEC Graph

:C(CONNECTIVITY MATRIX OF GRAPH) :

(ROWS AND COLUMNS OF MATRIX ARE VERTICES OF GRAPH)

	1	2	3	4	5	6	7	8	9	10	11	12	13	14	15	16
1	0	1	0	0	1	1	0	0	0	0	0	0	0	0	0	0
2	0	0	1	0	1	0	0	1	0	0	0	0	0	0	0	0
3	0	0	0	0	0	0	1	1	0	0	0	0	0	0	1	0
4	0	0	0	0	0	0	1	0	0	0	1	0	0	0	0	0
5	0	0	0	1	0	0	0	0	1	0	0	1	0	0	0	0
6	0	0	0	1	0	0	0	0	1	0	0	0	0	1	0	0
7	0	0	0	0	0	0	0	0	0	1	0	0	0	0	0	0
8	0	0	0	0	0	0	0	0	0	0	0	1	0	0	0	0
9	0	0	0	0	0	0	0	0	0	0	0	1	1	0	0	0
10	0	0	0	0	0	0	0	0	0	0	1	0	1	0	0	0
11	0	0	0	0	0	0	0	0	0	0	0	0	0	0	1	1
12	0	0	0	0	0	0	0	0	0	0	0	0	1	0	0	1
13	0	0	0	0	0	0	0	0	0	0	0	0	0	1	0	0
14	0	0	0	0	0	0	0	0	0	0	0	0	0	0	0	1
15	0	0	0	0	0	0	0	0	0	0	0	1	0	0	0	0
16	0	0	0	0	0	0	0	0	0	0	0	0	0	0	0	0

Figure 2. Connectivity Matrix

:R(REACHABILITY MATRIX OF GRAPH) :

(ROWS AND COLUMNS OF MATRIX ARE VERTICES OF GRAPH)

	1	2	3	4	5	6	7	8	9	10	11	12	13	14	15	16
1	1	1	1	1	1	1	1	1	1	1	1	1	1	1	1	1
2	0	1	1	1	1	0	1	1	1	1	1	1	1	1	1	1
3	0	0	1	0	0	0	1	1	0	1	1	1	1	1	1	1
4	0	0	0	1	0	0	1	0	0	1	1	1	1	1	1	1
5	0	0	0	1	1	0	1	0	1	1	1	1	1	1	1	1
6	0	0	0	1	0	1	1	0	1	1	1	1	1	1	1	1
7	0	0	0	0	0	0	1	0	0	1	1	1	1	1	1	1
8	0	0	0	0	0	0	0	1	0	0	0	1	1	1	0	1
9	0	0	0	0	0	0	0	0	1	0	0	1	1	1	0	1
10	0	0	0	0	0	0	0	0	0	1	1	1	1	1	1	1
11	0	0	0	0	0	0	0	0	0	0	1	1	1	1	1	1
12	0	0	0	0	0	0	0	0	0	0	0	1	1	1	0	1
13	0	0	0	0	0	0	0	0	0	0	0	0	1	1	0	1
14	0	0	0	0	0	0	0	0	0	0	0	0	0	1	0	1
15	0	0	0	0	0	0	0	0	0	0	0	1	1	1	1	1
16	0	0	0	0	0	0	0	0	0	0	0	0	0	0	0	1

Figure 3. Reachability Matrix

: Z (FINAL COVERING MATRIX BEFORE REDUCTION) :

Z = Q : P
ROWS OF Z ARE EDGES OF GRAPH
COLUMNS OF SUBMATRIX Q ARE EDGES OF GRAPH
COLUMNS OF SUBMATRIX P ARE PAIRS OF VERTICES FROM SET V

	1	2	3	4	5	7	10	11	12	18	19	20	22	28	30	1	2	3	4	5	6	7	8
1	1	1	1	0	0	0	0	0	0	0	0	0	0	0	0	0	0	0	0	0	0	0	0
2	1	1	1	0	0	0	0	0	0	0	0	0	0	0	0	0	0	0	0	0	0	0	0
3	1	1	1	0	0	0	0	0	0	0	0	0	0	0	0	0	0	0	0	0	0	0	0
4	0	1	1	1	1	0	0	0	0	0	0	0	0	0	0	1	0	0	0	0	0	0	0
5	0	1	1	1	1	0	0	0	0	0	0	0	0	0	0	1	0	0	0	0	0	0	0
6	0	1	1	1	1	0	0	0	0	0	0	0	0	0	0	1	0	0	0	0	0	0	0
7	0	1	1	0	1	1	0	0	0	0	0	0	0	0	0	1	1	1	1	0	0	0	0
8	0	1	1	0	1	1	0	0	0	0	0	0	0	0	0	1	1	1	1	0	0	0	0
9	0	1	1	0	1	1	0	0	0	0	0	0	0	0	0	1	1	1	1	0	0	0	0
10	0	0	0	1	0	0	1	1	0	0	0	0	0	0	0	0	1	1	1	0	0	0	0
11	0	0	0	1	0	0	1	1	0	0	0	0	0	0	0	0	1	1	1	0	0	0	0
12	0	0	1	1	0	0	0	0	1	0	0	0	0	0	0	1	0	1	0	1	0	0	0
13	0	0	1	1	0	0	0	0	1	0	0	0	0	0	0	1	0	1	0	1	0	0	0
14	0	0	1	1	0	0	0	0	1	0	0	0	0	0	0	1	0	1	0	1	0	0	0
16	1	1	0	0	0	0	0	0	0	0	0	0	0	0	0	1	0	0	1	1	0	0	0
17	1	1	0	0	0	0	0	0	0	0	0	0	0	0	0	1	0	0	1	1	0	0	0
18	0	0	0	0	0	0	0	1	0	1	0	0	0	0	0	0	0	0	0	0	0	0	0
19	0	1	1	0	1	1	0	0	0	0	1	0	0	0	0	1	1	1	1	0	1	1	0
20	0	0	0	1	0	0	0	0	1	0	0	1	0	0	0	0	0	1	1	0	1	0	1
22	0	0	0	0	0	0	0	1	0	0	0	0	1	0	0	0	0	0	0	0	0	0	0
24	0	0	0	0	0	0	0	0	0	0	0	0	0	0	0	0	0	0	0	0	0	0	0
26	0	0	0	0	0	0	0	0	0	0	0	0	0	0	0	0	0	0	0	0	0	0	0
28	0	0	0	0	0	0	0	0	0	0	0	0	0	1	0	0	0	0	0	0	0	0	0
30	0	0	0	0	0	0	0	0	0	0	0	0	0	0	1	0	0	0	0	0	0	1	1

Figure 4. Z Matrix

TEST POINTS MUST BE INSERTED ON THE FOLLOWING EDGES

EDGE NO:	FROM VERTEX	TO VERTEX	
18	7	10	
19	8	12	
20	9	12	
22	10	11	
28	13	14	
30	15	12	
11	4	11	OR ANYONE OF EDGES: 10,

MONITORING OF THE ABOVE 7 EDGES REQUIRES 32 TEST POINTS

IN ADDITION
TEST POINTS MUST BE INSERTED ON ANYONE OF THE FOLLOWING SETS OF EDGES

SET 1

EDGE NO:	FROM VERTEX	TO VERTEX	
3	1	6	OR ANYONE OF EDGES: 2, 1,
14	5	12	OR ANYONE OF EDGES: 13, 12,

MONITORING OF THE ABOVE 2 EDGES REQUIRES 4 TEST POINTS

Figure 5. Test Points for 1 - Distinguishability

FAULT SIGNATURE TABLE

MONITORED EDGES (WHERE 1:FAULT OBSERVED 0:FAULT NOT OBSERVED)

FAULTY VERTEX	18	19	20	22	28	30	11	3	14	25	27	29
1	1	1	1	1	1	1	1	1	1	1	1	1
2	1	1	1	1	1	1	1	0	1	1	1	1
3	1	1	0	1	1	1	0	0	0	1	1	1
4	1	0	0	1	1	1	1	0	0	1	1	1
5	1	0	1	1	1	1	1	0	1	1	1	1
6	1	0	1	1	1	1	1	0	0	1	1	1
7	1	0	0	1	1	1	0	0	0	1	1	1
8	0	1	0	0	1	0	0	0	0	0	1	1
9	0	0	1	0	1	0	0	0	0	0	1	1
10	0	0	0	1	1	1	0	0	0	1	1	1
11	0	0	0	0	1	1	0	0	0	1	1	1
12	0	0	0	0	1	0	0	0	0	0	1	1
13	0	0	0	0	1	0	0	0	0	0	0	1
14	0	0	0	0	0	0	0	0	0	0	0	1
15	0	0	0	0	1	1	0	0	0	0	1	1
16	0	0	0	0	0	0	0	0	0	0	0	0

Figure 6. Fault Signature Table

It is possible to express design requirements in quantitative terms to ensure that diagnostics are taken into account in the actual system design. It is usual to include a requirement for both fault detection and isolation. For instance a minimum system diagnostic requirement might be to:

(a) detect 95% of system failures and to isolate to a single unit 90% of system failures,
(b) have detection and isolation circuitry to contribute not more than 5% of the failure rate of the system,
(c) have a false alarm probability of less than 1%.

DESIGNING TO MEET THE REQUIREMENTS

General

Several levels of diagnostics may be required because operational constraints, for example memory size, do not permit the use of only one level of diagnostics. The term Built-In Test Equipment (BITE) is used here to define the several levels of diagnostics generally required in a complex system, i.e., BITE1, BITE2, BITE3.

BITE1 - This level is of the continuous monitoring type and does not interfere with the operation of the system. No test initiate signal is required but is initiated automatically when a system function is energized. Hardware or software may be involved but in any case this level is transparent to the operator. Examples: power supply overvoltage protection circuitry, background interface integrity checks.

BITE2 - This is a partially interruptive level of testing to provide a confidence check for the operator should he consider that the system is malfunctioning and not providing any BITE1 failure indication. The operation of the system is not significantly compromised other than the man-machine interface, such as displays, which may be used to exhibit test patterns that would attempt to give indications of malfunctions. Circuitry internal to the equipment will sequence the tests, generate the stimuli, and perform and evaluate the measurements. The circuitry is capable of being commanded into various test modes by means of operating controls, and a manual test initiation. Provision is made to ensure that output data cannot be interpreted as prime system data, during such testing by the system or operator.

BITE3 - This is a totally interruptive level of testing. It is used together with BITE1 and BITE2 to form a first line test facility. Diagnostic tapes would normally be available from magnetic tape equipment to fully exercise the system, making extensive use of a Central Processing Unit (CPU) aided by operator controls. Table 2 summarizes the diagnostic level characteristics discussed above.

Table 2. System Diagnostic Characteristics

Diagnostic Level	Major Characteristics	Purpose
BITE1	(a) Automatically initiated. (b) Non-interruptive.	Continuous monitoring.
BITE2	(a) Manually initiated. (b) Partially or temporarily interruptive.	Confidence check.
BITE3	(a) Manually initiated. (b) Totally interruptive.	Maintenance facility.

The object of BITE1 is to establish to necessary confidence that the system will operate when required to do so. Since the operator is interested in the performance at a particular time then it is important to be able to carry out the performance monitoring at the time, or immediately before, so as to obtain a reasonable assurance of completing the mission satisfactorily.

When the operator is not confident that the system is functioning correctly although there has been no indication from BITE1 that anything is wrong, he may then use BITE2, which is partially or temporarily interruptive, i.e., may temporarily interrupt processing but does not destroy existing data, but which in no way will destroy the prime equipment program, nor will it be possible for the results to be interpreted as prime data.

BITE3 is essentially a maintenance function where the objective is to diagnose the faulty element so that the fault may be cleared. It may be used as an addition to performance monitoring to enable a fault to be isolated. Since fault detection normally takes place at a time devoted to maintenance, it is not necessary to maintain the normal operational performance while it is taking place, although there must be no loss of performance. Usually a family of diagnostic program tapes will be made available from magnetic tape equipment to fully exercise the system making extensive use of the CPU and front panel controls.

A System Diagnostic scheme using the concepts discussed above is illustrated in Figure 7.

Approaches to Error Recovery

At the system level, assuming that adequate BITE1 and BITE2 hardware and software exists, it should be possible to detect the presence of system errors. This being so, the next question arising is to define the strategy in the event that a failure is announced. Several strategies are available and the appropriate one will depend upon the operational requirements and the type of system used (3). They are outlined as follows:

Bring the System to a Complete Halt - this strategy is simple and economical because the expense and trouble of redundant hardware is avoided. However, this strategy provides no possibility for recovery except within the system Mean Time To Repair (MTTR).

Fall Back to a Less Efficient Processing - depending where the error has occurred in the system, it may be possible to reduce processing by shutting down the affected hardware or bypassing it. This will permit the system to operate in a reduced or degraded mode. A good example of this, is assigning processing to certain memory areas in the event of failure and reducing system throughput. This is also known as a fail-soft system arrangement.

Switch to a Standby System - if high operational performance is of extreme concern, then a system may be completely duplicated both in hardware and software. This technique is more expensive than the other

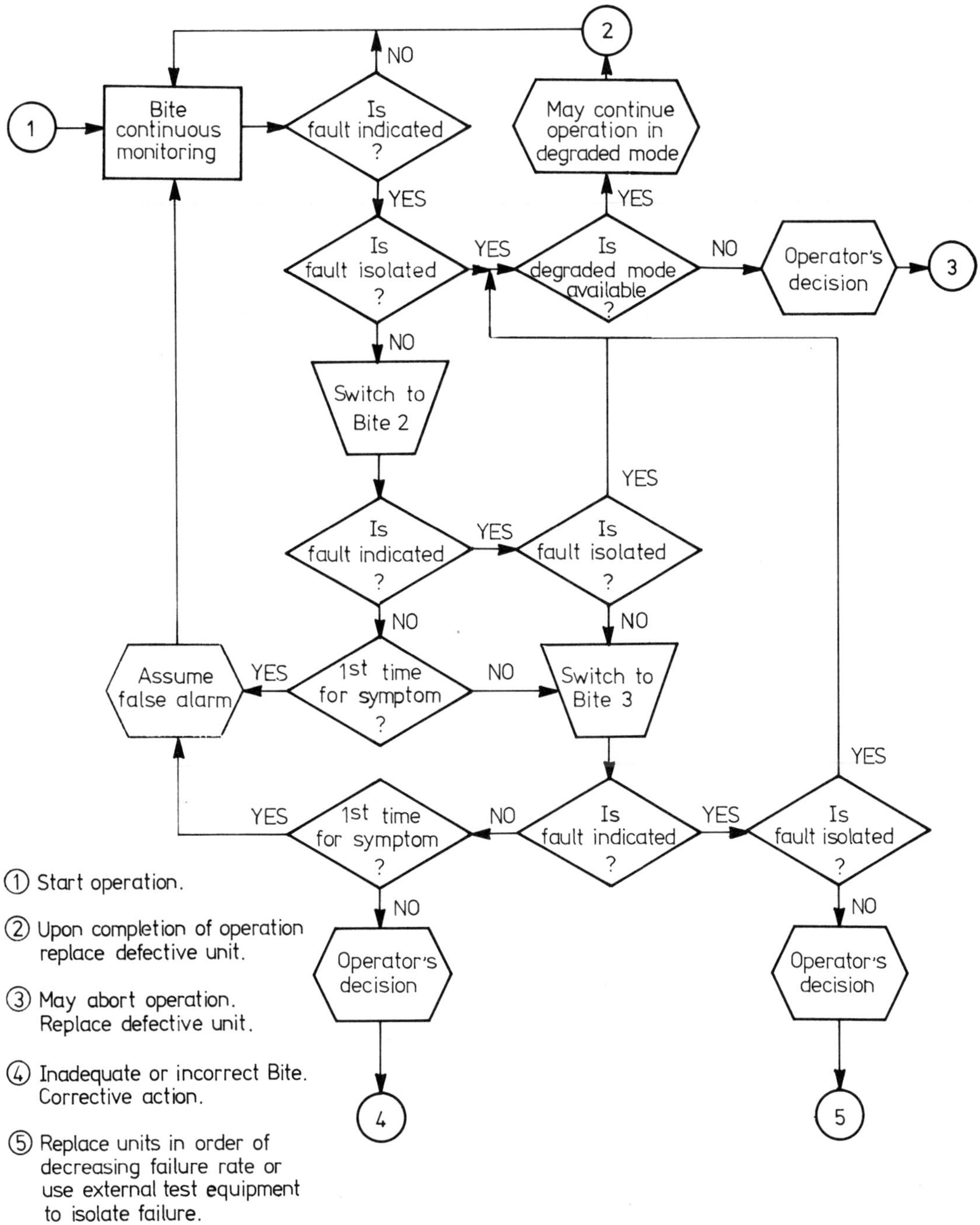

Figure 7. System Diagnostic Concept

strategies but is effective where a high Mean Time Between Failure (MTBF) is demanded. This arrangement is often called a fail-safe type of system.

System Diagnostic Hardware

Redundancy - a standard method of handling automatic error detection is through the use of redundancy. Comparators are located at critical interfaces to detect errors between the redundant paths. Redundancy can be used on all levels, i.e., part, assembly, unit, system. The general idea is illustrated in Figure 8.

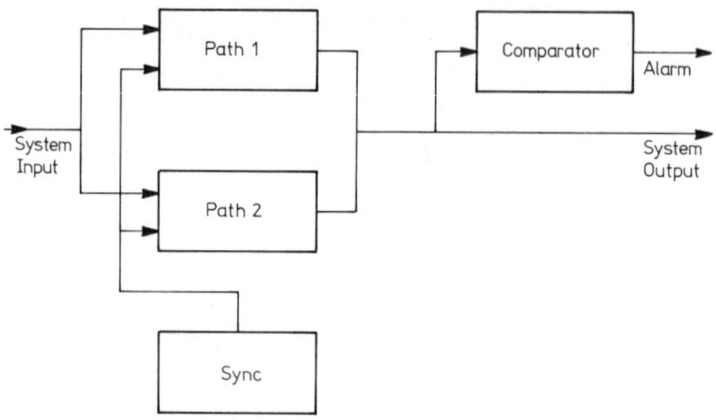

Figure 8. Use of Redundancy

Parity Checking - this technique is very common for ensuring data integrity during transmission along communications lines or during computer memory operations. The basic idea is shown in Figure 9.

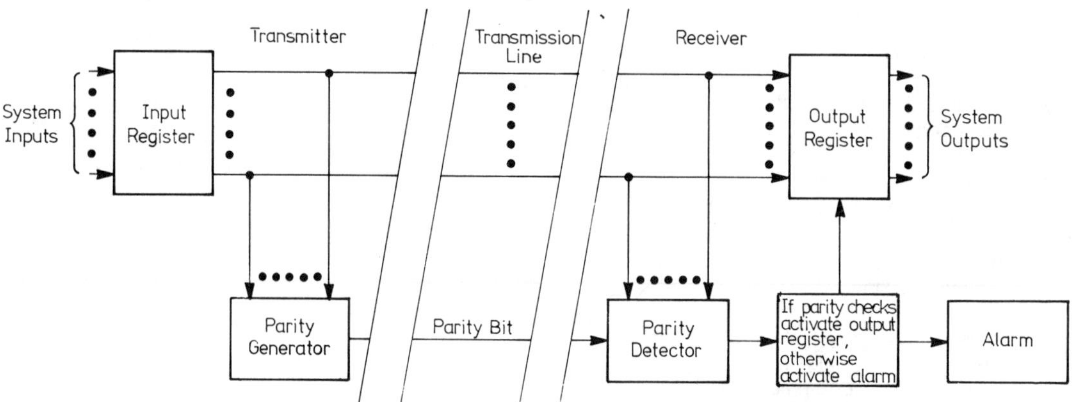

Figure 9. Parity Bit Encoding/Decoding

A parity bit is generated to ensure that the total number of logical ones in the word being transmitted or received is odd or even. On the receiving end a parity check is performed on the transmitted data to ensure that no word bit has been dropped (11). The system must be implemented using either odd or even parity.

Majority Voting - an arrangement can be designed (12) whereby the outputs of redundant paths are continuously compared for agreement by a voter. If all outputs agree the system output is correct. If one element fails the voter will choose as the system output, the output of the elements that agree. Therefore the system output will be correct. A minimum arrangement is shown in Figure 10.

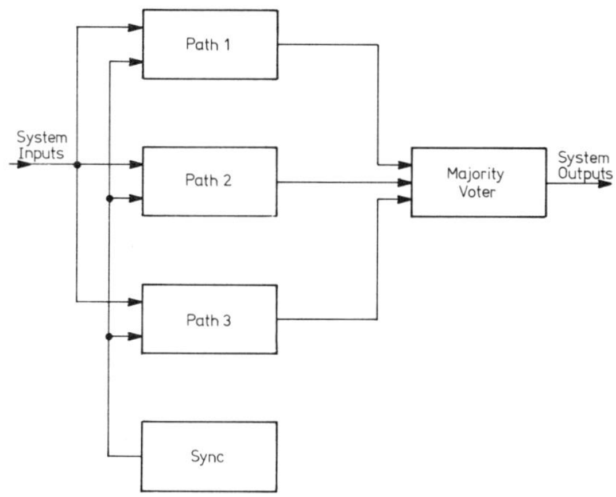

Figure 10. Majority Voter Concept

Status Reporting - this technique is very commonly used to provide information on the status of the various elements in a system. In most applications critical circuit values or conditions are monitored continuously and placed in a register which is interrogated under program control and the result is compared to known or correct conditions. The basic idea is shown in Figure 11.

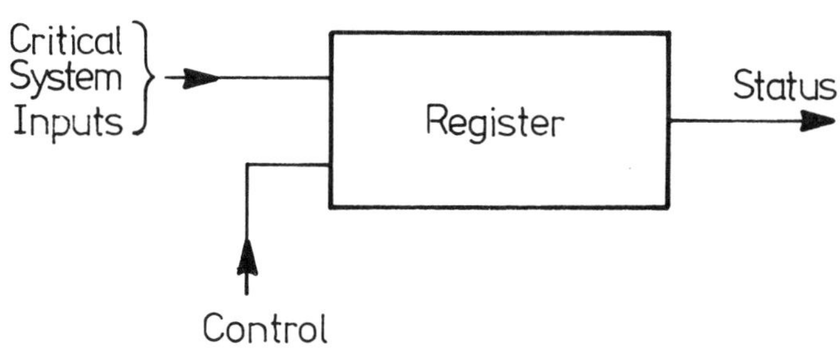

Figure 11. Status Register

Test Generation - this technique is often used to determine if a fault is in the system or is from some source outside the system. Usually normal system inputs are substituted with locally generated inputs (13). The basic concept is illustrated in Figure 12.

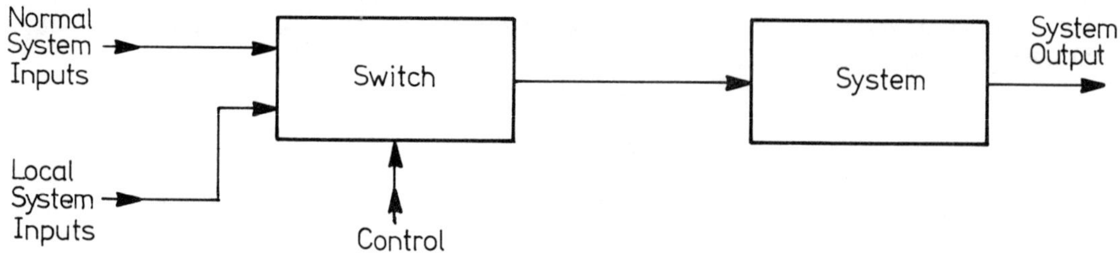

Figure 12. Test Generation

System Diagnostic Software

Because of the enormous number of machine states that a large system may assume, it is hardly ever practical to test a system 100 percent. Consequently provisions must be made to take care of those cases in which the right combination of data and processing steps cause a system failure to occur. Frequently occurring conditions are looping, deadlocks, etc. To provide for these situations a program is usually monitored in execution, i.e., once it is loaded in the system and is running. This program is usually referred to as the monitor program although it is known by other names as well. The interrupt mechanisms cause control to jump to predetermined memory locations occupied by the monitor program when system failures are detected. Depending on the specific memory location the monitor program will employ other programs to isolate the cause of the failure and report it to the system operator. Monitor programs usually will have to be able to set a timer just prior to execution and control is transferred to it when a process exceeds a specified time. The monitor program is stored in a protected area of memory in order that it may reference memory anywhere, as this allows the monitor to complete its analysis of the problem. In general, any attempt to reference protected memory will result in an interrupt to the monitor program (14, 15).

Diagnostic aids used for error isolation in software are generally concerned with the examination of system data structures for consistency (16). Examples of this technique are:

(a) tracing address entries in tables,
(b) using redundant information and,
(c) using checksum methods.

Tracing Address Entries - The set of addresses to which a pointer may link is fixed, relatively small and each branch of a 'tree' structure has some identifying information which may be identified. Examples are (14):

(a) Dump - is a listing of all or part of the program area of memory at the end of execution. It is often in octal or hexadecimal, although it is possible to provide a more useful symbolic listing,

(b) Post-mortem - an analysis of the contents of memory (after execution has been terminated) to gather as much information as possible that may be of use to the programmer,

(c) Snapshot - a dynamic dump of part or all of the memory that can be requested by the user's program. It is essentially a library program that will conveniently print sections of the memory,

(d) Trace - an automatic snapshot feature. It may be possible to request, for example, that certain registers or memory locations be printed every time that a branch is executed. This allows the user to get a picture of the flow of a program, since he knows that flow continues through increasing locations between branches.

Use of Redundant Information - During normal processing several copies of the same information are maintained in different forms. A resource data structure may contain allocation details, such as the name of each process holding some of the resource, while the process descriptors may redundantly contain a list of each resource allocated to the process.

Use of Checksum Methods - The idea is to combine all of the basic data into a table, or file into a single number by summing, exclusive 'ORing' or some other operation. The summing is done whenever a table is first created or changed. When the table or file is accessed, the checksum is computed to ensure that all of the data is present.

DIAGNOSTIC COMMISSIONING

Commissioning is the stage during which the system hardware and software are combined to:

(a) test the system known to be functioning correctly and pass all tests,
(b) test a system known to be outside specified limits, and show a test failure,
(c) detect system faults which have been deliberately introduced and display correct repair action to be taken. Figure 13 shows a typical commissioning procedure.

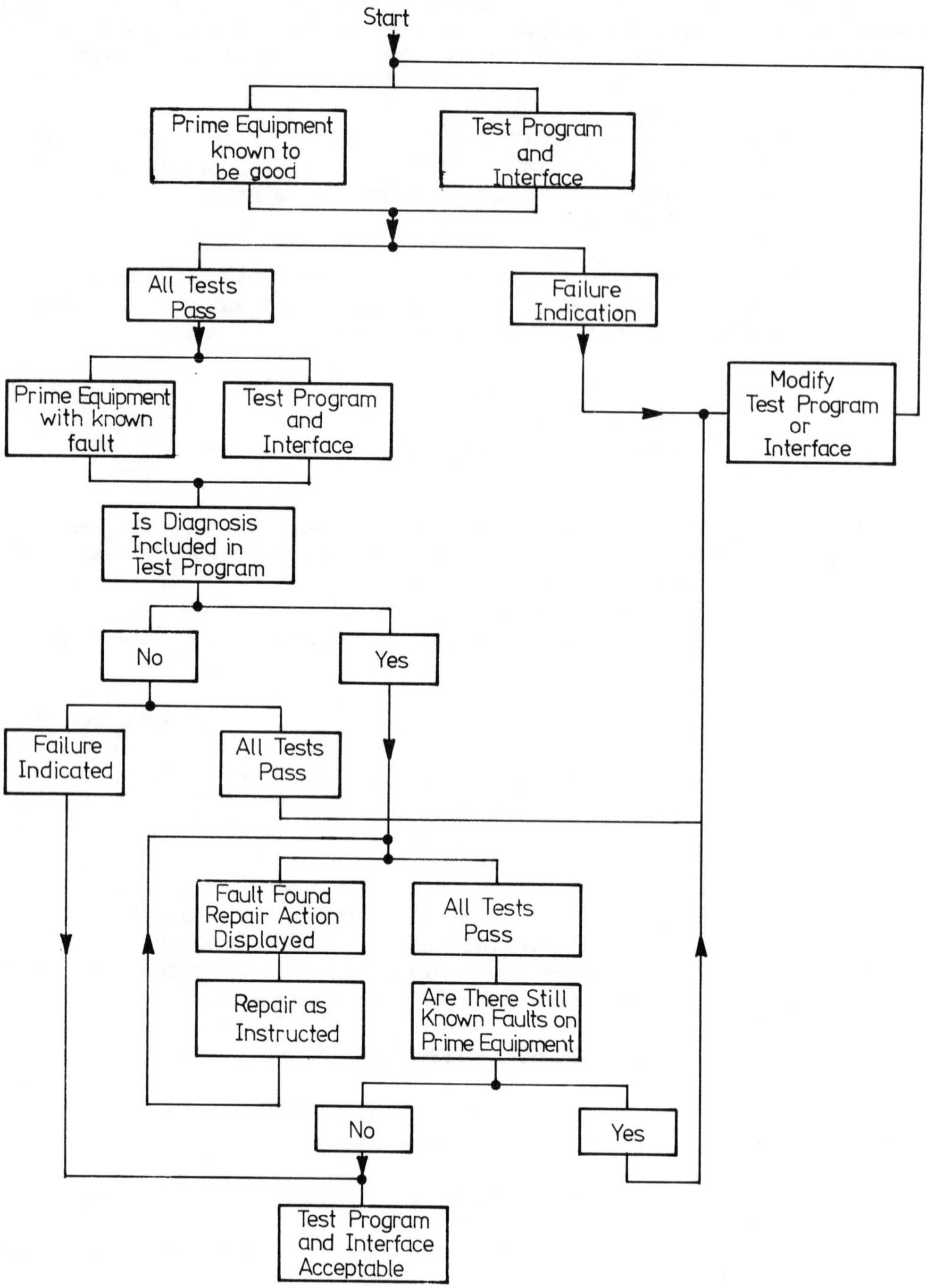

Figure 13. Diagnostic Commissioning Diagram

All parties concerned, i.e., customer, prime contractor, system contractors, must come to an agreement defining the modification state of the system used before commissioning and acceptance of the test program. Careful prior agreement between the interested parties is essential in order to support the commissioning activity so that it may proceed smoothly.

CONCLUSION

Design effort is required if systems are to have adequate levels of diagnostics. When these levels are specified to high requirements, the effort can represent a large proportion of project time. The diagnostic concept for a system must be defined early and given high visibility or specified requirements will not be met. The diagnostic area represents a field of endeavour where progress must continue at high level to keep pace with growing system complexity.

REFERENCES

1. ______________, Digital Computer Basics, Bureau of Naval Personnel, Dover Publications Inc., New York, 1969, p201-206.

2. MacGuire, B.W., Handbook of Computer Maintenance and Troubleshooting, Keston Publishing Inc., Reston, Virginia, 1973, p257-259.

3. Ramamoorthy, C.V., "A Structural Theory of Machine Diagnosis", Spring Joint Computer Conference, AFIPS, Conference Proceedings, Vol. 30, Washington, D.C., 1967, p743-756.

4. ______________, and Mayeda, W., "Computer Diagnosis Using the Blocking GAte Approach", IEEE Transactions, on Computers, Vol. C-20, No. 11, November 1971, p1294-1299.

5. Schertz, D.R. and Metze, G., "A New Representation for Faults in Combinational Digital Circuits", IEEE Transactions on Computers, Vol. C-21, No. 8, August 1972, p858-866.

6. DesMarais, P. and Krieger, M., "Fault Simulation and Digital Circuit Testability", Microelectronics and Reliability, Proceedings of Symposium on Reliability Engineering, Vol. 15 Supplement, 5 June 1976, p5-13.

7. Patel, M., "Fault Diagnosis by Inserting the Minimum Number of Test Points in System Graphs", NTIS Report No. AD 785272, Springfield, Va., 1974.

8. Torng, H.C., Switching Circuits, Theory and Logic Design, Addison-Wesley Publishing Co., Reading, Mass., 1972.

9. McMillan, C., Mathematical Programming, J. Wiley & Sons Inc., New York, N.Y., 1970.

10. Yourdon, E., Design of On-Line Computer Systems, Prentice-Hall, Englewood Cliffs, N.J., 1972, p534-542.

11. Blakeslee, T.R., Digital Design with Standard MSI and LSI, John Wiley and Sons, New York, N.Y., 1975, p251-253.

12. Shooman, M.L., Probabilistic Reliability - An Engineering Approach, McGraw-Hill, New York, N.Y., 1968, p293-324.

13. Peatman, J.B., The Design of Digital Systems, McGraw-Hill, New York, N.Y., 1972, p443.

14. Gear, W.C., Computer Organization and Programming (2 Ed.), McGraw-Hill, New York, N.Y., 1974, p137-140.

15. Kurzban, S.A., Heines, T.S. and Sayers, A.P., Operating Systems Principles, Petrocelli/Charter, New York, N.Y., 1975, p111.

16. Shaw, A.C., The Logical Design of Operating Systems, Prentice-Hall, Englewood Cliffs, N.J., 1974, p272-276.

22 UNIT DESIGN

J. E. Arsenault

In some systems diagnostics may be designed for fault isolation to the assembly level while others are designed to allow fault isolation to the unit only. Typical of the latter would be systems where repair is effected by replacement of faulty units comprised of assemblies. This approach ensures rapid repair and high operational readiness at the start of a mission. Having removed the faulty unit from this type of system the next step is to fault isolate to the assembly level using some diagnostic scheme on the unit as a stand-alone item.

The following discussion involves designing the diagnosis for a complex unit by replacement or repair of an assembly or part at a site removed from the operational system. In some cases this will involve the return of the unit to the manufacturer for repair; in others the system operator may maintain his own repair site or he may choose to sub-contract out the repair of the unit. The maintenance concept will determine the exact policy to be applied for unit repair.

REQUIREMENTS

It is possible to generalize unit diagnostic requirements as in the case for system diagnostics. They are:

(a) permit end-to-end unit tests and confidence tests in all modes of operation and provide the operator with a GO/NO GO and DEGRADE indication of unit readiness,
(b) isolate failures to individual assemblies and provide operator messages to locate the assembly.

The general requirements may be expressed quantitatively as follows:

(a) detect 98 percent of all faults,
(b) isolate 85 percent of detected faults to one assembly,
(c) isolate 90 percent of detected faults to two or fewer assemblies,

(d) isolate 95 percent of detected faults to three or fewer assemblies,
(e) isolate 98 percent of detected faults to four or fewer assemblies.

DESIGNING TO MEET THE REQUIREMENTS

Designing a complex unit, with the above requirements in mind usually leads to the conclusion that Automatic Test Equipment (ATE) will be necessary for efficient fault detection and isolation. A typical ATE arrangement is illustrated in Figure 1.

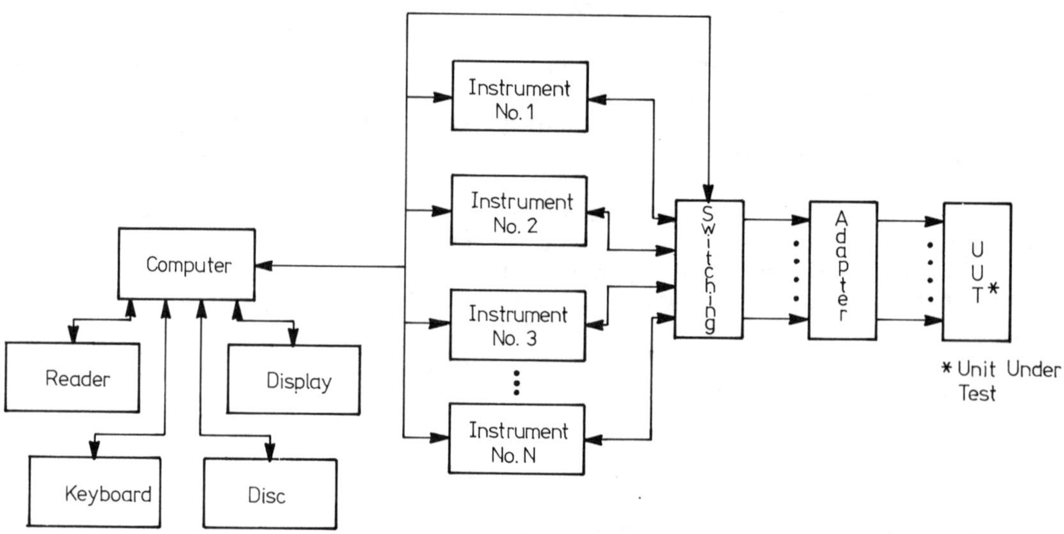

Figure 1. Typical ATE Complex

For the unit designer the problem is largely one of ensuring unit compatibility with the specified ATE and that the fault isolation requirements can be met. The problem is further complicated, however, by the fact that in many cases the ATE is supplied separately to a prime contractor and a great deal of documentation is required for both hardware and software to enable the prime contractor to integrate the unit with the ATE. More specifically, the unit designer usually will be responsible for the preparation of test data and diagnostic information for the unit/ATE integrator, to enable a test program to be generated and the associated coupling unit and any other special hardware required to be designed. In order that this may be accomplished, the ATE supplier should provide the unit designer with a complete specification of the ATE to be used. This will enable the unit supplier to prepare a maintenance program compatible with the ATE. Documentation requirements are summarized in Figure 2.

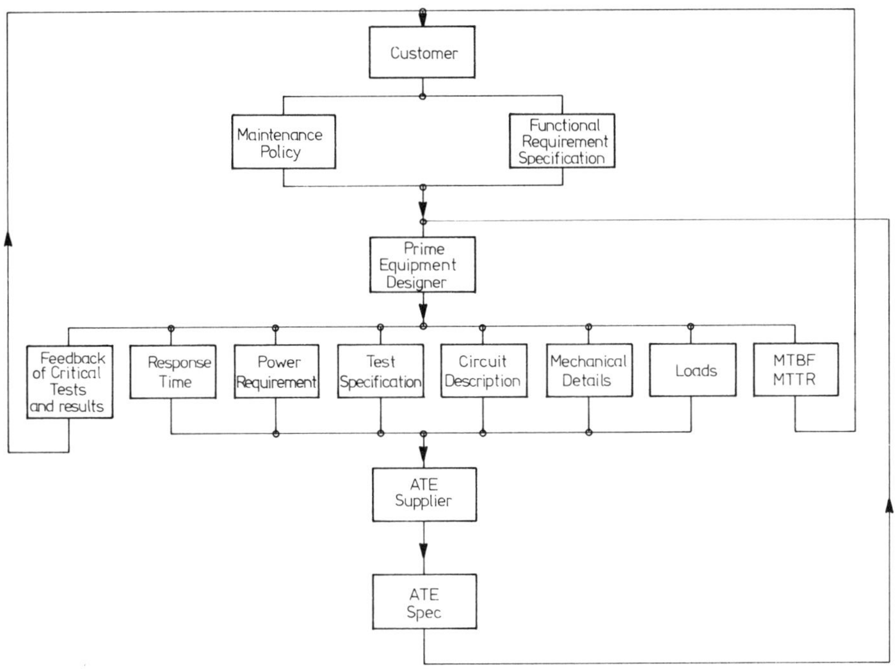

Figure 2. Documentation Requirements

Unit Hardware Design

Identification Signatures. The purpose of identification is to provide a fully automatic and positive means of assuring proper correlation between the Unit Under Test (UUT) and the ATE test program. Significant savings can be realized as a result of fewer manual operations and the prevention of equipment damage.

A method of automatic identification (1) uses an array of precision resistors arranged in a star configuration, as shown in Figure 3. These pins should be connected to the resistor network(shown in Figure 3) with R1 assigned to the lowest pin number or letter of the group reserved. Remaining resistors should be connected in ascending order of "R" numbers to the connector pins of ascending order. R1 and R2 are used for the unit manufacturers' code.

Test Connectors. For a complex unit, even if it is highly partitioned to the extent that there is only one function per circuit card, it is not usually possible to fault isolate to a single card because of inaccessibility of card interfaces. Consequently, it is often necessary to use test points at card interfaces and bring them out to dedicated test connectors for stimulation and monitoring purposes. These connectors are in addition to the operational connectors by which the unit interfaces with a system and are illustrated in Figure 4.

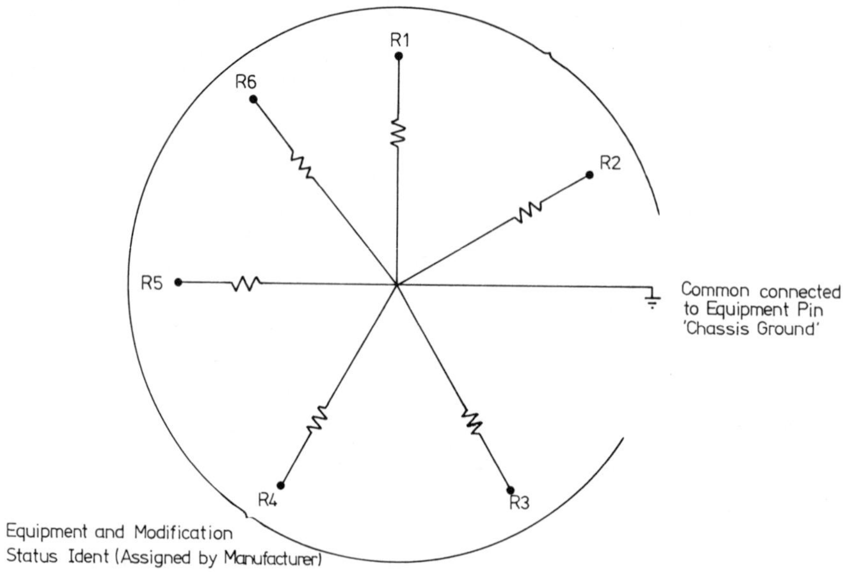

Figure 3. ATE Auto-Ident Resistor Network

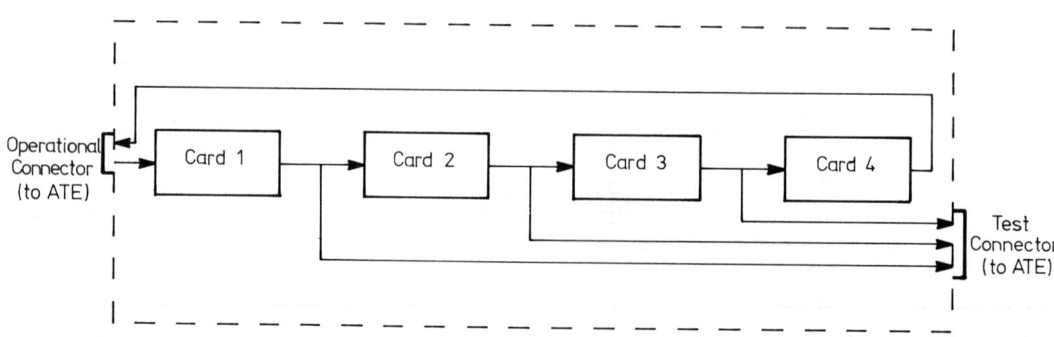

Figure 4. Test Points at Circuit Card Interfaces

It is generally assumed that at card interfaces the driving element is at fault rather than the driven element. This is especially true for digital circuits where the failure ratio is estimated to be about 10:1

relative to the driving card over the driven card, as shown in Figure 5.

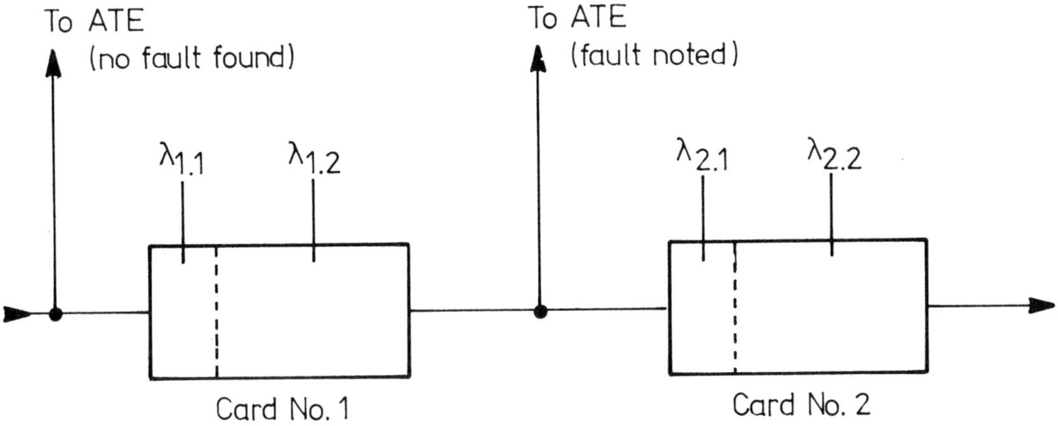

Figure 5. Circuit Card Interface Detail

Test Point Multiplexing. In many cases a unit may contain a large number of circuit card assemblies and attempting to observe unit behavior via test points at the card interfaces would result in an unacceptable number of test points. The solution is to be judicious in test point selection, in order to reduce the total quantity to a reasonable number, or to use a multiplexing scheme. Here a test card is inserted in the unit to multiplex test points from different locations in the unit onto the same connector. Figure 6 is an illustration of this idea.

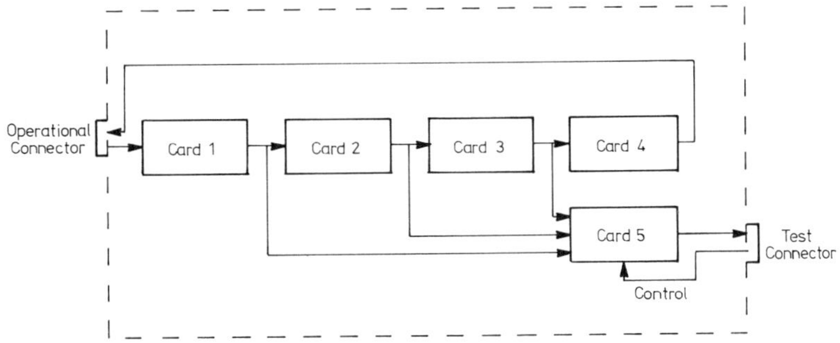

Card 5 is a Test Point Multiplex Card

Figure 6. Test Point Multiplexing

Where card space permits it is possible to add devices which permit multiplexing to the card itself, rather than by the use of a test point multiplexer card. This is illustrated in Figure 7.

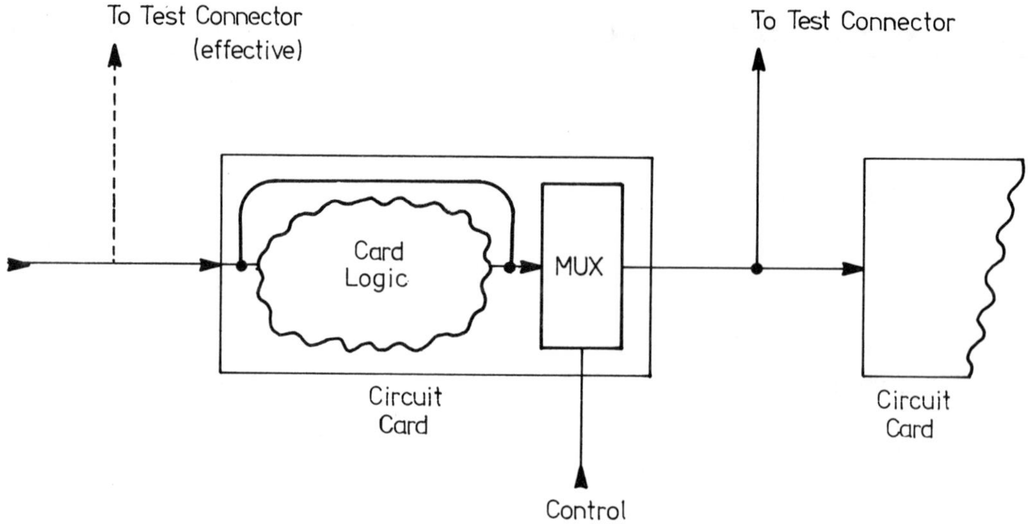

Figure 7. Test Point Multiplexing on a Card

Buffering. The use of test points at the unit level can lead to problems when testing with ATE. This is usually achieved by the addition of long cable runs which tend to be capacitative and may require buffering. Careful selection of test points will reduce the need for buffering. However, when it is required, a degree of buffering can be obtained by adding it to the card from where the test point originated. This can be done by adding gates and resistors. However, the card will often have unused gates and resistors available for buffering. If this cannot be accomplished an Interface Adapter (IA) must be used to provide the necessary buffering (2). A practical IA is used to house special circuits and loads as well as buffering circuits and to provide cooling for the unit. This is illustrated in Figure 8.

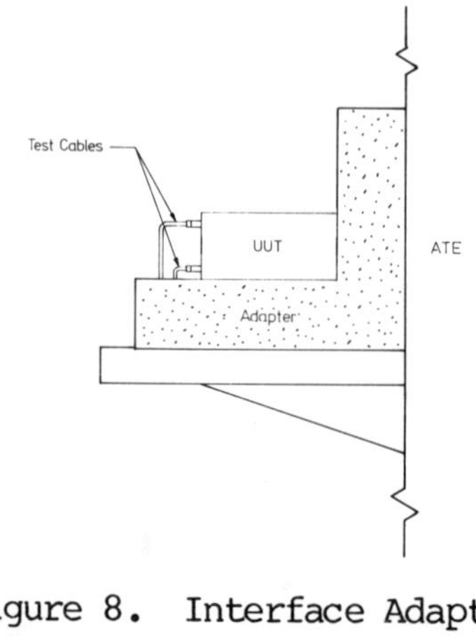

Figure 8. Interface Adapter

External Clock Control. Another critical problem which the unit designer must take into account is the fact that the ATE may not be capable of running at the same speed as the unit. Usually, the ATE will run a good deal slower. In this case, control of the unit clock(s) should be given to the ATE so that it controls unit operation completely, as shown in Figure 9.

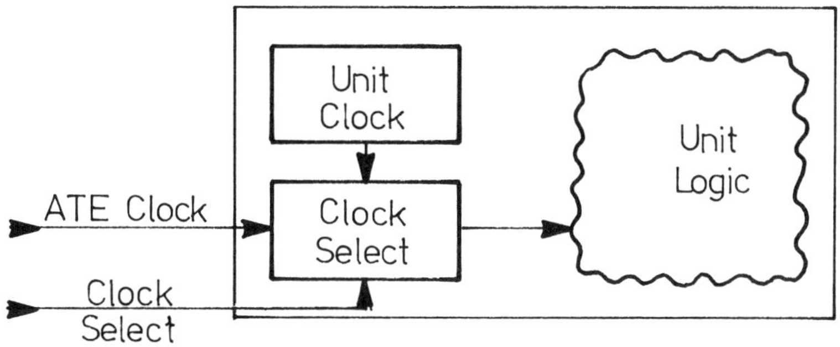

Figure 9. External Clock Control

Unit Diagnostic Software. Figure 10 shows the typical elements involved in programming ATE equipment. The engineer programming the ATE equipment utilizes the basic background material provided by the source documents in conjunction with a basic test plan to write the actual ATE program, utilizing a standard language. This program is then transcribed to a medium (such as punched tape or cards) for input to the compiling process.

In turn, the ATE equipment designer will have provided a compiler program for use with his own computer or some other standard machine. If this is a fully automated compiler the customer would read-in the punched cards, punched tape, or magnetic tape which contain all of the detailed test instructions translated into the ATE machine language. Alternatively, the ATE compiler may be designed in such a manner that the information fed through the standard interface must be manipulated first by "people" before being read-in to the computer for the automated compiling operation. Finally, the ATE system may include a generator (resident compiler located within the test station controller) which upon interpretation of the input source program, generates machine codes enabling the Test Station to execute the program immediately.

Diagnostic Software Characteristics. Assuming that the unit and ATE are compatible the problem remaining is to generate the diagnostic software to permit fault isolation to the assembly level. Carefully designed diagnostics will normally be based upon performance tests where only the unit performance is checked. In most cases, this will permit isolation to a group of cards. The diagnostic effort will be primarily directed toward breaking up the group to the single card.

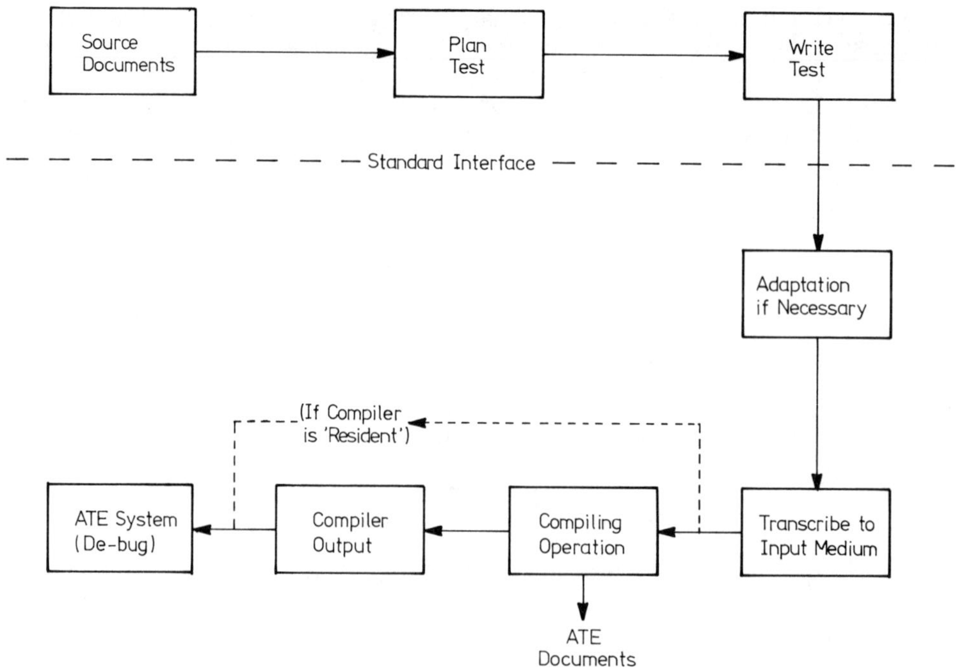

Figure 10. Typical Test Program Preparation

This work must be documented in the most convenient possible manner if the job of integration of the unit, ATE, and diagnostic information is to take place in an efficient manner. The usual procedure is to select a unit and an ATE independent language and generate diagnostic flowcharts and/or listings.

The following general set of guideline rules should be applied when designing test programs:

(a) there should be no ambiguity between the flowchart and the corresponding program listing,

(b) tests should be designed which detect faults that degrade operation beyond specified limits,

(c) fault isolation at the unit level will be accomplished by automatic means and the replacement of one or a group of cards,

(d) tests should be designed to isolate faults to the next lower replaceable item(s),

(e) tests, for reasons of program efficiency and physical access, should not be designed specifically to test wire paths or shorts between connector pins,

(f) static tests should be performed on the power input circuits and other circuits where a short or open circuit would be destructive,

(g) performance, i.e., 'go-chain', tests should be designed to exercise all functional circuitry in the Unit Under Test (UUT) within practical limits,
(h) the 'go-chain' should be designed for minimum execution time.
(j) Operator intervention should be minimized. Automatic fault isolation routines shall generally precede manual tests,
(k) where applicable, a resistor code is used to identify the UUT fully, so that the test program can be called up automatically. Consequently, UUT identification should be established before any tests are run,
(m) resistance tests should be conducted prior to application of power to ensure that there are no abnormal conditions. These tests should be performed to protect the ATE and the UUT against the consequences of failures such as a short circuit,
(n) the UUT should be initialized, i.e., all memory elements should be driven to known states before beginning performance tests,
(p) tests are performed to prove that the UUT is operating within performance specification. Each test essentially checks outputs against inputs. Generally, prime power inputs are programmed at nominal values. 'GO' results combine to form an 'end-to-end' test. 'NO GO' results branch the test program to a fault isolation 'sub-routine'.

Diagnostic Flowcharts. A Diagnostic Flowchart (DFC) is developed to show, in a graphic manner, the diagnostic program sequence and data for diagnostic testing at any functional level. The DFC diagrams should be developed at the first level by defining the significant parameters required to execute a UUT 'go-chain' or 'end-to-end' test. When appropriate, diagnostic 'sub-routine' branches should be developed from the main 'end-to-end' test locating the faults to reasonable repairable levels; for example in the case of a unit, to functional assembly.

Flowcharts are preferable to tabular listings for diagnostic test programs. The tabular listing is ideal for performance testing when there is no fault isolation. In this case, tests are executed in a fixed step-by-step sequence as they appear on the list. In diagnostic test programming, the sequence of test execution is dependent on evaluation of measured values at each step and is not a fixed step-by-step sequence. This sequence can be quite complex including loops, subroutine calls, macros, and semi-automatic sequences which require operator actions or decisions. The flowchart is the best format for describing these complex test sequences.

The DFC technique described herein is as straightforward as can be devised and has proven to be fully adequate through long usage on large projects. Typical symbology authorized for use in constructing DFCs is shown in Figure 11. A numbering system is used so that the information flow is referenced and easier to follow. Figure 12 is the 'go-chain' or the 'end-to-end' test. Figure 13 illustrates 'sub-routine' construction.

FLOWCHART SYMBOL	DEFINITION	EXAMPLE
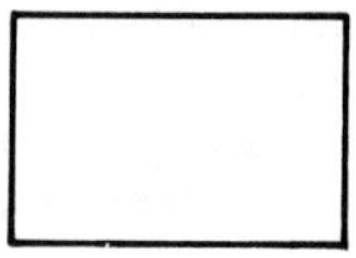 PROCESSING	Represents the Processing function, i.e. the process of executing an operation or groups of operations Key: G — 'Go' N — 'No-Go'	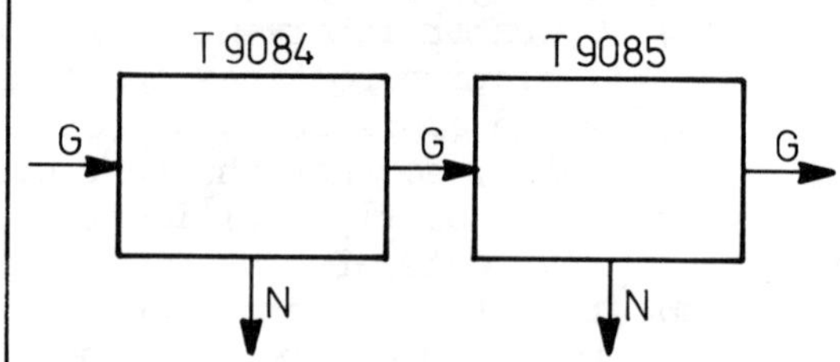
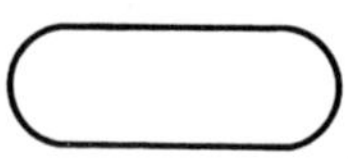 SUBROUTINE BRANCH	Represents the subroutine to which the 'go-chain' is directed in the event of a 'no-go' Key: SR — Subroutine OP — Operator Message R&R — Remove and Replace	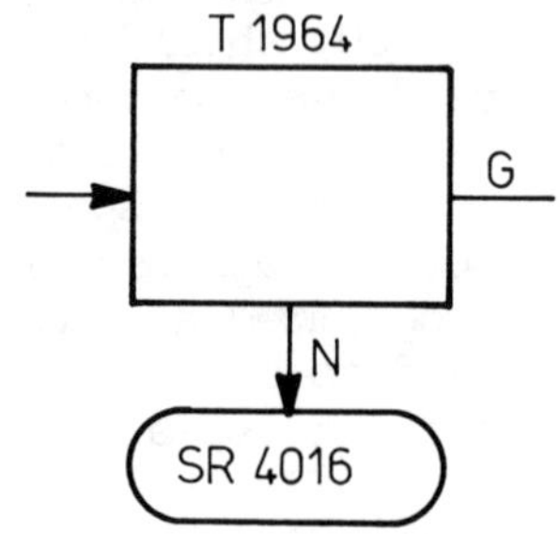 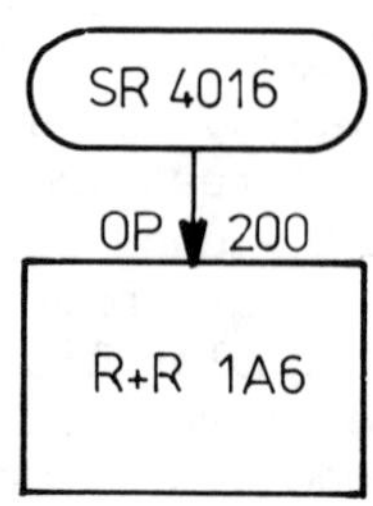
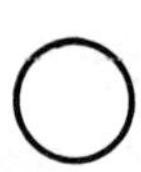 CONNECTOR	Represent a junction in the line of flow	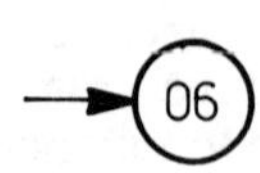
△ ▽ ▷ ◁ FLOW	Represents the direction of processing	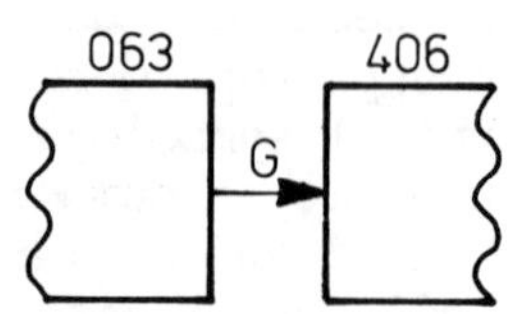

Figure 11. Flowchart Symbols

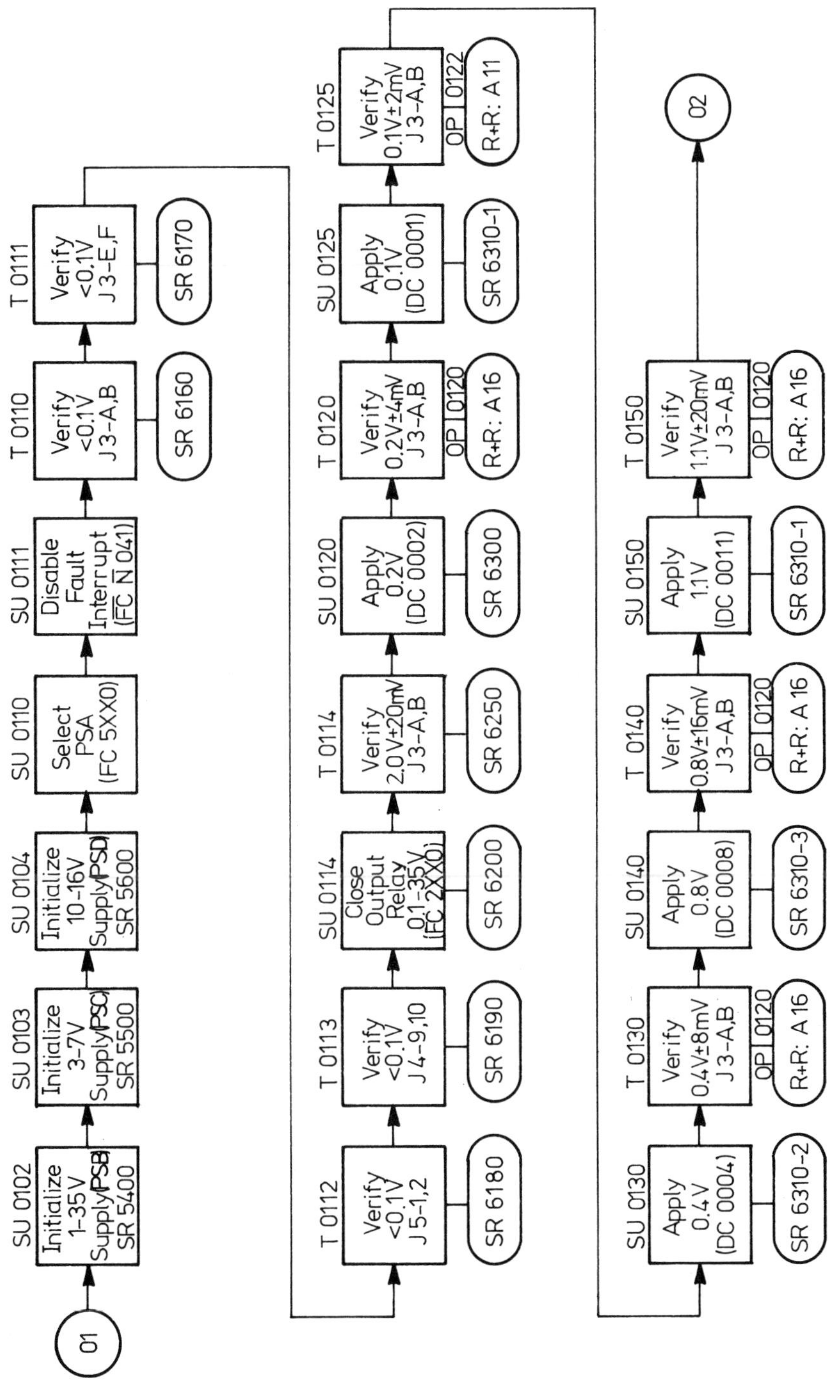

Figure 12. End-To-End Test

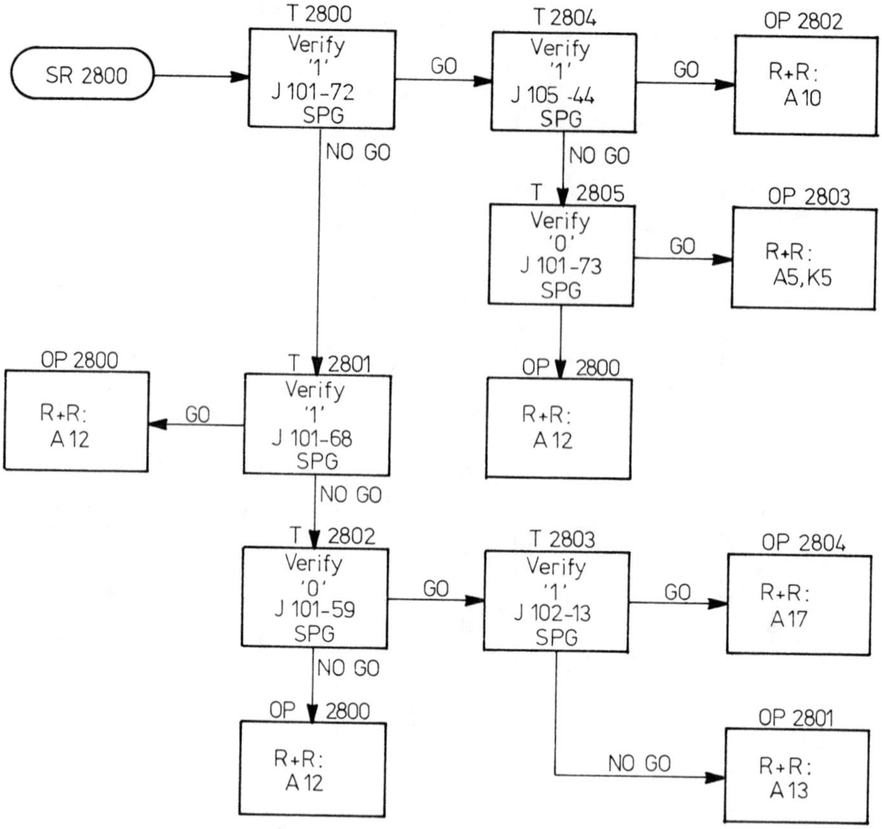

Figure 13. Sub-routine Construction

Language. Several languages have been designed specifically to serve as test specification languages for use by the unit designer. The characteristics of these languages are that they have been designed to be understood by programmers, technicians and engineers as well as a computer and their grammar is well structured to guarantee unambiguous description of a test procedure so that they can be translated by a computer into instructions that control ATE. Through standardization, they are fairly ATE independent with the use of ATE compatible compilers. Consequently, these test procedures can be employed by many users who may have different test systems.

The best known test procedure language, which is becoming a worldwide standard, is ATLAS (1,3). ATLAS stands for Abbreviated Test Language for Avionics Systems. It was originally designed for use in testing avionics systems but is now gaining acceptance as a universal test specification language. The official standard for ATLAS is maintained by Aeronautical Research Incorporated (ARINC) under document ARINC-416. An example of an ATLAS statement is:

```
APPLY, DC SIGNAL,
VOLTAGE 10V ERRLMT + -0.1V,
CNX HIJ4-1 LO J5 $
```

The English equivalent of this statement is: apply 10 ± 0.1VDC between UUT pins J4-1 and J5. In this example, APPLY defines what to do with the signal, the phrase DC SIGNAL, VOLTAGE 10V ERRLMT + -0.1V describes the type and characteristics of the applied signal, and CNX HIJ4-1 LO J5 names the UUT input connections.

Another example of an ATLAS statement is:

```
MEASURE, (FREQ ERRLMT + -0.004MHz), AC SIGNAL,
FREQ RANGE 4MHz TO 8 MHz).
VOLTAGE RANGE 1V to 1.5V
CNX HIJ1-1 LO J1-6 $
```

The words MEASURE (FREQ) describe the required action, AC SIGNAL specifies the type of signal, the characteristics for frequency and voltage provide the signal description, and the CNX field specifies the UUT output pins. This statement defines the test function as: measure the frequency, with the required measurement accuracy of ±0.004MHz, of an AC signal having expected amplitude and frequency characteristics of 1 to 1.5VAC at 4 to 8MHz present at UUT output connector pins J1-1 and J1-6.

The first example above is classified as a stimulation-type, signal-oriented statement. The statements in this class provide the stimulus function. The second example is a sensor-type, signal-oriented statement. These provide the response measurement function. A sequence of ATLAS statements must describe the test procedure in sufficient detail to allow the ATLAS compiler to generate an executable program for an automatic test system. The readability of ATLAS statements makes it possible for the same test procedure to be performed manually using bench instruments.

ATLAS uses commas to delineate fields within a statement. Usually there will be statement numbers preceding each statement and the statement can continue on separate lines at any point except where ending the line would split a word. The symbol $ is the statement terminator.

Other ATLAS statements, called procedure-oriented statements, are used to analyze and output test results and control the execution sequence of a test procedure. These include facilities for mathematical calculations, branching, looping, block structure, test operator interaction, and computer program control functions similar to other high-level languages.

Figure 14 shows an example of an adapted ATLAS listing (ATLAS-C) produced by one company and representative of typical ATLAS programs. Table 1 is an abbreviated example of a set of ATLAS statements used by another company (4). The Table is sub-divided into verbs, nouns and modifiers and is by no means complete.

```
E000000 BEGIN,ATLAS-C PROGRAM, FAUP00 ,$
C
C       FAUP00
C       SYSTEM GROUND CHECK.
C$
C       FAU PTS MODULE 00  THIS TEST CHECKS THE ADAPTOR
C       SIGNATURE AND THAT THE THREE PHASES AND
C       NEUTRAL ARE ISOLATED FROM SYSTEM GROUND.
C$
C       D.ASHWORTH      01 NOVEMBER 1977    VERSION 001
C
C
C
C$
C ADAPTOR CHECK ,$
 000100 MEASURE,(VALUE),LOGIC DATA,FORMAT BNR 6 BITS,
        CNX J8-F5 J8-F6 J8-E1 J8-E2 J8-E3 J8-E4 ,$
     05 COMPARE,'MEASUREMENT',EQ B'000101' ,$
     10 GO TO,STEP 012274 IF GO ,$
     15 PRINT,MESSAGE,
        TESTPROGRAM-ADAPTOR INCOMPATABILITY ,$
     20 INDICATE,REPLACEMENT,1 ,$
     25 GO TO,STEP 012399 ,$
     30 DISPLAY,RESULT,'MEASUREMENT' ,$
C       GROUND SYSTEM CHECK,$
 012274 MEASURE,(UUT-IMP),IMPEDANCE,
        UUT-IMP NOM 1 OHM,CNX PL6-66,$
     76 COMPARE,'MEASUREMENT',LT 10,$
     78 GO TO,STEP 012294 IF GO,$
     80 DISPLAY,MESSAGE,IN PO SIGNAL
        EARTH IN TEST CABLE 6 NOT
        CONNECTED TO SIGNAL EARTH IN
        BS CONNECTOR,$
     81 INDICATE,REPLACEMENT,2 ,$
     82 GO TO,STEP 012399,$
     94 MEASURE,(UUT-IMP),IMPEDANCE,
        UUT-IMP NOM .9 MOHM,CNX PL1-03 ,$
     96 COMPARE,'MEASUREMENT',GE 1 MOHM,$
     98 GO TO,STEP 012304 IF GO,$
 012300 DISPLAY,MESSAGE,IN PO PHASE A
        NOT ISOLATED FROM EARTHS,$
     01 INDICATE,REPLACEMENT,3 ,$
     02 GO TO,STEP 012399,$
     04 MEASURE,(UUT-IMP),IMPEDANCE,
        UUT-IMP NOM .9 MOHM,CNX PL1-06 ,$
     06 COMPARE,'MEASUREMENT',GE 1 MOHM,$
     08 GO TO,STEP 012314 IF GO,$
     10 DISPLAY,MESSAGE,IN PO PHASE B
        NOT ISOLATED FROM EARTHS,$
     11 INDICATE,REPLACEMENT,4 ,$
     12 GO TO,STEP 012399,$
     14 MEASURE,(UUT-IMP),IMPEDANCE,
        UUT-IMP NOM .09 MOHM,CNX PL1-08 ,$
     16 COMPARE,'MEASUREMENT',GE 1 MOHM,$
     18 GO TO,STEP 012324 IF GO,$
     20 DISPLAY,MESSAGE, IN PO PHASE C
        NOT ISOLATED FROM EARTHS,$
     21 INDICATE,REPLACEMENT,5 ,$
     22 GO TO,STEP 012399,$
     24 MEASURE,(UUT-IMP), IMPEDANCE,
        UUT-IMP NOM .09 MOHM,CNX PL1-02 ,$
     26 COMPARE,'MEASUREMENT',GE 1 MOHM,$
     28 GO TO,STEP 012398 IF GO,$
     30 DISPLAY,MESSAGE, IN PO NEUTRAL
        NOT ISOLATED FROM EARTHS,$
     32 INDICATE,REPLACEMENT,6 ,$
     34 GO TO,STEP 012399 ,$
     98 INDICATE,1 ,$
     99 TERMINATE,ATLAS-C PROGRAM,$
```

Figure 14. Adapted ATLAS Listing

Table 1. ATLAS Language Words

Procedure Oriented Verbs	
BEGIN	Used to start test program or BLOCK.
TERMINATE	Used with BEGIN. Last statement of test procedure.
DEFINE	Used to define sources, sensors, loads, messages, and procedures.
END	Delimiter for a PROCEDURE, BLOCK, IF/THEN/ELSE sequence, or FOR/WHILE loops.
DECLARE	Used to declare variable types.

Executable Verbs	
CALCULATE	Specifies computation.
COMPARE	Relates a labeled value to a specified limit or limits. Sets GO, NO GO flags.
DELAY	Used to delay execution of a test program for a specified interval.
DISPLAY	Used to present a temporary message or value to operator.

Signal Oriented Verbs	
CLOSE	Used to gate a source, sensor, or load function to the unit under test.
CONNECT	Used to connect pin connectors of unit under test to source, sensor, or load.
OPEN	Used to open switch at output of power or signal source or at input to measuring instrument.
SETUP	Used to set up power or signal sources or measuring instruments.
DISCONNECT	Used to open a specific connection through a switching matrix.
READ	Used to read present value of sensor function and retain that value, labeled 'measurement', until a new reading is taken.

Nouns		
AC SIGNAL	LOGIC LOAD	SHORT
AM SIGNAL	LOGIC REFERENCE	SQUARE WAVE
COMMON	MANOMETRIC	STEP SIGNAL
DC SIGNAL	PAM	SUP CAR SIGNAL

Table 1 (Cont'd)

Modifiers		
AC-COMP	IMP	PWR-SP-DENS
AC-COMP-FREQ	IND	Q
ALT-RATE	IN-PHASE	QUAD
AMPL-DENS	MOD-AMPL	REF-VOLT
AMPL-MOD	MOD-AMPL-PP	RES

CONCLUSION

The complementary and interactive nature of hardware and software in achieving fault detection and isolation at the unit level using ATE has been discussed. It is imperative that both hardware and software considerations be addressed early in the conceptual design and detailed design stages if desired system fault detection and isolation characteristics are to be achieved. There have been some recent attempts to standardize in the ATE area in both hardware and software, with the appearance of (5).

REFERENCES

1. ____________, Abbreviated Test Language for Avionics Systems (ATLAS), ARINC Spec 416-10, Aeronautical Radio Inc., Annapolis, Md., May 1975.

2. Matthews, N.O., ed., Introduction to Automated Testing (ATE), Network, Newport Pagnell, Bucks., England, 1974, p93.

3. ____________, A Guide to ATLAS for Test Specification Writers, ARINC Report 418, Aeronautical Radio Inc., Annapolis, Md., May 1969.

4. Finch, W.R. and Grady, W.B., "ATLAS: A Unit Under Test Oriented Language for Automatic Test Systems", Hewlett-Packard Journal, September, 1975, p2-13.

5. ____________, ATLAS Test Language, IEEE Std 416-1976, IEEE, New York, N.Y. 1976.

BIBLIOGRAPHY

1. ____________, Test Program Sets, General Requirements for AR-9, U.S. Navy, Naval Publications And Forms Center, 5801 Tabor Avenue, Philadelphia, Pa. 19120.

23 ASSEMBLY DESIGN

P. J. Des Marais

In the midst of ever-increasing digital logic complexity and density, particularly with the growing use of microprocessors and other LSI devices, testing has grown into a formidable problem at all levels, from component and circuit card assembly testing to unit and system testing. Fortunately, Automatic Test Equipment (ATE) and general-purpose software simulators have emerged as a cost-effective means of coping with test application and test program generation, respectively. However, the savings and benefits made possible by these tools can be fully realized only if circuit card assemblies are designed and laid out so as to provide a high degree of testability, i.e., the ability to detect the presence of and identify (locate) faulty network components to the desired level of completeness and accuracy.

The purpose of this chapter is basically two-fold: first, to review the role of software simulation in automated testing as applied to circuit card assemblies and outline capabilities and techniques implemented in currently available digital logic simulators; second, to outline a set of practical design guidelines aimed at enhancing testability of digital circuit card assemblies.

The above considerations may be viewed in a better perspective with Figure 1, illustrating the various test development activities typically encountered for a circuit card assembly. Of particular significance is the performance of a review as early as possible during the design phase to ensure that testability is designed into the circuitry such that maximum effectiveness/efficiency may be enjoyed from software simulation and ATE testing.

DIGITAL LOGIC SIMULATION

Simulation may be defined generally as "the representation of a system by mathematical relationships and the manipulation of those relationships for the purpose of envisioning the system's performance" (1). As applied to electronic systems, simulation offers a practical approach, not only to predicting correct network response but also to characterizing circuit behaviour in the presence of faults. To that end, a wide variety of general-purpose software simulators have been developed to address both analog circuitry (2) and digital logic (3), although only the latter class of simulators will be discussed in this chapter.

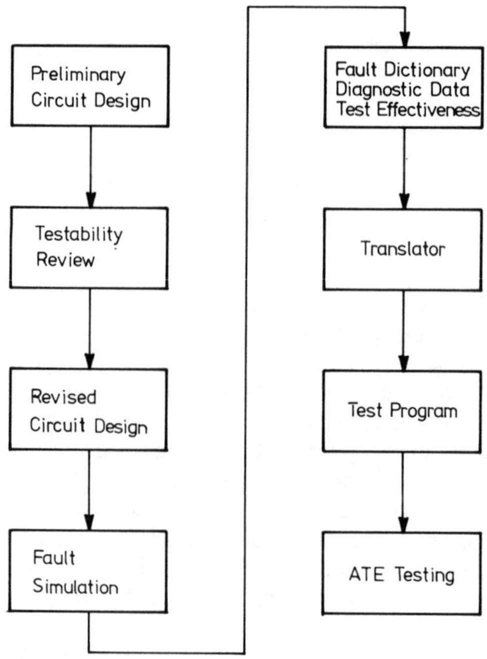

Figure 1. Test Development Process

This section proposes to describe the various software simulation models, techniques and capabilities that provide both a highly effective design aid to the development of digital electronic systems and a powerful diagnostic tool with which to assure adequate maintainability.

Upon stating the need for simulation, the basic software structure that essentially characterizes all simulators will be described. Techniques and capabilities currently available will then be outlined under four main considerations: modelling, simulation techniques, fault processing techniques, and test pattern generation. In conclusion, some of the more desirable features to look for in simulators will be identified.

The Need For Software Simulation

Simulation can be extremely beneficial to the most important phase of the system development process: design. Of particular significance is the fact that software simulation can verify logic design before any hardware is produced, which proves to be very attractive when circuit prototypes are expensive, as in the case of LSI devices, or when fast design turnaround is required. Software simulation thereby provides a cost-effective approach to evaluating a number of alternative designs before engaging in hardware development. In the process it not only eases the difficult task of anticipating the overall performance of increasingly complex circuits and systems, but also provides a strong basis for evaluating the effect of design changes and making design compromises.

Software simulation also plays a very important role in the diagnosis of digital networks. Present-day circuit complexity has brought about the development of two major automated test methods with which to replace the technician and his oscilloscope. On the one hand lies the computer guided probe which, in minutes, can isolate the faulty component with good accuracy. Essentially an operator successively probes IC nodal states, under computer directives issued on the basis of comparing observed IC pin signal values to predetermined correct responses. On the other hand, we find the significantly faster computer directory look-up (dictionary) approach which, on the basis of considerably more precompiled diagnostic information, isolates the fault without any manual assistance, but with generally poorer resolution. With increasing circuit complexity, there appears to be a trend to combine the two methods, namely to identify the several possible faulty ICs quickly using the dictionary approach, and then to zero in on the culprit under guided probing, possibly in the form of an automatic clip drive, in order to fully automate the process. Yet another approach would involve the use of online simulation whereby, upon observation of the first faulty response, only the corresponding set of faults (indistinguishable at that stage) would be further simulated and isolated to the desired level interactively, with subsequently observed responses. This last method would prove more appropriate when probing cannot be performed. It would improve on the resolution achieved by the dictionary approach while keeping the amount of prior information down to a more reasonable level. In any case, with the large variety of ATE hardware spanning a wide range of test capabilities, the problem of achieving efficient testing appears to be well in hand, as related to both testing time and data volume.

However, the task of generating the diagnostic information needed to program ATE appears to be the more challenging problem now facing the Test Engineer. While manual methods are no longer feasible, the hardware generation approach of physically inserting faults on a known-good-board and recording their effect on circuit response tends to become impractical for large, complex boards. The process of breadboarding a prototype, together with the need for manually testing and verifying that the circuit is indeed good, can prove to be fairly expensive, time-consuming and prone to error. Software simulation has thus emerged as the superior alternative to generating diagnostic information, in that it can be extremely fast and quite accurate, depending upon the complexity of the simulation program. Its major disadvantage, thus far, has been the high cost of processing on large computers, but a growing number of cheaper minicomputer-based systems are being introduced, making use of minicomputers already present in logic testers.

Basic Simulator Structure

Although a fairly wide variety of digital logic simulators are currently available (3), they all display the software structure shown in Figure 2. Three basic program modules are generally encountered.

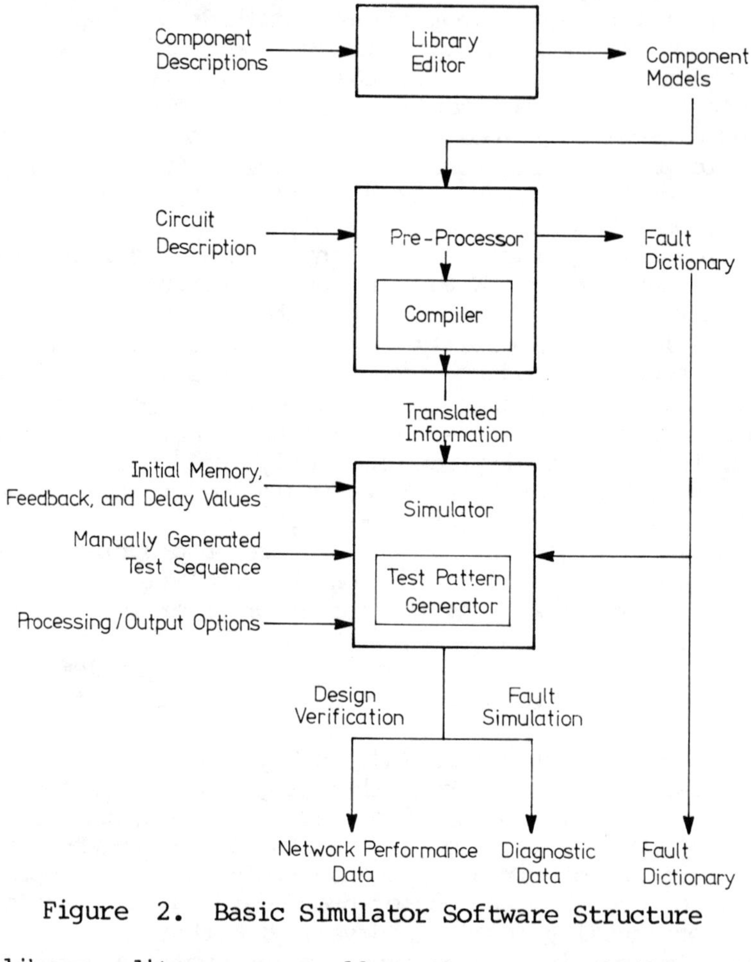

Figure 2. Basic Simulator Software Structure

The library editor program allows the user to build a library file of logic component simulation models, each of which needs to be described in detail only once. From an input description, the library program will usually generate structural information as well as a list of standard faults to be simulated for that component, as illustrated in Figure 3. The major advantage of a library facility lies in the fact that, while a component may be repeatedly encountered in various networks to be simulated, the user is then required only to refer to it by its given name, the key to retrieving the appropriate information from the library file.

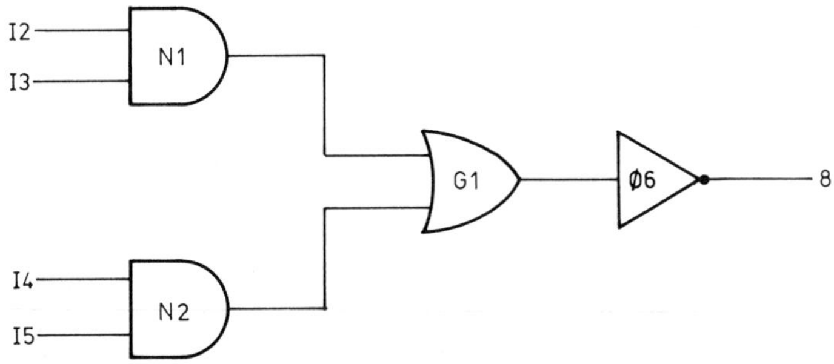

(a)

LOGIC LIBRARY VERSION 3.3.1 CONTRALL FAMILY 2/77 15.10.06. PAGE 1

BLOCK TYPE NAME = H51 , SECTION NAME = A , FAMILY NAME = TTL

THIS BLOCK CONTAINS 4 NODES, 7 INPUTS, AND 0 MEMORY ELEMENTS.

NODE TYPE	NODE NUMBER	OPERATION	INPUT TYPE	INPUT NUMBER
N	1	*	I	2
			I	3
N	2	*	I	4
			I	5
G	1	+	N	1
			N	2
0	6	-	G	1

FAULT NUMBER	FAULT	FAULT TURNED
1	I 2 TO N 1 STUCK AT 0	ON
2	I 2 TO N 1 STUCK AT 1	ON
3	I 3 TO N 1 STUCK AT 0	ON
4	I 3 TO N 1 STUCK AT 1	ON
5	I 2 TO N 1 SHORTED	ON
6	I 3 TO N 1 SHORTED	ON
7	N 1 STUCK AT 0	ON
8	N 1 STUCK AT 1	ON
9	I 4 TO N 2 STUCK AT 0	ON
10	I 4 TO N 2 STUCK AT 1	ON
11	I 5 TO N 2 STUCK AT 0	ON
12	I 5 TO N 2 STUCK AT 1	ON
13	I 4 TO N 2 SHORTED	ON
14	I 5 TO N 2 SHORTED	ON
15	N 2 STUCK AT 0	ON
16	N 2 STUCK AT 1	ON
17	N 1 TO G 1 STUCK AT 0	ON
18	N 1 TO G 1 STUCK AT 1	ON
19	N 2 TO G 1 STUCK AT 0	ON
20	N 2 TO G 1 STUCK AT 1	ON
21	N 1 TO G 1 SHORTED	ON
22	N 2 TO G 1 SHORTED	ON
23	G 1 STUCK AT 0	ON
24	G 1 STUCK AT 1	ON
25	G 1 TO 0 6 STUCK AT 0	ON
26	G 1 TO 0 6 STUCK AT 1	ON
27	G 1 TO 0 6 SHORTED	ON
28	0 6 STUCK AT 0	ON
29	0 6 STUCK AT 1	ON

(b)

Figure 3. (a) Logic Component (b) AFS Library Model

The pre-processor module basically translates a user-generated network description into an internal format more suitable for simulation. The user is required to specify interconnection of components in the network and its external inputs and outputs while the pre-processor refers to the library file for detailed information on each component. The main purpose of this latter program is to provide the user with an input description format which promotes simplicity, flexibility and ease of use.

The program also includes syntax-checking capabilities to ensure that the user has described his network properly, issues appropriate diagnostic messages in cases of error, and generally outputs (upon request) various cross-reference tables with which the user may double-check the accuracy of his description. Finally, for the purpose of fault simulation, the pre- processor generates a dictionary file of all faults to be analyzed by the simulator program, as illustrated in Figure 4, with a printout from the Automated Fault Simulation (AFS) program (4). Note that, in the case of a compiler-driven simulator, the pre-processor includes an additional module, usually referred to as the compiler, which further translates pre-processed information into simulation code in the computer's assembly language.

The third and last module consists of the simulator program itself which, for most simulation systems, can perform both design verification and fault simulation. The user is generally required to specify options pertaining to the simulation mode of processing and type of output information desired and, with some systems, the logic states to be given initially to memory and delay elements as well as feedback signals. In addition, the user must specify a sequence of input operands whose effect on circuit response the program is to simulate.

For the purpose of design verification, the simulator generates network performance data ranging, in format, from coded output vectors to timing charts describing network response, as illustrated in Figure 5 with a printout from the ASSIST program (5).

In the case of fault simulation, the program simulates not only the faultless behaviour of the network but also the effect of each fault in the fault dictionary on its response. The resulting diagnostic data generated by the program identifies the particular faults detected by each test input operand, as illustrated in Figure 6 with a printout from the AFS program (4). Note that different faults, producing identically erroneous output responses, are grouped into indistinguishability classes, leading to ambiguity for the purpose of fault isolation.

Finally, note that some simulators include test pattern generating routines that select input operands automatically, according to various strategies, for the purpose of detecting and isolating faults. These aim to relieve the user of the tedious and complex task of designing test sequences.

Modelling Techniques

Modelling of logic circuit behaviour covers four basic aspects: faults, components, timing, and signal levels.

FAULT DICTIONARY FOR MCM144422C (CODE=00160) 76/09/03. 09.01.38. PAGE 1

MACH.	FLT	---------ORIGIN---------			-----DESTINATION-------			FLT	---------ORIGIN---------			-----DESTINATION-------		
NO.	TYP	MODULE	BLOCK	PIN	MODULE	BLOCK	NODE	TYP	MODULE	BLOCK	PIN	MODULE	BLOCK	NODE
2	INT	MCM144422C	U03C	14										
3	SA0	MCM144422C	U03C	12				SA1	PRIM. INP.	C15		MCM144422C	U03C	1
	INT	MCM144422C	U03C	11				INT	MCM144422C	U03C	13			
4	SA1	MCM144422C	U03C	12				SA0	PRIM. INP.	C15		MCM144422C	U03C	1
	INT	MCM144422C	U03C	10				INT	MCM144422C	U03C	12			
5	SA1	PRIM. INP.	B33		MCM144422C	U16A	1	SHT	PRIM. INP.	C33		MCM144422C	U16A	1
6	SA1	PRIM. INP.	C33		MCM144422C	U16A	1	SHT	PRIM. INP.	B33		MCM144422C	U16A	1
7	INT	MCM144422C	U16A	11										
8	SA1	PRIM. INP.	A49		MCM144422C	U18D	1	SHT	PRIM. INP.	A50		MCM144422C	U18D	1
9	SA1	PRIM. INP.	A50		MCM144422C	U18D	1	SHT	PRIM. INP.	A49		MCM144422C	U18D	1
10	INT	MCM144422C	U18D	11										
11	SA0	MCM144422C	U18D	11				INT	MCM144422C	U18D	8			
	INT	MCM144422C	U18D	10										
12	SA1	MCM144422C	U18D	11				SA0	PRIM. INP.	A49		MCM144422C	U18D	1
	SA0	PRIM. INP.	A50		MCM144422C	U18D	1	INT	MCM144422C	U18D	7			
	INT	MCM144422C	U18D	9										
13	INT	MCM144422C	U20A	11										
14	INT	MCM144422C	U21A	8										
15	SA0	MCM144422C	U21A	2				SA1	PRIM. INP.	C23		MCM144422C	U21A	1
	INT	MCM144422C	U21A	5				INT	MCM144422C	U21A	7			
16	SA1	MCM144422C	U21A	2				SA0	PRIM. INP.	C23		MCM144422C	U21A	1
	INT	MCM144422C	U21A	4				INT	MCM144422C	U21A	6			
17	INT	MCM144422C	U21B	8										
18	SA0	MCM144422C	U21B	4				SA1	PRIM. INP.	C19		MCM144422C	U21B	1
	INT	MCM144422C	U21B	5				INT	MCM144422C	U21B	7			
19	SA1	MCM144422C	U21B	4				SA0	PRIM. INP.	C19		MCM144422C	U21B	1
	INT	MCM144422C	U21B	4				INT	MCM144422C	U21B	6			
20	INT	MCM144422C	U21C	8										
21	INT	MCM144422C	U21D	8										
22	SA0	MCM144422C	U21D	8				SA1	PRIM. INP.	A20		MCM144422C	U21D	1
	INT	MCM144422C	U21D	5				INT	MCM144422C	U21D	7			

Figure 4. Partial AFS Fault Dictionary Listing

```
**********RUN STATISTICS**********
   PEAK NO. OF EVENTS           32
   TOTAL NO. OF EVENTS         393
   NO. OF EVENTS REMOVED         6
   EVENT COUNTER AT END          8
**********************************
```

```
* 02/12/75    *
*             *
*PAGE 00001   *
*             *
*TIME       0*
*  TO      78*
*             *          11111222223333344444555556666677777
*GROUP 01     *     0246802468024680246802468024680246802468

   REG XINPUT        0000000000000000000001111111111111111111
   REG XINPUT        0000000000000000000001111111111111111111
   REG XINPUT        0000000000000000000001111111111111111111
   REG XINPUT        0000000000000000000001111111111111111111

   REG YINPUT        0000000000000000000000000000000000000000
   REG YINPUT        0000000000000000000000000000000000000000
   REG YINPUT        0000000000000000000000000000000000000000
   REG YINPUT        0000000000000000000001111111111111111111


   REG OUTPUT        0000000000000000000000000000000000000000
   REG OUTPUT        0000000000000000000000000000000000000000
   REG OUTPUT        0000000000000000000000000000000000000000
   REG OUTPUT        0000000000000000000000000000000000000000
```

Figure 5. ASSIST Output Listing

Fault Models All simulators basically use the classical "stuck-at" logic fault model which assumes that faults occur as component inputs or outputs being permanently stuck at one of the two logic states. Some simulators also model the shorted fault whereby a component output responds to only one of its input signals. Furthermore, all simulators analyze the effect of faults on the basis that, at any time, only a single fault may occur in the logic circuit. However, most simulators allow the user to manually define specific multiple faults whose combined effect on circuit response is to be simulated. While still very little is known about LSI failure modes, the stuck-at model appears to cover most circuit failures fairly well, while keeping the magnitude of the fault analysis problem down.

Component Modelling Component modelling ranges from gate to element to functional models. Gate level modelling offers the best structural accuracy, as components are described internally in terms of simple

```
BEGIN ANALYZER RUN
DATE - 76/09/03.
TIME - 16.51.11.
NETWORK - MCM144422C  (CODE=0016014064)       2-VALUE SIMULATOR
DETECT MODE

INPUT EQUIVALENCE CLASS  -  FIRST MACHINE =   1 ,   NO OF MACHINES = 995
 INITIAL RESETS
             FEEDBACK STATES - 4
             MEMORY STATES - 4646464646464646464646
  INPUT OPERAND  1 - 3625214373

             OUTPUT EQUIVALENCE CLASS  -  FIRST MACHINE =  1 ,  NO OF MACHINES = 787
               OUTPUT STATES - 04574527170332514420060

             OUTPUT EQUIVALENCE CLASS  -  FIRST MACHINE =  2 ,  NO OF MACHINES =  2
               OUTPUT STATES - 04503652570332514420060

             OUTPUT EQUIVALENCE CLASS  -  FIRST MACHINE =  7 ,  NO OF MACHINES =  3
               OUTPUT STATES - 04574527174126515420060

             OUTPUT EQUIVALENCE CLASS  -  FIRST MACHINE = 10 ,  NO OF MACHINES =  2
               OUTPUT STATES - 04574527170332514421460

             OUTPUT EQUIVALENCE CLASS  -  FIRST MACHINE = 13 ,  NO OF MACHINES = 12
               OUTPUT STATES - 03574527170332514420060

             OUTPUT EQUIVALENCE CLASS  -  FIRST MACHINE = 14 ,  NO OF MACHINES =  2
               OUTPUT STATES - 04574527170332503220060

             OUTPUT EQUIVALENCE CLASS  -  FIRST MACHINE = 20 ,  NO OF MACHINES =  3
               OUTPUT STATES - 04574527173631254420060

             OUTPUT EQUIVALENCE CLASS  -  FIRST MACHINE = 21 ,  NO OF MACHINES =  2
               OUTPUT STATES - 04575775170332514420060

             OUTPUT EQUIVALENCE CLASS  -  FIRST MACHINE = 24 ,  NO OF MACHINES =  2
               OUTPUT STATES - 04574527170332534420060

             OUTPUT EQUIVALENCE CLASS  -  FIRST MACHINE = 27 ,  NO OF MACHINES =  2
               OUTPUT STATES - 04574127170332514420060

             OUTPUT EQUIVALENCE CLASS  -  FIRST MACHINE = 30 ,  NO OF MACHINES =  2
               OUTPUT STATES - 04574527170332514420070

             OUTPUT EQUIVALENCE CLASS  -  FIRST MACHINE = 33 ,  NO OF MACHINES =  2
               OUTPUT STATES - 04574527170332514420064

             OUTPUT EQUIVALENCE CLASS  -  FIRST MACHINE = 36 ,  NO OF MACHINES =  2
               OUTPUT STATES - 04574527170332514520060

             OUTPUT EQUIVALENCE CLASS  -  FIRST MACHINE = 39 ,  NO OF MACHINES =  3
               OUTPUT STATES - 04574527170332514424060

             OUTPUT EQUIVALENCE CLASS  -  FIRST MACHINE = 40 ,  NO OF MACHINES =  3
               OUTPUT STATES - 04574527170332514422060

             OUTPUT EQUIVALENCE CLASS  -  FIRST MACHINE = 41 ,  NO OF MACHINES =  3
               OUTPUT STATES - 04574527170332514430060

             OUTPUT EQUIVALENCE CLASS  -  FIRST MACHINE = 42 ,  NO OF MACHINES =  2
               OUTPUT STATES - 04574527170332514437300
```

Figure 6. AFS Output Listing

single-output gates, but the approach becomes impractical in both simulator time and memory requirements for large boards involving complex MSI and LSI devices. Element modelling of SSI devices such as flip-flops, can cope better with these problems, but functional modelling with its more generalized method of expressing component output response as a function of input signals, provides the most practical approach to modelling complex components such as ROMs, RAMs, and microprocessors. Nevertheless, this higher level approach to describing component logic behaviour generally results in reduced internal fault modelling accuracy.

Timing Models As regards timing, modelling ranges from zero and unit delay models to more complex models whereby components may be assigned different nominal delay values or even min-max delay ranges.

The zero-delay model assumes that all components are delayless and hence fails to model any timing characteristics of a logic circuit. On the other hand, the unit-delay model assumes that all gates of a circuit each contribute a single-unit propagation delay. Although this latter model recognizes that longer circuit paths lead to longer propagation delays, its timing accuracy remains fairly limited.

The nominal delay model acknowledges the fact that propagation delays vary widely from one component to another, by allowing for the assignment of different delay values to different components. However, even better timing accuracy is provided by those models which allow for the specification of a minimum and maximum delay time within which a component may change from one logic state to the other.

Very precise timing models are essential to design verification applications where circuit hazards and races must be analyzed accurately. However, for fault simulation, the need for precision greater than that of simple assignable delay models becomes questionable, as processing time rapidly increases with timing mode complexity.

Signal Level Models Similar to timing models, signal levels are modelled to various degrees of accuracy ranging from the two-, to three-, to multiple-value models. The two-value model suffers some serious limitations from the fact that all signals must be either at the HIGH or LOW level. The three-value model, using the additional unknown (or indeterminate) X level, was introduced to cope with the problems of race and hazard detection; generally it proves adequate for fault simulation and is essential for specifying the initial state of feedback signals and memory elements that are only indirectly initialized. However, for design verification, the use of additional level values, to model, for example, rising and falling signal edges, becomes necessary as related to the need for greater timing accuracy.

Simulation Techniques

Two main simulation techniques are currently implemented: the compiler-driven approach (6, 7) developed in earlier simulators, and the table-driven approach (8) inherently providing greater timing accuracy.

Compiler-Driven Space Levelled Technique The compiler-driven technique is characterized by the translation of the topological logic description of the circuit into machine code, representing a levelization of the logic, generally assumed to be delayless, although not necessarily so. The simulation process consists of first assigning initial values to all memory, delay, and feedback signals (which must be carefully identified) and then sequentially determining new logic values at each successive level and repeating, for each new input pattern, as many of these simulation passes as may be required for all signals to stabilize. A certain limit is set on the number of simulation passes that may be processed which, when reached, results in termination of the analysis for the input operand and subsequent declaration of an oscillating condition. Upon discarding that particular input operand, the program then proceeds to simulate the next operand. The technique suffers some serious limitations, such as failing to recognize hazards or properly analyzing races when, in some cases, unrealistic oscillating conditions may be declared, and has been largely replaced by Table Driven Simulation.

Table-Driven Time-Based Technique More recent simulators use the table-driven approach which operates, not on compiled code, but directly upon the topology of the circuit, in the form of a set of tables storing information such as component interconnections, element functions, fan-in, fan-out, and signal values. This latter technique is basically event-directed, in that gates or elements are simulated only when at least one of their inputs change. It is also time-based in that those events are scheduled using some time-flow mechanism, usually in the form of a time queue, which reflect the precise delays specified for each component. This second technique can handle the problems encountered with the first one much more effectively.

Fault Processing Techniques

Here again three main techniques are implemented: parallel fault simulation (9) and deductive fault simulation (10) and concurrent simulation (15). The first approach calls for the simulation of not only the good circuit but also the effect of each fault on circuit response. Recognizing that signal values can be packed at the bit level of the computer memory, the method provides for the simulation of as many faults in parallel as there are bits in the computer word. Hence for each new input operand, simulation of m faults on an n-bit word computer would require m/n sets of passes.

By contrast, the deductive or fault list propagation approach requires that only the good circuit be simulated. By associating with each circuit line a list of faults which can produce a wrong signal, only a single pass is needed whereby lists of faults are propagated down the various circuit paths.

Concurrent simulation has many of the advantages of deductive simulation but does not require extensive list computations and hence may be faster.

Recent studies have demonstrated the second approach to be more cost-effective in processing a large number of faults, although requiring much more memory as fault lists tend to grow very large. On the other hand, the parallel approach has been shown to perform faster on smaller "highly sequential" circuits.

Test Pattern Generation

This appears to be the area in greatest need of improvement as regards to fault diagnosis applications. Most simulators generate test patterns automatically using purely heuristic strategies that fail to consider, in any way, the structure of the specific circuit analyzed. Test operands are evaluated and selected on the basis of how well they perform under simulation, a process that can become very demanding of computer time. Hence, this trial-and-error approach generally proves to be fairly expensive because of extensive processing time spent on evaluating generated operands. Furthermore, resulting test sequences are usually not as effective as manually generated ones. There is fortunately a trend towards developing more structure oriented generation techniques, such as D-LASAR (11) with its path sensitizing method of tracing output lines back to input lines, and TEGAS (12) with its extension of the D-algorithm. Further significant developments are, however, needed to handle complex sequential circuits effectively.

Desirable Features

In conclusion, let us single out some of the more desirable features that should be included in a software simulation system:

(a) The overall system should be modular in design for ease of expansion and modification,

(b) Fault modelling should allow for the specification of multiple faults and the user-definition of more complex logic faults, e.g. a NAND-gate behaving as an AND-gate,

(c) For more cost effective use of the system, the input coding format should be as simple as possible, incorporate extensive syntax-checking, and offer a macro-capability whereby large components can be readily described in terms of simpler ones,

(d) For more accurate results, especially for races, hazards and oscillating conditions, component descriptions should include timing parameters and three-value, time-based simulation should be provided,

(e) The simulation system should be capable of addressing both design verification and fault diagnosis, providing for those two different applications full input compatibility with regard to component modelling and network descriptions, thereby improving the cost-effectiveness of the system. Ideally the system should offer two basic processing modes, one which provides complex modelling capabilities in order to achieve accurate design verification, and a second mode where simpler models, while assuring a minimum level of accuracy, favor better processing efficiency,

(f) For better processing efficiency and reduced memory requirements, functional modelling should be available for complex MSI and SSI devices, selective trace should be implemented (i.e., simulating only those parts of the fault-free circuit that are changing under the current stimulus pattern), and an effective fault-collapsing algorithm provided in order to group, as completely as possible, indistinguishable faults under one representative,

(g) To maximize the cost-effectiveness of fault simulation, effective automatic test pattern generation should be implemented.

DESIGN GUIDELINES FOR TESTABILITY

The scope of testability as discussed here is limited essentially to functional (static) digital logic testing of circuit boards. The following set of practical guidelines is aimed at maximizing test efficiency and the effectiveness of software fault simulation for that generation. Testability is considered under three major aspects: initializability, or the ability to drive all memory elements and feedback signals to known initial values; observability, or the ability to monitor, directly or indirectly, the specific response of every circuit component; and controllability, or the ability to transmit to each component the set of input signals needed to test them fully.

Initializability

Generating diagnostics essentially consists of predicting (in particular by using a fault simulator) both the correct and faulty circuit responses to pre-determined input stimuli and, on that basis, of locating the faulty component. However, digital circuits are generally sequential in nature, i.e. their response depends not only on the current input stimulus but also on internal memory and/or feedback states. In particular, a circuit's response to the first test input of a diagnostic sequence will depend on memory/feedback states prior to the application of the test sequence. Hence, it becomes imperative that, upon powering up a circuit board, a capability be provided to initialize each and every memory and feedback element to a known state, even in the presence of a fault.

All memory elements, from simple latches and flip-flops to counters and shift-registers, should be directly initialized from either edge connector pins or a power-up circuit. In the case of internal components, more direct initialization should be inserted, as illustrated in Figure 7 for a J-K flip-flop, by using an additional gate (b,c) or preferably providing an open-collector gate whose output can be wire-ORed with an external pin (d). With 2-value simulators, this becomes essential if a fault in the initializing logic is not to be incorrectly diagnosed as being located in the memory element which consequently failed to be properly initialized. With 3-value simulators, the problem of indirect initialization is handled by assigning the X (unknown) value to corresponding memory elements, but complete diagnosis of the circuit will not be achieved until all X's have been eliminated. Hence, direct initialization contributes to more efficient testing, particularly for complex sequential circuits.

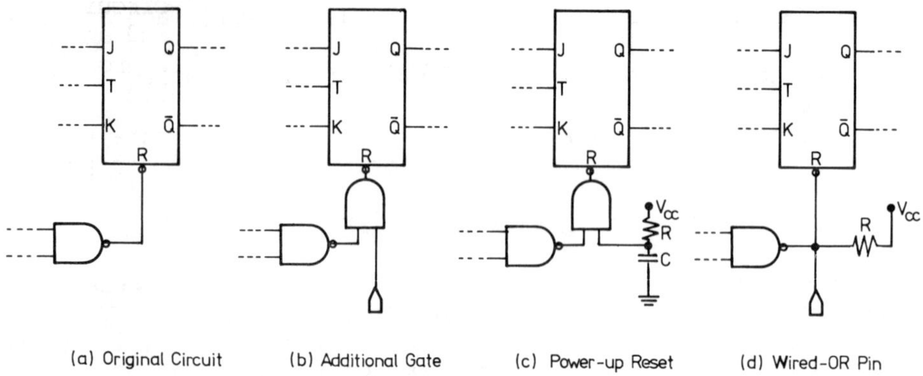

Figure 7. Inserting More Direct Initialization Control

Observability

The key to maximizing the effectiveness of using ATE to diagnose circuit assemblies is to minimize the human intervention needed to identify faulty components. In that respect proper monitoring of a circuit is essential to attain the desired degree of diagnostic resolution, and that is generally achieved by providing test points to a set of judiciously chosen internal circuit board nodes. The selection is based upon the structure of the circuit, but as a rule of thumb, test points are usually associated with large fan-in points, outputs of memory elements, and internal points in feedback loops. Unfortunately, in many cases, not enough pins are available to accommodate the required monitoring fully. An additional test connector should then be provided, or otherwise a funnelling of test points, achieved by using multiplexers or preferably parallel-in, serial-out shift- registers (to ensure that pulses are observed), as illustrated in Figure 8.

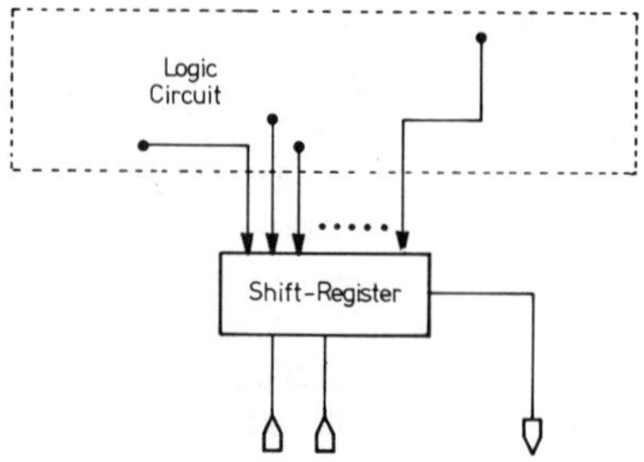

Figure 8. Achieving Desired Observability with Fewer Test Points

In addition, logic redundancy (where a line in a circuit is said to be redundant if the circuit response is independent of the logic signal at that line for all input combinations) should be avoided as it tends to mask out other detectable faults. However, redundancy is used sometimes to eliminate race problems. In those cases a test point should be provided to the redundant connection.

Controllability

Good controllability calls for designing into a circuit the capability to readily and properly exercise every component on the board. Such a capability promotes efficient use of ATE and, in particular, fault simulation, where the name of the game is to detect and isolate as many faults, as early as possible, in the test sequence in order to minimize simulation time.

It is imperative that full clocking control of the board be provided at its edge connector if ATE is to test the circuit at a rate compatible with its own. Free-running internal clock sources, such as oscillators, must therefore be isolated either via a jumper at the edge connector or by multiplexing external clock control, as illustrated in Figure 9.

Circuits employing one-shots and asynchronous sequential logic are often characterized by a relatively complex dynamic behaviour due to critical timing. On the one hand, accurate fault simulation of such circuits becomes difficult to achieve while, on the other, some faults may be impossible to detect and/or isolate through static testing. In particular, some output changes may go undetected if the latency period of the tester (i.e. period of time between stimulus application and response sampling) is long in relation to the operating speed of the circuit. Hence, synchronous logic should be preferred as it leads to more precise fault simulation and more adequate testing.

Feedback paths should be broken for more efficient testing of complex sequential circuits. If insufficient pins are available to jumper the loops at the edge connector, as shown in Figure 10 (a), a 3-state buffer should be inserted as shown in Figure 10 (b), where test points at the output pins become test inputs controlling the feedback signals when the buffer is disabled. If, however, this second alternative cannot be implemented, test inputs should at least be multiplexed into the loop, as illustrated in Figure 10 (c).

Long counter chains, requiring thousands of clock cycles for full testing, should be provided with the capability to exercise every counter chip simultaneously, in order to achieve more efficient testing and obtain more effective fault simulation with a reasonable number of test operands. A typical 12-stage counter (implemented with three 9316's) is depicted in Figure 11(a), requiring 2^{12} clock pulses to be exercised fully. With the count/load control available on the 9316 counter, there are several methods by which the upper part of a counter chain may be simulated more rapidly.

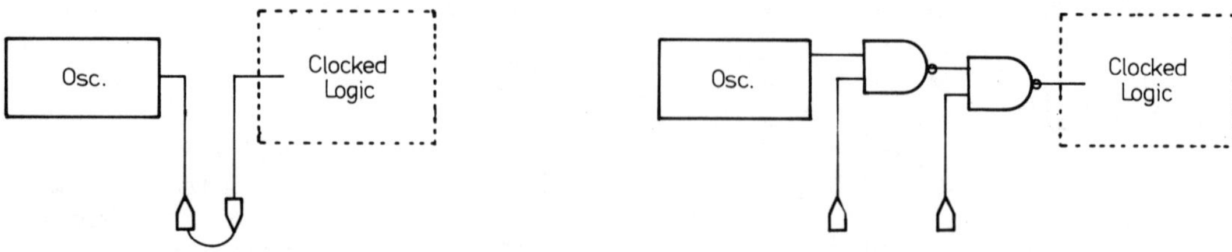

(a) Broken at Edge Connector (b) Multiplexed Control

Figure 9. Isolating Internally Generated Clocks

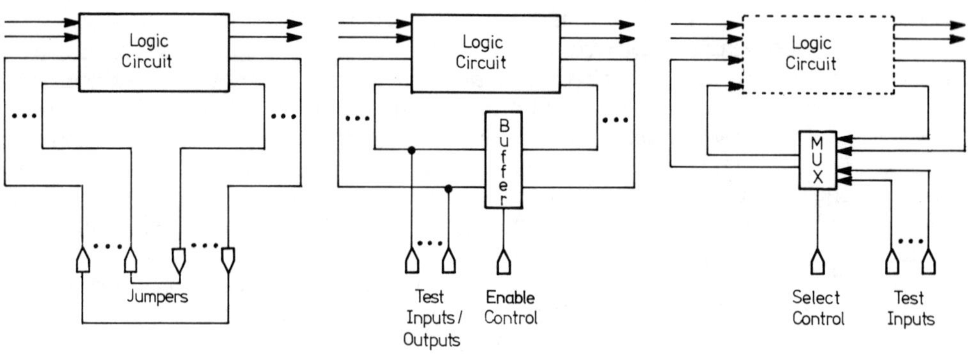

(a) At the Edge Connector (b) Using 3-State Buffer (c) Multiplexing Test Inputs

Figure 10. Controlling Feedback Loops

As illustrated in Figure 11 (b), the chaining could be broken at the edge connector via jumpers. Simultaneous operation of all three counters may also be gated into the chain using only one input pin, as shown in Figure 11 (c). Alternatively, efficient testing may be achieved by providing independent counting control to each counter, as shown in Figure 11 (d), whereby the chain can be fully exercised with 48 clock pulses. Finally, Figure 11 (e) illustrates the capability to load the chain with a terminal count and thereby exercise most of the counting hardware on the next clock pulse. However, this last scheme should be considered only as a minimal approach, since 2" clock pulses would still be required to test the highest stage of the chain fully.

Although the above examples considered 4-stage counter groups, note that as a rule of thumb, no more than 8 states (requiring 256 clock pulses) should go unbroken. Similarly, long chains of serially connected shift-registers should be monitored at intermediate outputs in order to achieve more efficient diagnosis of the low end of the register chain.

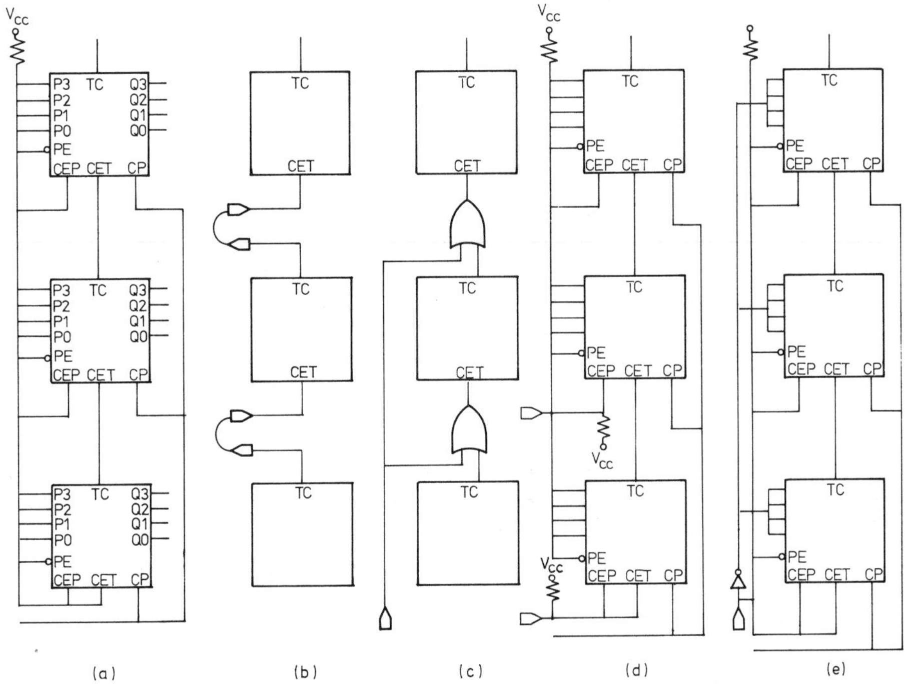

Figure 11. Breaking Long Counter Chains

Analog circuitry should be completely isolated from digital circuitry on a board, if logic fault simulation is to be applied effectively to the digital part of the circuit. In addition, such isolation will simplify the actual testing of the board.

Complex components, such as ROMs, RAMs, and microprocessors, all of which require extensive sets of input patterns to be fully tested, should be isolated to the extent that their input and output pins can be accessed, while neighbouring logic can be tested independently in order to ensure efficient diagnosis. In this respect, tri-state devices can be readily isolated. Test points used to monitor their outputs, while the devices are activated, can serve as test inputs to the logic that they drive, whenever they are disabled into their high impedance state.

Physical Considerations

Board layout and other physical assembly aspects should be considered while attempting to promote good maintainability. Factors such as package design and physical partitioning of logic functions between boards can affect test equipment and labor costs significantly.

A design should be partitioned functionally such that the various basic functions are each realized on a different board. Large complex boards should be partitioned physically into independent subsections whenever possible, although breaking up a logic function over several boards should be avoided. Furthermore, the number of levels of combinational logic on a board should be reduced as much as possible, since, as a general rule, the lower the number of logic levels, the lower the number of tests and internal test points needed to achieve good diagnostic resolution.

On the other hand, package standardization should be strived for on items such as card edge connectors and power supply pin assignment in order to minimize the number of adapters required and reduce setup time. The number of different logic types appearing at the edge connector of a card should also be minimized as mixed logic adds to tester cost in the form of logic level translators.

CONCLUSION

In conclusion, it should be emphasized that a greater awareness of the various techniques proposed in the literature (13) for the design of self-diagnostic and more testable structures is warranted. The more classical methods of analysis, which attempt to produce optimal solutions for combinational networks and small sequential circuits, fail to address the increasingly complex circuits currently produced. By contrast, recent design techniques indicate a significant trend toward a structured approach to design, at the expense of additional hardware. Although their general applicability appears to be somewhat limited at present, they may very well provide in the near future, within the context of lifecycle costs, a cost-effective approach to satisfying increasingly demanding maintainability requirements as circuit complexity and testing costs keep rising and hardware costs continue to drop. This fact is indeed confirmed by H. Falk (14) who reports that "in many computer systems, there is a designed increase of from 15 to 25 percent in system hardware to provide improved testability".

REFERENCES

1. Appel, A. "Digital Simulation", Machine Design, 10 July 1976, p74-77.

2. Bowers, J.C. et al., "A Survey of Computer-Aided Design and Analysis Programs", Report no. AFPAL-TR-76-33, Air Force Aero Propulsion Laboratory, April 1976.

3. Breuer, M.A., "Survey of Digital Logic Simulators and Automatic Test Generation Systems", available from Breuer and Associates, 16857 Bosque Dr., Encino, Calif., 91316, 1974.

4. ______________________, Automated Fault Simulation (AFS), Reference Manual, Control Data Corp., Minneapolis, Minn.

5. ____________________, Asynchronous Simulation System (ASSIST), User's Guide, Control Data Corp., Minneapolis, Minn.

6. Breuer, M.A., "The Design of Gate Level Simulation Systems for Fault Detection and Diagnosis of Digital Circuits", available from Breuer and Associates, 16857 Bosque Dr., Encino, Calif., 91316, 1974.

7. Seshu, S. and Freeman, D.N., "The Diagnosis of Asynchronous Sequential Switching Systems", IEEE Transactions on Electronic Computers, Vol. EC-11, August 1962, p459-465.

8. Szygenda, S.A. and Thompson, E.W., "Digital Logic Simulation in a Time-Based, Table-driven Environment, Part 1. Design Verification", Computer, Vol. 8, No. 3, March 1975, p24-35.

9. Szygenda, S.A. and Thompson E.W., "Digital Logic Simulation in a Time-Based, Table-driven Environment, Part 2. Parallel Fault Simulation", Computer, Vol. 8, No. 3, March 1975, p38-49.

10. Chang, H.Y. and Chappel S.G., "Deductive Techniques for Simulating Logic Circuits", Computer, Vol. 8, No. 3, March 1975, p52-59.

11. Thomas, J.J., "Automated Diagnostic Test Programs for Digital Networks", Computer Design, August 1971, p63-67.

12. ______________, Test Generation and Simulation System (TEGAS) User Information Manual, Control Data Corp., Minneapolis, Minn.

13. Bennetts, R.G. and Scott, R.V., "Recent Developments in the Theory and Practice of Testable Logic Design", The Radio and Electronic Engineer, Vol. 45, No. 11, November 1975, p667-679.

14. Falk, H., "Design for Production", IEEE Spectrum, October 1975, p48-53.

15. Ulrich, E.G. and T. Baker, "Concurrent Simulation of Nearly Identical Digital Networks", Computer, Vol. 7, April 1974, p39-44.

24 LIMITED LIFE ITEMS

J. A. Roberts and J. E. Arsenault

Components which require special maintenance consideration during operation or storage of a system will be classified for our purpose as "Limited Life Items". There are three states in which a piece of equipment may be ordinarily arranged, (a) Storage, (b) Dormancy, (c) Operation.

It cannot be assumed that in storage the equipment will be subject to zero failures. Although there is zero electrical stress, chemical and mechanical changes occur which may degrade components. Chemical changes progress even at preferred storage temperatures. Preservation and packing techniques described by (1 - 3), for example, although comprehensive, cannot mitigate against the internal chemical degradation of components. The storage deficiencies of the most critical components have been formally recognized (4, 5). Aluminum capacitors and electromechanical relays are acknowledged to be prime problems after storage.

The fact of the matter is that there is a good case for suspecting that predicted system reliability performance, defined in terms of MTBF, is not being achieved because the effects of storage and/or dormancy are neglected. During operation, electrical stress and mechanical wear contribute additional failure causes. For some components the wear out phase of the life cycle of a component follows almost directly upon the Stage I infant mortality phase, as shown in Figure 1.

This is the classic relationship observed for mechanical components subject to continuous wear (7, 8). Clearly, an extended "burn-in" or "wear-in" will, in such a situation, degrade, rather than enhance reliability. Electrical components such as large aluminum capacitors have been categorized as having similar life cycle characteristics. The "wear out" phase of electronic components is, in general, less well recognized than that of mechanical items. Thus the term "limited life" item is more applicable.

A list of components which are hazards during storage is given in Table 1 and a list of components which are "wear-out" risk or "limited life" items during operation is given in Table 2. For many pieces of equipment, these lists are identical. This relationship has been recognized formally in some procurement contracts (9). For such purposes the ratio of stand-by to operating failure rate of a given component is designated K_s,

$$K_s = \frac{\lambda_s}{\lambda_o}, \quad 0 \leq K_s \leq \infty \qquad (1)$$

where,

λ_o = operating failure rate,

λ_s = stand-by (storage) failure rate.

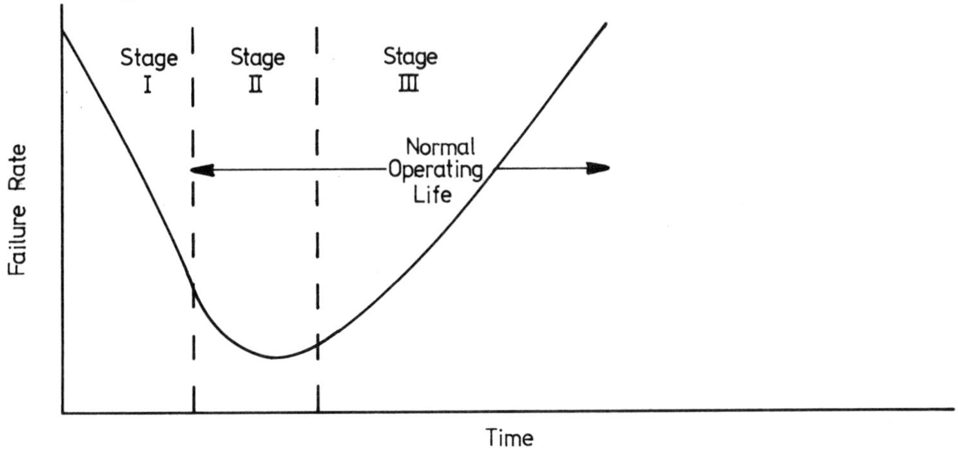

Figure 1. Mechanical Component Life Cycle

For general practice it has been customary to use a low value for K_s, typically between 0 and 0.10. TRW presents data in (9) to suggest that for spacecraft applications a value of 0.1 is appropriate. The early reports of Martin Marietta Corporation (10) suggest that values of K_s range from 0.077 to 0.007 for military standard parts using MIL-HDBK-217B (11) to compute λ_o. An average value for K would be 0.01. Since these computations relate to components displaying constant failure rates it is necessary to be cautious in applying such data to recognized "wear out" items. Actual life performance of components must be the basis of a maintenance action. The cause of failure of, for example, a motor or a relay, will be quite different during storage to that found during use. However, the fact that a relay or motor dominates both areas should suggest that massive efforts should be made to design out such parts. Recently issued Requests For Proposals (RFP) and Requirement Specifications have clearly demanded that solid state techniques should be used to the maximum extent. Much can then be done to eliminate the mechanical complexity of tape drives, for example. The pinch roller and capstan approach of audio cassette drives has now been eliminated in digital tape units by using electronic speed control.

Non-Operating Failures per Billion Part Hours	Resistors (1)	Capacitors (2)	Semiconductors and Integrated Circuits (3)	Transformers and Inductors (4)	Electromechanical and Rotating Devices (5)	Switches and Relays (6)	Connectors per Male and Female Connectors (7)	Hydraulic Equipment (8)	Hardware (9)
10,000						Acceleration timer switch		Electrohydraulic servos	
5,000					Brush type synchro devices	Thermal timer		Pumps, motors, and dynamic seals	
2,000			Discrete hybrid integrated circuits	Variable auto-transformers	Electromechanical timers AC and DC D'Arsonval type panel motors	Thermal switch			
1,000			Thin film hybrid integrated circuits High power alloy transistors	Variable air core RF coils	DC torquers	Ledex and telephone types of stepper switches Electronic timers		Electrohydraulic transducers	
500			High power Mesa transistors		AC motor tachometers	Dry circuit relays and sensing relays operating at less than 100mw			
200		Aluminum electrolytics	High power diodes		AC servos	Circuit breakers, push button switches, precision limit switches, and load contactors over 10 amps			
100				Variable iron core coils	DC generators and power motors	Rotary and toggle switches	Tape	Fluids and lubricants Solenoid operated valves	
50			Analog and digital monolithic circuits. High power planar transistors and medium power Mesa transistors.	Air core RF chokes.		One-sixth and one-half size crystal can relays	Coaxial cable Signal connectors which are high density crimped rectangular, minimum crimped rectangular, crimped edge, solder or crimped pin socket, or soldered edge. Rectangular power connectors which are blind mate soldered, blind mate crimped, screw lock soldered, or screw lock crimped.		
20		Wet tantalum foil and slug.	Medium power planar transistors.		AC generators and induction and synchronous power motors.	Reed relays	High density soldered circular and rectangular signal connectors. Soldered or crimped bayonet and threaded power connectors.		
10	Variable, non-hermetically sealed carbon composition, wire-wound, wire-wound trimmer, and metal film trimmer.	Ceramic general purpose high K, CV variable, and tubular variable.	Medium power planar diodes. Low power transistors.	Non-hermetically sealed pulse transformers under 1000 volts and power reactors.		Microminiature crystal can relays. Squib switches	Circular signal connectors which are minimum soldered, minimum crimped, or high density crimped.	Accumulators, adjustors, static seals, regulators, and self-operating valves.	
5	Non-hermetically sealed carbon composition	Variable air or tubular glass	Low power diodes						Bearings and rubber isolators.
2	Hermetically sealed carbon composition and non-hermetically sealed tin oxide	Metallized paper		Non-hermetically sealed magnetic amplifiers, audio transformers, power transformers and pulse transformers over 1000 volts.		Wetted mercury relays.		Relief and check valves.	

Table 1. Non-Operating Failure Chart

Table 2. Mechanical and Electromechanical Wearout Items Listed by Risk-RAC Survey (6)

Relays	23%
Switches	19%
Bearings	13%
Motors	10%
Gears	7%
Pumps & Valves	7%
Connectors	6%
Actuators	4%
Blowers	3%
Seals	2%
Solenoids	2%
Pots	2%
Servos & Resolvers	2%

Where such changes are not feasible other approaches are required. Reference (12) requires, in Section 5.2.6, that items which are hazards after storage should be identified. The U.S. Army (13) provides identification codes for components which are high deterioration risks throughout storage, handling or use. This data item requires the contractor to obtain shelf-life information. Fortunately a data base for the formulation of a maintenance plan, particularly for components of limited shelf or storage life, has been developed (10, 14, 15).

While the body of information on shelf-life has improved,less is known about dormancy. There are two forms of dormancy which can be considered, i.e., (a) shelf/retest/shelf/retest cycle, (b) very low but near continuous operating stress. In both cases the average stress is very low. This might suggest that the convenient reliability factors given by MIL-HDBK-217B (11) could be projected to ultra-low operating stresses. This does not accord with the statements contained within the document. What is required is a physical understanding of the oxidization rate of copper slip rings, of the rate a movement of surface charges in MOS transistors or the rate of decay of insulating oxides from electrolytic capacitors. With such information the retest/storage cycle, given in Dormancy Plan (a) and shown in Figure 2, could be implemented with some confidence. Clearly the reliability of such equipment could be very high and dramatically superior to simple shelf storage.

There are greater risks with the proposal offered in Dormancy Plan (b). It is difficult to build complex equipment that will, on command, lightly stress a large variety of interconnected components. Indeed, in the case of aluminum capacitors, a small voltage stress, of say less than 20% of the rated stress, will cause the oxide to reform eventually too thin to support higher voltages. This results in failure at normal rated operation. Components, such as some light current relays, require that currents through their contacts have a certain minimum value to prevent oxides forming. The operation of equipment at low continuous stress should proceed, therefore, only from a clear understanding of the effects of light and ultra light stress on individual components. Since little may be known about the effect of long term low stress operation, a false sense of security can be engendered.

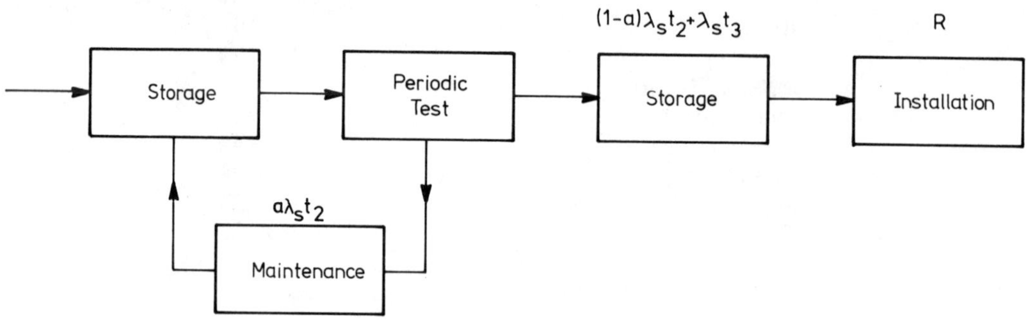

(a) Periodic Test Approach

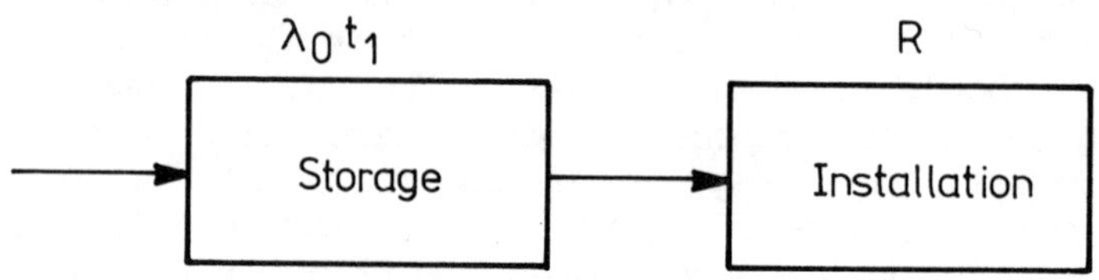

(b) Low Stress Approach

a = 90% (portion of failures detected by test),

t_4 = total non-operating time prior to installation,

t_2 = non-operating time up to last periodic test,

t_3 = non-operating time between last periodic test and installation,

R = probability of no failures.

Figure 2. Dormancy Plans

LIFE TESTING

In recent years changes have been made in the methods used for analyzing life-test, overstress and accelerated life test data. A variety of probability distribution functions have been considered. From this work it has been found that the most useful of these distributions is the Weibull distribution.

The Weibull distribution is closely related to the exponential, but has two additional parameters, the shape parameter and the location parameter. Instead of a single constant failure rate λ, as in the exponential case, a variety of hazard situations can be treated. For a given Weibull distribution, the failure rate can be either continually increasing, continually decreasing or else constant. If the data exhibit all three phenomena in what is commonly called the "bathtub" curve ("burn-in", constant failure rate with continued usage and "wear-out" towards end of life) different Weibull distributions can be obtained for different segments of the lifetime. A further convenience of the Weibull distribution is the relative ease of probability plotting to estimate the parameters, detect outliers and perform crude goodness-of-fit analysis. A rather complete discussion of the Weibull Probability Function is given in (16).

It is comparatively simple to perform graphical model fitting and goodness-of-fit analysis for Weibull time-to-failure data. The straight line plot which approximates the distribution also can be used to find crude estimates of the parameters, σ and β, hence an estimated value of the reliability. Weibull probability paper is constructed with double logarithmic scales in the vertical direction, corresponding to ln (-ln R(t)) and logarithmic scales in the horizontal, corresponding to ln t. The most convenient Weibull plotting paper was designed by Nelson-Thompson (17). An auxiliary scale is provided for the estimated shape parameter and a circular marker is used at the 0.6321 percentile, to help denote the estimated characteristic lifetime. In this plotting paper, the horizontal (time-stress) axis has one logarithmic cycle. Therefore, each number on the bottom line can be multiplied by the same constant, or else raised to the same power to correspond to any scale of values of time or stress level you might use. The vertical scale corresponds to the distribution function.

LIMITED LIFE COMPONENTS

In the following sections components which have been of concern to the writers, from the limited life aspect, are discussed. The treatment cannot be extensive in the space available. The reliability engineer, when facing unavoidable limited life items, has to gain knowledge of the technology of the component. Life testing together with data collection are the prime ways to reach an understanding of the problems peculiar to each component type.

SEMICONDUCTORS

It is not common practice to consider semiconductors as limited life items. However, there are defined failure modes and certain semiconductor devices which must be regarded as wearout risks. Defined failure modes include, (a) electromigration, (b) electrolytic-corrosion, (c) localized current concentration, (d) thermal cycling degeneration, (e) oxide charge instabilities, (f) intermetallic compounds. While efforts are made to reduce failures from these causes by manufacturing improvements, they often re-appear at different times, from different manufacturers or as new devices appear.

The electromigration problem is due to the current induced mass transport of metallization due to high temperature and current density. In conventional low power ICs this problem does not usually appear. In R.F. power devices, problems are still evident unless gold based or other special metal systems are used. Voids in aluminum interconnect patterns occur above 150 degrees C and with current densities of the order of 10 A^6/cm^2. A simplified theory (18, 19) provides insight into the variables influencing the Median Time to Failure (MTF).

$$MTF = \frac{(W\ t)}{(CJ^2)} \exp(\phi/KT) \quad (2)$$

where,

W = stripe width of the metallization,

t = stripe thickness of the metallization,

J = current density,

ϕ = activation energy for diffusion of the metal,

K = Boltzman's constant,

T = temperature in °K,

C = constant depending on degree of film crystallinity, resistivity, ion mass, density and stripe geometry.

Figure 3 (18) shows the expected best case MTF for a small-grained aluminum stripe 1μ thick x 2.54 μ wide, with a current density of 10^6 A/cm^2. MTF decreases rapidly as temperature and current density increase as shown on this log scale.

Electrolytic-corrosion of die metallization by moisture and applied voltage occurs in improperly sealed hermetic packages and conventional plastic devices. Improper sealing procedures, used for ceramic glass sealed packages, are known to be a prime cause of failure for all MOS technologies that use phosphorous doped, glass passivation methods (20). Failures can occur within days of initial operation. Corrosion is a constant problem for plastic MOS components. The median life is largely determined by (21):

(a) matching of thermal expansion coefficients of the leadframe, wire, die and plastic,
(b) purity of the encapsulant,

(c) adhesion of the encapsulant,
(d) length and width of the lead frame interface,
(e) integrity of the final die passivation layer.

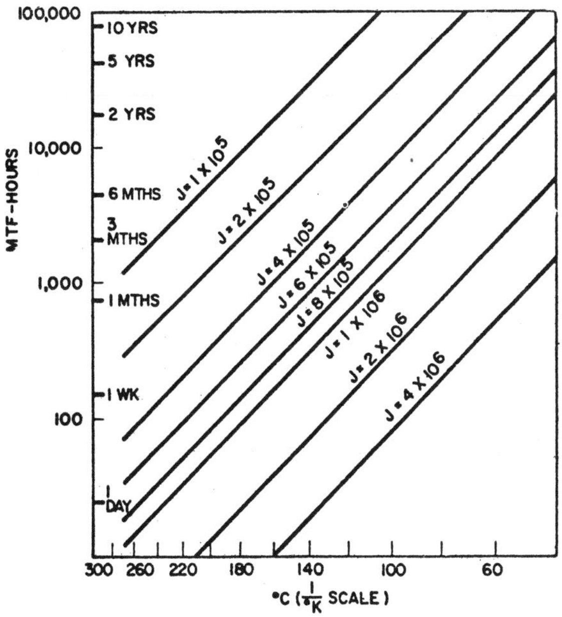

Figure 3. MTF vs. Temperature

These factors are clearly a hazard even in fairly good storage environments. Motorola has developed a model which substitutes vapour pressure for temperature in the Arrhenius equation. This model allows actual application environments to be related to the results of conventional tests. The plot shown in Figure 4 shows how the model relates various conditions to median life and provides a data point based on current data. Reference (11) of September, 1976 shows that commercial plastic ICs are given a Q of 300 while commercial hermetic ICs are given a Q of 150. The ratio being only 2 is probably optimistic in high humidity environments, particularly for MOS devices.

Of special interest is the performance of LED devices and displays. The availability and cost of plastic encapsulated LED components is considerably better than hermetic versions. There is, therefore, pressure to use such devices even in a high reliability application. Two approaches can be used:

(a) Arrange for the LED device to be front panel, user replaceable. Thus the LED device is treated in the same manner as an incandescent lamp,

(b) Test the LED device over the environmental limits of the equipment. This will eliminate a large percentage of vendor devices. Some products, however, can provide excellent performance even under high humidity and temperature cycling conditions. These tests should be qualification and not screening tests.

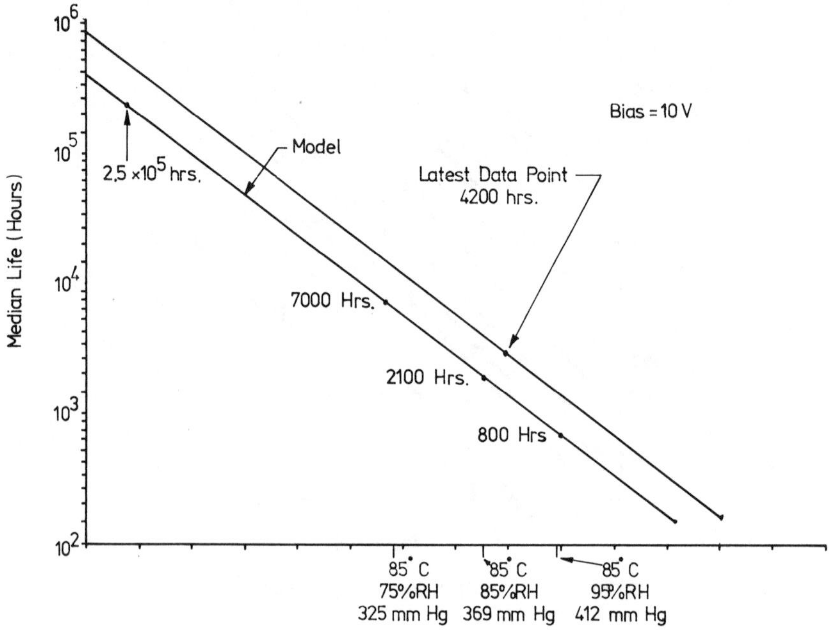

Figure 4. Vapour Pressure Model

The published data on LED devices is relatively sparse but the interest in this problem is likely to result in far more data and better devices.

Localized current concentrations can occur when power devices lack emitter ballasting or when any device is reverse biased to the point of breakdown. There is also the possibility that any high power device will display thermal failure due to hot spots. The quality of the die bonding to the header is a prime factor in preventing this latter defect. Thermal imaging under actual operating conditions is used to uncover these problems. Improper design, application and poor production control can lead to high failure rates. It is notable, however, that gradual deterioration of performance to a measurable degree, is a feature of thermal problems. The position is not unlike mechanical wear out and is preventable. The thermal cycling problem illustrated in Figure 5 is related largely to the difficulty of matching the temperature coefficients of silicon to those of conventional header materials. During each hot/cold, power on/power off cycle the mechanical bond between the silicon and the header is stressed.

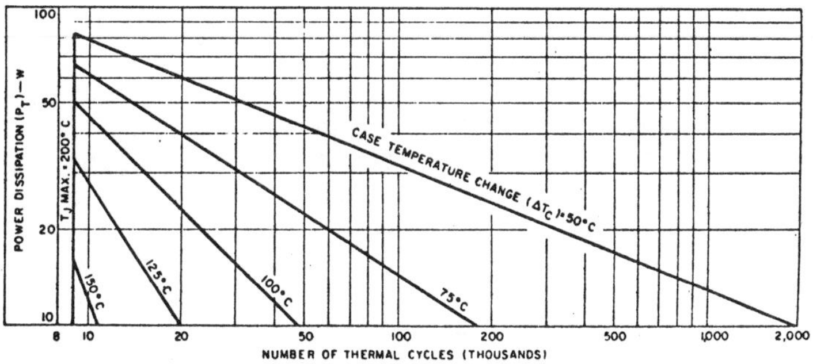

Figure 5. Thermal-cycling Rating Chart (Unique)

RCA has provided (22) thermal cycling data which allows this mode of failure to be estimated for each equipment requirement. It is possible, in some rare cases, that this mode of failure will dominate equipment failure rate. Maintenance and replacement action may then be determined by this factor. A flashing lamp or printer solenoid driver application may require special derating or maintenance action.

Oxide charge instabilities account for failures in MOS ICs and life instabilities in lower power transistors. The MOS gate oxide quality is a major contributor to the stability of the MOS threshold voltage (23). Contamination, in the form of mobile ions in the gate oxide, is a cause of threshold voltage instability. The instability appears as a long term drift which increases with increased temperature. Device failure occurs when the threshold voltage falls out of the design tolerance for the IC. This mechanism represents an insidious mode of failure peculiar to MOS devices having inadequate process controls. MOS devices that dissipate high power levels are notable for their susceptibility to this failure mode. While the problem is largely understood and controlled, a constant watch for this factor is required. The life instabilities of transistors are often due to mobile ion drift in oxides. In addition to life instabilities large noise increases and leakage current changes are observed. Occasionally the problem is increased by poor testing techniques, which force low power devices into unnecessary avalanche (breakdown) tests. In analogue circuits the devices affected show a continual pattern of performance degradation or instability, which leads to continual replacement problems. The device then becomes classified as a limited life item and remains a problem for as long as the original causes are overlooked.

Intermetallic compound generation occurs when gold-aluminum bonding is employed. In recently manufactured hermetic IC devices this couple is present only at the post bonding site. Noticeable intermetallic formation occurs even in good devices after 64 hours at 250 degrees C in high temperature operating tests (HTOT) (21). Since bond strength is not greatly degraded such formation is not considered serious. However, this factor is clearly a wearout element and in severe cases (poor devices) has been a prime failure cause.

Screening techniques, such as those described in MIL-STD-883, are capable of reducing the limited life occurrences of otherwise reliable devices. The question of what is the value of screening as a means of preventing early failures is often raised. Equally important is the possible damage that can occur to a reliable device during screening. By definition, there is no way that accelerated tests and screens can detect random failures. The fall-out experienced during testing must therefore arise from the infant failure (Stage I of Figure 1) and wear out failure (Stage III of Figure 1) sections of the life characteristic. There is ample evidence (24) that testing removes potential field failures accurately. The use of accelerated tests (25) has clarified the fact that the wear-out stage can follow directly upon the infant mortality stage. The degree of derating adopted for the design and application of the device plays a vital part in determining what occurs in practice. The mechanisms of failure are the same as those established from field failure sources and can apply to the entire population. The mechanical failure rate curves are possibly more representative of semiconductor failure patterns than has been supposed. Since, however, the failure rates are generally very low for good devices, a pattern of rapidly accelerating failure rate (at say the 15 or 20 year equipment life period) should not be of concern.

INCANDESCENT LAMPS

Incandescent lamps are used extensively for indicating equipment status in electronic systems. Rated average life is that obtained in closely controlled laboratory testing of lamps on 60 Hertz alternating current at their design voltage (26). Very long life lamps are generally rated on the basis of extrapolated laboratory test data. Service conditions such as shock, vibration, voltage fluctuations, temperature, etc. may contribute to shorter average life.

The life of a properly designed sub-miniature lamp is affected by each of the following, (a) filament color temperature which determines filament evaporation rate and is directly related to lamp life, (b) filament wire diameter which affects lamp life as a direct function of the square of the filament wire diameter for a given temperature, (c) electrical conditions under which the lamp is operated, (d) environmental conditions under which the lamp must perform and (e) particular processing techniques used in the manufacture of the lamp. Items (c), (d) and (e) cause actual life performance to vary from rated life. Lamp life ratings are mathematically determined as the point at which the filament wire weight has been reduced by 10%, then confirmed through actual laboratory testing.

In recent years another life-influencing lamp filament factor has come into use. This is commonly referred to as "filament notching"; its prominance is due to at least three factors, primarily associated with sub-miniature type lamps, (a) low temperature filament operation, less than that for significant normal tungsten evaporation. (Long life lamp designs, such as 10,000, 25,000, 50,000 and 100,000 hour designs. This does not apply to filament temperatures below 1600 degrees C), (b) small filament wire sizes, less than one mil (0.001") diameter in many cases, (c) increased use of D.C. voltage operation (generally resulting from advances in solid state technology).

Notching is the appearance of step-like or sawtooth irregularities, appearing on all or part of the tungsten filament surface, after some burning. These notches reduce the filament wire diameter at various points. In some cases, especially in fine wire diameter lamps, the notching is so severe as to penetrate almost the entire wire diameter. Thus accelerated spot evaporation due to this notching (as well as reduced filament strength), becomes the dominant mechanism for influencing lamp life. Because of its abnormal evaporation and/or reduced strength effects, lamp life-times due to notching may be one-half or less of so called ordinary or normal, predicted lamp lifetimes.

Ordinarily for still-rack operation, normal tungsten filament evaporation is the basic force or mechanism controlling incandescent lamp life. Where normal filament evaporation is the dominant failure mechanism, lamps should reach their design-predicted life-times.

Rerating. Generally, data included in lamp catalogs will show how amperage, luminance, and life change with voltage changes. Three basic equations used for rerating subminiature incandescent lamps are:

$$\text{Rerated MSCP} = (V/V_1)^{3.5} \times (\text{rated MSCP}) \quad (3)$$

$$\text{Rerated life} = (V_1/V)^{12.0} \times (\text{rated life}) \quad (4)$$

$$\text{Rerated current} = (V/V_1)^{0.55} \times (\text{rated current}) \quad (5)$$

where,

V is the application voltage, and V_1 is the rated voltage,

MSCP = Mean spherical candlepower.

These equations apply to ideal conditions and should be used only as guides. The greater deviation from rated voltage, the greater percentage of error.

Maintainability Design Note. Because lamps are limited life items they should be designed into the system in such a way that they are replaceable from the front panel without unsoldering. A press-to-test switch should be provided to illuminate ordinarily extinguished lamps for scheduled maintenance purposes.

CONTACTS

Switches, relays, circuit breakers, sockets and plugs are subject to possible degradation during storage and also suffer an established wear out risk. However, despite recent advances in gaining an understanding of the physical factors involved in contact problems the highly variable nature of applications and environmental factors preclude the generation of widely useful failure data.

The complexity of the problem is illustrated in Figure 6 which addresses only the environmental damage (27). On the other hand, contacts which are subject to near permanent closure and which are substantially gas tight at the point of closure, are largely free of these degradation factors. Corrosion effects between dissimilar metals should be considered, but the performance of, for example, tin plated ICs plugged into gold plated sockets is reassuring (28, 29). Contacts which are open for long periods, particularly periods appropriate to the storage of spares, are severely at risk from tarnish films which act as insulators. Even when the voltage employed is sufficient to break the film, the subsequent life of the contact is reduced due to poor arc characteristics and more rapid erosion. It should not be concluded that hermetically sealed or covered relays are necessarily a lower risk than non-sealed types. Many of the contamination problems of relays can be traced to the organic vapors from the coil construction which are trapped in the cover or can.

The problem of entrapped contaminants is not confined to conventional relays. Reed contacts which initially show low resistance in the tens of milliohms can, after several million operations, increase their contact resistance to 5 ohms or more. This is a wearout factor in some critical applications. The wearout factor of contacts can, for convenience, be divided into four areas (30):

(a) mechanical scoring, mass transfer due to arcing (pitting),
(b) mechanical fatigue of contacts and support items,
(c) heating, leading to thermal degradation or welding,
(d) deposit of conductive material on associated insulation.

The solution to all of these problems lies in correct contact choice (30), derating or the employment of a solid state switch (in the case of relays, circuit breakers and keyboards). Failing these possiblities, it is necessary to engage in planned maintenance activities. An example (28) of the form of a reliability replacement plan for keyboards is shown in Table 3. For one shift operation the 'E' letter is used about ten times more often than the letter 'U'. Table 4 illustrates a manufacturer's life estimation for a reed contact keyboard based on manufacturers' data, life test data and circuit experience.

Table 3. Estimate of Keyboard Life Requirements
(Worked-Out Example)

a. Statistical distribution in key usage for proposed application-derived (somewhat arbitrarily) from Lincoln's Gettysburg Address

Key	Percent to Total Keystrokes
Space bar	17
Carriage return	3
Vowel "A"	6.7
Vowel "E"	10.7
Vowel "I"	4.4
Vowel "O"	6.4
Vowel "U"	1.3
	49.5%

a. Estimate typing time in 8-hour shift, 8 hr x 60 min = 480 min-downtime Downtime = coffee breaks, startup and shutdown, personal, copy editing, and miscellaneous = 138 min. Estimated typing time = 342 min.

c. Estimate typing speed, assume average of 40 words/min x 5 chars/word = 68,000 keystrokes/shift total.

d. Estimate 252 average workdays/yr.

e. Estimated key usage/yr. depending on number of shifts. Relative numbers of total keystrokes for selected keys over a 1-yr period are as shown below:*

Key	1 shift	2 shifts	3 shifts
Space bar	2,913,120	5,826,240	8,739,360
Carriage return	514,080	1,028,160	1,542,240
"A"	1,148,112	2,296,114	3,444,336
"E"	1,833,552	3,667,104	5,500,656
"I"	753,984	1,507,968	2,261,952
"O"	1,096,704	2,193,408	3,290,112
"U"	222,769	445,536	668,304

* Adapted from a bulletin by Maxi-Switch Co.

Table 4. Estimating Keyboard Reliability* - MTBF
(Typical Illustration)

Assumptions (based on manufacturers' data, life test data and circuit experience). Reed failure 0.1%/10^6 operations. Electronic component failure rate of 0.1%/1000 hr (0.01 to 0.02%) is usually used. Solder joint failure rate of 0.001% is usually used. 25 components (MOS encoded); 300 solder joints max. Keying Rate - 30 words/min x 5 char/word = 150 char/min 150 char/min x60 min/hr = 9000 operations/hr (round off to 10,000). 10^6/10,000=100hr. to accumulate 10^6 operations. Reed Operations - Failure rate due to reeds. 0.1%/100 hr =1%1000 hr. Keyboard Survival Rate - $(0.99)(0.999)^{25}(0.999)^{300}$ = (0.9733)(0.9973) = 0.9610 = 96.1% survive the 1000-hr period. Failure rate is 3.98/1000 hrs. MTBF = 1/FR=1/ 0.039/1000 hr. = 25,000 hrs.

*Courtesy Keytronic Corp.

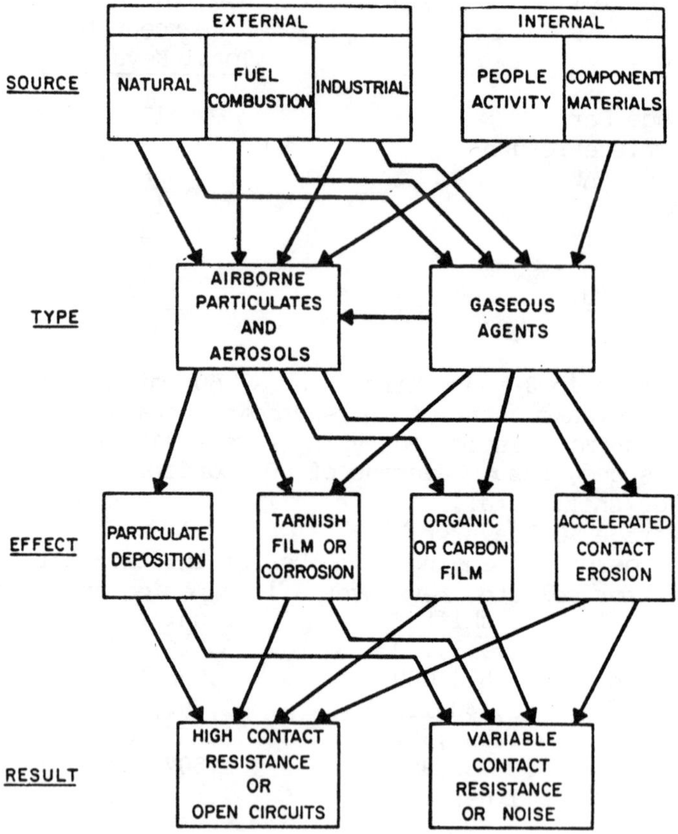

Figure 6. Environmental Damage to Electrical Contacts

BEARINGS

Providing that factors such as lubrication contamination, extreme temperatures, excessive internal clearances due to wear, etc., are eliminated from consideration, a bearing may be judged on the basis of its Rating Life.

Rating Life. The rating life of a group of apparently identical ball bearings is defined as the number of revolutions (or hours at some given constant speed) that 90% of a group of bearings will complete or exceed before the first evidence of fatigue develops. As presently determined, the life which 50% of the group of ball bearings will complete or exceed is approximately five times the rating life.

The rating life (for ball bearings) can be expressed by (31):

$$L = \left(\frac{C}{P}\right)^3 \qquad (6)$$

where,

L = rating life in millions of revolutions,

C = basic dynamic load rating (lb.),

P = the equivalent load (lb.).

By transposition equation (6) can be converted to a more practical form for calculating life as

$$L_h = \frac{16,666}{n} \frac{C}{P} \qquad (7)$$

where,

L_h = rating life in hours,

n = revolutions per minute (rpm).

In less stringent applications, the calculated rating life may seem unrealistically long but may serve at least as a "yard-stick" in approximating the life to be demanded of, or rectified with, other possible failure causing factors, such as the lubricant or wear properties of the bearing components, etc. The rating life serves as a more direct indication of the life to be anticipated for bearings subjected to higher C/P ratios or higher speeds.

Critical applications exist where even the standard fatigue life assumption of 90% survival provides an insufficient degree of service reliability. In such cases, bearings are often chosen whose calculated rating lives are longer than would normally be required, so as to provide an extra margin of safety. The rating life may be determined by inserting the basic dynamic load rating and the equivalent load in the above formula. Detailed bearing life calculations can be readily accomplished by using manufacturer's data provided in (31, 32).

BEARING LUBRICATION

Large, complex electronic systems normally will have critical units associated with them that depend upon bearings. These bearings require suitable lubrication for optimum performance (33). Lubrication is one of the primary factors governing the service life of rolling contact bearings. Theoretically, it might be assumed that where only pure rolling contact is involved, no lubrication is needed. However, in a ball or roller bearing there is a certain degree of sliding or skewing, in addition to the rolling action, due to the different peripheral velocities at various points on the curved surfaces of the contact areas. Sliding also occurs between the various component parts of the bearing.

In a rolling contact bearing, the lubricant performs the following functions: (a) lubricates the load carrying surfaces, (b) lubricates the sliding contact between the rolling elements and the cage, (c) lubricates the sliding contact between land riding cages and rings, (d) lubricates the sliding contact between rollers and guide lips or flanges of roller bearings, (e) excludes dirt and foreign matter and (f) prevents corrosion.

The limiting speed of a bearing, depends to a great extent, on the ability of the lubricant to reduce and disperse the heat generated in the sliding contact areas. Prevention of corrosion is of major importance in all applications. Such protection must be provided in storage and after mounting in the application. Two types of bearing lubrication are normally used, i.e. (a) oil, (b) grease. The relative merits of oil and grease lubrication are summarized in Table 5.

Table 5. Relative Merits of Oil and Grease

Lubrication	Lubrication Property
Oil	Suitable for operation at extremely high speeds. Easy to check for performance and reliability. Easy to feed to the rolling and sliding contact areas. Suitable for extremely high temperatures. Also acts as a coolant.
Grease	Can be retained without elaborate closure. Seals out contamination. Enclosed shielded and sealed bearings are prelubricated at factory with correct amount of clean grease. Long service without relubrication.

Oil

When oil lubrication is used, the following points should be observed: (a) high quality, approved oil should be used, (b) the oil should have the correct viscosity at the operating temperature and (c) the correct quantity of oil should be applied.

The correct viscosity of an oil depends on the load, speed, temperature and type of bearing. High speed, lightly loaded bearings should have a relatively low viscosity oil, while heavily loaded, low speed bearings would require a high viscosity oil. As a general rule, for normal operating conditions, the following minimum viscosities at operating temperatures are recommended as per Table 6.

Table 6. Recommended Oil Viscosity

Type of Bearing	Viscosity
Ball	70
Cylindrical Roller	100
Spherical Roller	100
Spherical Roller Thrust	150

Allowance should be made for the fact that in operating equipment the temperature of the oil will be several degrees higher than the housing temperature. As the temperature of the oil increases, the viscosity decreases (the oil thins out). In order to describe this change in viscosity the American Standard Test Method (ASTM) has provided a Viscosity Index Scale. Using a formula based on the viscosity of an oil at 100 degrees F and at 210 degrees F, the scale was derived ranging from 0 to 100. An oil having a viscosity index of 100 has the least change in viscosity per degree change in temperature. Where bearings have to operate over a wide range of temperatures, an oil with a high viscosity index should be selected; that is, one which will have a minimum change in viscosity with changes in temperature. Pure paraffinic oils have the highest viscosity index.

Quantity of Oil. Too much oil can be as detrimental as too little oil. For splash lubrication the oil level should be maintained at the middle of the lowest ball or roller in the bearing. For various oil feed systems such as jet oil and air-oil mist, the correct quantity of oil can be calculated or can be determined by running the machine and adjusting the oil rate to obtain the lowest operating temperature.

Oil Storage Life. Considerable variations in storage life exist between various types of lubricants. Shelf life assumes that a bearing is maintained in its original, clean, waterproof wrapping in a dry area and at fairly constant room temperature. Reference (33) provides an indication of the shelf lives of some commonly used oils.

Grease

Rolling bearing greases are a mixture of an oil and a soap base thickener. In a sense, the grease serves as an oil system, holding the oil in suspension and feeding it to the rotating parts. One of the more common ways of classifying bearing greases is by the type of soap base, or thickener.

Calcium Soap. Because of their excellent water resistance, these greases were very popular for paper mills, water pumps and chassis lubrication and were commonly referred to as "cup greases". Poor internal cohesion and poor stability at high temperatures make these greases unsuitable for operation at high speeds or at temperatures over 170 degrees F.

Sodium Base. Sodium greases (sometimes referred to as soda greases) have good load carrying ability, good high speed characteristics and a temperature range suitable for most bearing applications. Their structure allows them to absorb small quantities of moisture without detriment. However, they are not effective where appreciable amounts of water are present.

Lithium Base. Lithium greases combine some properties not found together in either of the above greases. They have good water wash out resistance, good internal cohesion and a high temperature range, almost equal to that of the sodium greases. Because of these characteristics, lithium soap greases are commonly referred to as "multi-purpose greases".

Synthetic Thickeners and Oils. In order to extend the operating characteristics of greases, various synthetic oils and thickeners are used. Diester oils are used for low temperature, low torque applications while silicone oils are used for both extreme low and extreme high temperature operations. Certain characteristics which are taken for granted in organic greases may be lost in a synthetic grease. For example: certain greases composed of synthetic thickeners and silicone oils have an operating temperature range to 450 degrees F. However, the load carrying ability is about 4% of the bearing capacity and very high temperatures are induced when parts run at high speeds.

Penetration. One indication of the suitability of grease for an application is its worked penetration number. This is based on a test where a standard cone is dropped from a standard height onto the surface of the grease and the penetration measured. The greater the penetration the higher the penetration number and the softer the grease. A channeling grease is preferred for higher speeds.

Water and Corrosion Resistance. Good water resistance does not necessarily imply good corrosion resistance. A water resistant grease may not sufficiently "wet" the surface of the bearing to protect it from corrosion. Conversely, a grease which has good corrosion resistance may not have good water wash out resistance. The combination of these factors in a grease should be carefully checked.

Grease Life. In many applications the life of the bearing will be determined by the life of the grease. Lightly loaded grease packed bearings running at moderate speeds with effective seals have been known to run for over 20 years. Load, speed and temperature will affect grease life. Of these, temperature usually has the greatest effect. Depending on the grease used, grease life is cut in half for every 18 degrees to 25 degrees F increase in temperature. Also, a sealed bearing packed 1/3 full is the grease quantity of optimum life.

Grease Storage Life. Oil tends to escape from a bearing, hence grease lubrication is often preferred when a storage period longer than approximately one year is anticipated. A gradual oxidation process of the grease will progress over prolonged storage periods and the oil can gradually escape eventually, leaving the soap medium in a hardened condition. Bearing demand requirements should be carefully controlled so that shelf lives exceeding two years can be avoided. Most instrument bearing greases will allow controlled storage for two to five years. Some greases are guaranteed for considerably longer periods, but longer storage periods should not be planned unless more elaborate precautions are undertaken, such as hermetically sealing each bearing in a plastic capsule filled with oil. Reference (33) gives a summary of the shelf life for some commonly used greases.

Spindle Life. A spindle life test is an excellent means of evaluation of the oxidation resistance of grease under conditions which offer close correlation with those experienced in operating service. In this test a measured quantity of grease is applied to a standard bearing assembly which is operated at 10,000 RPM and any desired temperature until the bearing is destroyed. Life over 1,000 hours at 300 degrees F would be considered very good while a life of 100 hours only, at this temperature would be considered poor. Similarly, a spindle life of 6000 at 200 degrees F would be considered very good, a life of 2000 hours fair, and a life under 1000 hours poor.

Lubricant Operational Life. The life of a grease is related to the bearing size, speed, load and temperature. The equation taken from (34) indicates that bearing grease life is

$$\log L = -2.60 + 4420/(460+T) \; -0.301S \qquad (8)$$

where,

L = time at which 10% of greased bearings fail in house,

T = temperature of bearing,

S = half-life reduction factor.

Detailed information for calculating grease life is given in (34).

Lubrication Schedules. Generally, regreasing should be scheduled at or before the calculated 10% grease life L. For critical equipment, regreasing should be scheduled at 1/2 L; this estimate corresponds to a 1% failure rate. If the indicated regreasing schedule is impractically short, design or operating changes that decrease S should be considered. The soundest basis for setting the regreasing schedule is past experience with similar applications.

GEARS

In general, gears still find their way into electronic systems only because they cannot be replaced in an economical fashion. Gears are widely used in electronic systems employing servomechanisms and stepper motors where performance depends on relative position.

For rapid calculation of gear life the following equation derived from (35) can be used, i.e.,

$$K_1 = \frac{4W}{D_1 F Q \sin\phi} \tag{9}$$

where,

K_1 = load stress factor, psi,

W = gear load, lb driver,

D_1 = pitch diameter of driver, in,

F = gear face width, in,

Q = tooth ratio factor,

ϕ = gear pressure angle, deg,

$$Q = \frac{2T_2}{T_1 + T_2} \text{ for spur gears} \tag{10}$$

where,

T_1 = number of teeth in driver,

T_2 = number of teeth in follower.

K_1 represents the load that will cause failure at a number of stress cycles and is derived from engineering tests. Engineering judgment requires that predictions should be based on 75% of the K_1 test data so that:

$$K_1' = \frac{K_1}{0.75} = 1.333\ K_1 \tag{11}$$

where,

K_1' = adjusted load stress factor, psi.

When K_1 has been obtained it is then possible to calculate a value for expected gear life also derived from (35) as:

$$L = \frac{\text{alog}\ (B-A\ \log K_1')}{60N} \tag{12}$$

where,

A = gear material constant,

B = gear material constant,

N = gear speed, rpm.

Required values of both A and B are taken from (36) as given in (35).

CONCLUSION

Limited Life Items are prime problems for the users of complex equipment. Users of equipment, that is subject to degradation in long term storage, have demanded storage tests and data generation to provide confidence of the operating success of the equipment (37). Equipment that employs accepted wear out items are subject to vigorous analysis in the design cycle in order to remove such components. The cost of ownership of poorly conceived equipment has grown to the point where it now exceeds the cost of purchase. As a result, increased emphasis on Limited Life Item study can be expected.

REFERENCES

1. ________________, MIL-P-116, Preservation - Packaging, Methods of, U.S. Department of Defense, Washington, D.C.

2. ________________, MIL-STD-107, Preparation and Handling of Industrial Plant Equipment for Shipment and Storage, U.S. Department of Defense, Washington, D.C.

3. ________________, MIL-E-17555, Electronic and Electrical Equipment Accessories, and Repair Parts, Packaging and Packing of, U.S. Department of Defense, Washington, D.C.

4. ________________, MIL-STD-1131, Storage Shelf Life and Reforming Procedures for Aluminum Electrolytic Fixed Capacitors, U.S. Department of Defense, Washington, D.C.

5. ____________________, MIL-R-6106, General Specification for Electromagnetic Relays, U.S. Department of Defense, Washington, D.C.

6. Fulton, D.W., "Nonelectronic Reliability", Reliability Workshop RAC, November 1976.

7. Jones, E., "A Guideline to Component Burn-In Technology", Wakefield Engineering Inc., 1972.

8. McCall J., "Renewal Theory", Machine Design, 25 March 1976.

9. Neuner G. and Barnett E., "An Investigation of the Ratio Between Stand-by and Operating Part Failure Rates", TRW Systems Group, November 1972, (GIDEP D7286).

10. ____________________, RADC-TR-67-307, Martin Marietta, 1967.

11. ____________________, MIL-HDBK-217, Reliability Prediction of Electronic Equipment, U.S. Department of Defense, Washington, D.C.

12. ____________________, MIL-STD-785, Reliability Program for Systems and Equipment Development and Production, U.S. Department of Defense, Washington, D.C.

13. ____________________, DI-L-1412, Shelf Life Data, U.S. Army, Washington, D.C.

14. Ficchi, R., "How Long-Term Storage Affects Reliability", Electronic Industries, March 1966.

15. ____________________, T-70-48891-007, "Handbook of Piece Part Failure Rates", GIDEP REF E031-1273.

16. Locks, M.O., Reliability, Maintainability and Availability Assessment, Hayden Book Company, Rochelle Park, N.J., 1973, p147.

17. Nelson, W. and Thompson, V.C., "Weibull Probability Paper", Journal of Quality Technology, 3, 1971, p45-50.

18. Flahie, M., "Reliability and MTF", Microwaves, July 1972, p36-40, 1972.

19. Black, J., "Electromigration", IEEE Transactions on Electron Devices, April 1969, p338-347.

20. Peattie, G. et al., "Elements of Semiconductor - Device Reliability", Proceedings of the IEEE, February 1974, p149-168.

21. ____________________, "McMos Reliability", Motorola B Series, CMOS Data Book, 1976, p6-2 to 6-18, 1976.

22. Lukach, V. et al., "Thermal-Cycling Ratings of Power Transistors", RCA Application Note AN-4793.

23. Colbourne, D. et al., "Reliability of MOS LSI Circuits", Proceedings of the IEEE, September 1973, p244-259.

24. Stick, M. et al., "Microcircuit Accelerated Testing Using High Temperature Operating Tests", IEEE Transactions on Reliability, October 1975, p238-250.

25. Gallac., L. and Whelan, C., "Accelerated Testing of COS/MOS Integrated Circuits", RCA ST-6379.

26. Curran, W., "Pick the Proper Lamps-Subminiature Lamps", Evaluation Engineering, January/February 1974, p4-10.

27. Russell, C.A., "How Environmental Pollutants Diminish Contact Reliability", Insulation/Circuits, September 1976, p43-46.

28. Davis, S., "Keyswitch and Keyboard Selections for Computer Peripherals", Computer Design, March 1973, p69-79.

29. Holbrook, W., "Environmental Test Reports 42259-1", Wyle Labs, (for Robinson-Nugent).

30. ____________________, Insulation Circuits Desk Manual, June/ July 1976, p45.

31. ____________________, Precision Instrument Bearings, 3rd Edition, Canada, FAG Bearings Limited, Stratford, Canada, p 23.

32. ____________________, Barden Precision Miniature and Instrument Ball Bearings, Catalog # M6, The Barden Corporation, Danbury, Conn., 1971.

33. ____________________, FAG Precision Bearings Catalog No. 41400, FAG Bearings Corporation, Stratford, Canada, 1974.

34. Booster, E.R., "When to Grease Bearings", Machine Design, August, 1975, p70-73.

35. Buckingham, E.K., "Accelerated Gear-Life Tests", Machine Design, 1974, p91.

36. Morrison, R.A., 'Load-Life Curves for Gear and Cam Materials', Machine Design, 1 August 1968, p102-108.

37. Gagnier, T.R., 'XM-712 COPPERHEAD (CLGP) Projectile Storage Reliability Verification Test Program', Proceedings of the IEEE, Annual Reliability and Maintainability Symposium, 1977, p353-360.

25 SPARING

R. L. Baglow

We cannot do better than to introduce our subject with a quotation from a basic text (1):

"Two separate classes of problems are involved in managing real property - management of inventory and management of durable equipment. Inventories supply items to the productive process; durable equipment supplies services."

In this chapter we discuss the management of durable inventory; but as we do so it is extremely important to understand the relationship of inventory management to the level of service provided by the durable equipment, and to the optimum method of maintenance which will provide the service at least cost. This formulation is the basis of a large and important part of economic theory both at the macro and micro levels. Here we will be considering inventory at the micro economic level, as it relates to the use of equipments in manufacturing, transportation, utilities and other services, and in military operations.

Another basic text in inventory management (2) reviews the fundamental theory. But of course a great deal of work has been published since these two important books. We will not attempt to give a complete survey of the literature here - that would take too much space. A review up to 1965 is given in (3). A forthcoming book (4), will cover more recent work.

There are a variety of motives for holding inventory. These are discussed fully in (2). In our context the motive is to counteract the unreliability of equipment. We can do this by having more equipments than are necessary to provide the required service, or by having parts, over and above those installed in the equipments, which can be used to restore them to operation when the installed part fails. We commonly encounter three or four indent levels of parts, but of course the hierarchy may be longer or shorter, e.g.,

(a) complete equipment (e.g. airplane),
(b) subsystem (e.g., weather radar),

(c) assembly (e.g. card in the radar),
(d) component (e.g. part of the card).

Any of the first three of these are candidates for spares. The lowest level of consumable spare parts is an important inventory item but for management purposes is on a different footing from spares which are durable items.

The equipment may have redundant parts built into it to improve the reliability, but we will not consider these to be spares, although their function is similar. Usually equipments will be operated intermittently or continuously over periods of time. A special problem arises when equipments are held on "alert" to be ready for operation under specified conditions, at times which cannot be predicted ((1), Section 2.3). The distinction becomes important when we are deciding how to measure the level of service provided by the equipments.

The "failure" of an item is a function both of the inherent physical behaviour of the piece of hardware and of the behaviour that is required if the hardware is to supply the desired service. Equipments may fail to give the "service" which is aimed at as a result of diverse physical and chemical effects, i.e.,

(a) random defects in manufacture,
(b) random defects induced by physical changes over time (e.g. oxidation, corrosion, crystallization, evaporation, migration of constituents, etc.),
(c) random defects induced in use by operation and maintenance (e.g. heat, vibration, contamination, thermal, electrical and mechanical shock, wear, etc.).

Essentially the time of a "failure" cannot be predicted and a statistical approach is forced. With few exceptions there is no physical theory leading to a statistical description of the failure process, especially as in practice the events which are identified as "failures" arise from the combined effects of many separate processes in a complex structure.

In most cases, therefore, we represent the failure process by a more or less arbitrarily chosen statistical function which at best may have been shown to be consistent with failure data. Appropriate functions are the well known reliability functions, e.g., Exponential, Weibull, Gamma and Normal. All these are discussed thoroughly elsewhere in this book (Chapters 7, 19). We have remarked immediately above that the observed failure process is the aggregated result of a number of separate processes. When we are interested in spares calculation the process may be even more complex, as we sum the failure processes over a set of equipments operating in parallel - machines in a factory, fleets of vehicles, and so on. It is fortunate that in these circumstances we can appeal to a general statistical theorem which gives the exponential distribution a privileged position ((5), Theorem 3.1). By this theorem, under plausible assumptions, the failure law of complex equipments tends to the exponential. This will be reasonable when we are considering a stream of failures from a set of equipments. An important practical exception occurs when we deal with small fleets of equipments exhibiting increasing failure rate.

MODELLING THE SERVICE PROVIDED BY DURABLE EQUIPMENT

We have indicated above that a basic goal of management is to relate the inventory of spares to the level of service provided by the durable equipment. Many measures of service have been proposed and used, but it is useful to distinguish two general types.

The first is based on economic considerations and is appropriate when the equipment, or equipments, are producing an item or a service for sale. The tradeoff in the analysis is between the costs of the inventory, and the costs of lost or delayed sales induced by failures and removals. The costs of inventory are grouped into two: an ordering, or set-up cost, and a holding cost. We should note that the ordering cost, in our case, may depend on the maintenance policy. Intangible factors, such as customer good will, may be included in the costs of lost or delayed sales. The basic problem is discussed in (2), Chapter 9. It has been shown that under rather general assumptions the optimal method of managing inventory is to specify two levels of inventory - a reorder point, and an order-up-to level. This approach to inventory management is called (s, S), or (r, R), depending on which textbooks are read. An order is placed (for repair, or, in the case of consumables, for purchase) whenever the inventory level falls to or below s (or r). The size of the workorder (or purchase order) is the difference between the level to which inventory has fallen, at the time it is observed to be below the order point, and S. It can be seen that both the order quantities and the times at which orders are placed are irregular. It is possible to order fixed quantities, and/or at regular intervals for convenience, but this is usually not the best, in the sense that the sum of the costs of ordering and holding inventory and of the costs of lost sales are presumably greater than if an (s, S) policy is followed. For calculation of the parameters see (6).

Figure 1 gives an illustration of the history of inventory with a demand for items (spare parts) which is random. Whenever the inventory falls to, or below, the reorder level, s, an order is placed for a quantity equal to the difference between the inventory level at the time the reorder is placed and the order-up-to level S. Occasionally the inventory falls to 0 before the order is received and a stock out occurs. If the unsatisfied demands are filled after the inventory level comes back up, the unsatisfied demands are converted into back orders. Otherwise the demands may be cancelled (the lost sales case). The lead time between placing an order and receiving it is itself usually variable. In most inventory systems the lead time is taken as fixed, as an exact treatment of random demand combined with random lead time is quite complicated. In the special case s=S-1 the theory allows the lead time to be a random variable.

The calculation of the two control parameters depends on the probability distribution of demands (i.e. failures, in our case) and the three cost factors. The calculations may be too onerous if a large range of inventory is involved, then approximations may be used. A common one is to set the difference S-s to be equal to the economic order quantity ((7), Section 2.2). In the management of repair this would be the economic batch quantity.

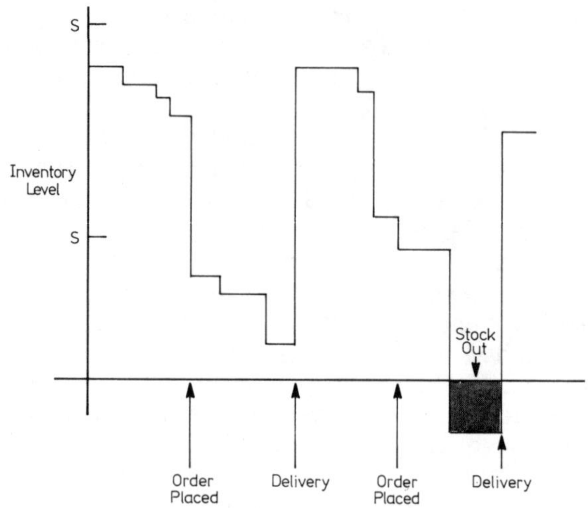

Figure 1. Typical History of Inventory

Intuitively it can be seen that the reorder level, s, will be larger the greater the penalty cost of lost or delayed sales, and the greater the rate of demand for the item. Also if the set up cost for an order is large we will want to place orders infrequently and so the difference S-s will tend to be greater. On the other hand if the cost of the item is large we will want to keep the inventory small and so the difference S-s will tend to be smaller. For items which are very expensive (relative to the order cost) we will initiate repair (or purchase) in quantitites of 1, that is whenever a failure occurs. This leads to useful mathematical simplification, of which many inventory models take advantage. Having selected the batch quantity for repair it remains to choose the reorder point, S, so that the total average annual cost is minimized - that is, the sum of ordering costs, holding costs and penalty costs (of lost or delayed sales). A complete discussion is given in (7), Chapter 4.

If the equipment is not being used to produce an item or service for sale, or if the operation of the equipment contributes only indirectly to such production, it becomes impractical to determine inventories of spares on cost factors alone. Such is the case, for example, for military equipments. When this is the situation it is necessary to measure the service provided by the equipment, not by a cost function, but in some other way. It seems best to base such a measure on fundamental reliability concepts: primarily the average availability of equipments and their reliability. The availability A is defined in the classical way by:

$$A = \frac{MTBR}{MTBR + MTTR} \qquad (1)$$

where MTBR is the mean time between removals and MTTR is the mean time to replace (or repair). These innocent looking parameters conceal some complex factors. If an equipment or piece is subjected to both unscheduled removals (failures) and scheduled removals (preventive maintenance) the calculation of MTBR is slightly involved ((1), Section 4.4). If the equipment is operated intermittently the "time" in MTBR is operating time, not clock time. Also in this case, the simple formula gives only an approximate value for the availability. The MTTR term is made up of a number of factors, some of which can also be complicated - particularly the contributions of time waiting for technicians and time waiting for spares. In many situations the time waiting for spares is the largest contributor to MTTR. Then the determination of the inventory of spares is fundamental for planning the effective employment of the equipments. The reliability, of course, measures the probability of operating without failure for a specified time, which we might call a mission. In some cases an appropriate measure of effectiveness is the product of the availability (i.e., the probability that an equipment is serviceable when it is wanted) and the reliability (i.e., the probability that if it is available it will complete the mission). One must be sure, however, of the independence of these two probabilities, or take account of the conditioning of the second probability. Having selected one or another measure of effectiveness, the inventory problem becomes one of attaining some specified level of effectiveness at minimum costs; that is, cost minimization under a constraint on effectiveness. (Alternatively, the problem is sometimes posed as maximizing effectiveness under a cost constraint).

MODELLING OF MAINTENANCE POLICIES

The modelling of maintenance is dealt with in Chapter 19 of this book. Since the selection of a maintenance policy has a significant impact on spares, these two pillars of support should not be considered independently of one another. Selection of a maintenance policy determines the locations at which stocks of spares for replacement and repair are required. The depth of maintenance will determine the indent level of the spares. The location of the repair facilities relative to the location of the operating equipments will determine the delay and cost of transportation. In some cases it may be advantageous to carry out a stock of spares at a location other than the repair facilities, that is at a central depot warehouse. Central warehousing of spares should not be decided upon unless it is shown to be advantageous by a logistic model.

Intuitively it is clear that it is simpler and faster to replace at the highest level in the hierarchy of spares. This will give a high availability of the equipment, and not require as highly skilled a technician as needed for repair at a greater depth. However, the high indent level spares will be much more expensive. The cost of spares will be reduced if replacement and repair can be carried out to a greater depth. Counterbalancing this, however, is the need for a more highly trained technician, more expensive test equipment, and a longer time for diagnosis and repair/replacement. If the operating equipments are dispersed at several locations there is a tradeoff between repair at each location (more technicians) and repair at a central point (more spares). It can be seen that the analysis leading to optimum repair level is a relatively complicated matter. In many cases a mixed policy can be shown to be most economical overall: that is, replacement at a relatively high indent level, and repair of these by replacement at a lower indent level at a second or third echelon of repair, as illustrated in Figure 2.

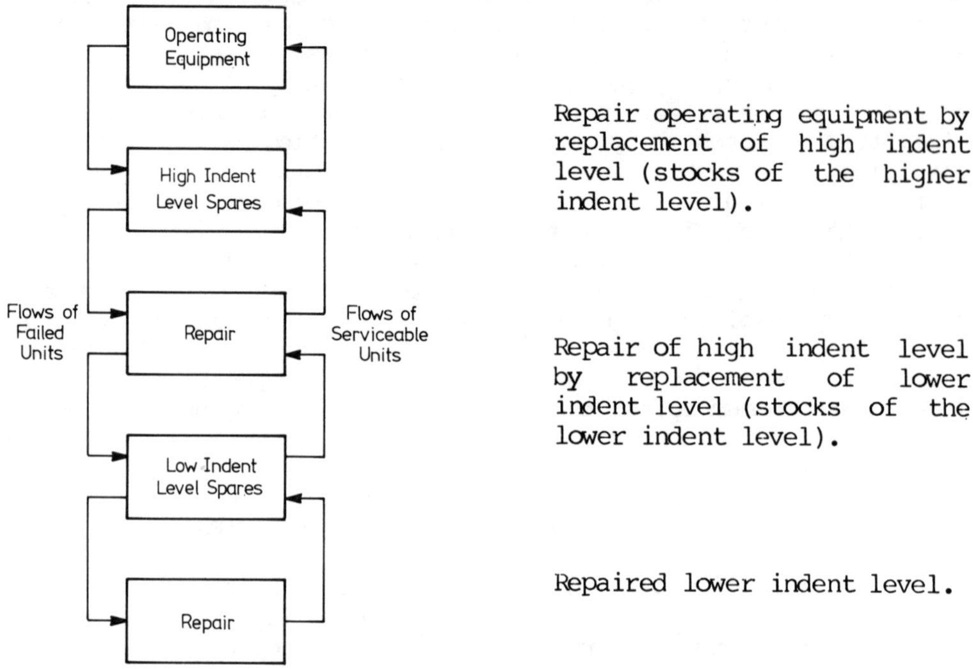

Figure 2. Equipment Repair Flow - Two Echelons of Repair

Downtime of the equipments resulting in non-availability of service results from, (a) the time to replace a failed unit; and (b) the delay incurred if a serviceable spare is not available on demand from the spares stock. It is this latter factor which is controlled by the quantity of spares. If there are no spares, every failure will result in downtime of an equipment for the length of time to complete the repair and return the unit. This time may include transportation delays, scheduling delays in the repair facility, as well as the active repair time. The more spares available, the shorter the wait will be, though it can never be reduced to zero in practice. If the repair facility can accept and work immediately on any number of units then the delays are reduced to a minimum. Usually this will not be the case; there will be a scheduling delay whose magnitude will depend on the capacity of the repair facility and on the inflow rate of failed units. The rate of flow of failed units, in its turn, will depend on the number of equipments which are operating; that is, those which have not failed and are not waiting for a replacement spare. To construct a complete mathematical model of such a situation is quite an undertaking and is usually only attempted for very important cases ((5), Section 5.3).

Some simplifications are often appropriate. One is to assume that the availability of the operating equipments does not vary greatly, so that the rate of occurrence of failed units is constant. This will be a reasonable assumption if the availability is high. If the rate of flow of units to the repair facility does not vary then the scheduling delay also can be assumed to be constant. The final simplification is to assume that the control policy is (S, S-1); that is no batching for repairs. The analytic advantage is that in this case the number of units in resupply depends only on the average delay required for the failed units to go to the repair facility, through repair and back to serviceable stock, and not on the distribution of these delays ((6), Section 4-13). The final simplification, which is almost always made, is to neglect the transient solutions to the mathematical models and consider only the steady state solutions to which the system is assumed to tend in the long run.

The simple model sketched in the above paragraph is the basis for many inventory management systems. Given the assumptions, it is easy to calculate various measures of effectiveness. If we are interested in effectiveness measured in economic terms, the most appropriate statistic to calculate is the average number of backorders. If effectiveness is measured in terms of availability, the appropriate statistic is the probability of 0 backorders. A number of extensions are easily made to this model.

We will illustrate this model with simple examples. If the total failure rate of the assemblies is assumed to be Poisson with mean λ , then the probability of observing n failures in a time period t is

$$(\lambda t)^n \exp(-\lambda t)/N! \qquad (2)$$

It can be shown that if we have a stock of spares, S, and the average repair time is T, then the probability of no backorders, i.e., that not one of the parent equipments will be down waiting for a spare of the item in question, is

$$\begin{aligned} P(\text{no backorders}) = {} & P(0 \text{ demands in period } T) + P(1 \text{ demand in period } T) \\ & + \ldots + P(S \text{ demands in period } T) \qquad (3) \end{aligned}$$

which is equal to

$$\sum_{X=0}^{S} (\lambda T)^X \exp(-\lambda T)/X! \qquad (4)$$

The repair time, T, is the time from the removal of the faulty item, to its return to stock in a serviceable condition. It includes all sources of delay - administration, scheduling, transportation, active repair, etc. If the availability of individual equipments is wanted we take the n th root of the quantity calculated in equation (4). Tables for calculating (4) are available in a number of texts; a rather complete set is given in (8). Apart from tables, however, Figure 3, can be used to estimate requirements as in the following example.

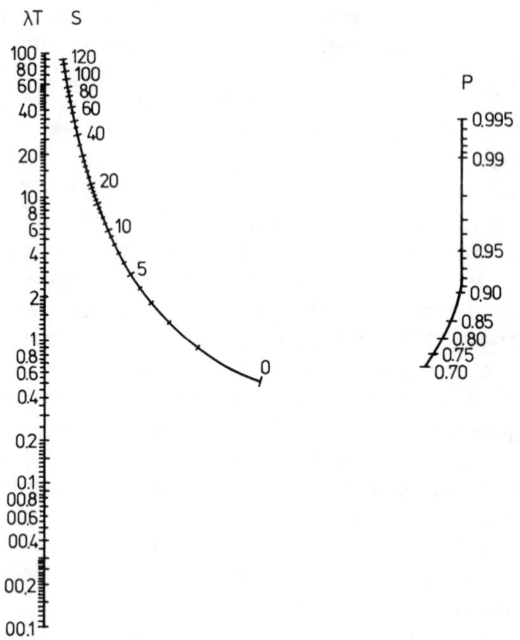

Source: NAVSHIPS 94324, *Maintainability Design Criteria Handbook for Designers of Shipboard Electronic Equipment*, Naval Ship Systems Command, Navy

Figure 3. Spares Requirement Nomograph

Consider the problem of sparing circuit cards for 4 systems, operated continuously at one site, where it is required to estimate the spares quantity knowing that the re-stocking time is 90 days and provided that 95% probability of no backorders is an acceptable risk. Each system has 2 memory cards of one type and a failure rate of $50F/10^6$ hrs each and therefore $\lambda = (4x2x50) = 400F/10^6$ hrs. T= 90x24 = 2160 hrs and $\lambda T=(400x2160)=$ 0.864. Entering Figure 3 with $\lambda T=0.864$ and P=0.95, draw a line from these two points and the spares quantity, S, is given at the point of intersection with the S line. In this case the result is 3 and the exercise is repeated for all items to be spared. Of course the same result can be obtained with a programmable, hand calculator of the type which has replaced the engineer's slide rule.

To continue the discussion of the example we can see that the probability of no backorders with a spares stock of 3 is 0.988. That is, 98.8% of the time none of the 4 systems will be unserviceable because of lack of a spare. Thus the availability of the 4 systems is $(0.988)^{1/4}$ = 0.997.

The average number of backorders is easily calculated:

Average backorders = 1 x P((S + 1) demands in time T)
+2 x P((S + 2) demands in time T)
+3 x P((S + 3) demands in time T)
+ ...

= 1 x 0.0098 + 2 x 0.0017 + ...

= 0.014.

The average duration of a backorder is:

0.014/0.0004 = 35 hrs.

This way of calculating downtime is appropriate when the replacement time is negligible compared to the downtime which accumulates waiting for spares. If this is not the case then the availability calculated from expression (4) will have to be multiplied by the factor 1 - R where R is the mean replacement time. Suppose, for example, λ = 0.01 per hour and T is 100 hours. Then Table 1 shows the availability, average backorders and average wait for spares which may be expected for various quantities of spare stock.

Table 1. Effects of Sparing

Spares	Availability	Average Backorders	Average Wait (Hours)
0	0.360	1.000	100.0
1	0.736	0.368	36.8
2	0.920	0.104	10.4
3	0.981	0.023	2.3
4	0.996	0.004	0.4
5	0.999	0.001	0.1

The reader should reflect on the extreme limitations of this model. Apart from the strong hypothesis of constant failure rate, it deals with only one item, with a single level of repair, and takes no account of the cost of the item.

If the parent equipment contains a number of replaceable/repairable items the availability of the parent equipment (as far as spares are concerned) is given by the product of the availabilities of the individual items. This product can become small unless the number of spares of each individual item is sufficiently great to give an availability close to 1. This view of the problem implicitly assumes that the failures of the items are independent of one another. This will certainly not be true if more than one indent level of spares is carried. This is because the failure rate of the higher level assembly will depend in whole or in part on the failure rate of the parts it contains. In some circumstances (which fortunately seem to be rare) secondary failures may occur.

One important generalization is to consider all the replaceable items for a given machine, or equipment, take account of their cost and use marginal economic analysis so as to make an optimal division of investment among the items, on either economic or availability considerations ((6), Chapter 6). This leads in a natural way to equipment system management of the spares inventory.

Another generalization, described in (9) and (10), is to take account of many stockage points and more than one echelon of repair. A further generalization allows two indent levels of spares (11). As long as we retain the simple Poisson hypothesis it is not difficult to generalize to more than two echelons of repair, or more than two indent levels of repair.

In principle, it is easy to allow batching of repairs and to generalize to other demand distributions than Poisson, but the calculations become much more onerous. Such models are also usually limited to fixed lead times for repair, unlike the basic Poisson model. A rather flexible approach is described in (12).

We will illustrate these ideas with some simple examples calculated with the aid of the METRIC model. This model, which was originally developed by the Rand Corporation, is a multi-item model; that is it considers a group of items all belonging to a single system. It is a two-echelon model, that is, it considers the possibility of repair at a (single) central repair facility, or at a number of secondary repair facilities. Finally it is a multi-location model, that is, it considers equipments operating at a number of different locations, some or all of which may have secondary repair facilities. The model considers the cost of each spare item and seeks to minimize the total cost of spares to give a specified availability (or average number of backorders). Conversely if the total budget for spares is filled, the model will determine the quantity of spares of each item in such a way as to maximize the overall availability. In its present form the METRIC model only considers one indent level of spares, but generalizations to more than one indent level are available (11).

The first example illustrates a basic case of one equipment, with five items, all items cost the same and are equally reliable. There is one echelon of repair, i.e., all repairs can be effected at the single repair facility, which we will represent as BRFR = 1.00. The repair time at this echelon we will call BRT. The repair time at a second echelon is DRT. There are ten equipments, and we set a target for equipment availability not less than 0.95. The results are shown in Table 2.

Table 2. METRIC Model Example 1

Item	Failures Per Year	Cost $	BRFR	BRT (Max)	DRT (Max)	Stock
1	10.08	300	1.0	0.5	3.0	2
2	10.08	300	1.0	0.5	3.0	2
3	10.08	300	1.0	0.5	3.0	2
4	10.08	300	1.0	0.5	3.0	2
5	10.08	300	1.0	0.5	3.0	1
Investment		\$2700				
Equipment Availability		0.9525				

If the items have different prices it is possible to obtain the same availability at less cost. This is the advantage gained by marginal analysis. The results are shown in Table 3.

Table 3. METRIC Model Example 2

Item	Failures Per Year	Cost (\$)	BRFR	BRT (Max)	DRT (Max)	Stock
1	10.08	100	1.0	0.5	3.0	3
2	10.08	200	1.0	0.5	3.0	2
3	10.08	300	1.0	0.5	3.0	2
4	10.08	400	1.0	0.5	3.0	2
5	10.08	500	1.0	0.5	3.0	1
Investment		\$2600				
Equipment Availability		0.9570				

If the prices are the same but the failure rates are different, we achieve the results shown in Table 4. If only one of the items can be repaired at the second echelon of repair, we obtain the results shown in Table 5.

SPARES UNDER PREVENTIVE REPLACEMENT POLICIES

We have mentioned above that most inventory models assume that the process has gone on for a sufficiently long time to reach a steady state.

In many practical situations the time required may be so long that it is questionable whether the steady state is reached at all. A particular case of this kind may be presented by durable items subject to wearout. Typically such items may be subject to age replacement, or more complicated policies ((1), Chapter 4). In the steady state the removal rates are easily calculated ((1), Section 4.4). If a fairly large population of operating equipments is involved, we may assume a Poisson distribution of removals with the above parameter. If the population is small, or if the steady state has not been reached this will be an inappropriate model. In most cases it will be too pessimistic.

Table 4. METRIC Model Example 3

Item	Failures Per Year	Cost $	BRFR	BRT (Max)	DRT (Max)	Stock
1	3.76	300	1.0	0.5	3.0	1
2	6.72	300	1.0	0.5	3.0	1
3	10.08	300	1.0	0.5	3.0	2
4	13.44	300	1.0	0.5	3.0	2
5	16.80	300	1.0	0.5	3.0	3

Investment $2700
Equipment Availability 0.9600

Table 5. METRIC Model Example 4

Item	Failures Per Year	Cost $	BRFR	BRT (Max)	DRT (Max)	Stock
1	10.08	300	1.0	0.5	3.0	2
2	10.08	300	1.0	0.5	3.0	4
3	10.08	300	1.0	0.5	3.0	2
4	10.08	300	1.0	0.5	3.0	2
5	10.08	300	1.0	0.5	3.0	2

Investment $3600
Equipment Availability 0.9585

The simplest method of meeting this situation is to set up a discrete Markov chain model ((5), Chapter 5). The transition probabilities for the matrix are readily calculated ((5), Chapter 5, (1), Section 7.4). Such a model can be used for forecasting when there is an arbitrary initial distribution of the ages of the items, that is, for non-steady state conditions. With it, we can calculate effectiveness functions, for a given stock of spares. These functions will, of course, vary with time.

SOME TYPICAL PROBLEM AREAS

Notwithstanding the utility of the logistics models we have identified above, it is all too apparent that drastic simplifications have been made for analytic convenience. Anyone using such models should be alert to the possibility that in some situations the assumptions may not be realistic. If historical data are available, statistical tests can be performed to check this.

A common difficulty is the lack of a precise specification of the parameters of the statistical distributions used in the logistics models. This difficulty is particularly acute when new equipments are being introduced. Notwithstanding the skill of reliability engineers the common experience is that forecasts of failure rates show considerable variances from those experienced after the equipments come into service. These forecast errors have serious consequences - either they result in buying too large an inventory of spares of some items, or, which is worse, too few. Experience has shown that in the total inventory of spares required for the support of a major equipment, relatively few - typically 2% to 5% - of these items account for most of the cost of inventory, and most of the downtime of the equipment. Therefore, they warrant particular attention. A rough rule of thumb for identifying such a class of items is to pick them out on the basis of unit cost. All too often these same items are found to show high failure rates. Errors in forecasting and reliability of these "key" items will have serious economic and operational consequences. Special effort should be devoted to determining their reliability, special attention should be paid to the procurement and the maintenance of these items, and they should be the object of selective management when they come into service.

A particularly valuable feature of the model described in (7), is the use of a Bayesian method for representing the failure rates. Judiciously used, this technique can improve confidence in achieving the desired level of support effectiveness. Naturally this can be achieved only at a certain economic penalty. Nevertheless this is a useful approach (see also (13)).

CONCLUSION

Responsibilities for procurement, inventory management, and maintenance are divided in some organizations and it may be difficult to adopt a unified 'system' approach to planning production and support. Unless a system approach is used, considering the whole useful life of the equipment, from introduction to disposal, the value of logistics models will not be fully realized. In short, the concept of support planning in this chapter is based upon the idea of life cycle management. A useful survey of logistics models for support planning has been published by the Rand Corporation (14).

The success of any of these models will depend on the correctness of the values of the parameters. In most practical situations the economic and operational consequences of relying on poor data are serious, so effort devoted to obtaining good data is well spent. The reliability parameter is usually the weakest, despite the best efforts of the reliability engineer. Failure data obtained from actual use are to be preferred, of course; but even when such data are available the user should verify completeness and correspondence with the planned use. An insidious feature of information systems is the hidden aggregation which tends to take place. Unsuspected aggregation can make the predictions of logistics models look ridiculous. This is because the probability functions are not independent of the units in which demands or failures are measured. The danger signal is an excessive ratio of variance to mean - a ratio greater than 5 is suspicious. In some cases a change in unit of measurement will enable more reliable predictions to be made (15). The well-known moral is: garbage in, garbage out.

REFERENCES

1. Jorgenson, D.W., McCall, J.J. and Radner, R., Optimal Replacement Policy, Rand-McNally, Chicago, Ill. 1976.

2. Arrow, K.J., Karlin, S. and Scharf, H., Studies in the Mathematical Theory of Inventory and Production, Stanford University Press, Stanford, Ca., 1958.

3. Veinott, A.T., Jr., "The Status of Mathematical Inventory Theory", Management Science, Vol. 12, No. 11, 1966, p 745-777.

4. Peterson, R. and Silver, E.A., Decision Systems for Inventory Management and Production Planning, John Wiley and Sons Inc., New York, N.Y., (to be published).

5. Barlow, R.E. and Proschan, E., Mathematical Theory of Reliability, SIAM Series in Applied Mathematics, John Wiley and Sons Inc., New York, N.Y., Chapter 2, Section 3.

6. Snyder, R.D., "Computation of (S,s) Ordering Policy Parameters", Management Science, Vol. 21, No. 3, October 1974.

7. Hadley, G. and Whitin, T.M., Analysis of Inventory Systems, Prentice Hall, Englewood Cliffs, N.J., Chapter 4.

8. Molina, E.C., Poisson's Exponential Binomial Limit, Van Nostrand, New York, N.Y., 1942.

9. Sherbrooke, C.C., "METRIC: Multi-Echelon Technique for Recoverable Item Control", Operations Research, Vol. 16, January-February 1968.

10. Clark, A.J., "An Informal Survey of Multi-Echelon Theory", Naval Research Logistics Quarterly, Vol. 19, No. 2, December 1972.

11. Muckstadt, J.A., "Model for Multi-Item, Multi-Echelon, Multi-Indenture Inventory System", Management Science, Vol. 20, No. 4, December 1973, Part I of II.

12. Wagner, H.M., Statistical Management of Inventory Systems, John Wiley and Sons Inc., New York, N.Y., 1962.

13. Brown, Jr., G.F. and Rogers, W.F., "A Bayesian Approach to Demand Estimation and Inventory Provisioning", Naval Research Logistics Quarterly, Vol. 20, No. 4, December 1973.

14. Paulson, P.M., Waina, R.B. and Zacks, L.H., Using Logistics Models in System Design and Early Support Planning, R-550-RR, Rand Corporation, February 1971.

15. Burgin, T.A., "The Gamma Distribution and Inventory Control", Operational Research Quarterly, Vol. 26, No. 3, 1 September 1975, p 507-525.

26 MANUALS

D. T. Ingram

In today's environment of ever increasing complexity, and changing techniques in electronics, the role of documentation is more important than ever. Besides containing the more usual attributes such as clear and simple English, non-ambiguity, clear illustrations, etc., documentation must now contain that which is really relevant to the end task. This is certainly the case with equipment manuals, particularly those directed toward the maintenance aspects of the industry. The time is past when a few well chosen words on circuit operation coupled with the odd schematic, wiring diagram, and spare parts list, would suffice to point the Service Technician or Field Engineer in the right direction. Such individuals are today professionals and specialists in their own right and as such, expect to have at their disposal professionally written, adequate technical documentation to aid them in their mammoth task of cost effective maintenance for large and complex electronic systems.

However, the Field Engineer's requirements are not the only points to be considered when producing manuals. In certain areas, such as the military aspect of the electronic business, the guidelines for the content of manuals of all descriptions, are well defined. This is true irrespective of which side of the Atlantic the work is directed. It is necessary, therefore, to consult with all the relevant Military Specifications and Standards.

In giving consideration to the Maintenance Manuals, the more important points for attention fall into such categories as:

(a) the general style in which the material is written, or illustrated,
(b) the section of the industry for which the final product is aimed,
(c) fundamental aim of the publication,
(d) the style of contents required,
(e) the validation required.

GENERAL

Technical Writing

There exist on the market today, numerous publications which delve into the art and profession of technical writing, and it is not the object of this section to teach technique. However, there are a few ground rules which are worthwhile noting:

(a) irrespective of the language used, only simple terms and phrases conveying the intended meaning should be utilized,
(b) language should be free from ambiguous and vague terminology. There are standards which set out recognized definitions, terminology and abbreviations,
(c) the text should contain all relevant and essential information pertinent to the topic under discussion, either directly or by reference,
(d) terminology used should be consistent throughout the manual or set of manuals,
(e) sentences should be short, concise and to the point,
(f) punctuation should be kept to a minimum to aid reading,
(g) with regard to the content, always include sufficient information to ensure peak performance of equipment or system,
(h) aim to keep cross-referencing to other paragraphs, sections, chapters, etc., to a minimum,
(i) preparation of all draft work should be double spaced. Subsequent correction or additions are then much more easily discernible.

Observance of the above when writing original material will aid the generation of text that has continuity, is easy to read, and is not open to mis-interpretation.

Technical Illustrations

The only intent here is to discuss under what circumstances illustrations should be used, and when produced, the attributes that the finished item should possess. Apart from the use of the more obvious types of illustrations such as existing engineering drawings, schematics and the like, illustrations can be considered as useful tools, which:

(a) describe and enlarge upon an item or an idea,
(b) call attention to details which would be difficult or lengthy to describe by text alone,
(c) clarify points in the text,
(d) present sequences or phases that would be difficult to describe by text alone,
(e) describe tools and parts.

When generating illustrations the aim must be to produce an original sketch or schematic, with layout and detail as near as possible to that expected in the finished article. Problems can arise when utilizing extracts from engineering schematic diagrams. To illustrate a point in circuit functions, Figure 1 (a) details an area of logic as it appeared in the production schematic. Figure 1 (b) shows the rearrangement of the portion of the circuit when redrawn to aid in the unit description contained in a hardware manual. The point to be understood is that a cursory glance at the logic of Figure 1 (a) leaves an impression of parallel functioning of the flip flops U2, U3, and U4. Valuable time is wasted in deciphering the true intent of the circuit, which is a sequential operation, as shown in Figure 1 (b).

In producing an illustration the following general guide lines should be observed:

(a) ensure that the scale of reproduction is adequate for the amount of detail to be shown,
(b) line drawings must be capable of reproduction. Inks and other materials used must be capable of maintaining high density tonal values,
(c) when determining the proximity of lines, consideration must be given to the amount of reduction to which the final illustration might be subjected,
(d) colors should be used only as a last resort,
(e) if engineering drawings are used, all borders, title blocks and irrelevant material should be removed,
(f) symbols used should conform to the relevant standard in force. For example, in North America the accepted standard for logic symbols is the American National Standards Institute (ANSI), ANSI-Y32.14 (1), while in Great Britain the same area of symbology is covered by the British Standards Institue document BS 3939 (2) Section 21,
(g) if a photograph is used, aim for covered exposure, and ensure that the picture is free of dark areas which could mask out detail or create false impressions. Any retouching should be kept to a minimum.

Positioning of illustrations with respect to the relevant text is an important point which must be given careful consideration. There are differing ideas, that range from inserting illustrations on the same or following page after the first reference is made, to grouping all illustrations, particularly fold outs, at the end of each chapter. Whatever the positioning, the primary idea to be kept in mind, particularly when a technical description is involved, is continuity of reading. There is nothing more disconcerting for the reader than continual interruption of reading to find the relevant illustration. Although equipment design is seldom influenced by maintenance manual requirements, manuals which are difficult to follow lead to increased equipment downtime.

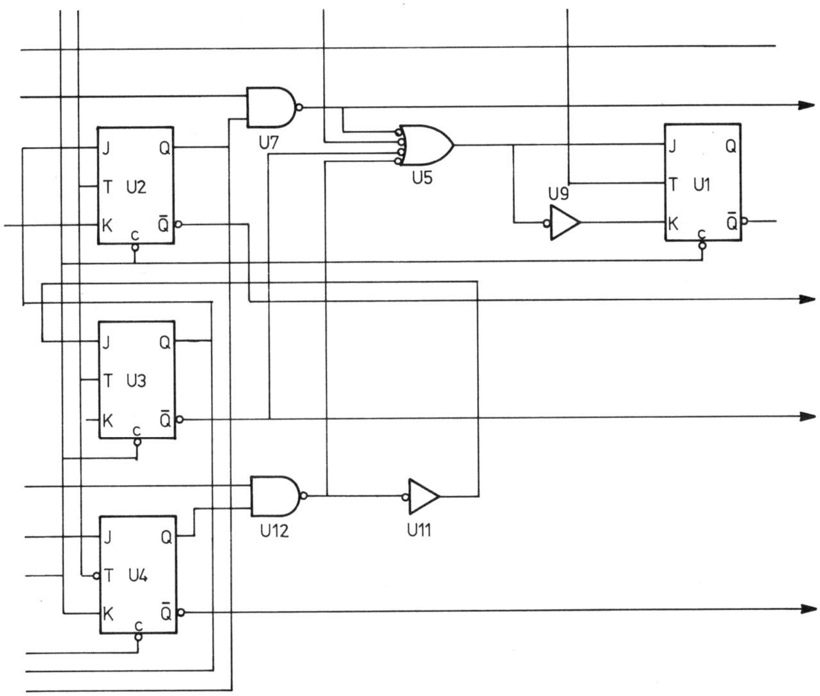

(a) Production Schematic

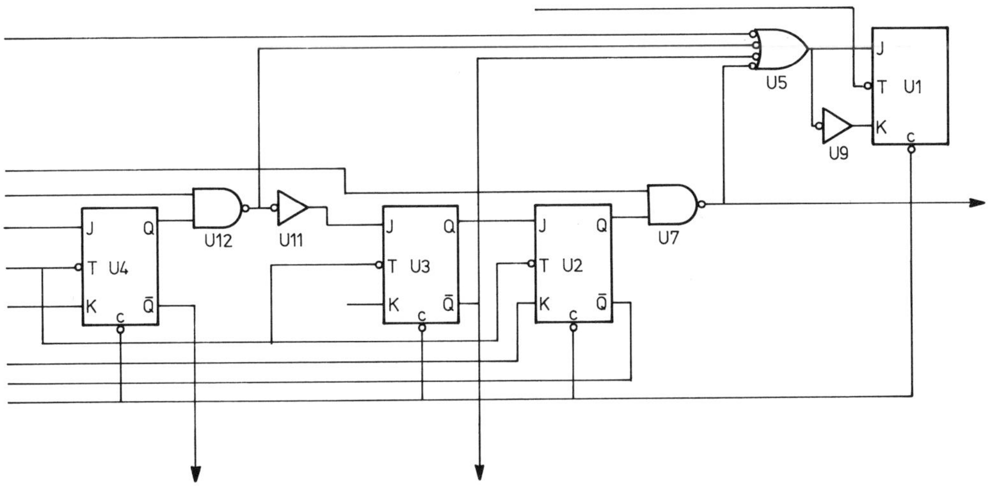

(b) Schmatic Redrawn to show Flow

Figure 1. Diagramatical Flow

SPECIFIED MANUAL REQUIREMENTS

Introduction

A customer's requirements for technical documentation to cover the equipment he is about to purchase, can vary from an obscure statement in an equally obscure clause of the job contract, to the point of being defined almost to the last comma or full stop. Generally, the situation is influenced by such factors as: the customer's field of operation, military, space or commercial, and whether the equipment is 'one off', or a production run. It is obvious, for instance, that one would not allocate a large budget for the production of manuals, when only one equipment is to be produced. At the other end of the scale, the military have a whole series of standards and specifications that define the production and acceptance of technical documentation such as maintenance manuals. Certain areas of the commercial market, such as civil aviation, also impose stringent requirements. Generally speaking, the areas where requirements are less stringent are those where the equipment is purely experimental, a prototype, or a 'one off' special. Even so, equipment in these categories have a habit of turning into a more substantial production run. It is therefore prudent, when preparing documentation for this group, to use as a base one of the more well known standards.

Usually requirements can be divided into three main areas, i.e.:

(a) General Requirements,
(b) Requirements for Contents,
(c) Requirements for Verification, Submission and Approval.

Although these areas are somewhat interrelated, each is worthy of some individual discussion.

General Requirements

At the outset of the task, the general areas and guidelines for the production of the manual must be defined. The establishment of an official channel of communication between the contractor and the customer department responsible for final acceptance is essential, and in military work, will almost certainly be mandatory. From this liaison, and the controlling documents, will emerge a finalized version of the General Requirements. The channel of communication will also provide a useful connection throughout the duration of the task, putting those responsible for production in touch with those who will accept the documentation, for the resolution of such queries that will almost certainly arise on both sides.

What are these general requirements? Perusal of MIL-M-38784 (3), or such British documents as TPN(L) 114 (4) or AVP 70, Spec. 0, (5) will certainly leave the reader in no doubt. The earlier sections of ATA 100 (6) for civil aviation also contain facts about requirements. For the writer seeking information, these volumes are certainly to be recommended.

Because of the detail contained in the aforementioned references, the following is offered as a précis of the subject:

(a) liaison, details of planning and progress meetings, schedule of progress reports to be submitted, solution of queries,
(b) the overall book plan, detailing the volume, part and chapter layout,
(c) type of copy to be presented, i.e., manuscript and/or reproducible,
(d) copy layout: an almost endless string of details, stating page layout, margining, typescript areas; paragraphing, numbering systems, table and illustration numbering; types and sizes of illustrations, etc.,
(e) the versions of reproducible copy which are required, i.e. preliminary and final manuals, changes, revisions,
(f) security classifications to be enforced,
(g) the allowable terminology,
(h) the system of change implementation to be adopted,
(j) the type of binding of copy,
(k) quality assurance provisions.

And so the list could continue. In view of the depth of detail it is important to be fully aware of the controlling documents and their contents.

Requirements For Contents - Hardware Manuals

Until the last decade the contents of manuals had always contained what have now become known as the traditional contents. Recently the somewhat revolutionary content format, known generally as Symbolic Integrated Maintenance Manual (SIMM), has been developed and is discussed in more detail later in this chapter. Not everyone agrees that SIMM is an economical proposition, and there is a polarization of opinion, between the diehards for the retention of tradition and those who would sweep the manual business with SIMM. Whatever the contents and their format, it is again important to know just what are the controlling documents. The following is aimed as an insight into the two content format systems.

Conventional contents. Typical documents for the definition of contents and their format are MIL-M-1501 (7), ATA101(8), (4), (5) and (6). It is interesting to note that their contents are fairly similar, and fall into the topics of: general information, operation, functional description, maintenance, troubleshooting, parts list and installation. Table 1 is a comparison of the topics and their order as specified by four different control documents using (7) as the base. As can be seen from the comparison, titles are not always identical or associative, yet the items that appear as oddmen out will undoubtedly fit into one topic or another. Typically the contents for each topic are outlined in the following paragraphs.

Table 1. Comparison of Manual Contents

TOPIC NO.	CONTROL DOCUMENTS			
	MIL-M-15071 Equipment and Systems	MIL-M-38797 Operating and Maintenance	ATA 101 Ground Equipment	TPN(L) 114 General and Technical
1	General Information	Introduction and General Information	Maintenance	General and Technical Information
2	Operation	Special Tools and Test Equipment	Maintenance	General Orders and Modification
3	Functional Description	Preparation for Use and Shipment	Overhaul	Equipment and Spares Schedules
4	Schedule Maintenance	Operation Instructions	Parts List	Progressive Servicing (Aircraft) Schedule
5	Trouble Shooting	Maintenance Instructions	Manufacturers Appendices	Basic Servicing Schedules
6	Corrective Maintenance	Diagrams	- - -	Repairs and Re-conditioning
7	Parts List	Parts Breakdown	- - -	10 Servicing Diagrams and Data
8	Installation	Difference Data Sheets	- - -	- - -

General information. The content shall be such as to permit management or command personnel rapid assimilation of equipment purposes, functions, physical size, operational characteristics, etc.

Operation. This part of the chapter should include such items as operating instructions and procedures, including emergency procedures, descriptions of all controls and indicators, protective devices and jacks.

Functional description. This is a description in simple technical terminology, supported by equally simple line diagrams. It should not be restricted to electrical operations but should also include mechanical operations. An ideal technique is to use a building block type of system, commencing with an overall block diagram, and then working down to detailed description of the various blocks. A three level system is generally sufficient, viz,

(a) overall (block diagram),
(b) major functions (block or functional),
(c) circuit level (schematics)

Scheduled Maintenance. Scheduled maintenance includes such items as preventative maintenance schedules and performance test instructions. Planned maintenance using official forms is permissible. Periodicity is another important topic for discussion. Description of tools and special test equipments should also be included.

Troubleshooting. This topic should include information guides and diagrams to lead to the logical isolation of faults, utilizing indications and readings available from the equipment. Diagrams of signal flow, power distribution, control and, if applicable, piping and hydraulic flow schemes.

Parts list. This topic should identify all repair parts and hardware. Information on part manufacturers should also be included. The requirements of certain areas of industry and service dictate that parts lists shall be illustrated. This requirement is usually implemented with line diagrams of exploded views of assemblies or sub-assemblies which enumerate those parts included. Normally pages opposite diagrams are tabulations of the parts with all relevant information.

Installation. All relevant information and drawings concerned with the installation of the equipment should be included under this topic heading. Diagrams should be included to cover cabling, wiring, piping, and outline and mounting. Information concerning unpacking, foundations, power requirements and any special environmental considerations should be discussed in detail. Post-installation verification test details may also be a topic for inclusion, though this could well be the subject of a separate document.

In conclusion, it should be remembered that the above discussion is only a brief summary of the detailed information that will be required, and only the controlling document can define the exact nature of contents.

Integrated Maintenance Manuals

During the last decade, a form of manual, new in concept but still retaining traditional elements of contents, has appeared. The Integrated Maintenance Manual, or to give its full title, The Symbolic Integrated Maintenance Manual (SIMM), was the subject of an investigation by the Systems Effectiveness Laboratory Technical Operations Inc. (9) in the early 1960's, and was subsequently further developed by their British interest, Tec-Script Division of Joyce Loeble Ltd. The effectiveness of the SIMM is disputed but, authorities have adopted, or even adapted the basic concept for their own use. The US Military have implemented the Manual for certain of their missile systems, and have produced specification MIL-M-24100(10), covering Functionally Oriented Maintenance Manuals (FOMM), for equipments and systems. The British Royal Navy have adopted a modified version, Functionally Identified Maintenance System (FIMS), as the basis for naval maintenance documentation. On either side of the Atlantic industrial corporations have, or are modifying SIMMS, FIMS and FOMM's to their own requirements. Since the original concept, papers have appeared in the technical journals of various institutes, an example being reference (11).

The somewhat revolutionary concept of SIMM is based on the supposition that one good illustration is worth 10,000 words of text. Its approach is purely pictorial, keeping the written word to a minimum. Figure 2 illustrates the SIMM approach to the overall maintenance manual associated with an electronic equipment, Figure 3 illustrates typical data and parts information. Maintenance data can take the form of comprehensive waveform and timing diagrams, or a maintenance dependency chart; yet another development of SIMM.

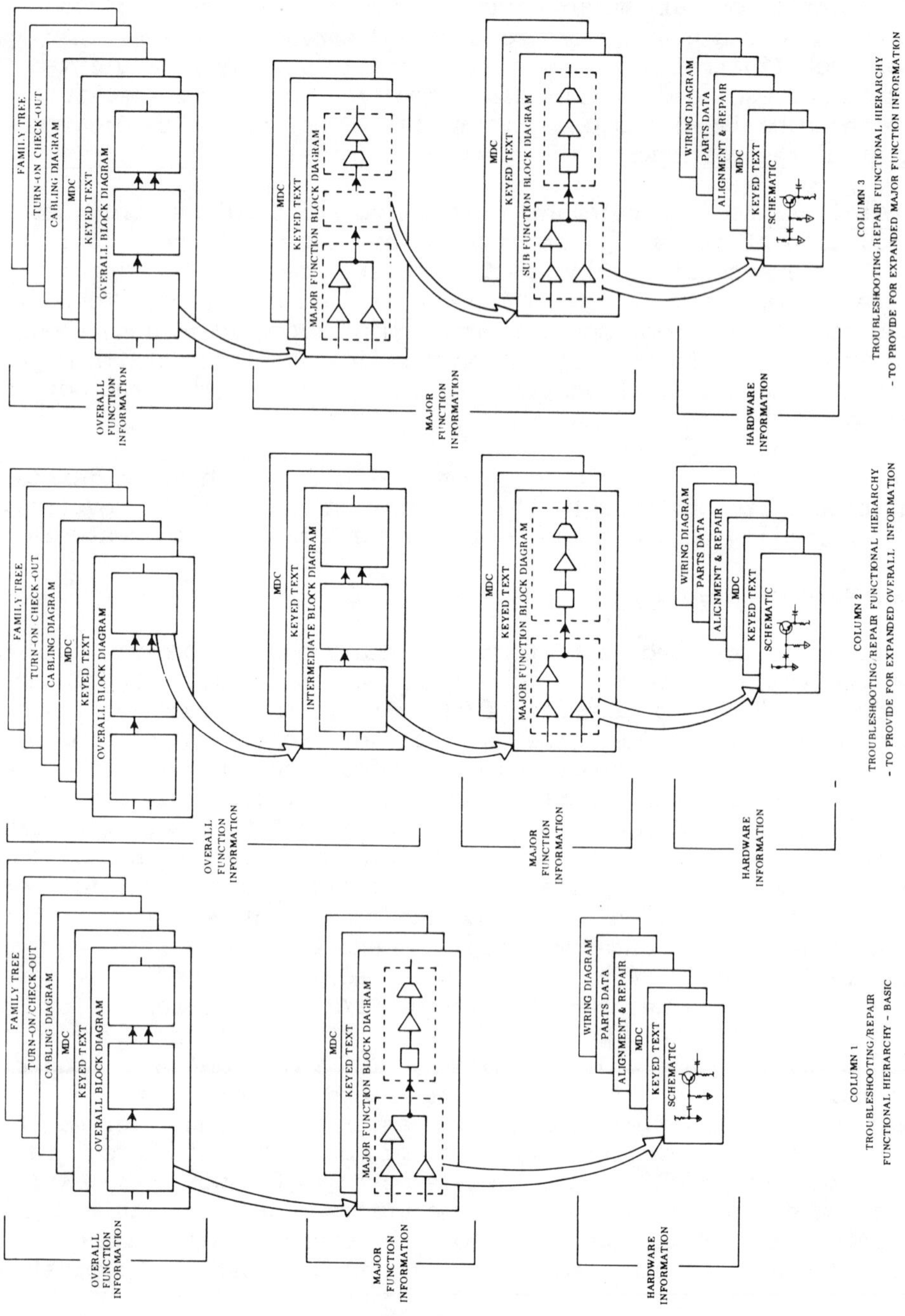

Figure 2. Troubleshooting/Repair Volume Organization

The maintenance dependency chart is a detailed diagrammatic representation of signal flow through the subject material part of such a chart, illustrated in Figure 3, bottom left hand side. It should be noted, that although the examples shown here are electronic in nature, the principles can be easily applied to other fields of technology.

In broader terms, at least to growing sections of industry, the SIMM system approaches the ultimate in maintenance manuals. There is evidence, extracted from tests carried out by various military and commercial organizations, to suggest that use of SIMM or one of its off-shoots, can cut equipment 'down' times in half, aiding cost effective maintenance. Whatever its merits, the SIMM is only as good as the information it contains. Accuracy and detail are essential. The inability of the maintenance dependency chart to cover all of the failure modes for any given piece of equipment, is perhaps the only real failing in the system.

One big disadvantage of SIMM is the cost aspect. The average SIMM printing costs are about 75% to 100% more than a conventional publication, while authors and illustrators need training in special techniques associated with this type of work. These considerations are apparently sufficient to kill the interest of a large number of equipment manufacturers and their customers, whose aim is to keep the cost of documentation to a minimum.

Requirements for Contents - Software Manuals

The preceding paragraphs have dealt with manuals specifically oriented toward hardware maintenance. However, in this age of computers and data processing, documentation of the software aspects of system design implementation and maintenance hold an equal placing in the maintenance hierarchy. From a global viewpoint, although format may differ, manuals will normally dispense similar information. A typical standard which sets out the hierarchy for software documentation as a whole is FIPS PUB 38 (12). The following is a summary of the manuals required under this standard.

User Manual. The scope of the user manual is such as to describe the functions of the software. The information is intended to permit the user to determine when and how to use it. The user manual should contain information on the following topics:

(a) general information,
(b) application,
(c) procedures and requirements,
(d) outputs,
(e) error and recovery,
(f) file query.

Operations Manual. Provides operations personnel with details of the software and its environment, thus permitting its use. Normally it contains information on the following topics.

SYMBOL	EXPLANATION OF SYMBOLS
[A]	AVAILABLE SIGNAL WHICH IS WITHIN SPECIFICATION
IND	INDICATION (DIGITAL READOUT, AUDIO INDICATION, ETC.)
LIT	INDICATOR LAMP LIT
OUT	INDICATOR LAMP NOT LIT
[ON]	STATE OF A DEVICE (EXAMPLE: ANTENNA BRAKE APPLIED, ETC.)
[OFF]	STATE OF A DEVICE (EXAMPLE: ANTENNA BRAKE NOT APPLIED, ETC.)
[A1]	SIGNAL MEETS FIRST SET OF REQUIRED CONDITIONS
[A2]	SIGNAL AVAILABLE AT SAME MEASUREMENT POINT UNDER A SECOND SET OF CONDITIONS.
[L]	LOW LOGIC LEVEL, 0 VOLTS
[H]	HIGH LOGIC LEVEL, +5 VOLTS
[NA]	NOT AVAILABLE
[NM]	NOT MEASURABLE
[EN]	RELAY ENERGIZED
[I/H EN]	RELAY ENERGIZED I INITIALLY ENERGIZED H HELD ENERGIZED
[DEN]	RELAY DEENERGIZED
●	INDICATES FUNCTIONAL ENTITY OR CIRCUIT ELEMENT THAT MUST BE PERFORMING ITS FUNCTION PROPERLY FOR THE EVENT ([A]) ON SAME LINE TO OCCUR WITHIN SPECIFICATIONS.
◑	INDICATES FUNCTIONAL ENTITY WHICH IS PROVEN PARTIALLY GOOD IF EVENT ([A]) ON SAME LINE OCCURS WITHIN SPECIFICATON. TO PROVE FUNCTIONAL ENTITY FULLY GOOD, EVENTS ON AT LEAST TWO DEPENDENCY LINES, OF WHICH FUNCTIONAL ENTITY IS A PART, MUST BE OPERATING WITHIN SPECIFICATION.
●̸	INDICATES OUTPUT OF FUNCTIONAL ENTITY IS ADJUSTABLE.
•/	INDICATES RELAY CONTACT SET THAT PROVIDES CONTINUITY IN THE DEENERGIZED STATE.
/•	INDICATES RELAY CONTACT SET THAT PROVIDES CONTINUITY IN THE ENERGIZED STATE.
▲	INDICATES THAT EVENT OR EVENTS DIRECTLY ABOVE DEPENDENCY MARKER MUST BE AVAILABLE (AND WITHIN SPEC) FOR EVENT ON SAME LINE AS DEPENDENCY MARKER TO OCCUR (WITHIN SPEC) PROVIDED OTHER REPRESENTED FUNCTIONAL ENTITIES ALONG THE LINE ARE PERFORMING PROPERLY.
▼ ↞	INVERTED DEPENDENCY MARKER USED IN CONJUNCTION WITH FEEDBACK ARROW (↞) TO DENOTE FEEDBACK LOOP.

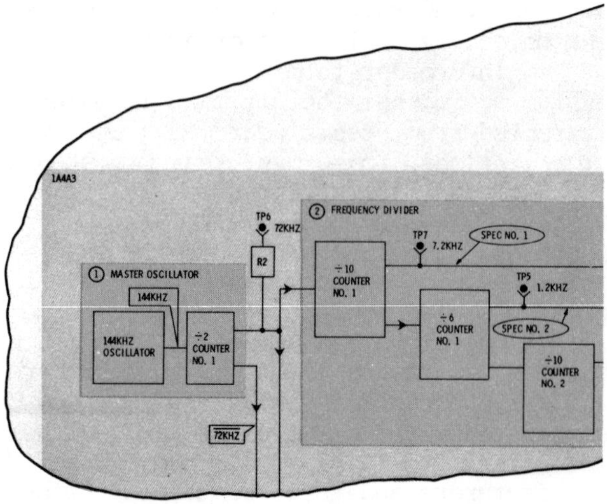

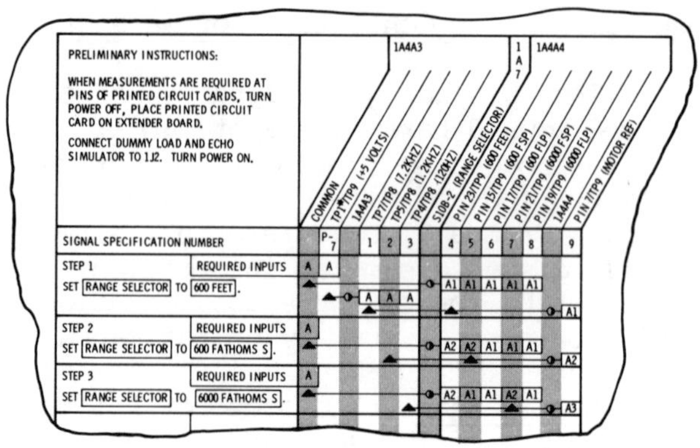

Figure 3. Typical SIMM Data

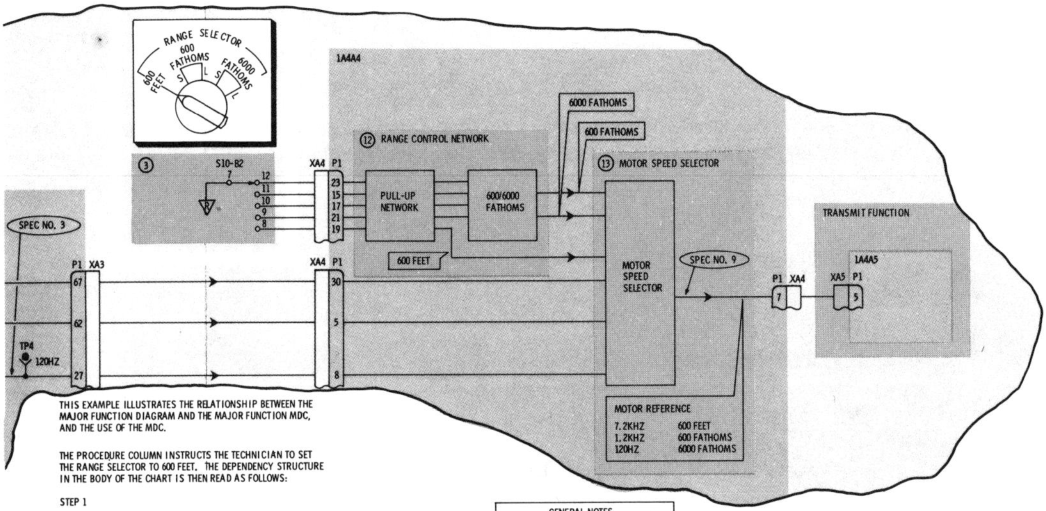

THIS EXAMPLE ILLUSTRATES THE RELATIONSHIP BETWEEN THE MAJOR FUNCTION DIAGRAM AND THE MAJOR FUNCTION MDC, AND THE USE OF THE MDC.

THE PROCEDURE COLUMN INSTRUCTS THE TECHNICIAN TO SET THE RANGE SELECTOR TO 600 FEET. THE DEPENDENCY STRUCTURE IN THE BODY OF THE CHART IS THEN READ AS FOLLOWS:

STEP 1

1. DEPENDENT ON THE AVAILABILITY ([A]) OF COMMON AND IF S10 (RANGE SELECTOR) IS PARTIALLY GOOD (◑), AVAILABILITIES ([A]) WILL BE MEASURABLE AT PINS 23, 15, 17, 21, AND 19 OF 1A4A4. THE SIGNAL SPECIFICATION TABLE DEFINES THESE MEASUREMENTS AS 0 VOLTS AT PIN 23 AND +5 VOLTS AT THE OTHER FOUR PINS.

2. DEPENDENT UPON THE AVAILABILITY OF THE +5 VOLTS AT TP1 AND IF 1A4A3 IS PARTIALLY GOOD, THREE AVAILABILITIES WILL BE PRESENT. ONE AT TP7, ONE AT TP5, AND ONE AT TP4 OF 1A4A3. PHOTOGRAPHS OF OSCILLOSCOPE PRESENTATIONS ARE USED TO DESCRIBE THE WAVEFORMS AT THE TEST POINTS.

3. DEPENDENT ON THE AVAILABILITY OF A CORRECT SIGNAL AT 1A4A3 TP7 AND THE SIGNAL AT PIN 23 OF 1A4A4, AND IF 1A4A4 IS PARTIALLY GOOD, A SIGNAL WILL BE AVAILABLE AT PIN 7 OF 1A4A4.

   NOTE THAT THE MDC DOES NOT ATTEMPT TO DESCRIBE WHAT THE SIGNAL AT PIN 7 WILL BE IF 1A4A4 IS FAULTY OR THE REQUIRED INPUTS ARE NOT PRESENT. A LIST OF ALL POSSIBLE WRONG SIGNALS OR VOLTAGES ALONG WITH A LIST OF ALL PROBABLE CAUSES WOULD BE OVERWHELMING.

   WHAT IS GIVEN, IS THE CORRECT SIGNAL, THE CIRCUITRY THAT PRODUCED THE SIGNAL, AND THE INPUTS THAT WERE REQUIRED BY THE CIRCUITRY TO PRODUCE THE SIGNAL. SIMPLE LOGIC DEMANDS THAT IF THE INPUTS ARE CORRECT AND THE OUTPUT IS INCORRECT, THE CIRCUIT MUST BE AT FAULT. THE SAME LOGIC ALSO IMPLIES THAT IF THE INPUTS ARE NOT CORRECT, THE PROBLEM LIES IN SOME PREVIOUS CIRCUITRY.

STEP 2

4. AFTER THE TECHNICIAN HAS SET THE RANGE SELECTOR TO THE 600 FATHOMS S RANGE THE SIGNAL AVAILABILITIES AT PINS 23 AND 15 CHANGE VALUE. PIN 23 IS NOW +5 VOLTS AND PIN 15 IS 0 VOLTS. THE OTHER THREE PINS RETAIN THEIR ORIGINAL +5 VOLT SIGNALS.

5. A DIFFERENT SIGNAL WILL BE AVAILABLE AT PIN 7 OF 1A4A4 UNDER THE EQUIPMENT CONDITIONS SET UP BY STEP 2.

GENERAL NOTES
UNLESS OTHERWISE SPECIFIED:
A ALL SCOPE PHOTOS WERE TAKEN WITH INPUT COUPLING SET TO DC.
B MEASUREMENTS ARE ±10%.

REQUIRED TEST EQUIPMENT
MULTIMETER AN/USM-183
OSCILLOSCOPE AN/USM-281A

SPEC NO.	SIGNAL SPECIFICATION	SPEC NO.	SIGNAL SPECIFICATION
P-7	+5 VOLTS, ±5%	4A1 A2	0 VOLTS +5 VOLTS
1	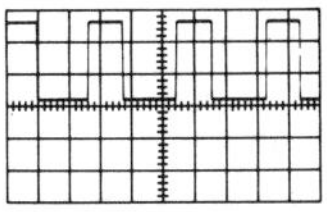 PROBE: X10 VERT: 0.2V/CM HORIZ: 50USEC/CM SYNC: AC + INT	5A1 A2	+5 VOLTS 0 VOLTS
		6A1 A2	+5 VOLTS 0 VOLTS
		7A1 A2	+5 VOLTS 0 VOLTS
2	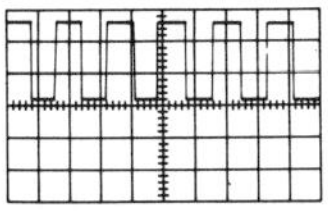 PROBE: X10 VERT: 0.2V/CM HORIZ: 0.5MSEC/CM SYNC: AC + INT	8A1 A2	+5 VOLTS 0 VOLTS
		9A1 A2 A3	COMPLEMENT OF SPEC NO. 1 COMPLEMENT OF SPEC NO. 2 COMPLEMENT OF SPEC NO. 3
3	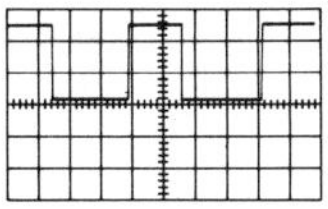 PROBE: X10 VERT: 0.2V/CM HORIZ: 2MSEC/CM SYNC: AC + INT		

(a) general information,
(b) overview,
(c) description of runs,
(d) non routine procedures,
(e) remote operations.

Program Maintenance Manual. Provides the maintenance programmer with sufficient information to promote an understanding of the program, operating environment and maintenance procedures. Information will be listed under the following topics:

(a) general information,
(b) program description,
(c) operating environment,
(d) maintenance procedures.

Validation and Verification

Validation and Verification fall into the area of Quality Control. Validation is the process whereby the contractor proves that the publication is technically accurate and meets the requirements in all other respects. Verification, normally called for when an outside procuring agency is involved, is the customers opportunity to test the publication for accuracy. The two areas may interlock; for example, the customer may choose to witness the contractors validation test of any maintenance material, instead of carrying out their own isolated verification tests.

As far as the contractor is concerned the process of validation forms an integral part of the overall plan defined by the general requirements. All material changes or corrections, made as a result of engineering tests, service tests and similar checks, are normally subject to revalidation.

Records are an important part of validation, particularly where outside agencies are concerned. These should show dates of validation reviews, the material reviewed, the findings and any corrective action to be taken. They will be subject to review by the procuring agency or customer.

CONCLUSION

One word, above all others, must be borne in mind by the potential technical writer. That word is 'criticism'. If the writer's personality is such that he or she cannot accept criticism, forget about writing. This is not advice which is given glibly; it is offered for the following reasons.

Almost any work which involves writing, like a good many other arts and crafts, is subject to comment by others. In the case of technical manuals, the comments arise during verification and validation. No matter how well the writer completes the task, there will be comment. Most of it will appear to the writer as destructive criticism, although probably it is not intended as such by designers, engineers and other professional employees, or by procuring agencies employed for just such a purpose, in an effort to build into the manual what they believe are the true criteria.

Derogatory criticism can be avoided by the writer only by employing common sense guidelines such as:

(a) full understanding of the requirements,
(b) full understanding of the topics or subjects,
(c) suppression of the creative genius temperament; normally the writer is just another member of a team.

REFERENCES

1. __________, ANSI-Y32.14 Graphic Symbols for Logic Diagrams (Two State Devices), American National Standards Institutes, Institute of Electrical and Electronic Engineers, 345 East 47th Street, New York, NY., 10017.

2. __________, BS3939 Graphic Symbols for Electrical and Electronics Diagrams, British Standards Institute, 2 Park St, London, Wl.

3. __________, MIL-M-38784, Military Specification Manauls Technical General Requirements, US Department of Defense, US Government Printing Office, Washington, D.C., 20402.

4. __________, TPN(L) 114. Technical Procedures Notice (Electronic). UK Ministry of Defence, Procurement Executive, 77-91 New Oxford St., London, WC1A 1DT.

5. __________, AVP 70, Specification for Air Technical Publications, UK Ministry of Defence, Procurement Executive, 77-91 New Oxford St, London, WC1A 1DT.

6. __________, ATA 100, Specification for Manufacturer Technical Data, Air Transport Association of America, 1000 Connecticut Ave, Washington, D.C., 20036.

7. __________, MIL-M-15017, Military Specification, Manuals Technical Equipments and Systems Content Requirements For, US Department of Defense US Government Printing Office, Washington, D.C., 20402.

8. __________, ATA 101 Specification for Ground Equipment, Air Transport Association of America, 1000 Connecticut Ave, Washington, D.C., 20036.

9. __________, Symbolic Integrated Maintenance Manuals Versus Conventional Technical Manuals, Technical Operations Inc., 1968.

10. ____________, MIL-M-24100, Military Specification Manuals, Technical, Functionally Oriented Maintenance Manual (FOMM) For Equipment and Systems, US Department of Defense, US Government Printing Office, Washington, D.C., 20402.

11. Plata, E.F., Feasibility of Preparing a Symbolic Integrated Maintenance Manual for Complex New Electronic Systems, IEEE, Transactions on Engineering Writing and Speech, No. 4, 1971, p34.

12. ____________, Federal Information Processing Standards Publication 38-FIPS PUB 38, National Bureau of Standards, U.S. Government Printing Office, Washington, D.C., 20402.

27 TRAINING

M. B. Darch

More than any other industry, electronics is characterized by rapid changes in technology. Accompanying the changes in technology are concerted efforts to incorporate reliability and maintainability into electronic systems. The rapid changes in technology result in a continuing need for maintenance personnel to up-date their skills, to keep abreast of new equipment. The increased emphasis on reliability and maintainability results in shifting maintenance policies, which are reflected in the skill requirements of the maintenance personnel.

The process of altering skill levels rests with the training function within an organization. Training has generally been regarded as a necessary but not a critical factor in organization planning. For the reasons mentioned above, training must play an important role if maintenance personnel are to retain their ability to keep systems operating.

The necessity of keeping abreast of the altering skill levels impacts directly on the cost competitiveness of the organization. Failure to up-date the skill levels can lead to excessive downtime or the purchase of sub-optimal quantities of spares. Training necessitates a considerable expenditure of funds. Consequently, cost savings from training can only arise when training requirements are clearly defined and a training plan is developed to meet these requirements, preferably in the most cost effective manner.

A MODEL FOR DEVELOPING A TRAINING PLAN

When training is not a critical cost factor, ad hoc methods for developing a training plan will generally meet the organization's requirements. As training becomes more critical, the formulation must be done in a more organized manner. The following is a suggested flow diagram for demonstrating how the training plan evolves.

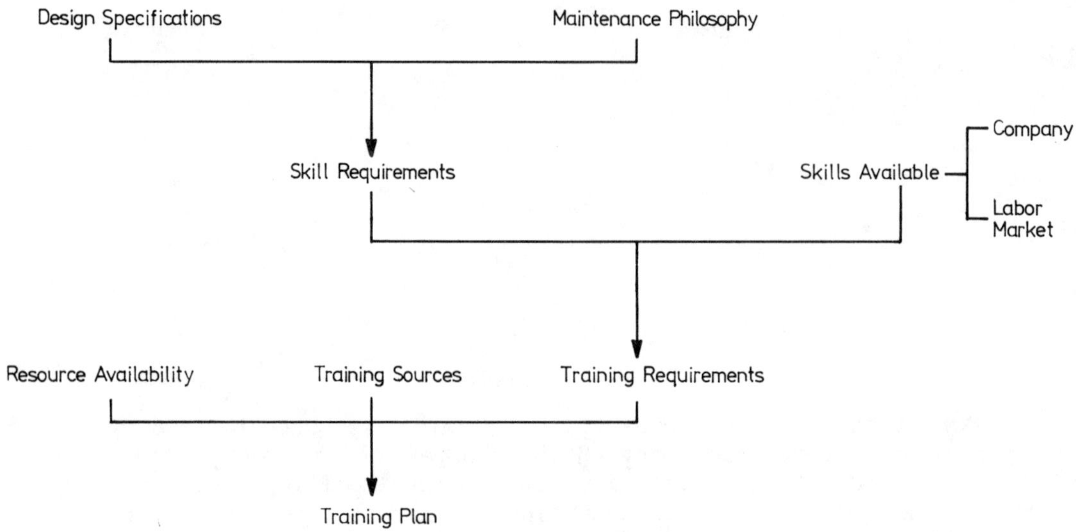

Figure 1. Evolution of the Training Plan

The training requirements within an organization are primarily influenced by three factors; the design specifications of the equipment, the philosophy being used to maintain the equipment and the skill levels of the personnel. The first two factors determine the skill levels necessary to maintain the equipment properly. The difference between these levels and those levels present within the organization represent the areas to be addressed by the training plan.

The Influence of Design Specifications

Design specifications determine the physical form of the equipment to be maintained. Training requirements will be most directly influenced by four design features; the technology level employed, the complexity of the design, the provisions for interface and the incorporation of features to enhance maintainability.

The technology level employed determines many of the basic skills required by the maintainers. Different types of skills are required to troubleshoot vacuum tube type systems as opposed to systems using microcircuit chips. Newer technology attempts to have systems that are as small and lightweight as possible. With circuits becoming smaller and smaller, specialized equipment is needed to isolate and correct faults in individual components. Modern technology usually employs components that are very specialized. Part substitution is therefore a complex subject that demands specific knowledge of the part, its function, and its environment within the system.

The complexity of the design will affect skill levels in fault isolation and in fault correction. As a system becomes more complex, it becomes increasingly difficult to isolate failures. Fault isolation by techniques such as component replacement are prohibitively expensive. The maintainers must be taught how to isolate even the most obscure failure. Complex systems usually imply a large dependence between components and a difficulty in accessing many components. Higher skill levels are, therefore, required to properly correct a fault without damaging other components.

The trend toward miniaturization, standardization and modularization is creating systems which are composed of relatively independent and highly specialized subsystems. Many of these subsystems are designed as individual entities rather than as components of a larger system. The provisions for interface, therefore, become an important determinant in maintaining the overall systems. A well integrated system will ease interface problems and simplify fault isolation, while a poorly integrated system will demand technicians with high skill levels.

The cost of maintaining complex systems has become a major consideration in system design. Previous discussion has indicated that as designs become more complex and employ a higher level of technology, and as interface points increase, maintenance skill levels must increase. To counter the need for better trained and, therefore, more expensive technicians, design features aiding maintainability can be added. These features may include the use of modules, automatic test features, buit-in test points, ease of access to major components and simplified component correction and thereby skill level requirements can be reduced.

The Influence of the Maintenance Philosophy

The maintenance philosophy of the organization will decide a number of factors including repair levels, the repair philosophy at a particular level and what repair will be done in-house.

Repair levels are used to separate the maintenance tasks into discrete packages. Accompanying each package is a requirement for technicians with given sets of skills. In developing the training plan, these levels must be examined to determine the manpower required at each level and the skill requirements.

Certain faults will be designated for repair at each level. Fault correction can range from a detailed analysis to find and replace an individual component, or fault isolation to a module and then replacement of the entire module. Repair to the component level will require extensive knowledge in the areas of both fault isolation and fault correction, whereas a replace philosophy will require only skills in fault isolation.

The trade-off between in-house and contractor maintenance will also have a significant impact on the training plan. Photo copier owners generally do no maintenance at all on their equipment, resulting in no need for trained maintenance personnel. Maintenance is totally contracted out. Most military organizations desire to keep a self-contained operation so that they will perform all maintenance internally. Personnel must, therefore, be trained for all facets of maintenance. In between are a continuum of alternatives.

An analysis of the design specifications and the maintenance philosophy will lay out the skill requirements needed. To determine the necessary training, the required skills must be matched to the available skills.

Looking at available Skill Levels

When obtaining personnel possessing a set of skills, two extremes exist. One extreme is the use of a competitive labor market where the necessary personnel can be hired and put to work directly. The other extreme is the military where hiring occurs at only the bottom level and then skills are built-up. Between the two are various situations where a company will want to hire directly in some cases or train company personnel in others. To determine training requirements, an on-going process of examining the skill requirements, the availability of qualified personnel in the marketplace and the skills available internally, must exist.

Training Requirements

The above process will outline particular deficiencies within the organization that are to be met through training programs. The problem then becomes how to obtain the necessary training.

The training requirements should be recorded by giving a clear statement of what is required and why it is required. Furthermore, the skills should be linked to a general educational level, such as an engineer, a technologist or a technician. It should be stated whether a general knowledge is required or a specific knowledge of a skill or technique. A wide variety of training is available. To ensure that the training is cost effective, the individual deciding on training must be able to evaluate the objectives of a given course versus his requirements.

Length of course. The length will give a good indication of the objective and background required for a course. A course lasting one day to one week is designed to give a practical knowledge of a narrowly defined subject. Courses of this type should be used to obtain specific skills. Care must be taken in assigning personnel to these courses as their short length demands that attendees have the required background.

Courses lasting from two weeks to one month can have two objectives. The first is a general introduction to a field. Theory will generally be presented in a lecture format, interspersed with case studies or exercises to relate the theory to practice. The objective is to introduce the attendee to the subject matter. Therefore prerequisites are less strict. The second objective for a course of such length is to give an in-depth knowledge of a new technique or system. An example would be a course for technicians to maintain a new radio set. These courses are very practical and are designed to have the attendee leave as an expert. The attendee at such a course must have a thorough knowledge of basic principles.

Courses lasting one month to a year are used for general upgrading, for introducing new concepts or for an introduction to a new system or subsystem. The length of the course allows sufficient time to bring all attendees to a certain level before proceeding with new material. The wider subject area results in the attendee gaining a broader knowledge.

Courses in excess of one year are designed to impart a general education in a field. Prerequisites are usually broad. These courses are generally left to public or quasi-public institutions, such as universities. The knowledge is broad, permitting its application to a wide variety of circumstances.

Training Sources

Training can be obtained from five basic sources:

(a) public and quasi-public educational institutions,
(b) manufacturers,
(c) professinal associations,
(d) private training institutions, and
(e) an in-house training department.

Public and quasi-public educational institutions. This class would include state or provincially funded vocational institutions, colleges and universities. The prime objective of these institutions is to provide basic skill development through various certificate, diploma or degree programs.

Recently, there has been a greater tendency for these institutions to offer short intensive courses in very narrowly defined areas. These courses are generally offered in the evening or during the summer. The individuals offering the courses are generally well qualified and the costs are comparatively low. The individual in the company responsible for training should have available the list of such courses from local institutions. References (1, 2) published annually, can be used to obtain listings of summer courses available at universities in the United States and Canada.

Manufacturers. Most manufacturers offer training programs to acquaint their customers' maintenance personnel with their equipment. These courses are specific in nature giving the trainee knowledge directly related to maintaining individual equipments or systems.

Care must be taken in contracting for such courses to ensure that the course meets your training needs. The manufacturer generally develops a preferred maintenance structure and philosophy. The course or courses offered must be compatible with the client's maintenance organization and available skill levels. The client must ensure that course attendees possess the necessary basic skill level and that the course then brings them to the level necessary to perform their function.

Professional Associations. Professional associations offer two basic forms of training, annual symposia and seminars. The annual symposia usually consist of a series of papers and/or panel discussions relevant to the association. This permits the attendee to become acquainted with new ideas and the direction in which his field is heading. The papers and their references provide him with a good research base should any particular topic be directly relevant to his work.

Increasingly, the education committees of associations are developing training programs for their members. These programs are directed at maintaining the currency of their member's skills and providing training in peripheral subjects that provide a broader educational base. (An example would be management accounting courses for engineers).

A list of professional associations can be found in references (3, 4). Writing to the executive director should yield information on what is available.

Private Training Institutions. There are a number of private training firms which offer a variety of courses. The best known of this type would be business colleges and technical correspondence schools.

Historically, these schools have offered, in the technical area, basic trade training. The increasing demand for specialized courses has resulted in a considerable growth in this type of company. This growth has been accompanied by a diversification in the types of courses offered and a movement into courses designed not only for the technician but also the technologist and the engineer.

Information on these institutions can usually be found in the advertisements in trade and association publications. Very often a organization will receive direct mail solicitations outlining available services.

In-house. The previous sources of training lack one ingredient that can be a necessity in a course: a thorough understanding of how the client operates. It is often felt that for the contents to be of practical use, they must be directly relevant to the trainee's work environment.

In-house training courses are developed to meet specific organization needs. The training function evolves into a separate department which can handle a wide variety of courses. Because the attendees and course leaders are with the same group, feedback on the usefulness of any course is obtained. In-house training would not only consist of lectures or seminars but would also include on-the-job training.

Great care must be taken with in-house courses to ensure that course leaders are familiar with what they are teaching. Too often, the training staff becomes an entity unto itself and loses working contact with the subject of their lectures.

This source of training is best for familiarization type courses and basic courses on maintenance training. It is more cost effective to use outside sources for highly specialized training and for an introduction to new techniques or technology. It must be remembered that the maintenance of an in-house training section requires a fixed overhead expense.

The Training Plan

Emerging from the process will be the training plan. The training plan will embody the formal description of what training will be carried out, who will perform the training and its purpose. The training plan should therefore include, for each course.

(a) the course title,
(b) a description of the course,
(c) the purpose of the course (what skills will be developed);
(d) the length of the course,
(e) the target group for attendance,
(f) prerequisites, if any,
(g) frequency of the course,
(h) number of attendees per course,
(i) the source of the course and
(j) the cost.

A formalization of the training plan assists the maintenance organization in two important ways. First, the process of establishing the requirements necessitates a review of the maintenance philosophy and how that philosophy is put into practice. If the process is considered to be dynamic, the interchange relationship established between the training and maintenance organizations will help ensure a responsive, relevant training plan.

The second important purpose of the formal training plan is to provide a ready reference for the individuals who require the training. With the proliferation of educational courses, it is difficult for the individual, who must still peform his normal duties, to keep abreast of what is available. Career development of the individual is, therefore, facilitated if he has easy access to a comprehensive listing of available training.

PSYCHOLOGY OF LEARNING

With the increasing expense of training, it is necessary to ensure that resources for training are well utilized. When developing the training plan, it is, therefore, helpful to review some of the basic principles in the learning process of adults.

Learning Curves

Learning does not progress in a steady manner but rather is typified by two basic patterns. The learning curve for simple material is illustrated in Figure 2 while that for complex material is illustrated in Figure 3.

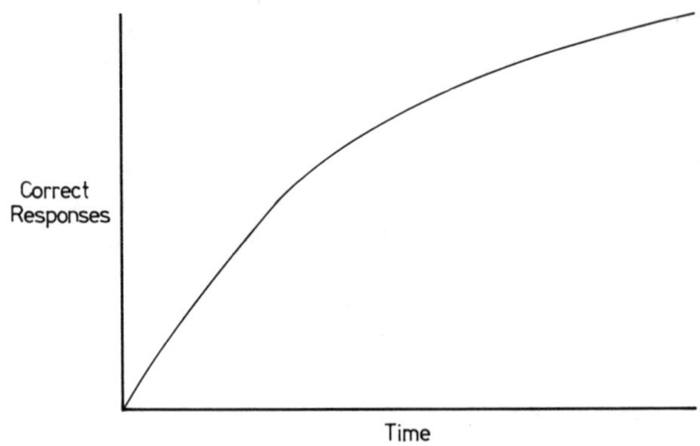

Figure 2. Learning Curve for Simple Material.

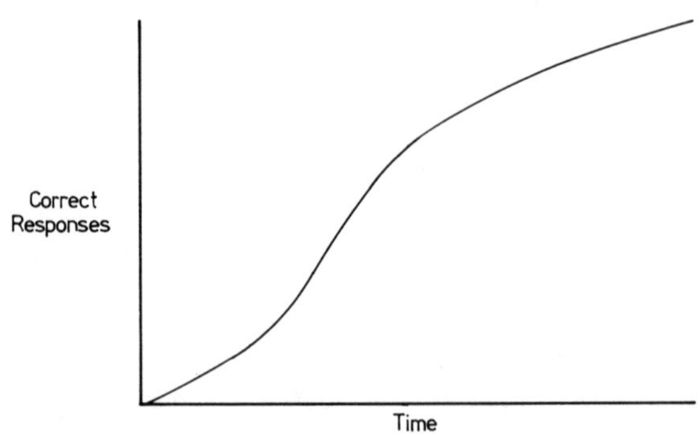

Figure 3. Learning Curve for Complex Material

For simple material, learning occurs first at a rapid rate and then levels off as the material is mastered. This curve would apply for training aimed at basic skill development. The student is able to absorb the material at a fairly constant rate until the subject is mastered.

Comparing Figures 2 and 3 illustrates that the learning curve of complex material is significantly different from that of simple material. Rather than being able to absorb the material quickly, initial progress is slow as the student develops an understanding of the material. The learning process accelerates as the matter is absorbed. Again, levelling off occurs as the subject is mastered. Figures 2 and 3 give a general view of the learning process but plateaus can exist where relatively little learning occurs. These plateaus arise due to factors such as fatigue, poor motivation, a loss of interest, or a pause to absorb the old material before progressing on to new material. Knowing that such plateaus exist, courses must be developed to ensure that new material is not thrust upon the student at such times.

The learning curves presented are universal in nature and do not consider individual differences in learning ability. Learning curves indicate to the course designer how the general format for the course should be laid out. A later section on learning performance goes into greater detail on the factors which influence individual learning capabilities.

Principles of Learning

If training is to be successful, it is necessary that the participants absorb the material and then are able to apply it in the work environment. Although individuals learn at different rates and, therefore, will each respond differently to a course, certain basic principles are common to adult learning. In developing the training plan, consideration of these principles increases the probability of achieving the desired results.

Meaningful material. Learning occurs at a much faster rate if the student perceives that the material is relevant and that the material is presented in a comprehensible manner. Training in most maintenance organizations is very goal directed. Skills are to be imparted which will assist the maintainer in doing his job. Adherence to this notion of goal direction ensures that the employee is sent on courses for which he sees immediate usefulness. Very little return is likely from material that may be rarely used, or material that is learned far in advance of the time that the skills will be used.

A major problem in adult education is linking the needs of the student to the abilities of the instructor. Too often, the instructor knows his subject very well but knows little of the students' working environment. The learning process will be impeded if the vocabulary of terms used by the instructor is unfamiliar to the students. The instructor must be able to organize the material so that new concepts are built upon prior knowledge, while familiarity with the students' working environment permits the instructor to relate new ideas to practical experience.

Appropriate practice. The probablility of retention of material is enchanced if the training process permits practice. Depending upon the type of course this can involve practical demonstrations or the performance of case studies. The practice is best achieved in a distributed manner, whereby material is presented and then consolidated through practical exercises.

Knowledge of results. Particularly in longer courses, it is necessary for the student to have some indication of his progress. Formal testing is not necessary but efforts must be made to relate to the student some level of achievement. This can be done through short quizzes, discussion of case studies, review of practical exercises, etc. Feedback of a less formal nature allows the student to progress without creating the anxieties inherent in formal testing.

Memory. The principles of learning are concerned with developing courses which present the material in a manner which encourages the learning process. Also of importance to the individual developing the training plan is the ability of the trainees to retain the material learned. Figure 4 gives an indication of recognition and recall of material over time.

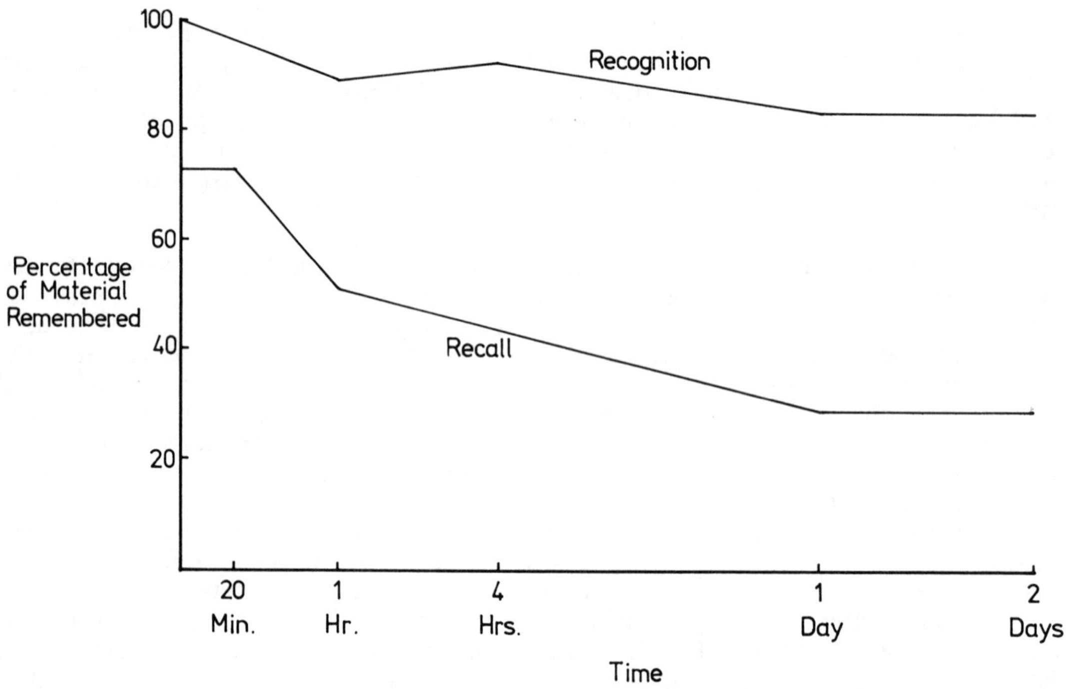

Figure 4. Long - Term Memory by Recall and Recognition

Figure 4 illustrates that the recall of material drops off significantly after the course but the recognition does not exhibit this phenomenon. During normal maintenance tasks, recognition rather than recall is required, since reference material is available. Again, it is important to emphasize that training which is not related to the individual's work is unlikely to be retained.

Influences on Learning Performance

There are three main influences on learning performance: type of material, previous education and occupation. As the material becomes more complex, the learning process becomes more difficult. In discussing training requirements earlier, it was stated that care must be taken to ensure that course participants have the proper educational background. Previous education assists learning performance through establishing learning skills and through the establishment of an education base. Occupation has an important influence on learning performance because of mathematical and verbal skills that are associated with various occupations. For the purposes of the training plan, the individuals to be sent on the courses will have similiar occupational backgrounds, but it is critical to ensure that outside training courses are aimed to that background. It must be realized, when deciding upon courses and who will attend, that previous education and occupation are more critical factors in learning performance than age.

CONCLUSION

Current trends toward complex electronic systems, employing state-of-the-art technology, demand a dynamic, responsive maintenance organization. To provide this responsiveness, training plays an increasingly important role in maintenance. Training should not be considered as a necessary evil but rather as a systematic method of developing an organization's human resources. The development of human resources must be considered as an investment; the return is realized in the continuing efficient and effective operation of the maintenance organization.

REFERENCES

1. ________________, Profiles of American Colleges, Volume 2, Barron Publishing Co., Woodbury, N.Y., published annually.

2. ______________, Directory of Summer School and Intersession Courses Offered By Canadian Universities, Statistics Canada Government of Canada, Ottawa, published annually.

3. Land, B., Directory of Associations In Canada, University of Toronto Press, Toronto, Ontario, published annually.

4. ______________, Encyclopedia of Associations, Gale Reseach Company, Detroit, published annually.

BIBLIOGRAPHY

1. Cross, K.P., Valley, J.R., and Associates, Planning Non-Traditional Programs, Jossey-Bass Publishers, San Francisco, California, 1975.

3. Deighton, L.C., editor, The Encyclopedia of Education, McMillan Co., New York, N.Y., 1971.

4. Ebel, Noll and Bauer, Encyclopedia of Educational Research, McMillan Co., New York, N.Y., 1969.

5. Kausler, D.H., Psychology of Verbal Learning and Memory, Academic Press, New York, N.Y. 1974.

6. Knowles, M.S., The Modern Practice of Adult Education, Association Press, New York, N.Y., 1970.

28 MAINTAINABILITY TESTING

A. B. Keith

The intention of this chapter is not to present ready made test plans for testing maintainability but to provide the basis for choosing and planning a demonstration method to suit particular maintainability requirements. The purpose of a Maintainability Demonstration is to provide a tangible measure of the elusive property of equipment known as 'maintainability'. It is to provide the customer with a measure of confidence that the contracted equipment has, in fact, the specified maintainability characteristics.

The maintainability requirements for a piece of equipment are the result of economic trade-offs, operational requirements and maintenance facility constraints. These maintainability requirements are to allow a piece of equipment to fit into an operational system and perhaps into an overall maintenance plan in the most efficient manner. These desired maintainability characteristics are described in terms of Mean Time To Repair (MTTR), Maintenance Man Hours per Operating Hour (MMH/OH), Maximum Maintenance Time (Mmax), Availability (A), etc., for each level of maintenance. The maintainability parameters include the extent of repair and maintenance facilities at each level of maintenance. Table 1 shows the relationships between maintainance level and associated maintenance tasks.

Once the parameters are satisfactorily defined, a maintainability demonstration plan is designed to ensure that the desired characteristics are designed into the equipment. This plan with its reject/accept criteria then becomes the specified maintainability requirement.

The demonstration plan can call for preliminary theoretical calculations, prototype verification tests, production model demonstrations and actual in-use operational verification. The important point is that all accept/reject criteria reflect the actual requirements, and acceptance of preliminary measures of maintainability indicate that the equipment has a reasonable chance of final acceptance.

Table 1. Maintenance Levels vs Maintenance Tasks

Maintenance Level	Maintenance Task
Organizational	Test tasks required to restore equipment to operational status.
Field	Test tasks required to repair assemblies removed from the Organizational level.
Depot	Test tasks required to repair sub-assemblies received from Field level.

MAINTAINABILITY DEMONSTRATION SPECIFICATIONS (MDS)

Impact of MDS

The specifications of maintainability parameters in equipment specifications do not take on meaning until a demonstration requirement is included. Before the inclusion of the requirement for demonstration, the maintainability requirements in a specification are only a statement of customer desires, which may or may not affect the design of an equipment.

The maintainability demonstration is the criterion for acceptance or rejection of the equipment with respect to maintainability parameters and is, therefore, the basis for the producers design to build equipment which will pass the demonstration test. It is important that demonstration requirements reflect the desired maintainability parameters, and will, in fact, result in the equipment having the desired maintainability characteristics.

Writing the MDS

The equipment specification must contain a clear definition of the maintainability demonstration test. This is more important, as previously pointed out, than statements of MTTR, MMH/OH, etc., which are desired characteristics resulting from trade-offs and are relatively meaningless until stated in terms of a demonstration test.

The maintainability demonstration section must state clearly the extent of repair to be accomplished at each level of maintenance, the support equipment to be used, the personnel training requirements and skill levels required. MTTR requirements, etc., are meaningless unless the extent of repair is clearly defined. For example, does 'field level' maintenance mean repair by removing and replacing subassembles or does it also include repairing subassemblies. If this division of maintenance is not possible at the time of specification, the requirements (MTTR, etc.)

must be left flexible, with perhaps some overall requirements in terms of MMH/OH, although the test methods should be specific.

The statistical plan with accept/reject criteria must leave no misunderstanding of the test results which will lead to acceptance or rejection. It is up to the producer and consumer to understand the implications of the accept/reject criteria in terms of producers and consumers risk, so that they can interpret the accept/reject criteria into design requirements and operational requirements respectively. The accept/reject criteria should result in a reasonable level of confidence that the equipment will be designed to meet operational requirements. The method of selecting tasks must be defined in the specification.

Penalties

The nature of penalties must vary with the stage of development or production of the equipment. If evaluation of theoretical estimates does not indicate the potential of the equipment for passing further demonstrations, redesign of hardware and/or software is indicated. This includes testing done in breadboards, brassboards, etc., during the design phase.

Failure of a laboratory environment test on prototypes or early production models may involve a redesign and possibly a retrofit type of penalty. However, price penalties might be considered depending on the severity of the deficiencies. It is possible that the producer might find redesign cheaper than price reduction if the deficiencies were severe. Figure 1 shows a typical penalty-incentive curve.

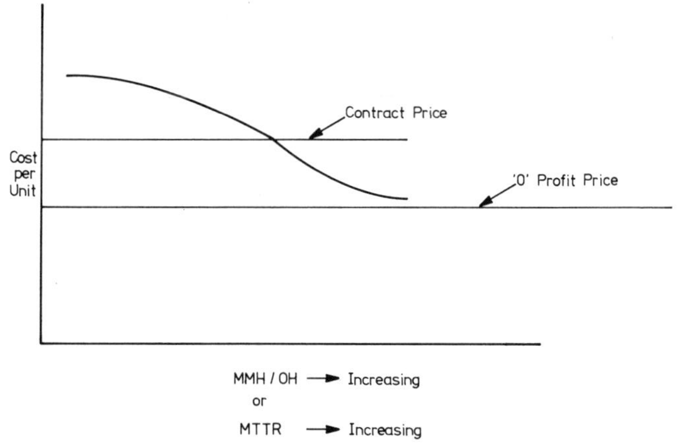

Figure 1. Penalty - Incentive Curve

It is possible that the only official acceptance testing performed is carried out under actual operations on production equipment. If this is the case, the manufacturer should have protected himself by extensive preliminary testing. By the time compliance or non-compliance could be determined, redesign and retrofit would be quite expensive.

A price reduction penalty could be applied but the producer must be careful not to be placed in a position of producing at no profit or even at a loss. Perhaps the consumer, in placing his acceptance decision so late in production, should accept some of the risk. This is done by having a penalty curve which approaches the no profit point, while never reaching it. The consumer then carries the risk of excessive maintenance cost while the producer carries the risk of reduced or no profit. It would be possible to combine this type of penalty with an incentive clause where the producer is paid extra for exceeding the maintainability requirements.

A similar approach is for the producer to pay a portion of the excessive maintenance costs. If the equipment is of a specified long life the cost implications could be staggering. In any case the penalties should not last longer than the specified life of the equipment. It would be possible, of course, to specify a limit on the period of time for which the penalty must be paid. Where penalties are applied late in production, a low producer risk, and consumer risk, plan should be chosen.

APPROACH TO TESTING

There are basically two approaches to testing maintainability. One approach is to select failures randomly from the universe of equipment failure modes and measure repair times for each level of maintenance. Statistics are then applied to the sampled task times, to accept or reject hypotheses with respect to the MTTR, Mmax, etc. The risks involved are twofold: the risk coming from choosing a non-representative set of long or short tasks on which to base the accept/reject decision and the risk involved with not measuring the 'true' task time for a given task.

Another approach is to base the testing on the maintainability estimate or apportionment. The tasks are still selected in the same random manner but the task times measurements are compared to the expected times according to the estimate, thus eliminating the risk associated with choosing a non-representative sample. This method tests the hypothesis that the maintainability estimate is correct. The maintainability estimate provides the measure of MTTR, Mmax, MMH/OH, etc., and acceptance of the estimate means acceptance of the estimated parameters. Table 2 provides a comparison of the two basic approaches.

Testing of production equipment under operational conditions can apply either principle. It may be that reliability testing is combined with maintainability testing in the form of availability or MMH/OH. In operational testing the observed failure rate is used to select the maintenance task. A sufficient sample size must be accumulated to prove boththe maintainability and reliability before reaching an accept/reject decision. Corrective actions must be directed properly in the case of rejection. The rejection could be due to either an unforeseen reliability or maintainability problem. A reliability problem area might not be severe enough to cause rejection on the basis of reliability, but it might bias the task selection toward some larger tasks, thus causing rejection.

Table 2. Comparison of the Two Approaches to Testing

First Approach

Task No.	Estimated Task Time (min)	Measured Task Time (x_i)	Task Frequency (f_i)	MMM/OH (f_i x_i)
3	10	12	0.01	0.12
10	30	35	0.02	0.70
32	37	30	0.05	1.50
40	20	25	0.03	0.75
45	5	7	0.14	0.98
70	50	40	0.07	2.80
89	80	70	0.02	1.40
		= 219	= 0.34	= 8.25

Notes:

1. Without accounting for maintenance rate MTTR = $\Sigma x_i/n$ = 219/7 = 31.3.
2. Accounting for maintenance rate MTTR = $\Sigma f_i x_i/\Sigma f_i$ = 8.25/0.34 = 24.3.
3. Acceptance or rejection is then based on this observed MTTR. It is advantageous to take the task maintenance rate into account when calculating the observed MTTR.

Second Approach

Task No.	Estimated Task Time (min)	Measured Task Time (min)	Diff. (min)
3	10	12	2
10	30	35	5
32	37	30	-7
40	20	25	5
45	5	7	2
70	50	40	-10
89	80	70	-10

Table 2 (Cont'd)

Notes:

1. Acceptance or rejection could be based on the average weighted difference or by counting passed or failed tasks.

DEMONSTRATION TEST OUTLINE

Choosing Task Sample

The method of choosing tasks for demonstration will vary considerably with types of equipment. Mechanical equipment will have a highstress on remove, replace, repair and adjust tasks. Electronic equipment will place greater stress on fault isolation tasks.

Electronic equipment packaged on printed circuit boards may not require a remove and replace action for each fault isolation task. The important function is to verify the diagnostics. If the diagnostics are carried out by automatic test equipment, the fault isolation time becomes a function of the test equipment operation (provided that the diagnostic software is correct). In this case the task sample should contain relatively few remove and replace tasks but sufficient faults to demonstrate the diagnostics.

It is important to apportion the tasks according to failure rate, in order to ensure that high failure rate items are verified. If the demonstration is based on verification of a maintainability estimate there is no need to have the tasks in direct proportion to the failure rate, and excessive testing of similar tasks should be avoided. The sample size will be governed by the desired confidence levels and the measured or assumed deviations of the sample.

Ideally a Failure Modes and Effects Analysis (FMEA) will be used to choose the sample, as it lists all of the failures which theoretically could occur. The probability of choosing a failure mode as part of the sample should be proportional to its probability of occurrence. Demonstration of field level maintenance where the repair is to 'circuit card asembly' level would use 'circuit card assembly' failure modes, demonstration of depot level maintenance would use part failure modes. This, of course, can become very complex with LSI. Perhaps computer aided sample selection could be used. With complex 'circuit card assemblies' the major effort should be to test the diagnostics thoroughly.

Equipment may contain mechanical, electro-mechanical and electronic portions or analog and digital portions. A task sample of a predetermined size should be selected from each portion. The sample from each portion should be proportional to the maintenance rate of that portion. If the approach taken to testing is that of testing the maintainability estimate, the number of tasks concerning simple mechanical operations, or electronic packages containing many similar assemblies, may be reduced. If sequential testing is being used, the samples will not be of predetermined size but of predetermined proportions, amongst the portions of the equipment.

Table 3 (Task Selection) shows the listing of tasks from which to choose a sample. Fault isolation tasks are chosen from a list of part failure modes, as each part failure mode could result in different fault isolation problems. Repair task selection ignores failure modes, and groups similar items together where the remove and replace action is the same. This avoids unnecessary repetition of similar tasks. Adjust tasks should be selected from their own list to ensure that they are properly represented.

Each level of maintenance should have its own separate demonstration plan and selection of tasks. In order to avoid duplication of effort in fault simulation, fault isolation testing could be co-ordinated for the three levels of maintenance. 'Organizational level' tasks could be a subset of the 'field level' tasks which could be a subset of the 'depot level' tasks. All aspects of task time determination should be covered in the specification of the demonstration test.

Table 3. Task Selection
(Each level of maintenance will have its own task selection)

Fault Isolate Tasks		Repair Tasks		Adjust Tasks	
Failure Mode	Frequency	Assembly Subassembly Part (1)	Frequency	Adjustment	Frequency
Q open short drift	0.01	PC Board(2)	120	X	1.7
Q open short drift	0.1	R1	0.1	Y	0.5
A o/p high low	1.2	C2, C3, C4(3)	1.5	Z	5.2

Notes:

1. Identify part assembly, etc., which is removed in order to effect repair.
2. Printed circuit board removals identical so all grouped as one item.
3. Parts with identical mountings grouped together.

Determination of Task Time

Assurance that each observed task time reflects the 'real' task time is a necessary part of maintainability demonstration. It is perhaps more important when testing the validity of an estimate as the variability in task lengths becomes overriding when selecting failures randomly from the equipment failure mode universe.

At least three different technicians should be used. They should each carry out a set of tasks identical to the other technicians. Significant differences should be noted, correction factors should be agreed upon for each technician. A wide variation in technician performance should cause a review of the selection of technicians.

Remove and Replace Tasks

High maintenance rate remove and replace tasks should have three or four trials to determine their task times. There is no need to report the remove and replace for each fault isolation action. Complex mechanical tasks may require more trials if a large variance of time is found in the initial trials. Where there are large numbers of similar mechanical tasks the number of trials may be reduced. In cases where the tasks consistently conform closely to the estimated times, the number of trials may also be reduced.

Fault Isolation Tasks

If fault isolation is carried out by automatic test equipment, only the actions after the automatic diagnostic need be measured. The time and accuracy are a function of the test equipment and software, and these are the subject of a different acceptance test.

Fault isolation times for any given task are liable to vary more than remove and replace times and hence may require a greater number of trials. Large numbers of similar fault isolate tasks could take the place of repeating the same tasks. Consistent, small variations from estimated times should also reduce the number of trials. If tasks are marginal and there are significant differences between technicians, extra trials and additional technicians may be needed.

Determining Compliance

The method of determining compliance is governed by the approach to testing. The two methods discussed here are 'Maintainability Estimate Verification' and 'Testing Maintainability Parameters Directly'.

Maintainability Estimate Verification

There is a degree of flexibility in the determination of compliance with this approach of verification. Basically the estimated parameters,

MTTR, M max, etc., are accepted or rejected through the testing of the hypothesis that the estimate is accurate. The tasks are sampled and the measured task times are compared to the estimated time for the task. If estimated parameters are less than the specified parameters an apportionment could be used rather than the estimate, thus reducing the producer's risk. Another approach would be to use the estimate and allow a positive deviation of the measured times over the estimated times.

A simple approach to testing is to accept only if all measured task times are below the apportioned task times. If during testing some tasks are failed and others passed with a large margin, a reapportionment should be possible. The number of reapportionments should be limited during a testing session, before testing is suspended pending a new maintainability estimate. A large number of reapportionments would indicate a marginal situation. An accept decision would be reached when sufficient tasks have been measured to state that the apportionment times are accurate (or pessimistic) to a predetermined confidence level.

The use of a computer program for purposes of the maintainability estimate of apportionment would facilitate reapportionments during the demonstration. Figure 2 illustrates the maintainability estimate reapportionment method.

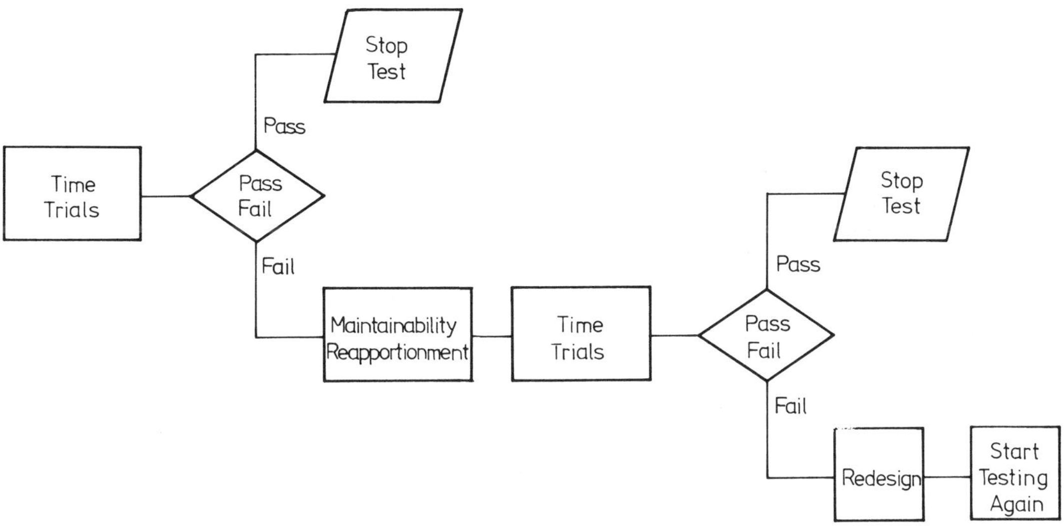

Figure 2. Maintainability Estimate Reapportionment Method

When a new set of tasks is selected after reapportionment, the already demonstrated tasks do not constitute part of the sample unless they are part of the new randomly chosen sample of tasks. Use of the previously tested tasks would constitute a biased sample of known accurate task times. The reapportionment would have assured that all tested tasks were accurate in the reapportionment.

The use of deviations of measured task times from estimated task times will be further explored here, providing a suggested method of approach. The principle of this method is to calculate a mean deviation weighted by task maintenance rates and to test the hypothesis that this deviation is zero or negative.

Table 4 shows a simple calculation of the weighted mean deviation ($\overline{D}$) which is given by:

$$\overline{D} = \frac{\Sigma f_i D_i}{\Sigma f_i} \qquad (1)$$

where,

f_i = task frequency,

D_i = task time deviation.

The value for $\overline{D}$ can be tested to a given confidence level by accepting if

$$\overline{D} \leq 0 + Z_\alpha \hat{d} / \sqrt{n} \qquad (2)$$

where,

Z_α = standardized normal deviate corresponding to the desired confidence level,

$\hat{d}$ = calculated sample standard deviation

n = sample size.

The distribution of deviations between measured and estimated task times is assumed to be a normal distribution about the average deviation, as shown in Figure 3. An accurate estimate would exhibit an average deviation close to '0' and a small 'standard deviation' when tested.

The 'standard deviation' expected from testing by deviation measurements is much smaller than that expected from measuring task times with no reference to the maintainability estimate. Because of this higher standard deviation a much larger sample is required for acceptance or rejection at a given level of confidence, for testing by measured task times rather than by deviations. This is illustated by comparing typical distributions in Figure 4.

A sample of measured task times will result in an average observed deviation, which may be larger or smaller than the actual average deviation of the entire population of tasks. The 'average observed deviations' from randomly sampled sets of tasks is assumed to have a normal distribution about the actual average deviation.

Table 4. Demonstration by Sample Task Differences

Measured-Estimated Deviation (D_i)	Failure Rate $x10^{-6}$ (f_i)	MMH/OH $x10^{-6}$ Deviation ($D_i f_i$)
5	0.1	0.5
-5	1.1	- 5.5
7	0.05	0.35
8	1.2	9.6
-7	3.1	-21.7
2	2.3	4.6
-1	0.01	- 0.01
0	5.2	0.0
1.125	= 13.06	= - 9.01

From equation (1) $\bar{D} = -9.01/13.06 = -0.69$ and from equation (2) $-0.69 < 0 + 1.96\ (5.44/2.83) = 3.77$.

Figure 5 illustrates this curve and shows the producer's risk if the actual average deviation was '0'. The points on the X axis represent observed average deviations of randomly sampled sets of tasks. The equipment will be rejected if the sample exhibits an average deviation to the right of point 'D_r'. The area under the curve to the right of 'D_r' represents the producer's risk, that is, the equipment meets requirements but is rejected.

The producer's and consumer's risks may be controlled by specifying an MTTR less than that actually acceptable and allowing a positive deviation. This is the same basic approach as used in (1) for reliability testing, which uses a 'minimum acceptable MTBF' and a 'specified MTBF' in order to control producer's and consumer's risks. Figure 6 illustrates this assignment of risk. The curve to the left represents the 'sample observed average deviation' probability distribution for an estimate with a '0' average deviation for the entire population. The curve to the right represents the 'sample observed average deviation' for an estimate with a positive average deviation for the entire population. An accept/reject point 'D_r' provides a producer's risk based on the specified (0 deviation)

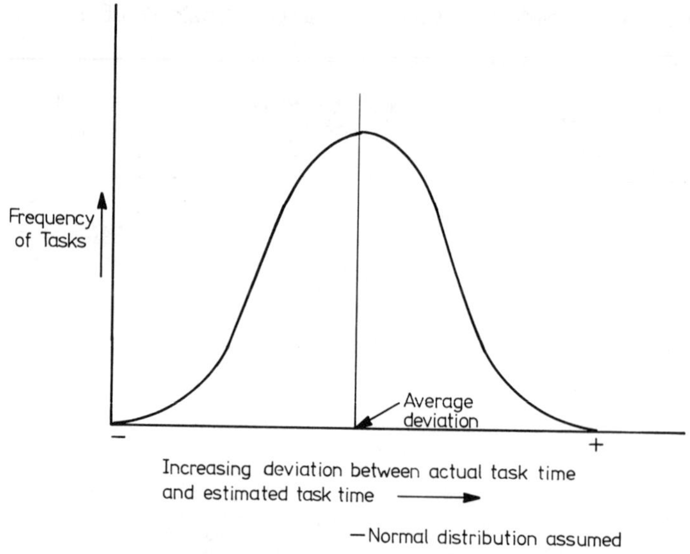

Figure 3. Assumed Distribution of Task Time Difference Between Actual and Estimated Times

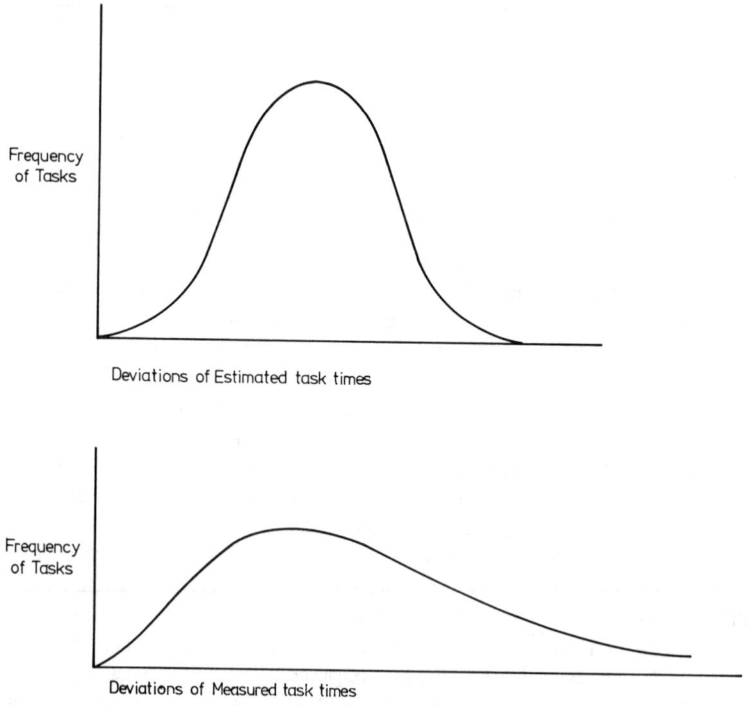

Figure 4. Comparisons of Samples For 'Testing by Deviation' vs 'Measuring Task Times'

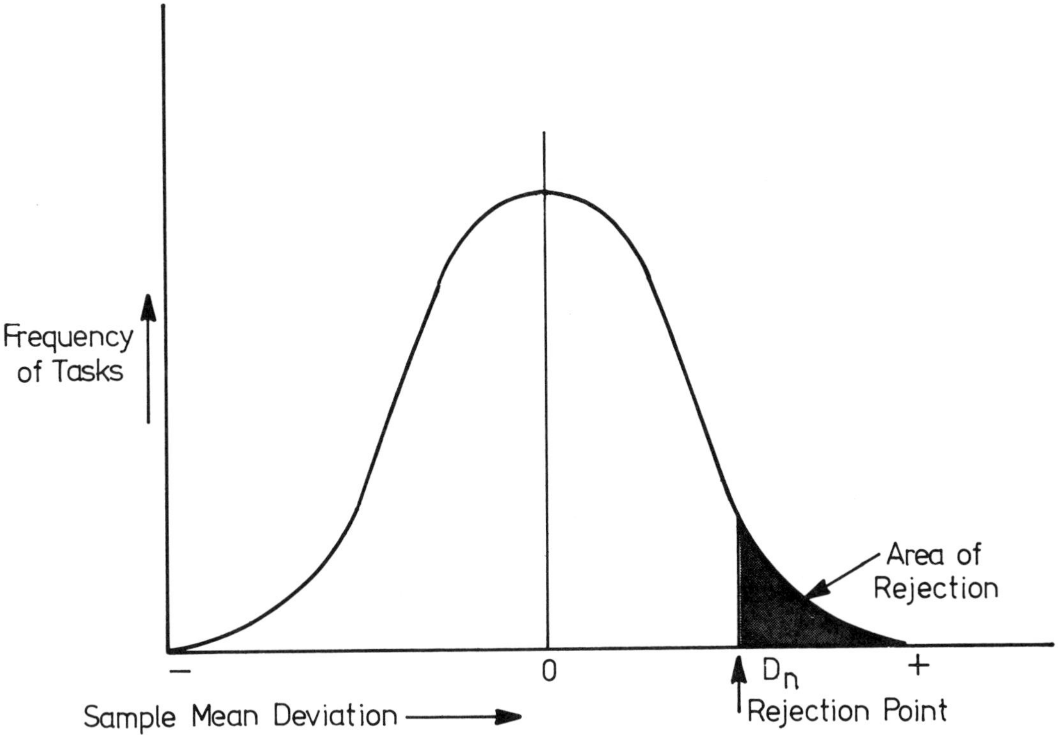

Figure 5. Producer's Risk

distribution and a consumer risk based on the distribution for the deviation acceptable to the consumer. The risks can be controlled by the choosing of the point 'D_r', the difference between the '0' deviation and the deviation acceptable to the consumer, and the sample size.

Figure 7 shows the 'operating characteristics' curve for relating consumer's and producer's risks to the difference between the specified MTTR ('0' deviation) and the inherent deviation of the equipment MTTR from the specified. As the mean deviation of actual repair times from the estimated repair times decreases, the probability of acceptance increases. The theoretical produced risk is based on the maintainability estimate being accurate and showing an MTTR equal to the specified. The producer's risk is reduced if the equipment has lower repair times than shown in the estimate. The producer can lower his risk by designing to a lower MTTR and providing a maintainability apportionment to the specified MTTR for testing. This will move the mean deviation to the right along the X-axis increasing the probability of acceptance. It is possible to develop sequential variable length testing as is used in (1). Figure 8 shows what would be a typical accept/reject graph.

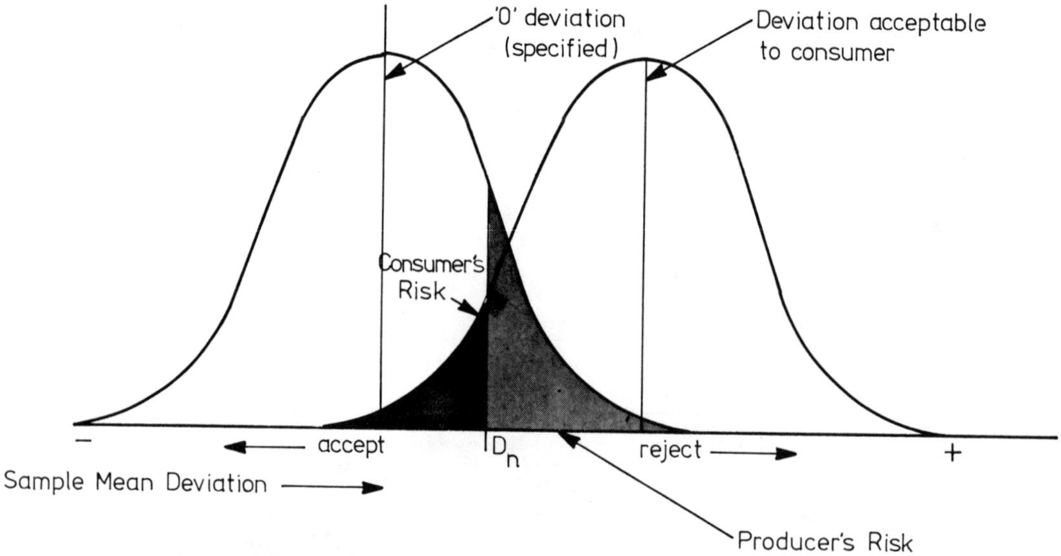

Figure 6. Risk Assignment

Where variations are large, a reapportionment should be carried out even if the hypothesis that the deviation is not '0' cannot be rejected. There are two basic reasons for this statement: (a) the measured average deviation from the MTTR is not necessarily the actual deviation from the MTTR. It is the deviation based on the sample, which may not be proportional by task type to failure rate, particularly where the number of samples of task types has been reduced because of similarities to other tasks, (b) if a maintainability estimate with large deviations were accepted, it would not be valid for maintenance planning.

Field Testing

Field usage demonstration of maintainability can rely on the same techniques of averaging deviations as lab testing. The main objective is the control and recording of the maintenance actions under field conditions. The supplier is liable to require some field representation.

The demonstration could be used as an availability demonstration if sufficient operating hours were accumulated to demonstrate the reliability. This is not likely to be a precise test, as there is no assurance that the short term failure distribution will be the same as the long term failure distribution. Even if there are sufficient operating hours to demonstrate

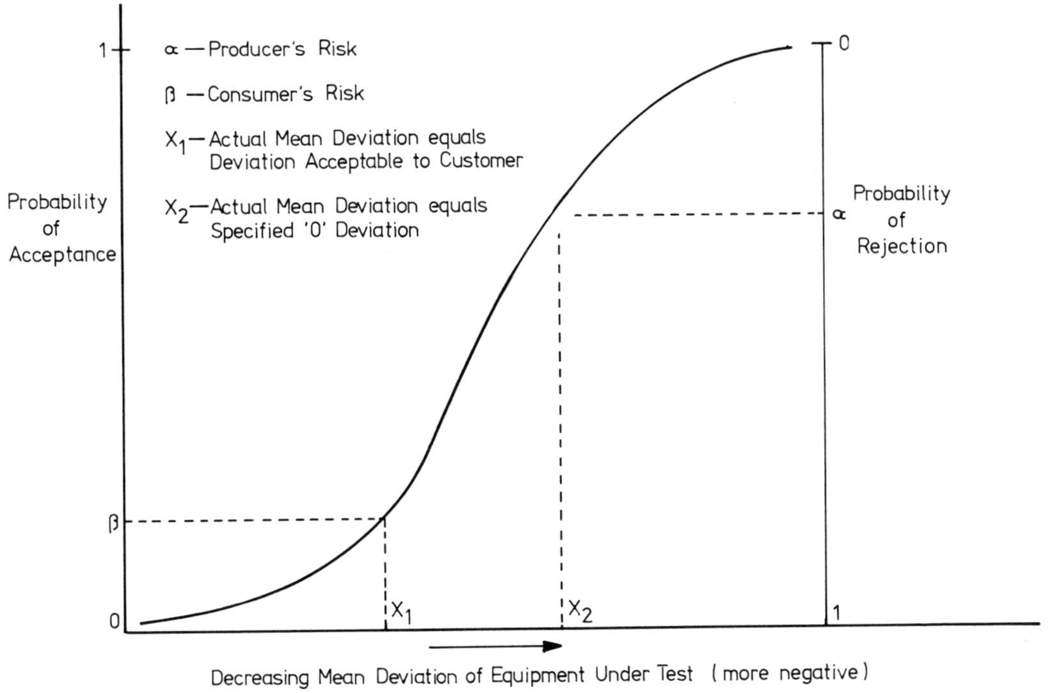

Figure 7. Operating Characteristics Curve

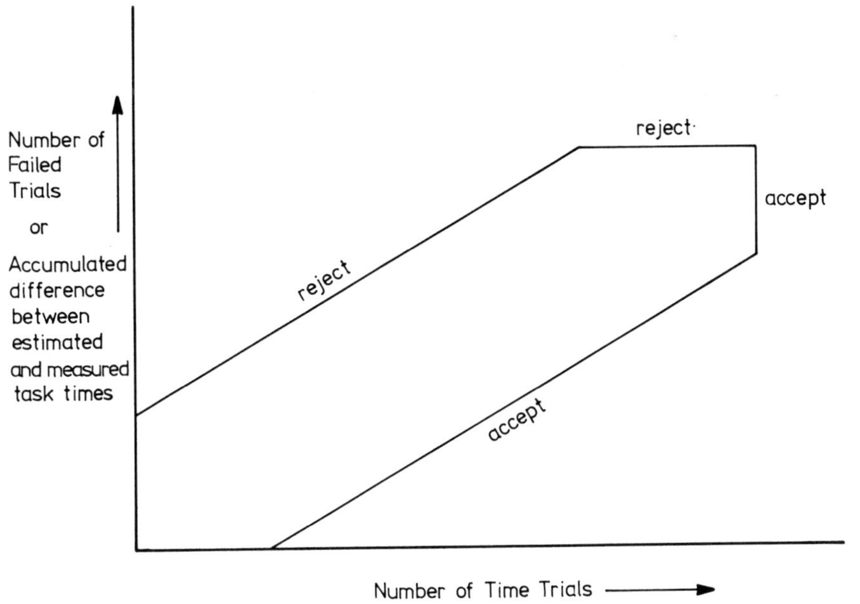

Figure 8. Sequential Testing

the reliability, the failure rate distribution is not demonstrated. The testing of availability should involve sufficient operating hours to demonstrate the failure rate of each replacement subassembly or major failure mode. This means that field demonstrations should, at least in the initial phases, rely on the reliability estimate for maintenance rate distribution.

Prolonged maintenance times, due to inadequate facilities or spares, must be assessed taking into consideration the following points, (a) are the spares available in accordance with the recommended spares level?, (b) are maintainance facilites in accordance with recommendations?, (c) is the inadequacy of spares or facilites due to a reliability problem?, (d) is the sparing confidence level too low (perhaps the spares plan is inadequate)?

Testing Maintainability Parameters Directly

It is possible to demonstrate maintainability by sampling maintenance tasks directly with no reference to the maintainability estimate. The idea here is to sample an unknown universe and build a model of the maintainability characteristics from statistical analysis of the sample. In this type of testing it is extremely important to have the sample proportional to the failure rate, that is, each type of failure is represented in the sample by a number of tasks in direct proportion to the failure rate for that type of failure.

This type of testing has the risk that the tasks chosen are not representative of the task time for the type of failure, in addition to the possibility that the observed task time may be longer or shorter than the 'true' task time. There is also the risk that the test could be passed through extremely short times on low failure rate items, or failed through extremely high times on low failure rate items.

Reference (2), Method 1 provides a test of the mean MTTR using this sampling procedure. It tests the hypothesis that the mean repair time is equal to or less than a predetermined specified mean. The sample mean ($\overline{X}$) is calculated from:

$$\overline{X} = \frac{1}{n} \sum_{i=1}^{n} X_i \tag{3}$$

where,

n = number of tasks,

X_i = task time of task i,

i = task number.

The hypothesis is accepted if

$$\overline{X} \leq \mu_o + Z_\alpha \hat{d}/\sqrt{n} \tag{4}$$

where,

μ_o = the specified mean,

Z_α = standardized normal deviate corresponding to the desired confidence level,

$\hat{d}$ = calculated sample standard deviation.

The probability of acceptance will vary with the sample size and the ratio of the true mean repair time to the specified mean repair time. An Operating Characteristic (OC) curve would provide the probability of acceptance for a given ratio and sample size.

This test could be transformed into a variable length test, similar to the reference (1) tests, by having a reject hypothesis to test. The test would reach a reject decision if,

$$\overline{X} \geq \mu_1 - Z_\beta \, \hat{d}/\sqrt{n} \qquad (5)$$

The test would continue until a reject or accept decision was achieved or until arrival at a predetermined point.

This type of test requires timing of induced faults through fault isolation, repair and check out. The individual tasks such as remove and replace, cannot be tested individually as with testing an estimate. Reference (2), Method 2, tests the critical percentile assuming a log normal distribution.

CONCLUSION

Two basic approaches to the demonstration of maintainability, sampling maintenance times directly and that of sampling the deviations between estimated and measured maintenance task times have been discussed in this chapter. The difference in the approaches is not in the statistics but in the use of available information (the maintainability estimate) to reduce the sample variance and hence the amount of testing required to achieve a desired confidence level. The distribution of maintenance times is also less important in the testing of deviations, as the distribution being tested is the distribution of estimated deviations to measured maintenance times, not the deviations of all maintenance times.

The maintainability demonstration determines compliance or non-compliance with a maintainability requirement and therefore should be carried out according to well defined, contractually agreed rules. These rules should be in the form of a written detailed plan showing: accept/reject criteria, methods of measurements, methods of recording, supervision, task sampling, maintenance procedures, diagnostics, test equipment, tools, skill levels, penalties, corrective action procedures, and any other items of particular importance to the product or maintenance environment.

REFERENCES

1. ___________, MIL-STD-781B Reliability Tests: Exponential Distribution, U.S. Department of Defense, Washington, D.C., 15 November 1967.

2. ___________, MIL-STD-471A Maintainability Verification/Demonstration/ Evaluation, U.S. Department of Defense, Washington, D.C., 27 March 1973.

BIBLIOGRAPHY

1. Blanchard, B.S., Logistics Engineering and Maintenance, Prentice-Hall, Englewood Cliffs, N.J., 1974.

2. Cunningham, C.E. and Cox, Wilbert, Applied Maintainability Engineering, John Wiley and Sons, New York, N.Y., 1972, Chapter 12.

3. Smith, D.J. and Babb, A.H., Maintainability Engineering, Pitman Publishing, New York, N.Y., 1973, Chapter 7.

4. ___________, AR-10 Maintainability of Avionics Equipment and Systems, General Requirements for, Naval Air Systems Command, Department of the Navy, DOD, Washington, U.S.A.

5. Dixon, W.J., and Massey, F.J., Introduction to Statistical Analysis, McGraw-Hill, New York, N.Y., 1969.

Appendix A1

RELIABILITY AND MAINTAINABILITY DOCUMENTS

H. Reiche and J. A. Roberts

The following is a list of documents that provides a basic source of reference for an initial study of Reliability and Maintainability Standards, Specifications and related material. The list has been developed mainly from that originally produced by B.A. MacDonald and published in 'Quality Progress', September 1976. The mailing address of the issuing organization is given at the end of the list.

1. ANSI Standards

C83.86-1966(R1972)	- Highly Reliable Soldered Connections in Electronic and Electrical Applications, Criteria for Inspection (includes 2x2 Color Transparencies) (reaffirmation and redesignation C99.1-1966)
C83.93-1973	- Solderability Test Standard (EIA RS-178-Jan. 1973)
C83.79-1973	- Tools, Crimping, Solderless Wiring Devices, Procedures for User Certification (EIA RS-270-Nov. 1972)
C83.107-1975	- Acceptability of Printed Wiring Boards (IPC-A-600B)
G35.25-1973	- Straight-Beam Ultrasonic Examination of Plain and Clad Steel Plates for Special Applications, Specification for (ASTM A578-71b)
Z1.1-1958(R1969)	- Guide for Quality Control (ASQC B1-1958)
Z1.2-1958(R1969)	- Control Chart Method of Analyzing Data (ASQC B2-1958)

Z1.3-1958(R1969)	- Control Chart Method of Controlling Quality During Production (ASQC B3-1958)
Z1.4-1971	- Sampling Procedures and Tables for Inspection by Attributes (MIL-STD-105D)
Z1.5-1971	- Definitions, Symbols, Formulas and Tables for Control Charts (ASQC A1-1971)
Z1.6-1971	- Definitions and Symbols for Acceptance Sampling by Attributes (ASQC A2-1971)
Z1.7-1971	- Glossary of General Terms Used in Quality Control (ASQC A3-1971)
Z1.8-1971	- A Quality Program, General Requirements for (ASQC C1-1968)
Z1.9-1972	- Sampling Procedures and Tables for Inspection by Variables for Per Cent Defective (MIL-STD-414, with Notice 1)
Z1.10-1973	- Probability Sampling of Materials, Practice for (ASTM E105-58 1970)
Z1.11-1973	- Choice of Sample Size to Estimate the Average Quality of a Lot or Process, Practice for (ASTM E122-72)
Z1.12-1973	- Acceptance of Evidence Based on the Results of Probability Sampling, Practice for (ASTM E141-69)
Z1.13-1973	- Use of the Terms Precision and Accuracy as Applied to Measurement of a Property of a Material, Practice for (ASTM E177-71)
Z1.14-1973	- Dealing with Outlying Observations, Practice for (ASTM E178-68)
Z34.1-1947(R1959)	- Practice for Certification Procedures
Z34.2-1969	- Practice for Certification by Producer or Supplier

2. ASQC Documents

ASQC Std A1-1971 (ANSI Std Z1.5-1971)	- Definitions, Symbols, Formulas and Tables for Control Charts
ASQC Std A2-1971 (ANSI Std Z1.6-1971)	- Definitions and Symbols for Acceptance Sampling by Attributes
ASQC Std A3-1971 (ANSI Std Z1.7-1971)	- Glossary of General Terms Used in Quality Control
ASQC Std B1-1958 (ANSI Std Z1.1-1958, R1969)	- Guide for Quality Control

ASQC Std B2-1958 (ANSI Std Z1.2-1958,R1969)	- Control Chart Method of Analyzing Data
ASQC Std B3-1958 (ANSI Std Z1.3-1958, R1969)	- Control Chart Method Controlling Quality During Production
ASQC Std C1-1968 (ANSI Std Z1.8-1971)	- Specification of General Requirements for a Quality Program
ASQC #66 1973	- Glossary of Tables for Statistical Quality Control
ASQC #70 1973	- Matrix of Nuclear Quality Assurance Program Requirements
ASQC #158 1969	- Procurement Quality Control (Vendor-Vendee Handbook)
ASQC #165 1971	- Quality Costs - What and How
ASQC #170 1967	- Quality Motivation Workbook
-----	- Handbook of Product Maintainability (available from Reliability Division)

3. American Society for Testing and Materials Standards

ASTM D2906-74	- Statements on Precision and Accuracy, Recommended Practice for
ASTM E29-67 (1973)	- Indicating which Places of Figures are to be Considered Significant in Specified Limiting Values, Recommended Practice for
ASTM E268-68	- Electromagnetic Testing, Definition of Terms Relating to
ASTM E431-71	- Interpretation of Radiographs of Semiconductors and Related Devices, Recommended Guide to
ASTM E433-71	- Liquid Penetrant Inspection, Reference Photographs for
ASTM F25-68 (1973)	- Sizing and Counting Airborne Particulate Contamination in Clean Rooms and Other Dust-Controlled Areas Designed for Electronic and Similar Applications

4. Federal Standards

FED-STD-209B 24 Apr. 73	- Clean Room and Work Station Requirements, Controlled Environment
FED-STD-358 10 Jan. 75	- Sampling Procedures

5. US Military (General)

MIL-STD-454D	- Standard General Requirements for Electronic Equipment
MIL-STD-785A	- Reliability Program for Systems and Equipment Development and Production
MIL-STD-757	- Reliability Evaluation from Demonstration Data
MIL-STD-1600	- Reliability Prediction of Monolithic Integrated Circuits
MIL-STD-756A	- Reliability Prediction
MIL-STD-790C	- Reliability Assurance Program for Electronic Parts Specifications
MIL-A-8866	- Airplane Strength and Rigidity Reliability Requirements, Repeated Loads and Fatigue
MIL-R-22732C	- Reliability Requirements for Shipboard Electronic Equipment
MIL-STD-839	- Parts, with Established Reliability Levels Selection and Use of

6. U.S. Military (Support Documents)

Environmental

MIL-STD-210B	- Climatic Extremes for Military Equipment
MIL-T152(2)	- Treatment Moisture and Fungus Resistant of Communications, Electronic and Associated Electrical Equipment

Quality

MIL-Q-9858A	- Quality Program Requirements
MIL-Q-21549B	- Product Quality Program Requirements for Fleet Ballistic Missile Weapon System Contractors
MIL-I-45208A	- Inspection System Requirements
MIL-S-52779	- Software Quality Assurance Program Requirements
MIL-HDBK-107	- Single-Level Continuous Sampling Procedures and Tables for Inspection by Attributes
DOD-HDBK-110	- Evaluation of Contractor Quality Control Systems

Test Methods

MIL-STD-781B	- Reliability Tests: Exponential Distribution
MIL-STD-750B	- Test Methods for Semiconductor Devices

MIL-STD-883B	- Test Methods and Procedures for Microelectronics
MIL-STD-810C	- Environmental Test Methods
MIL-STD-202E	- Test Methods for Electronic and Electrical Component Parts
MIL-STD-446B	- Environmental Requirements for Electronic Parts
MIL-STD-271E	- Nondestructive Testing Requirements for Metals
MIL-E-5272C	- Environmental Testing, Aeronautical and Associated Equipment, General Specification for
MIL-T-5422	- Testing, Environmental, Airborne Electronic and Associated Equipment
MIL-T-18303B	- Test Procedures; Preproduction, Acceptance, and Life for Aircraft Electronic Equipment, Format for
MIL-STD-109B	- Quality Assurance Terms and Definitions
MIL-STD-252B	- Wired Equipment, Classification of Visual and Mechanical Defects
MIL-STD-280A	- Definition of Item Levels, Item Exchangeability, Models and Related Terms
MIL-STD-410D	- Nondestructive Testing Personnel Qualification and Certification (Eddy Current, Liquid Penetrant, Magnetic Particle, Radiographic and Ultrasonic)
MIL-STD-1235A	- Single and Multilevel Continuous Sampling Procedures and Tables for Inspection by Attributes
MIL-STD-1246A	- Product Cleanliness Levels and Contamination Control Program
MIL-STD-1460	- Soldering of Electrical Connections and Printed Wiring Assemblies, Procedure for
MIL-STD-1520A	- Corrective Action and Disposition System for Nonconforming Material
MIL-STD-1535A	- Supplier Quality Assurance Program Requirements
MIL-STD-1556	- Government/Industry Data Exchange Program Contractor Participation Requirements
MIL-STD-1540A	- Test Requirements for Space Vehicles

Design

MIL-STD-439B	- Electronic Circuits

MIL-E-4158E	- Electronic Equipment Ground: General Requirements for
MIL-E-5400P	- Electronic Equipment, Airborne, General Specification for
MIL-E-8198H	- Electronic Equipment, Missiles, Boosters and Allied Vehicles, General Specification for
MIL-E-16400G	- Electronic, Interior Communication and Navigation Equipment, Naval Ship and Shore: General Specification for
MIL-E-19600A	- Electronic Modules, Aircraft, General Requirements for
Wiring	
MIL-T-713D	- Twine: Impregnated, Lacing and Tying
MIL-W-5088E	- Wiring, Aircraft, Selection and Installation of
MIL-W-8160D	- Wiring, Guided Missile, Installation of, General Specification for
Data	
MIL-D-8706B	- Data and Tests, Engineering: Contract Requirements for Aircraft Weapon Systems
MIL-D-26239A	- Data, Qualitative and Quantitative Personnel Requirements Information (QQPRI)
AFSCM 310-1	- Management of Contractor Data and Reports
MIL-STD-749B	- Preparation and Submission of Data for Approval of Nonstandard Electronic Parts
Safety	
MIL-STD-882	- System Safety Program for Systems and Associated Subsystems and Equipment: Requirements for
Interference	
MIL-E-6051D	- Electromagnetic Compatibility Requirements, Systems
MIL-I-6181D	- Interference Control Requirements, Aircraft Equipment
MIL-STD-461A	- Electromagnetic Interference Characteristics, Requirements for Equipment and Measurement of

Packaging

MIL-P-116F	- Preservation Packaging, Method of

Human Factors

MIL-STD-1472B	- Human Engineering Design Criteria for Military Systems, Equipment and Facilities
MIL-HDBK-759	- Human Factors Engineering Design for Army Materiel
MIL-H-46855A	- Human Engineering Requirements for Military Systems, Equipment and Facilities
MIL-H-81444	- Human Factors Engineering Systems Analysis Data

Vibration

MIL-STD-167	- Mechanical Vibrations of Shipboard Equipment (Type I - Environmental and Type II - Internally Excited)

Sampling

MIL-STD-690B	- Failure Rate Sampling Plans and Procedures
MIL-STD-105D	- Sampling Procedures and Tables for Inspection by Attributes
MIL-STD-414	- Reliability of Military Electronic Equipment
DOD-HDBK-106	- Multi-Level Continuous Sampling Procedures and Tables for Inspection by Attributes
DOD-HDBK-108	- Sampling Procedures and Tables for Life and Reliability Testing (Based on Exponential Distribution)

Test Equipment

MIL-I-21200L	- Test Equipment for Use with Electronic and Electrical Equipment, General Specification for

Maintainability

MIL-STD-470	- Maintainability Program Requirements (for Systems and Equipments)
MIL-STD-471A	- Maintainability Verification/Demonstration/Evaluation
MIL-M-24365A	- Maintenance Engineering Analysis: Establishment of, and Procedures and Formats for Associated Documentation; General Specification for

MIL-HDBK-472	- Maintainability Prediction
Reports	
MIL-STD-1304A	- Reports: Reliability and Maintainability Engineering Data
MIL-T-9107	- Test Reports, Preparation of
Enclosures	
MIL-STD-108E	- Definitions of and Basic Requirements for Enclosures for Electric and Electronic Equipment
MIL-E-2036C	- Enclosures for Electric and Electronic Equipment, Naval Shipboard
MIL-C-172C	- Interim Amendment Cases; Bases; Mounting; and Mounts, Vibration (For Use with Electronic Equipment in Aircraft)
Installation	
MIL-I-8700A	- Installation and Test of Electronic Equipment in Aircraft, General Specification for
MIL-E-25366D	- Electrical and Electronic Equipment, Guided Missile, Installation of, General Specification for
Management	
MIL-STD-483	- Configuration Management Practices for Systems, Equipment, Munitions, and Computer Programs
MIL-STD-499A	- Engineering Management
AFM-66-1	- Maintenance Management
AFSCM375-5	- Systems Engineering Management Procedures
AFLCM/AFSCM 400-4	- Contract and Data Requirements
Life Cycle Cost	
DOD LCC - 1	- Life Cycle Costing Procurement Guide
DOD 7041.3	- Economic Analysis and Program Evaluation for Resource Management
Logistics	
MIL-STD-1388-1	- Logistic Support Analysis
ILS DOD 4100.35 - G	- Integrated Logistic Support Planning Guide
ILS NAV MAT P4000	- Integrated Logistic Support Implementation Guide

References

PB - 121839 (Navy)	- Reliability Design Handbook, Navy
NAVSHIP 93820	- Handbook for the Prediction of Shipboard and Shore Electronic Equipment Reliability
MIL-HDBK-217B	- Reliability Prediction of Electronic Equipment
RADC-TR-69-458	- Non-Electronic Reliability Notebook
RADC-MRB-0474	- Microcircuit Reliability Bibliography
RADC-TR-73-403	- Reports on Maintainability
RADC-HMRD-0175	- Hybrid Microcircuit Reliability Data
MIL-STD-721B	- Definitions of Effectiveness Terms for Reliability, Maintainability, Human Factors and Safety

7. Canadian Forces

CA-G100, Issue 1-W/1(M)	- Climatic and Durability Testing of Army Electronic Equipment
CA-G101, Issue 2	- Parts, Materials & Processes for Army Electronic Equipment
PACK 2-5, Issue 4-w/4(M)	- Packaging Inspection and Testing for Packing of Material
SB 3 Revision 61	- List of Approved CAMESA Specifications and Related Documents
SB 5 Revision 7	- Approved Products List
SB13 Revision 4	- Canadian Military Preferred and Guidance Lists of Electronic Parts and Materials

8. (a) NASA Reliability and Quality Publications

NHB 5300.4(1A) Apr. 1, 1970	- Reliability Program Provisions for Aeronautical and Space System Contractors
NHB 5300.4(1B) Apr. 1969	- Quality Program Provisions for Aeronautical and Space System Contractors
NHB 5300.4(1C) July 1971	- Inspection System Provisions for Aeronautical and Space System Materials, Parts, Components and Services
NHB 5300.4 (1D-1) Aug. 1974	- Safety, Reliability, Maintainability and Quality Provisions for the Space Shuttle Program
NHB 5300.4(2B) Nov. 1971	- Quality Assurance Provisions for Government Agencies
NHB 5300.4(3A) May 1968	- Requirements for Soldered Electrical Connections

NHB 5300.4(3C) May 1971	- Line Certification Requirements for Microcircuits
NHB 5300.4(3D) May 1971	- Test Methods and Procedures for Microcircuit Line Certification
NHB 5300.4(3F) June 1972	- Qualified Products Lists Requirements for Microcircuits
NASA Sp-5002	- Soldering Electrical Connections
NASA SP-5113	Nondestructive Testing, A Survey

8. (b) NASA Reliability and Quality Publications

NASA SP-3079	- Nondestructtive Evaluation Technique Guide
NASA SP-6501	- Evaluation of Reliability Programs
NASA SP-6502	- Design Review for Space Elements
NASA SP-6503	- Introduction to the Derivation of Mission Requirements Profiles for System Elements
NASA SP-6504	- Failure Reporting and Management Techniques in the Surveyor Program
NASA SP-6505	- Parts and Materials Application - Review for Space Systems
NASA SP-6506	- An Introduction to the Assurance of Human Performance in Space Systems
NASA SP-6507	- Parts, Materials, and Processes Experience Summary (Vols. I and II)
NASA SP-6508	- Failure Analysis of Electronic Parts: Laboratory Methods
NASA SP-6509	- Techniques of Final Preseal Visual Inspection
NASA CR-2120 (N72-33482)	- Summary of Nondestructive Testing Theory and Practice
NASA CR-98433 (N69-25697)	- Development of Highly Reliable Soldered Joints for Printed Circuit Boards
NASA CR-123421 N72-12418)	- Applications of Aerospace Technology in Industry (Contamination Control)
NASA CR-124816 (N72-13417)	- Applications of Aerospace Technology in Industry (Welding)
NASA CR-126574 (N72-25482)	- Applications of Aerospace Technology in Industry (Nondestructive Testing)
NASA CR-127779 (N72-29892)	- A Case Study in Technology Utilization (Fracture Mechanics)
NASA TMX-2290	- Solder Circuitry Separation Problems Associated with Plated Printed Circuit Boards
NASA TMX-64706 (N73-14483)	- Assessment of and Standardization for Quantitative Nondestructive Testing

8. c. NASA Reliability and Quality Publication

MIL-STD-975 (NASA)	- Standard Parts List for Flight and Mission - Essential Ground Support Equipment

9. IEC Recommendations

68-1(1968)	- Basic Environmental Testing Procedures for Electronic Components and Electronic Equipment, Part 1: General
68-2	- Basic Environmental Testing Procedures for Electronic Components and Electronic Equipment, Part 2: Tests (describes over 20 different tests)
271(1974)	- List of Basic Terms, Definitions and Related Mathematics for Reliability
272(1968)	- Preliminary Reliability Considerations
300(1969)	- Managerial Aspects of Reliability
319(1970)	- Presentation of Reliability Data on Electronic Components (or parts), including Amendment 2 and Supplement 319A
352(1971)	- Solderless Wrapped Connections General Requirements, Test Methods and Practical Guidance
382(1971)	- Guide for the Collection of Reliability, Availability, and Maintainability Data from Field Performance of Electronic Items
409(1973)	- Guide for the Inclusion of Reliability Clauses into Specifications for Components (or parts) for Electronic Equipment
410(1973)	- Sampling Plans and Procedures for Inspection by Attributes
419(1973)	- Guide for the Inclusion of Lot-by-lot and Periodic Inspection Procedures in Specifications for Electronic Components (or Parts)
493(1974)	- Guide for the Statistical Analysis of Aging Test Data - Part 1: Methods Based on Mean Values of Normally Distributed Test Results

10. ISO Standards

R286-Part 1-1962	- System of Limits and Fits Part 1: General Tolerances and Deviations (Supersedes ISA 25) (B4.1-1967)
R1938-1971	- ISO System of Limits and Fits - Part II: Inspection of Plain Workpieces (Reprinted in Special Metric Edition)
2602-1973	- Statistical Interpretation of Tests Results - Estimation of the Mean-Confidence Interval
2859-1974	- Sampling Procedures and Tables for Inspection by Attributes
3058-1974	- Nondestructive Testing - Aids to Visual Inspection - Selection of Low-Power Magnifiers
3059-1974	- Nondestructive Testing - Method for Indirect Assessment of Black Light Sources
3207-1975	- Statistical Interpretation of Data: Determination of a Statistical Tolerance Interval
3301-1975	- Statistical Interpretation of Data - Comparison of Two Means in the Case of Paired Observations

11. British Standards

BS4891 1972	- A Guide to Quality Assurance
BS5179: Part 1 1974	- Guide to the Operation and Evaluation of Quality Assurance System Part 1 - Final Inspection System
BS5179: Part 2 1974	- Guide to the Operation and Evaluation of Quality Assurance System Part 2 - Comprehensive Inspection System
BS5179: Part 3 1974	- Guide to the Operation and Evaluation of Quality Assurance System Part 3 - Comprehensive Quality Control System

12. NATO Documents

AQAP-1 Edition #2 Dec 72	- NATO Quality Control System Requirements for Industry
AQAP-2 Edition #2 Dec. 72	- Guide for the Evaluation of a Contractor's Quality Control System for Compliance with AQAP-1
AQAP-3 Feb. 70	- List of Sampling Schemes used in NATO Countries

AQAP-4 Dec. 70	- NATO Inspection System Requirements for Industry
AQAP-5 Apr. 72	- Guide for the Evaluation of a Contractor's Inspection System for Compliance with AQAP-4
AQAP-6 Dec. 70	- NATO Calibration System Requirements for Industry
AQAP-9 Dec. 70	- NATO Basic Inspection Requirements for Industry

13.(a) Federal Aviation Administration Documents

FAA-STD-016 Aug. 27, 1975	- Quality System Requirements
FAA Advisory Circular 00-41A Nov. 3, 1975	- FAA Quality Certification Program

(b) OD (Navy) Documents

OD42565	- The Quality Control Chart Technique
OD39223	- Maintainability Engineering Handbook

(c) European Organization for Quality Control Document

-----	- General Guide to the Preparation of Specifications (EOQC Committee on Specifications in Quality and Reliability Clauses)

(d) DCAS Documents

DSAM 8200.1 Aug. 1973	- Procurement Quality Assurance

Copies of the above standards and specifications can be ordered by writing to the organization indicated below:

1.	ANSI Standards	American National Standards Institute 1430 Broadway New York, N.Y. 10018
2.	ASQC Documents	American Society for Quality Control 161 West Wisconsin Ave. Milwaukee, Wis. 53203
3.	ASTM Standards	American Society for Testing Materials 1916 Race St. Philadelphia, Pa. 19103

4. Federal Standards	General Services Administration Region 3, Federal Supply Service Special Programs Division Specifications Activity, Building 197 Washington Navy Yard Annex Washington, D.C. 20407
5. US Military (General)	The Naval Publications and Forms Center 5801 Tabor Ave. Philadelphia, Pa. 19120
6. US Military (Support Documents)	The Naval Publications and Forms Center 5801 Tabor Ave. Philadelphia, Pa. 19120
7. Canadian Forces	Department National Defence DDDS 45 Sacre Coeur Blvd. Hull Printing Bureau Hull, Que., Canada
8.(a) NASA Reliability and Quality Publications	Superintendent of Documents U.S. Government Printing Office Washington, D.C. 20402
(b) NASA Reliability and Quality Publications	National Technical Information Service 5285 Port Royal Rd. Springfield, Va. 22151
(c) NASA Reliability and Quality Publications	The Naval Publications and Forms Center 5801 Tabor Ave. Philadelphia, Pa. 19120
9. IEC Recommendations	American National Standards Institute 1430 Broadway New York, N.Y. 10018
10. ISO Standards	American National Standards Institute 1430 Broadway New York, N.Y. 10018
11. British Standards	British Standards Institution 2 Park St. London, W1A 2BS, England
12. NATO Documents	The Naval Publications and Forms Center 5801 Tabor Ave. Philadelphia, Pa. 19120

13.(a) Federal Aviation Administration Documents	Federal Aviation Administration ALG 380 Department of Transportation 800 Independence SW Washington, D.C. 20591
(b) OD (Navy) Documents	Superintendent of Documents U.S. Government Printing Office Washington, D.C. 20402
(c) European Organization for Quality Control Documents	European Organization for Quality Control P.O. Box 1976 Weena 734 Rotterdam 3003, The Netherlands
(d) DCAS Documents	Superintendent of Documents U.S. Government Printing Office Washington, D.C. 20402

Appendix A2

RELIABILITY AND MAINTAINABILITY BOOKS

J. A. Roberts

The following list of Reliability and Maintainability books was compiled largely by Ken Eagle, Dimitri Kececioglu and other members of the Society of Reliability Engineers(SRE). The list was originally included in the SRE Newsletter, International Edition of June, 1977 and has been augmented by a few publications which were not included at that time. Not all these books have been the subject of favorable reviews or are thought to provide adequate treatment of the subject. Knowledgeable and entertaining reviews of many of these books have been provided by R.A. Evans in the 'IEEE Transactions on Reliability' and by G.W.A. Dummer in the 'Microelectronics and Reliability' journals. An examination of the current issue of 'Books in Print' will provide sources for new books and will determine which of the listed books are still conveniently available.

1962 Abraham, L.H. "Structural Design of Missiles and Spacecraft", McGraw-Hill, New York, N.Y.

1971 Amstadter, B.L. "Reliability Mathematics", McGraw-Hill, New York, N.Y.

1976 Anderson, R.T. "Reliability Design Handbook", (Cat. RDH-376) IIT Research Institute (RAC) RADC, Griffiths Air Force Base, Rome, N.Y.

1963 Ankenbrandt, F.L. "Maintainability Design", Engineering Publishers Div. of A.C. Book Co., Elizabeth, N.J.

1960 Ankenbrandt, F.L. "Electronic Maintainability", Vol. 3 Engineering Publishers, Dist. Reinhold Pub. Corp. New York, N.Y.

1964	ARINC Research Corp.	"Reliability Engineering", Prentice-Hall, Englewood Cliffs, N.J.
1965	Barlow, R.E. and Proschan, F.	"Mathematical Theory of Reliability", Wiley & Sons, New York, N.Y.
1975	Barlow, R.E., and Proschan, F.	"Statistical Theory of Reliability and Life Testing Probability Models", Holt C HR&W, New York, N.Y.
1969	Barlow, R.E., and Scheuer, E.M.	"An Introduction to Reliability Theory", Ceir, Inc.
1962	Barlow, W.R. et. al.	"Probabilistic Models in Reliability Theory", Wiley & Sons, New York, N.Y.
1961	Bazovsky, I.	"Reliability Theory and Practice", Prentice-Hall, Englewood Cliffs, N.J.
1969	Belyayev, Gnedenko, and Solovyev	"Mathematical Methods of Reliability", (Trans.Ed.) - Barlow, Wiley & Sons, New York, N.Y.
1970	Benjamin, J.R. and Cornell, C.A.	"Probability Statistics and Decision for Civil Engineers", McGraw-Hill, New York, N.Y.
1970	Billington, R.	"Power System Reliability Evaluation", Gordon and Breach Science Pub., New York, N.Y.
1973	Billington, R. et.al.	"Power System Reliability Calculations", MIT Press, Cambridge, MA.
1978	Blanchard, B.S.	"Design to Manage Life Cycle Cost", M/A Press, Portland, Ore.
1969	Blanchard, B. and Lowery, E.	"Maintainability, Principles and Practice", McGraw-Hill, New York, N.Y.
1969	Bolotin, V.V.	"Statistical Methods in Structural Mechanics", Holden-Day, San Francisco, CA.
1973	Bompas-Smith, J.	"Mechanical Survival: The use of Reliability Data", McGraw-Hill, New York, N.Y.
1972	Bourne, A.J. and Greene, A.E.	"Reliability Technology", Wiley Interscience, New York, N.Y.
1970	Breipchu, A.M.	"Probabilistic Systems Analysis", Wiley & Sons Inc., New York, N.Y.

1972	Brook, R.H.W.	"Relability Concepts in Engineering Manufacture", Wiley & Sons, New York, N.Y.
1974	Bockland, William R.	"Statistical Assessment of the Life Characteristic, A. Biographical Account", Hafner, Collier-Distribution Center, Riverside, N.J.
1962	Calabro, S.R.	"Reliability Principles and Practice", McGraw-Hill, New York, N.Y.
1972	Caflen, R.H.	"A Practical Approach to Reliability", Business Books Limited.
1975	Carruepa, E.R., et al.	"Assuring Product Integrity", D.C. Heath & Co, Lexington, MA. 021373.
1973	Carter, A.C.S.	"Mechanical Reliability", Halsted Press, New York, N.Y.
1969	Chaney, F.B. and Harris, C.B.	"Human Factors in Quality Assurance", Wiley & Sons, New York, N.Y.
1968	Chapguille, P. et. al.	"Fiabilite Des Systemes", Masson & Cie., Editors, Paris.
1960	Chorofas, Dimitri N.	"Statistical Processes and Reliability Engineering", Van Nostrand, New York, N.Y.
1974	Cluley, J.C.	"Electronic Equipment Reliability", Mac Millan Press, London.
1962	Cox, D.R.	"Renewal Theory", Wiley & Sons, New York, N.Y.
1966	Cox, D.R. and Lewis, P.A.	"The Statistical Analysis of Series of Events", Wiley & Sons, New York, N.Y.
1972	Cunningham, C.E. and Cox, WIlbert	"Applied Maintainability Engineering", Wiley & Sons, New York, N.Y.
1975	Dhillon B.S.	"The Analysis of the Reliability of Multi-State Device Networks", U. of Windsor, Windsor, Ontario.
1950	Dummer, G.W.A.	"Electronic Equipment Reliability", Wiley & Sons, New York, N.Y.
1965	Dummer, G. and Winton, R.C.	"An Elementary Guide to Reliability", Pergamon, Elmsford, New York, N.Y.

1966	Dummer, G. and Griffin, N.B.	"Electronic Reliability: Calculation and Design", Pergamon, Elmsford, New York, N.Y.
1976	Eagle, K.H.	"Poisson Confidence Table; SRE-1201276", Society of Reliability Engineers, 2304 Rockwell, Wilmington, Del.
1968	English, J.M.	"Cost-Effectiveness - The Economic Evaluation of Engineered Systems", Wiley & Sons, New York, N.Y.
1966	Enrick, N.L.	"Quality Control and Reliability", Industrial Press, New York, N.Y., (5th ed. 1972)
1972	Enrick, N.L.	"Quality Control & Reliability", Indus. Pub. Edison, N.J.
1968	Gedye, G.R.	"A Manager's Guide to Quality and Reliability", Wiley and Sons, New York, N.Y.
1969	Gertsbakh, I.B. and Kordonskiy, K.H.	"Models of Failure", Springer-Verlag, New York, N.Y.
1969	Gilmore, H.L. and Schwaltz. H.C.	"Integrated Product Testing and Evaluation", Wiley & Sons, New York, N.Y.
1969	Gnedenko, B.V. et al.	"Mathematical Methods of Reliability Theory" Vol. 6, Academic Press, New York, N.Y.
1964	Goldman, A.S. and Slattery, T.B.	"Maintainability: A Major Element of System Effectiveness", Wiley and Sons, New York, N.Y.
1973	Gordon, F.	"Maintenance Engineering: Organization and Management", Halsted-Wiley, New York, N.Y.
1964	Gorgy, H.M., Myers, R.H. and Wong, K.L.	"Reliability Engineering For Electronic Systems", Wiley and Sons, New York, N.Y.
1972	Green, A.E. and Bourne, A.J.	"Reliability Technology", Wiley-Interscience, London, England.
1972	Grouchko, D.	"Operations Research & Reliability", Gordon and Breach, New York.
1965	Gryna, F.M.	"Reliability Theory and Practice", 6th Annual Workshop, ASEE Headquarters.
1970	Gyrna, F.M. and Juran, J.M.	"Quality Planning and Analysis", McGraw-Hill, New York, N.Y.

1960 Gyrna, McAfee, Ryerson, Zwerling — "Reliability Training Text", Institute of Radio Engineers

1967 Hahn, G.J. and Shapiro, S.S. — "Statistical Models in Engineering", Wiley & Sons, New York, N.Y.

1962 Haugen, E.B. — "Probabilistic Approaches to Design", Wiley & Sons, New York, N.Y.

1964 Havilanc, R.P. — "Engineering Reliability and Long Life Design", Van Nostrand, Princeton, N.J.

1956 Henney, K. — "Reliability Factors for Ground Electronic Equipment", McGraw-Hill, New York, N.Y.

1968 Hofman, W. — "Zuverlassigkeit Von Mess-, Stuer-, Regel-Und Sicherheitssytemen", Verlag Karl Thiemig KG.

1965 Hummitzsch, P. — "Zuverlassigkeit Von Systemen", Friedr.Vieweg & Sohn, Braunschweig, West Germany, RA28.

1966 Ireson, W.G. — "Reliability Handbook", McGraw-Hill, New York, N.Y.

1973 Jardine, A.K.S. — "Maintenance, Replacement and Reliability", Halsted Press-Wiley, New York, N.Y.

1970 Jardine, A.K.S. — "Operational Research in Maintenance" Halsted Press - Wiley, New York, N.Y.

1970 Jelen, F.C. — "Cost and Optimization Engineering", McGraw Hill, New York, N.Y.

1964 Johnson, L.G. — "The Statistical Treatment of Fatigue Experiments", American Elsevier Pub., New York, N.Y.

1964 Johnson, L.G. — "Theory and Techniques of Variation Research", American Elsevier Pub., New York, N.Y.

1966 Jowett, C.E. — "Reliability of Electronic Components", London Iliffe Book Ltd., England.

1971 Jowett, C.E. — "Reliable Electronic Assembly Production", Tab Books, Blue Ridge Summit, PA.

1976 Kececioglu, Dimitri — "Manual of Product Assurance Films...", 7340 N. La Oesta Ave., Tucson, AZ. 85704.

1966	Kenney, D.P.	"Application of Reliability Techniques", Argyle Pub., 605 3RD Ave, New York, N.Y., 10016.
1963	Kerawalla, J. & Lipson, C.	"Engineering Applications of Reliability", University of Michigan.
1971	Kivenson, G.	"Durability and Reliability in Engineering Design", Hayden Rock Co., Rochelle Park, N.J.
1975	Kogan, J.	"Crane Design: Theory and Calculations of Reliability", Halsted-Wiley, New York N.Y.
1970	Kozlov, B.A. and Ushakov	"Reliability Handbook", Trans:Lisa Rosenblatt, Holt, Rinehart & Winston Inc, New York York, N.Y.
1963	Landers, R.R.	"Reliability and Product Assurance", Prentice-Hall, Englewood Cliffs, N.J.
1960	Leake, C.E.	"Understanding Reliability", Pasadena Lithographers, Inc, Pasadena, CA.
1967	Lin, Y.K.	"Probabilistic Theory of Structural Dynamics", McGraw-Hill, New York, N.Y.
1973	Lifson, C. and Narencra, J.S.	"Statistical Design and Analysis of Engineering Experiments", McGraw-Hill, New York, N.Y.
1964	Litle, William A.	"Reliability of Shell Buckling Predictions", MIT Press, Cambridge, MA.
1962	Lloyd, Cavic K. and Lipow, Nyron	"Reliability: Management, Methods, and Mathematics", Prentice-Hall, Englewood Cliffs, N.J.
1973	Locks, Mitchell O.	"Reliability, Maintainability, & Availability Assessment", Hayden Book, Rochelle Park, N.J.
1962	Mann, W.C. and Wilcox, R.H.	"Redundancy Techniques for Computing Systems", Spartan Press, Washington, DC.
1974	Mann, N.R., Shafer, Esingpurwallar	"Methods for Statistical Analysis of Reliability and Life Data", Wiley & Sons, New York, N.Y.

1971	Messerschmitt-Bolkow-Blohm	"Technische Zuverlassigkeit", Gmbh, Munchen, Springer-Verlag, New York, N.Y.
1975	Myers, Glenford J.	"Reliable Software Through Composite Design", Petrocelli Books, New York, N.Y.
1976	Myers, Glenford J.	"Software Reliability Principles and Practice", Wiley-Interscience, New York, N.Y.
1971	Nixon, Frank	"Managing Costs to Achieve Quality and Reliability", McGraw-Hill, New York, N.Y.
1970	Osgood, C.C.	"Fatigue Design", Wiley and Sons, New York, N.Y.
1969	Peyrache, G. and Scwob, M.	"Traite de Fiabilite", Masson & Cie, Editeurs.
1965	Pierce, W.H.	"Failure-Tolerant Computer Design", Academic Press, New York, N.Y.
1963	Pieruschka, E.	"Principles of Reliability", Prentice-Hall, Englewood Cliffs, N.J.
1964	Polovko, A.M.	"Fundamentals of Reliability Theory", (Translation 1968: Pierce), Academic Press, New York N.Y.
1973	Pronikov, A.S.	"Dependability and Durability of Engineering Products", Halsted-Wiley, New York, N.Y.
1974	Proschan and Serfling	"Reliability and Biometry - Statistical Analysis of Life Length", Siam, Philadelphia, PA.
1966	Pugsley, Sir A.	"The Safety of Structures", Edward Arnold Ltd., London, England.
1970	Rau, J.G.	"Optimization and Probability in Systems Engineering", Van Nostrand, New York, N.Y.
1964	Roberts, N.H.	"Mathematical Methods in Reliability Engineering", McGraw-Hill, New York, N.Y.
1963	Sandler, G.H.	"System Reliability Engineering", Prentice-Hall, Englewood Cliffs, N.J.

1969	Seiler, K. III	"Introduction to Systems Cost-Effectiveness", Wiley-Interscience, New York, N.Y.
1968	Shooman, M.	"Probabilistic Reliability: An Engineering Approach", McGraw-Hill, New York, N.Y.
1961	Shyop, J. and Sullivan, H.	"Semiconductor Reliability", Engineering Publishers, Elizabeth, N.J.
1976	Smith, O.	"Introduction to Reliability in Design", McGraw-Hill, New York, N.Y.
1969	Smith, C.S.	"Quality and Reliability: An Integrated Approach", Pitman, New York, N.Y.
1972	Smith, C.J.	"Reliability Engineering", Barnes and Noble, Scranton, Pa.
1973	Smith, D.J. and Babb, A.H.	"Maintainability Engineering", Pitman Publishing, New York, N.Y.
1963	Society of Automotive Eng.	"Reliability Control in Aerospace Equipment Development", SAE Inc, New York, N.Y.
1970	Stormer, H.	"Mathematische Theorie Der Zuverlassickeit", R. Oldenborgh, Munich, Germany.
1960	Tangerman, E.J. (Ed)	"Manual CF Reliability", Product Engineering.
1969	Thomason, R.	"An Introduction to Quality and Reliability", Machinery Publishing Co., Brighton, England.
1976	Unipub	"Reliability of Nuclear Power Plants", (Paper) Unipub, New York, N.Y.
1972	Unipub	"Reliability Guidebook", (Paper) Unipub, New York, N.Y.
1971	Waller, F.	"Component Reliability", Giechenhaus.
1970	Weinberg, S.	"Profit Through Quality Management Control of Q & R Techniques", Cahners, Boston, MA.
1963	Zelen, M.S.	"Statistical Theory of Reliability", Univ. of Wisconsin Press, Madison, WS.

INDEX

A. J. B. SWAIN